**Gauge Integral Structures
for Stochastic Calculus
and Quantum Electrodynamics**

Gauge Integral Structures for Stochastic Calculus and Quantum Electrodynamics

Patrick Muldowney

Registered Office
John Wiley & Sons, Inc., 111 River Street, Hoboken, NJ 07030, USA

Editorial Office
111 River Street, Hoboken, NJ 07030, USA

For details of our global editorial offices, customer services, and more information about Wiley products visit us at www.wiley.com.

Library of Congress Cataloging-in-Publication Data
Names: Muldowney, P. (Patrick), 1946- author.
Title: Gauge integral structures for stochastic calculus and quantum electrodynamics / Patrick Muldowney.
Description: Hoboken, NJ : Wiley, [2020] | Includes bibliographical references and index.
Identifiers: LCCN 2020016333 (print) | LCCN 2020016334 (ebook) | ISBN 9781119595496 (cloth) | ISBN 9781119595502 (adobe pdf) | ISBN 9781119595526 (epub)
Subjects: LCSH: Stochastic analysis. | Henstock-Kurzweil integral. | Feynman integrals. | Quantum electrodynamics–Mathematics.
Classification: LCC QA274.2 .M85 2020 (print) | LCC QA274.2 (ebook) | DDC 519.2/2–dc23
LC record available at https://lccn.loc.gov/2020016333
LC ebook record available at https://lccn.loc.gov/2020016334

Set in 9.5/12.5pt STIXTwoText by SPi Global, Chennai, India

SKY10025659_031721

Dom' bean lán de rtuaim

Contents

III   Appendices   303

Preface

This book is about infinite-dimensional integration in stochastic calculus and in quantum electrodynamics, using the gauge integral technique pioneered by R. Henstock and J. Kurzweil.

A link between stochastic calculus and quantum mechanics is provided in a previous book by the author ([121], *A Modern Theory of Random Variation*, or [MTRV] for short), which establishes a mathematical connection between large scale Brownian motion on the one hand and, on the other, small scale quantum level phenomena of particle motion subject to a conservative external mechanical force. In [MTRV] each of the two subjects is a special case of *c-Brownian motion*.

The present book is a continuation of [MTRV], in the sense that it develops and extends some of the themes of that book. On the other hand this book is a stand-alone introduction to particular problems of integration in the probabilistic theory of stochastic calculus, and in the probability-like theory of quantum mechanics.

Between [MTRV] and this book there is a significant difference in style of exposition. Practically all the underlying mathematical theory is already set out in [MTRV]. The present book includes motivational explanation of the key points of the underlying mathematical theory, along with ample illustrations of the calculus—the routine procedures—of the gauge theory of integration.

But because the "mathematical heavy lifting" (or rigorous mathematical underpinning) is already accomplished in [MTRV], the present book can take a more gradual, relaxed, and discursive approach which seeks to engage the reader with the subject by exploring a much smaller range of chosen themes.

Thus there is hardly anything of the formal Theorem-Proof structure in this book. Instead the text is organised around Examples with accompanying introductions and explanation, illustrating themes from probability and physics which can be difficult and taxing. Particular areas of interest in the book can be selected and read without engaging with other topics. Its relatively self-contained component parts can easily be "dipped into".

In addition to [MTRV], two principal physics sources for this book are [39], *Space-time approach to non-relativistic quantum mechanics* (cited as [F1] for short) by R. Feynman; and [46], *Quantum Mechanics and Path Integrals* (cited as [FH] for short) by R. Feynman and A. Hibbs.

Certain modes of expression used by physicist R. Feynman are highly illuminating—but from a physics perspective. For instance: *Integrate* [some expression] *over all degrees of freedom* [all variables] *of the* [physical] *system.* This statement does not specify the mathematical domain Ω for this integration process, nor how $\int_\Omega \cdots$ (the integral on Ω) is to be actually calculated.

For quantum electrodynamics, and with $\mathbf{R}$ denoting the real line and $\mathbf{T}$ an interval of time, this book proposes a product space domain

$$\Upsilon = \left(\left(\mathbf{R}^3 \right)^{\mathbf{R}^3} \times \left(\mathbf{R}^3 \times \mathbf{R} \right)^{\mathbf{R}^3} \right)^{\mathbf{T}},$$

along with methods of calculating integrals on product spaces such as

$$\int_{\mathbf{R} \times \mathbf{R}} \cdots, \qquad \int_{\mathbf{R}^{\mathbf{T}}} \cdots, \qquad \int_\Upsilon \cdots.$$

The examples, explanations, and illustrations of this book, though prompted by issues in physics and finance, are primarily about underlying mathematical problems. A reader whose primary purpose is to investigate finance or physics as such should seek out other, more relevant sources.

Also, it is inadvisable to approach the gauge or Riemann sum theory of integration as if it were unchallenging or easy compared with, say, Lebesgue's integration theory. That said, the general idea of gauge integration is more accessible (initially, at least) than the mathematical theory of measure which underpins other approaches to integration. (Gauge integration theory includes measure and measurability—see section A.1 of [MTRV]—but these are outcomes rather than prerequisites of the theory.)

Another central theme of this book is to devise new and better versions of stochastic integrals. Functionals of the form $\int_{\mathbf{T}} f\left((x_t)_{t \in \mathbf{T}} \right) dx(t)$ appear in both the theory of stochastic processes (as stochastic integral) and quantum theory (as integral of lagrangian function, or system action).

Such integral functionals either do not actually exist, or are not easy to define. In their place, this book proposes replacement or equivalent functionals which involve Riemann sums rather than integrals. In the case of stochastic processes such sums are called *stochastic sums*. In quantum theory they are designated *sampling sums*. (The former are a special case of the latter.)

The general idea of infinite-dimensional integration on a Cartesian product domain can be found in chapter 3 of [MTRV]. The technical foundations of the subject are in chapter 4 of that book.

The core of this book consists of Chapters 6 and 9. Without tackling the companion book [MTRV], a reader who wants a quick introduction to basic gauge integral theory can get it in Chapter 10 below. This includes a re-definition of the standard Lebesgue integral (on an abstract measure space) as a Riemann-Stieltjes integral $\int_{\mathbf{R}} f \, dg$ on domain $\mathbf{R}$.

There is a website for commentary on technical issues:
`https://sites.google.com/site/StieltjesComplete/`
cited as [`website`] in this book. Technical communications are welcome and can be addressed to:
`stieltjes.complete.integral@gmail.com`

Pat Muldowney

Reading this Book

The term "gauge" in the title relates to gauge integration in mathematics (a generalized form of Riemann integration). It is not about gauge symmetry or gauge transformations in theoretical physics.

The following abbreviations are used because of frequent references to these sources:

- [F1] for [39], *Space-time approach to non-relativistic quantum mechanics*, by R.P. Feynman;

- [FH] for [46], *Quantum Mechanics and Path Integrals*, by R.P. Feynman and A.R. Hibbs;

- [MTRV] for [121], *A Modern Theory of Random Variation*, by P. Muldowney.

- [website] for [122], `https://sites.google.com/site/StieltjesComplete/` This is the website for this book, and for [MTRV].

References to chapters, sections, and figures in this book use a capital letter, "Chapter x"; but for material from other sources lower-case is used: "figure y".

This book develops themes in probability and quantum mechanics which were introduced in [MTRV]. The range of topics is smaller, and the range of notation is correspondingly smaller, with only a few new symbols. One such is the notation $\not{f}$ (denoting *stochastic sum*) on page 132 below. Section 13.4 has a list of the main symbols used.

The subject of the predecessor book [MTRV] is the role of Cartesian product spaces $\mathbf{R}^T$, $= \prod_{t \in T} \mathbf{R}$, in the theory of probability (including quantum mechanics), where $\mathbf{R}$ is the set of real numbers, and T is an interval of time.

The present book examines in more detail different kinds of Cartesian products of $\mathbf{R}$, as domains for integration of functions f,

$$\int_\Omega f(\cdot)\,d(\cdot), \qquad \Omega = \cdots \times \mathbf{R}^T \times \cdots \times \left(\mathbf{R}^\mathbf{R}\right)^T \times \cdots.$$

Though the symbol T will generally represent time, it is also used as an arbitrary finite or infinite set of labels t, depending on the context.

The notation $\int_{\mathbf{R}^T} f$ has two components: domain $\mathbf{R}^T$, and integrand f. When T is an infinite set, such as an interval of time, two different perspectives are present. These are the perspectives indicated in figures 3.1 and 3.2 on page 87 of [MTRV].

Figure 3.1 represents the graph of x_T where x_T is an argument of integrand $f(\cdots, x_T, \cdots)$. Figure 3.2 represents the domain $\mathbf{R}^T$ whose elements x_T are points, not graphs. Both of these perspectives should be kept in mind while using this book.

To see the significance of these alternate perspectives[1], suppose $T =]0, t[$ and $\mathbf{T} =]0, t]$, so $\mathbf{T} = T \cup \{t\}$. In this case, for integrands f which appear in this book, there may be little difference between the values of integrands $f(\cdots, x_T, \cdots)$ in domain $\mathbf{R}^T$, and $f(\cdots, x_{\mathbf{T}}, \cdots)$ in $\mathbf{R}^{\mathbf{T}}$ if f is continuous.

But domains $\mathbf{R}^T$ and $\mathbf{R}^{\mathbf{T}} = \mathbf{R}^T \times \mathbf{R}$ are very different; just as $\mathbf{R}^2$ differs geometrically from $\mathbf{R}^3$, $= \mathbf{R}^2 \times \mathbf{R}$. The latter difference is the vehicle for the 19th century satire *Flatland* by Edwin A. Abbott [1], in which two-dimensional beings struggle with the idea of a three-dimensional universe.

The book can be read as a collection of standalone accounts of topics which are suggested, or introduced, or touched upon, in [MTRV]. Equally, it can be read as a supplement to [MTRV], developing the ideas of stochastic sums and path integrals which were introduced in [MTRV].

The formal theorem-proof layout is avoided in favour of concrete illustrations. In almost cases, the relevant theorems and proofs are cited from [MTRV]. But exceptions are made for a few fundamental results such as analogues or versions of theorems 4, 62, 64, and 65 of [MTRV]. The theorem-proof format is applied in these few exceptional cases.

Also, some of the illustrative examples are "thought experiments", exercises of imagination intended to clarify, not finance or physics as such, but certain mathematical issues which arise in these subjects.

Notes and additional material for the book can be found at:

> `https://sites.google.com/site/StieltjesComplete/`

—also used to record typos, errors, and corrections; and cited as [`website`] in this book.

Sincere thanks to colleagues and readers who help by sharing their thoughts via: `stieltjes.complete.integral@gmail.com`

---

[1]In [MTRV], T usually denotes $]\tau', \tau]$, with $\mathcal{T}$ denoting $]0, t]$. Also, T^- is used in [MTRV] to denote $T \smallsetminus \{\tau\}$, or $]\tau', \tau[$.

Introduction

The *gauge integral* is a version of the Riemann integral, with much improved convergence properties. Convergence properties are conditions which ensure integrability of a function; in particular, integrability of the limit of a convergent sequence of integrable functions, with integral of the limit equal to the limit of the integrals—the limit theorems.

Another notable property of the gauge integral in one dimension is that, if a function possesses a corresponding derivative function, the derivative is integrable, with indefinite integral equal to the original function. Curiously, this "schooldays meaning"—integration as the reverse of differentiation—does not hold universally for the more widely used integration systems. See Section 10.2 of Chapter 10, which provides an overview of this subject.

The gauge integral (called -complete integral in [MTRV], and in this book) is non-absolute. Other kinds of integration, such as Lebesgue's or Riemann's, have restrictive requirements of absolute convergence. But existence of -complete integrals requires only that the Riemann sum approximations converge non-absolutely to the value of the integral; and this is central to the present book.

In [MTRV], instead of the more familiar term gauge integral, the term *-complete integral* (as in Riemann-complete) is found to be helpful.[2] This is because there are a great many different integration techniques—Riemann, Lebesgue, Stieltjes, Burkill, and others—which are used in different situations; most of which can be subsumed or adapted into a gauge integral system. But to assign indiscriminately the blanket designation "gauge integral" to each of the adapted versions is to ignore, firstly, their considerable difference in usages and origins; and, secondly, the fact that "gauge integral" has become practically a synonym for the one-dimensional generalized Riemann (or Riemann-complete) integral— also known as the Kurzweil-Henstock integral.

Also, the general or abstract integral—called Henstock integral in chapter 4 of [MTRV]—has diverged historically from the more mainstream gauge or Kurzweil integration which has "integral-as-antiderivative" as its driving force. This aspect of the subject is touched on in Chapter 10 below.

---

[2] The attachment "-complete" was introduced by R. Henstock in [70], the first book-length exposition of this kind of integration theory. A few of copies of this edition were printed in 1962. A replacement edition with different page size was printed and distributed in 1963. Up to that time J. Kurzweil and R. Henstock had worked independently on this subject from around the mid-1950s, without knowledge of each other.

The integral-as-antiderivative feature of one-dimensional Riemann-complete integration was mentioned in passing in Henstock's 1962–63 exposition [70], which concentrated on other aspects of integration (such as limit theorems[3] and Fubini's theorem).

As a student Henstock was attracted to the theory of divergent series. When in 1944 he applied to Paul Dienes to do research in this subject, he was steered towards integration theory [11]; and his subsequent work often focussed on the margins between divergence and convergence.[4]

The gauge idea made its first appearance in Henstock's published work in [69], in a scenario of extreme divergence in which the gauge method is "tested to destruction" in its first public outing. (This counter-example is rehearsed in pages 178–181 of [MTRV], section 4.14, *Non-Integrable Functions*. See also Example 13 below.) There is no mention in his 1955 paper [69] of the reversal of differentiation which many students of the subject have found so useful. Nor does it touch on the notion of random variation in which theories of integration and measure play a central role, and where integral convergence is much more important than differentiation.[5]

The emphasis on convergence is maintained in the present book, which can be read as a stand-alone, self-contained, or self-explanatory volume expanding on certain themes in [MTRV]. Like [MTRV] this book aspires to simplicity and transparency. No prior knowledge of the subject matter is assumed, and simple numerical examples set the scene. There is a degree of repetitiveness which may be tedious for experts. But experts can cope with that; more consideration is owed to less experienced readers.

For reasons demonstrated in [MTRV], and amply confirmed in the present volume, non-absolute convergence is one of the characteristics which, in comparison with other methods, makes the gauge (or -complete) integrals suitable for the two main themes of this book: stochastic calculus and Feynman integration.

Stochastic calculus is the branch of the theory of stochastic processes which deals with stochastic integrals, also known as stochastic differential equations. A landmark result is Itô's lemma, or Itô's formula.

Stochastic integration is part of the mathematical theory of probability or random variation. Broadly speaking, quantities or variables are **random** or **non-deterministic** if they can assume various unpredictable values; and they are **non-random** or **deterministic** if they can take only definite known values.

Classically, stochastic integrals are constructed by means of a procedure

---

[3]Henstock's introduction of the "-complete" appendage is suggestive of "enhanced integrability of limits" rather than "completeness of a domain with respect to a norm".

[4]As part of the College Prize awarded by St. John's College, Cambridge, on the results of the 1943 Mathematics Tripos Part 2 examination, Henstock received a copy of Dienes' book [23], which includes close analysis of convergence-divergence issues. In a late, unfinished work [78], c. 1992–1993, Henstock used some notable ideas from Dienes' book.

[5]The final chapter of Henstock's 1962-1963 book [70] has the title *Integration in Statistics*. It deals mostly with tests of significance, and touches on some questions of probability theory using the Riemann-complete method. The 1955 paper is concerned strictly with the nature of integration. But ancillary matters such as probability—and, indeed, differentiation—featured consistently in Henstock's subsequent work.

involving weak limits. The purpose of this book is two-fold:

- To treat **stochastic integrals as actual integrals**; so that the limit process which defines a stochastic integral is essentially the same as the limit of Riemann sums which defines the more familiar kinds of integral.

- To provide an alternative theory of **stochastic sums** which achieves the same purposes as stochastic integrals, but in a simpler way.

Mathematically, integration is more complicated and more sophisticated than summation (or addition) of a finite number of terms. It is demonstrated that stochastic sums can achieve the same (or better) results as stochastic integrals do. **In the theory of stochastic processes, stochastic sums can replace stochastic integrals.**

Examples of concrete nature are used to illustrate aspects of stochastic integration and stochastic summation, starting with relatively elementary ideas about finite numbers of things or events, in which there is no difference between summation and integration. A basic calculation of financial mathematics (growth of portfolio value) is used as a reference concept, as a vehicle, and as an aid to intuition and motivation.

In a review [145] of a book [31], Laurent Schwartz stated:

> *Each of us* [Schwartz and Emery] *tried to help the probabilists absorb stochastic infinitesimal calculus of the second order "without tears"; I don't know whether any of us succeeded or will succeed.*

This book is a further effort in this direction.

The action functionals of quantum mechanics (see (8.7), page 211 below) are analogous to stochastic integrals. They appear as integrands in the infinite-dimensional integrals used by R.P. Feynman in his theory of quantum mechanics and quantum electrodynamics.

In comparison with alternative approaches such as those of J. Schwinger ([147, 148, 149, 150]) and S. Tomonaga ([88, 89, 90, 91, 92, 164, 165]), Feynman's method is said to be physically intuitive. It contrasts with the mathematics-leaning approach of Paul Dirac [27]:

> *The present lectures, like those of Eddington, are concerned with unifying relativity and quantum theory, but they approach the question from a different point of view. Eddington's method is first to get the physical ideas clear and then gradually to build up a mathematical scheme. The present method is just the opposite—first to set up a mathematical scheme and then try to get its physical interpretation.*

In reading [FH] it can be helpful to bear in mind that "[Feynman was] *the outstanding intuitionist of our age ...*", (attributed to Schwinger in [32]).

Feynman's first published paper on path integrals was [F1], *Space-time approach to non-relativistic quantum mechanics* [39]. In a long tradition of the relationship between physics and mathematics it entailed problems of a pure mathematical kind:

There are very interesting problems involved in the attempt to avoid the subdivision and limiting processes [in Feynman's construction of path integrals]. *Some sort of complex measure is being associated with the space of functions* $x(t)$. *Finite results can be obtained under unexpected circumstances because the measure is not positive everywhere, but the contributions from most of the paths largely cancel out. These curious mathematical problems are largely side-stepped by the subdivision process. However, one feels as Cavalieri must have felt calculating the volume of a pyramid before the invention of calculus.* [39] (R.P. Feynman [F1]; also page 79 of [10].)

These are problems essentially of mathematics, not physics or quantum mechanics. And the solutions proposed in [MTRV], and here, are intended to be contributions to mathematics, not physics.

- The space of functions $x(t)$ (for $0 < t < \tau$) is $\mathbf{R}^T$ where $\mathbf{R}$ is the set of real numbers and $T =]0, \tau[$. It is likely that Feynman's reference to "measure" above relates to Lebesgue-type measure on measurable subsets of $\mathbf{R}^T$, which is not available in the form suggested by Feynman. Here are some mathematical issues:

- Instead of measurable sets and measure of sets, [MTRV] provides a solution based on a structure of interval-type subsets of $\mathbf{R}^T$, with a "natural" volume function for such subsets, and using the -complete system of non-absolute integration described in [MTRV].

- Feynman's statement that *"the contributions from most of the paths largely cancel out"* suggests a non-absolute convergence approach, and confirms the unsuitability of methods requiring absolute convergence.

- Stochastic integrals sometimes have the form $\int_T f(X(s)) \, dX(s)$ where $X(s)$ is a stochastic process. Feynman's integrals often include expressions involving the integral of kinetic energy $\frac{1}{2}mv^2$. These are *action functionals*, integrals such as

$$\int_T \frac{m}{2} \left(\frac{dx(s)}{ds} \right)^2 ds \quad \text{or} \quad \frac{m}{2} \int_T \left(\frac{dx(s)}{ds} \right) dx(s);$$

where the latter has the form of a stochastic integral $\int_T \left(\frac{dX(s)}{ds} \right) dX(s)$. Generally speaking, for $x \in \mathbf{R}^T$, $x(s)$ is non-differentiable. So none of these functionals actually exists as an integral and, in order to give mathematical meaning to them, various devices have to be used, such as the weak integrals of classical stochastic calculus, or Feynman's subdivision and limiting processes.

- Feynman's *"subdivision and limiting processes"* are described in [F1], and in [FH] (*Quantum Mechanics and Path Integrals* [46], by R.P. Feynman and A.R. Hibbs). They are also examined in section 7.18 of [MTRV], along with their relationship[6] to the -complete integral solution.

This book provides an alternative solution to these problems. Instead of integrals $\int_T \cdots dx(s)$, or $\int_T \cdots ds$, sample times $s \in T$ are used to form Riemann sums. These are called *stochastic sums* in the stochastic integral case, and *sampling sums* in the case of action integrals:

- These functionals are finite sums, not integrals;

- they are sample versions of stochastic integrals (or of action integrals in the case of quantum mechanics);

- they always exist;

- and their expected values and other properties are defined and calculated by a well defined system of -complete (or gauge) integration in $\mathbf{R}^T$.

- And, just as it is reasonable to estimate integrals by means of finite Riemann sums, it is equally reasonable to use finite samples to estimate the functional integrands by means of finite samples (or sampling sums).

This book considers mathematical aspects of the Feynman integral theory as it is expounded in [FH], which starts with

- a single **particle** μ interacting with a conservative mechanical force,

- and which progresses through to a system consisting of the interaction of a charged particle with an electromagnetic **field** $\mathcal{F}$.

For the latter system, [FH] declares that a certain action functional should be integrated over "all degrees of freedom" of the system—over all possible values of each of the variables.

This highly intuitive mode of expression is physically suggestive and resonant. But in mathematics a domain of integration must be defined and formulated as a definite mathematical set composed of definite mathematical elements or points.

In [FH] as in [MTRV] this is achieved for motion μ by translating "integration over all degrees of freedom" of the single particle motion into integration on a domain $\mathbf{R}^T$ consisting of elements or points $(x(t) : t \in T)$; or, simply, $x_T \in \mathbf{R}^T$. (This deals only with one-dimensional particle motion. For physical realism elements of the domain should be points $\mathbf{x}_T$ of $\left(\mathbf{R}^3\right)^T$, where

$$\mathbf{x}_t = (x_1, x_2, x_3)_t = (x_{1t}, x_{2t}, x_{3t})$$

and where x_{1t}, x_{2t}, x_{3t}, are the particle position co-ordinates in $\mathbf{R}^3$ for each $t \in T$.)

---

[6] In the terminology of [MTRV] and this book, Feynman's method consists of substituting *cylinder function* approximations in the action functional.

For system $\mathcal{F}$ the domain and its elements are less obvious. In this book the domain

$$\Omega = \left(\mathbf{R} \times (\mathbf{R} \times \mathbf{R})^{\mathbf{R}} \right)^{T}$$

is proposed. This involves a one-dimensional simplification (like the simplific-ation $\mathbf{R}^{T}$ instead of $\left(\mathbf{R}^{3}\right)^{T}$ for system μ), and also other simplifications which are contrary to physical reality but which make the mathematical exposition a bit easier to follow. An element of this domain is

$$\chi = \left(\left(x_t, \, (A_{yt}, \phi_{yt})_{y \in \mathbf{R}} \right)_{t \in T} \right)$$

where $x_t \in \mathbf{R}$ is particle position at time t; and, at time t, elements $A_{yt} \in \mathbf{R}$ and $\phi_{yt} \in \mathbf{R}$ correspond to electromagnetic field components[7] at a point y in space. (An element χ is called a **history** of the interaction.)

The reason for trailing advance notice of details such as these is to provide a sense of the mathematical challenges presented by quantum electrodynamics (system $\mathcal{F}$ above), further to the challenges already posed by system μ.

Feynman's theory of system $\mathcal{F}$—or interaction of $\mathcal{F}$ with μ—posits certain integrands in domain Ω, the integration being carried out over "all degrees of freedom" of the physical system. But how is an integral on Ω, $\int_{\Omega} \cdots$, to be defined? Is there a theory of measurable sets and measurable functions for Ω? (Even if such a measure-theoretic integration actually existed it would fail on the requirement for non-absolute convergence in quantum mechanics.) And if integrands f in "$\int_{\Omega} f$" involve action functionals of the form $f = \int_{T} \cdots ds$, we face the further problem of how to give meaning to "$\int_{T} \cdots ds$" as integrand in domain Ω.

This is reminiscent of the stochastic integrals/stochastic sums issue men-tioned above. The resolution in both cases uses the following feature of the -complete or gauge system of integration.

A Riemann-type integral $\int_{a}^{b} f(x)\,dx$ in a one-dimensional bounded domain $[a,b]$ is defined by means of Riemann sum approximations $\sum f(x)|I|$ where the subintervals $\{I\}$ of domain $[a,b[$ are formed from partitions such as

$$\mathcal{P} = \{x_1, x_2, \dots, x_{n-1}\}, \quad x_0 = a, \quad x_n = b, \quad x_0 < x_1 < \cdots < x_n.$$

Riemann sums can be expressed as Cauchy[8] sums $\sum_{j=1}^{n} f(\bar{x}_j)(x_j - x_{j-1})$ where $\bar{x}_j = x_j$ or x_{j-1}. In fact the Riemann-complete integral can be defined in terms of suitably chosen finite samples $\{x_1, \dots, x_{n-1}\}$ of the elements in the domain of integration, without resort to measurable functions or measurable subsets—or even without explicit mention of subintervals of the domain of integration.

---

[7]This simplification represents each of the variables x_t, ϕ_{yt}, and A_{yt} as one-dimensional. The electric field component is essentially vectorial, and one-dimensional A_{yt} is contrary to the physical nature of the system. A physically more accurate version can be arrived at by a careful reading of chapter 9 of [FH]. And even though it is a bit more complicated, it is not too difficult to adapt the mathematical theory presented in this book.

[8]A Cauchy sum has $\bar{x}_j = x_{j-1}$. But allowing $\bar{x}_j$ to be either of x_{j-1} or x_j makes a connection with the Riemann sums of -complete integration.

To define -complete integration in "rectangular" or Cartesian product domains such as Ω above—no matter how complex their construction—the only requirements are:

- Exact specification of the elements or points of the domain, and

- A structuring of finite samples of points consistent with Axioms DS1 to DS8 of chapter 4 of [MTRV].

In other words integration requires a domain Ω and a process of selecting samples of points or elements of Ω—without reference to measurable subsets, or even to intervals of Ω at the most basic level.

This skeletal structuring of finite samples of points of the domain provides us with a system of integration (the -complete integral) with all the useful properties—limit theorems, Fubini's theorem, a theory of measure, and so on. More than that, it provides criteria for non-absolute convergence (theorems 62, (∗),[9] 64, and 65 of [MTRV]) which are crucially important for Feynman integrals.

---

[9]Theorem 63 (page 175 of [MTRV]) is false. See Section 11.2 below; and also [**website**].

Part I

Stochastic Calculus

Chapter 1

Stochastic Integration

The idea or purpose of *stochastic integration* is to define a random variable $\mathcal{S}_t$ =

$$\int_0^t \mathcal{Z}(s)d\mathcal{X}(s), \quad \text{or} \quad \int_0^t f(\mathcal{X}(s))\,d\mathcal{X}(s)$$

where $\mathcal{S}_t$ is a random or unpredictable quantity, depending in a particular manner on unpredictable entities $\mathcal{X}$ and $\mathcal{Z}$; and where

$$\mathcal{X}, = (\mathcal{X}(s) : 0 < s \leq t), \qquad \mathcal{Z}, = (\mathcal{Z}(s) : 0 < s \leq t)$$

are stochastic processes and $\mathcal{S}_t$ depends on time t. In textbooks, the integrand is usually presented as $f(s)$, but $\mathcal{Z}(s)$ is used here in order to emphasise that the integrand is intended to be random.

The integrand $\mathcal{Z}(s)$ (or, when appropriate, $f(\mathcal{X}(s))$) is to be regarded as a measurable function—as is $\mathcal{X}(s)$—with respect to a probability space $(\Omega, \mathcal{A}, P)$.

If $\mathcal{Z}(s)$ is a deterministic or non-random function $g(s)$ of s, its value at time s is a definite (non-random) number which, whenever necessary, can be regarded as a degenerate random variable. If $\mathcal{Z}(s)$ is *the same* random variable for each s in $t_{j-1} \leq s < t_j$, each j, then the process $\mathcal{Z}$ is a *step function*. (In textbooks, the term *elementary function* is often applied to this.)

The most important kind of stochastic integral is where $\mathcal{X}, = (\mathcal{X}(s))_{0 < s \leq t}$, is standard Brownian motion, and this particular case (called the *Itô integral*) is outlined here. The main steps are as follows.

I1 Suppose the integrand $\mathcal{Z}(s)$ is a step function, with constant random variable value $\mathcal{Z}(s) = \mathcal{Z}_{j-1}$ for $t_{j-1} \leq s < t_j$, $0 = t_0 < t_1 < \cdots < t_n = t$. Then define

$$\mathcal{S}_t = \int_0^t \mathcal{Z}(s)d\mathcal{X}(s) := \sum_{j=1}^n \mathcal{Z}_{j-1}\left(\mathcal{X}(t_j) - \mathcal{X}(t_{j-1})\right).$$

In this case (that is, $\mathcal{Z}(s)$ a step function), the *Itô isometry* holds for expected values:

$$\mathbf{E}\left(\mathcal{S}_t^2\right) = \mathbf{E}\left(\int_0^t (\mathcal{Z}(s))^2 ds\right).$$

Gauge Integral Structures for Stochastic Calculus and Quantum Electrodynamics, First Edition. Patrick Muldowney.
© 2021 John Wiley & Sons, Inc. Published 2021 by John Wiley & Sons, Inc.

I2 Suppose the process $\mathcal{Z}(s)$ (not necessarily a step function) satisfies

$$\mathrm{E}\left(\int_0^t (\mathcal{Z}(s))^2\, ds\right) \;<\; \infty.$$

Then there exists a sequence of step functions (processes) $\{\mathcal{Z}^{(p)}(s)\}$, $p = 1, 2, 3, \ldots$ such that

$$\lim_{p\to\infty} \mathrm{E}\left(\int_0^t \left|\mathcal{Z}_j^{(p)}(s) - \mathcal{Z}(s)\right|^2 ds\right) \;=\; 0.$$

I3 For such $\mathcal{Z}(s)$, define its stochastic integral $\mathcal{S}_t$ with respect to the process $\mathcal{X}(s)$ as

$$\mathcal{S}_t = \int_0^t \mathcal{Z}(s)\, d\mathcal{X}(s) \;:=\; \lim_{p\to\infty} \mathcal{S}_t^{(p)} \;=\; \lim_{p\to\infty} \int_0^t \mathcal{Z}^{(p)}(s)\, d\mathcal{X}(s)$$

$$=\; \lim_{p\to\infty} \sum_{j_p=1}^{n_p} \mathcal{Z}_{j_p}^{(p)}\left(\mathcal{X}(t_{j_p}) - \mathcal{X}(t_{j_p-1})\right).$$

I4 If $\mathcal{X}$ is Brownian motion the latter limit exists.

An objective of this book is to provide an alternative to the classical theory, not develop it. Thus the commentary, interpretation, and speculation of this section can be safely omitted by anybody who is either already familiar with, or is not interested in, the standard theory of stochastic integration.

Regarding notation, many textbooks use the symbol B for Brownian motion, whereas $\mathcal{X}$ is used above. Textbooks also use the symbol $f(s)$ for the integrand, where $\mathcal{Z}(s)$ is used above. The reason for using notation $\mathcal{Z}(s)$ instead of $f(s)$) is to emphasise that the value of the integrand function is generally a random variable depending on s, and not generally a single, definite real or complex number (such as the deterministic function $g(s) = s^2$, for instance) of the kind which occurs in ordinary integration.

In classical probability theory, an underlying mathematical probability measure space $(\Omega, \mathcal{A}, P)$ is assumed, such that, for all random variables and processes, the probability that any random variable has an outcome in a particular set $A \in \mathcal{A}$ can be calculated using the appropriate technical calculation[1] relevant to each random variable. If the random variables or processes have a time structure, then mathematical properties of *filtration* and *adaptedness* ensure that sets A which qualify as $(\mathcal{A}, P)$-measurable events at earlier times will still qualify as such at subsequent times.

The integrator $d\mathcal{X}(s)$ is a random variable. The integrand function $f(s)$ or $\mathcal{Z}(s)$ is also a random variable. And the (stochastic) integral $\mathcal{S}_t, = \int_0^t \mathcal{Z}(s)\, d\mathcal{X}(s)$, is a random variable. This point is sometimes illustrated in textbooks by means of examples such as the following.

---

[1]The random variable could be normally distributed, or Poisson, or binomial, etc.

Example 1 *Suppose* $\mathcal{X}(t)$ *(a random quantity) is the price of an asset at time* *t. Then, for times* $t_{j-1} < t_j$, $\mathcal{X}(t_j) - \mathcal{X}(t_{j-1})$ *is the change in the price of the asset, the change or difference also being random. Suppose the quantity of asset holding* $\mathcal{Z}(t_{j-1})$ *(sometimes denoted as* $f(t_{j-1})$) *is unpredictable or random. The product of these two,*

$$\mathcal{Z}(t_{j-1})(\mathcal{X}(t_j) - \mathcal{X}(t_{j-1})),$$

is then a random variable representing the change in the value of the total asset holding. The stochastic integral $\mathcal{S}_t = \int_0^t \mathcal{Z}(s)\,d\mathcal{X}(s)$, *represents the aggregate or sum of these changes over the period of time* $[0,t]$; *and is a random variable.*

Here, use of the symbol $\mathcal{Z}(s)$ for the integrand (instead of the usual $f(s)$) indicates that while the integrand is a random variable dependent on s, it does not necessarily depend on the integrator random variable $\mathcal{X}(s)$. If, in fact, there is such dependence, then an appropriate notation[2] for the integrand is $f(\mathcal{X}(s))$.

The notation and terminology of ordinary integration is used in **I1, I2, I3, I4**, and they provide a certain "feel" for what is going on. But the various elements of the system are clearly different from ordinary integration. Can we get some more precise idea of what is really going on?

The "integration-like" construction in **I1** suggests that the domain of integration is $0 \le s \le t$, and that the integrand takes values in a class of functions (—random variables; that is, functions which are measurable with respect to some probability space, or spaces).

How does this compare with more familiar integration scenarios? Basic integration (" $\int_a^b f(s)\,ds$ ") has two elements: firstly, a domain of integration containing values of the integration variable s, and secondly, an integrand function $f(s)$ which depends on the values s in the domain of integration. The more familiar integrand functions have values which are real or complex numbers $f(s)$; and which are deterministic (that is, "definite", not approximate or estimated).

The construction in **I1, I2, I3** indicates an integration domain $[0,t]$ or $]0,t]$. (There is nothing surprising in that.) But in **I1, I2, I3** the integrand values are not real or complex numbers, but random variables—which may be a bit surprising.

But it is not unprecedented. For instance, the Bochner integration process in mathematical analysis deals with integrands whose values are functions, not numbers. The construction and definition of the Bochner integral [105] is similar in some respects to the classical Itô integral. What is the end result of the construction in **I1, I2, I3**?

In general, the integral of a function f gives a kind of average or aggregation of all the possible values of f. So if each value of the integrand $f(s)$ is a random variable, the integral of f should itself be a random variable—that is, a function which is measurable with respect to an underlying probability measure space.

---

[2]It is sometimes convenient to denote the integrand function by $\nu(s)$, where $\nu(s)$ can be $g(s)$ (deterministic), $\mathcal{Z}(s)$ (random, independent of $\mathcal{X}(s)$), or $f(\mathcal{X}(s))$ (random, dependent on $\mathcal{X}(s)$).

If the notation $\mathcal{Y} = \mathcal{S}_t(\mathcal{X}) = \int_0^t (\cdots)\, d\mathcal{X}(s)$ is valid or justifiable for the stochastic integral, it suggests that the Itô integral construction $\int_0^t (\cdots)\, d\mathcal{X}(s)$ derives a single random variable $\mathcal{Y}$ (or $\mathcal{S}_t$) from many jointly varying random variables, such as $\mathcal{X}(s)$, as s varies between the values 0 and t. This is reminiscent of Norbert Wiener's construction in [169], which is in some sense a mathematical replication in one dimension of Brownian motion; even though the latter is essentially an infinite-dimensional phenomenon with infinitely many variables. Without losing any essential information, a situation involving infinitely many variables is converted to a scenario involving only one variable.[3]

The proof of the Itô isometry relation (see **I1**) indicates that, as a stochastic process, $\mathcal{Z}(s)$ must be independent of $\mathcal{X}(s)$. Otherwise the construction **I1, I2, I3** would seem to be inadequate as it stands, whenever the process $\mathcal{Z}(s)$ is replaced by a process $f(\mathcal{X}(s))$.

In **I3** the integrand $\mathcal{Z}(s)$ does not have step function form; and, on the face of it, $\int_0^t \mathcal{Z}(s)\, d\mathcal{X}(s)$ indicates dependence of $\mathcal{Y}$ (or $\mathcal{S}_t$) on random variables $\mathcal{Z}(s)$ and $\mathcal{X}(s)$ for **every** s, $0 \le s \le t$. If the integrand were $f(\mathcal{X}(s))$ (which, in general, it is not), with joint random variability for $0 \le s \le t$, and if $(\mathcal{X}(s))$ is Brownian motion, then the joint probability space for the processes $(\mathcal{X}(s))$ and $(f(\mathcal{X}(s)))$ is given by the Wiener probability measure and its associated multi-dimensional measure space. (The latter are described in Chapter 5 below.)

Returning to **I1**, the Itô integral $\int_0^t \mathcal{Z}(s)\, d\mathcal{X}(s)$ of step function $\mathcal{Z}(s)$ is defined as

$$\sum_{j=1}^n \mathcal{Z}_{j-1} \left(\mathcal{X}(t_j) - \mathcal{X}(t_{j-1}) \right)$$

where the $\mathcal{Z}_j$ are random variable values of $\mathcal{Z} = (\mathcal{Z}(s))$. It is perfectly valid to combine finite numbers of random variables in this way, in order to produce, as outcome, a single random variable (—which may be a *joint* random variable depending on many underlying random variables).

This part of the formulation of the integral of a step function in **I1** corresponds to the integral of a step function in basic integration, and does not require any passage to a limit of random variables.

Now suppose each $\mathcal{Z}_j$ is a fixed real number α_j; so, for $t_{j-1} \le s < t_j$, $\mathcal{Z}(s) = \mathcal{Z}_{j-1} = \alpha_{j-1}$. (Accordingly, in **I1**, $\mathcal{Z}_j$ can be regarded as a "degenerate" random variable, with atomic probability value.) Suppose the integrator is the real-valued ds instead of the random variable-valued $d\mathcal{X}(s)$. Then[4]

$$\int_0^t \mathcal{Z}(s)\, ds = \sum_{j=1}^n \left(\mathcal{Z}_{j-1} \times (t_j - t_{j-1}) \right) = \sum_{j=1}^n \alpha_{j-1} (t_j - t_{j-1}).$$

Formally, at least, this looks like the definition in **I1** of $\int_0^t \mathcal{Z}(s)\, d\mathcal{X}(s)$ when $\mathcal{Z}(s)$ is a step function. The factor $t_j - t_{j-1}$ equals $\int_{t_{j-1}}^{t_j} ds$ for each j. This emerges naturally from the mathematical meaning of the length or distance variable s, and from the mathematical meaning of $\int_0^t$.

---

[3]This construction is also described in Muldowney [115].

[4]Recall also that **I1** and **I2** make reference to a construction $\int_0^t \mathcal{Z}(s)^2\, ds$.

Can this be replicated in **I1** when $\mathcal{Z}(s)$ is a step function, or when each $\mathcal{Z}_j$ is a fixed real number α_j? Is it the case that

$$\int_0^t \mathcal{Z}(s)\, d\mathcal{X}(s) = \sum_{j=1}^n \left(\mathcal{Z}_{j-1} \int_{t_{j-1}}^{t_j} d\mathcal{X}(s) \right) = \sum_{j=1}^n \left(\alpha_{j-1} \int_{t_{j-1}}^{t_j} d\mathcal{X}(s) \right)?$$

With each $\alpha_j = 1$, this would imply

$$\int_0^t d\mathcal{X}(s) = \mathcal{X}(t) - \mathcal{X}(0) = \mathcal{X}(t). \tag{1.1}$$

If this is unproblematical, it should be possible to deduce it from one or other of the various mathematical definitions of Brownian motion $(\mathcal{X}(s))$, along with some mathematical definition of the integral $\int_0^t$ in this context.

But it appears that there is no such understanding of $\int_0^t d\mathcal{X}(s) = \mathcal{X}(t)$. So, as in **I1**, it seems that this formulation is to be regarded as a basic postulate or axiom of stochastic integration.

Returning to the definition of the classical Itô integral, **I2** has the following condition on the expected value of the integral of the process $(\mathcal{Z}(s)^2$:

$$\mathrm{E}\left(\int_0^t (\mathcal{Z}(s))^2\, ds \right) < \infty.$$

The idea here is that, if $\mathcal{Y}$ is the random entity obtained by carrying out some form of weighted aggregation—denoted by $\int_0^t (\mathcal{Z}(s))^2\, ds$—of all the individual random variables $\mathcal{Z}(s)$ $(0 \le s \le t)$, then

$$\mathrm{E}(\mathcal{Y}) = \int_\Omega \mathcal{Y}(\omega)\, dP < \infty.$$

This formulation assumes that the aggregative operation $\int_0^t (\mathcal{Z}(s))^2\, ds$, involving infinitely many random variables $\mathcal{Z}(s)$ $(0 \le s \le t)$, produces a single random entity $\mathcal{Y}$ whose expected value can be obtained by means of the operation $\int_\Omega \cdots dP$.

Additionally, $\int_0^t \mathcal{Z}(s)^2\, ds$ is said to be a Lebesgue integral-type construction. The $\int_0^t \cdots ds$ part of this statement should be unproblematical. The domain $[0, t]$ is a real interval, and has a distance or length function, which, in the context of Lebesgue integration on the domain, gives rise to Lebesgue measure μ on the space $\mathcal{M}$ of Lebesgue measurable subsets of $[0, t]$. So $\int_0^t \cdots ds$ can also be expressed as $\int_0^t \cdots d\mu$.

However, the random variable-valued integrand $\mathcal{Z}(s)^2$ is less familiar in Lebesgue integration. Suppose, instead, that the integrand is a real-number-valued function $f(s)$. Then the Lebesgue integral $\int_0^t f(s)ds$, or $\int_0^t f(s)d\mu$, is defined if the integrand function f is Lebesgue measurable. So if J is an interval of real numbers in the range of f, the set $f^{-1}(J)$ is a member of the class $\mathcal{M}$ of measurable sets; giving

$$f^{-1}(J), = \{s : f(s) \in J\}, \in \mathcal{M}.$$

That is, for each J, $f^{-1}(J)$ is a Lebesgue measurable subset of $[0,t]$. This is valid if, for instance, f is a continuous function of s, or if f is the limit of a sequence of step functions.

How does this translate to a random variable-valued integrand such as $\mathcal{Z}(s)^2$? Two kinds of measurability arise here, because, in addition to being a μ-measurable function of $s \in [0,t]$, $\mathcal{Z}(s)$ is a random variable (as is $\mathcal{Z}(s)^2$), and is therefore a P-measurable function on the sample space Ω:

$$[0,t] \times \Omega \quad \overset{\mathcal{Z}}{\mapsto} \quad \mathbf{R},$$

$$(s,\omega) \quad \rightarrow \quad \mathcal{Z}(s,\omega) \in \mathbf{R}.$$

Likewise $\mathcal{Z}(s)^2$. For $\int_0^1 \mathcal{Z}(s)^2 \, ds$ to be meaningful as a Lebesgue-type integral, the integrand $\mathcal{Z}(s)^2$ must be $\mathcal{M}$-measurable (or μ-measurable) in some sense. At least, for purpose of measurability there needs to be some metric in the space of (Ω, P)-measurable functions f_s, $0 \leq s \leq t$, with $f_s(\omega) \in \mathbf{R}$, $\omega \in \Omega$:

$$\left\{ f_s = \mathcal{Z}(s)^2 \ : \ 0 \leq s \leq t \right\}.$$

For example, the "distance" between f_{s_1} and f_{s_2} could be

$$\int_\Omega |f_{s_1}(\omega) - f_{s_2}(\omega)| \, dP.$$

With such a metric at hand, it may then be possible to define $\int_0^t f_s \, ds$, or $\int_0^t \mathcal{Z}(s)^2 \, ds$, as the limit of the integrals of (integrable) step functions $f_s^{(p)}$ converging to f_s for $0 \leq s \leq t$, as $p \rightarrow \infty$.

Unfortunately, most standard textbooks do not give this point much attention. But for relatively straightforward integrands such as $\mathcal{Z}(s)^2$, it should not be too difficult.

Continuing the discussion of **I1, I2, I3, I4**, it appears that the output of this definition of stochastic integral is a random entity $\mathcal{Y}$; perhaps a process which is some collection of random variables $(\mathcal{Y}(s))$.

Again comparing this with basic integration of a real number-valued function $f(s)$, the integral $\int_0^t f(s) \, ds$ is some kind of average or weighted aggregate value for $\{f(s) : 0 \leq s \leq t\}$. This integral, if it exists, produces a single unique real number (depending on the value of t), denoted by $\int_0^t f(s) \, ds$.

For random variable-valued integrand $f(s) = \mathcal{Z}(s)$, suppose (for the purpose of speculation) that the stochastic integral

$$\int_0^t \mathcal{Z}(s) \, d\mathcal{X}(s)$$

(if it exists) is equivalent (in some unspecified sense) to a single, unique random variable $\mathcal{Y}$. Remember, a random variable is a function, usually real-valued[5],

---

[5]See [MTRV] for discussion of complex-valued random variables.

defined on a sample space Ω. Two such functions, $\mathcal{Y}$ and $\mathcal{Y}'$, are the same function if and only if

$$\mathcal{Y}(\omega) = \mathcal{Y}'(\omega) \quad \text{for each } \omega \in \Omega.$$

Does the definition of stochastic integral in **I1, I2, I3** yield such a unique value for $\int_0^t \mathcal{Z}(s)\, d\mathcal{X}(s)$? **I2** and **I3** do not guarantee uniqueness: there may be different sequences $\{\mathcal{Z}^{(p)}\}$ in **I2** which converge "in mean square" to $\mathcal{Z}$. In effect, **I4** asserts weak convergence of the integrals $S_t^{(p)}$ of the step functions $\mathcal{Z}^{(p)}$ to a value S_t for the integral of $\mathcal{Z}$, that value being not necessarily unique.

If the integral does not have a unique value, what connections may exist between alternative values? Suppose there is more than one candidate random variable, say $\mathcal{Y}$ and $\mathcal{Y}'$, for the value of the stochastic integral,

$$\mathcal{Y} = \int_0^t \mathcal{Z}(s)\, d\mathcal{X}(s), \qquad \mathcal{Y}' = \int_0^t \mathcal{Z}(s)\, d\mathcal{X}(s)$$

In that case, what is the relation between $\mathcal{Y}$ and $\mathcal{Y}'$? For instance, is it the case that, for each real number a, the probabilities of corresponding measurable sets are equal (such as $P(\mathcal{Y} \neq a) = P(\mathcal{Y}' \neq a)$):

$$P(\mathcal{Y} < a) = P(\mathcal{Y}' < a), \qquad P(\mathcal{Y} > a) = P(\mathcal{Y}' > a) \ ?$$

The framework outlined above does not include the important case $\int_0^t d\mathcal{X}(s)^2 = t$, where $(\mathcal{X}(s))$ $(0 < s \leq t)$ is Brownian motion. Broadly speaking, $\int_0^t d\mathcal{X}(s)^2 = t$ means that the random variables represented by finite sums

$$\sum_{j=1}^{n} (\mathcal{X}(t_j) - \mathcal{X}(t_{j-1}))^2$$

converge as $t_j - t_{j-1}$ tend to zero, each j. In fact the convergence is weak, not point-wise, with

$$E\left(\sum_{j=1}^{n} (\mathcal{X}(t_j) - \mathcal{X}(t_{j-1}))^2 \right) \ \to \ t,$$

and the weak limit t is a fixed real number which can be regarded as a degenerate random variable. This result is basic to the construction **I1, I2, I3, I4**.

A closer reading of source material may provide answers and/or corrections to some or all of the above comments and queries. Any misinterpretation, confusion, and errors may be dispelled by closer examination of the underlying ideas.

Aside from these issues, and looking beyond the classical mathematical theory, the general idea of stochastic integral is, in intuitive terms, a persuasive, natural and practical way of thinking about the underlying reality.

An alternative (and hopefully more understandable) mathematical way of representing this reality is presented in subsequent chapters of this book.

Example 2 *In order to focus on the underlying ideas, here is a simple illustration. Suppose, at different times t, Y_t (or $Y(t)$) is a random variable, with sample space Ω. For simplicity suppose, for each t, Y_t has the same sample space Ω and the same probability distribution. Suppose further that Y_t has only a finite number m of possible values $y_{t1}, \ldots, y_{tm}$, each equally likely. Then we can take*

$$\Omega = \{1, 2, \ldots, m\},$$

and probability $P\left(Y_t(\omega) = i\right) = \frac{1}{m}$. For $A \subseteq \Omega$,

$$P\left(Y_t \in A\right) = \frac{|A|}{m},$$

where $|A|$ is the number of elements in A. Then, for each t, Y_t is a (Ω, P)-measurable function and thus a random variable. (We may also suppose, if it is convenient for us, that for any t, t', the random variables $Y_t, Y_{t'}$ are independent.)

Now suppose that, for $0 < t \leq \tau$, Z_t is another indeterminate or unpredictable quantity; and that, for given t, the possible values of Z_t depend in some deterministic way on the corresponding values of Y_t, so

$$Z_t = f(Y_t)$$

where f is a deterministic function. For instance, the deterministic relation could be $Z_t = Y_t^2$, so if the value taken by Y_t at time t is y_t, then the value that Z_t takes is y_t^2. Provided f is a "reasonably nice" function (such as $(\cdot)^2$), then Z_t is measurable with respect to (Ω, P), and is itself a random variable.

This scenario is in broad conformity with $I1$, $I2$, $I3$, $I4$ above. So it may be possible to consider, in those terms, the stochastic integral of Z_t with respect to Y_t. Essentially, with $\Omega = \{1, \ldots, m\}$, then for each t, for $\omega = i \in \Omega$, and for $i = 1, \ldots, m$,

$$Z_t(\omega) = f\left(Y_t(\omega)\right), \quad \text{or} \quad Z_t(i) = f\left(Y_t(i)\right),$$

the two formulations being equivalent. If the stochastic integral "$\int_0^\tau Z_t \, dY_t$" is to be formulated in terms of Lebesgue integrals in $0 < t \leq \tau$ (as intimated in $I1$, $I2$, $I3$, $I4$), then some properties of t-measurability $(0 \leq t \leq \tau)$ are suggested. This aspect can also be simplified, as follows.

Just as Ω was reduced to a finite number m of possible values, $[0, \tau]$ can be replaced by a finite number of fixed time values $0 < \tau_1 < \tau_2 < \cdots < \tau_n = \tau$ if the family of random variables Y_t $(0 < t \leq \tau)$ is replaced by Y_{τ_j} $(1 \leq j \leq n)$; so there are only a finite number n of random variables $Y_j = Y_{\tau_j}$,

$$Y_t = Y_{\tau_j} = Y_j \quad \text{for} \quad \tau_{j-1} < t \leq \tau_j, \quad 1 \leq j \leq n;$$

and the random variables can be written

$$Z_t(\omega) = Z_t(i) = f\left(Y_t(\omega)\right) = f\left(Y_t(i)\right) = f\left(Y_j(i)\right)$$

for $1 \le i \le m$, $1 \le j \le n$. *(Below, $f(\cdot)$ will be taken to be $(\cdot)^2$.) Replacing the domains Ω and $[0,\tau]$ by $\{1,\dots,m\}$ and $\{1,\dots.n\}$, respectively, ensures measurability in ω and t. It also ensures measurability for the conditional cases of Y_t (or $f(Y_t)$) with $Y_{t'}$ already determined as known real numbers when $t' < t$.*

Let y_{ji} represent sample values (or potential occurrences) of the random variables Y_j. For any given j, i can have value

$$ i = i_j, \quad 1 \le i_j \le m, \quad 1 \le j \le n; $$

so $\{i_j\}$ is the set of permutations, with repetition, of the numbers $i = 1,\dots,m$ taken n at a time.

*The ideas in **I1, I2, I3, I4**) suggest the following sample values for the stochastic integral $\int_0^\tau f(Y_t)\,dY_t$ (or "$\int_0^\tau f(Y_j)\,dY_j$"):*

$$ \sum_{j=1}^{m} f\left(y_{ji_j}\right)\left(y_{ji_j} - y_{j-1,i_{j-1}}\right). $$

The subscript i_j labels the random variability in this calculation, and demonstrates that this version of the stochastic integral can take m^n possible values; though not all of the possible values are necessarily distinct.

For further simplification, take $m = 2$ and $n = 3$; so, at each of times τ_j ($j = 1,2,3$), the random variable Y_j can take one of two possible values, y_{j1}, y_{j2}. Then, by enumerating the permutations with repetition of $m = 2$ things taken $n = 3$ at a time, the 8 possible sample values of the stochastic integral $\int_0^\tau f(Y_t)\,dY_t$ are:

$$ f\left(y_{11}\right)\left(y_{11}-0\right) \;+\; f\left(y_{21}\right)\left(y_{21}-y_{11}\right) \;+\; f\left(y_{31}\right)\left(y_{31}-y_{21}\right), $$

$$ f\left(y_{11}\right)\left(y_{11}-0\right) \;+\; f\left(y_{21}\right)\left(y_{21}-y_{11}\right) \;+\; f\left(y_{32}\right)\left(y_{32}-y_{21}\right), $$

$$ f\left(y_{11}\right)\left(y_{11}-0\right) \;+\; f\left(y_{22}\right)\left(y_{22}-y_{11}\right) \;+\; f\left(y_{31}\right)\left(y_{31}-y_{22}\right), $$

$$ f\left(y_{11}\right)\left(y_{11}-0\right) \;+\; f\left(y_{22}\right)\left(y_{22}-y_{11}\right) \;+\; f\left(y_{32}\right)\left(y_{32}-y_{22}\right), $$

$$ f\left(y_{12}\right)\left(y_{12}-0\right) \;+\; f\left(y_{21}\right)\left(y_{21}-y_{12}\right) \;+\; f\left(y_{31}\right)\left(y_{31}-y_{21}\right), $$

$$ f\left(y_{12}\right)\left(y_{12}-0\right) \;+\; f\left(y_{21}\right)\left(y_{21}-y_{12}\right) \;+\; f\left(y_{32}\right)\left(y_{32}-y_{21}\right), $$

$$ f\left(y_{12}\right)\left(y_{12}-0\right) \;+\; f\left(y_{22}\right)\left(y_{22}-y_{12}\right) \;+\; f\left(y_{31}\right)\left(y_{31}-y_{22}\right), $$

$$ f\left(y_{12}\right)\left(y_{12}-0\right) \;+\; f\left(y_{22}\right)\left(y_{22}-y_{12}\right) \;+\; f\left(y_{32}\right)\left(y_{32}-y_{22}\right). $$

Now suppose that the deterministic function f is exponentiation to the power of 2 (so $f(y) = y^2$); and suppose the random variable Y_t (or Y_j above) has sample values -1 and $+1$ with equal probabilities 0.5. Calculating each of the above expressions, the 8 sample evaluations of the stochastic integral $X = \int_0^\tau Y_t^2\,dY_t$ are, respectively,

$$ -1, \ 1, \ -1, \ 1, \ -1, \ 1, \ -1, \ 1, $$

each having equal probability; so the distinct sample values of the random variable X are -1, $+1$ with equal probabilities 0.5. Thus it happens, in this case, that the stochastic integral has the same sample space Ω, and the same probabilities, as each of the random variables Y_t.

This example gets round the technical problem of measurability by discretizing the domains Ω and $]0, \tau]$, and using only function $Y_t(\omega), = Y(t, \omega)$, which are step functions in respect of their dependence on t and ω.

This discrete or step function device is a fairly standard ploy of mathematical analysis. Following through on this device usually involves moving on to functions $Y_t(\omega)$ which are limits of step functions; and this procedure generally involves use of some conditions which ensure that the integral of a "limit of step functions" is equal to the limit of the integrals of the step functions.

Broadly speaking, it is not unreasonable to anticipate that this approach will succeed for measurable (or "reasonably nice") functions—such as functions which are "smooth", or which are continuous.

But the full meaning of measurability is quite technical, involving infinite operations on sigma-algebras of sets. This can make the analysis difficult.

Accordingly, it may be beneficial to seek an alternative approach to the analysis of random variation for which measurability is not the primary or fundamental starting point. Such an alternative is demonstrated in Chapters 2 and 3, leading to an alternative exposition of stochastic integration in ensuing chapters.

The method of exposition is slow and gradual, starting with the simplest models and examples. The step-by-step approach is as follows.

- Though there are other forms of stochastic integral, the focus will be on $\int_\Omega g(Z) \, dX$ where Z and X are stochastic processes.

- The sample space Ω will generally be $\mathbf{R}^\mathbf{T}$ where $\mathbf{R}$ is the set of real numbers, $\mathbf{T}$ is an indexing set such as the real interval $]0, t]$, and $\mathbf{R}^\mathbf{T}$ is a cartesian product.

- Z and X are stochastic process (Z_s), (X_s) $(s \in \mathbf{T})$; and g is a deterministic function. More often, the process Z is X, so the stochastic integral[6] is $\int_\Omega g(X) \, dX$.

- The approach followed in the exposition is to build up to such stochastic integrals by means of simpler preliminary examples, broadly on the following lines:

  - Initially take $\mathbf{T}$ to be a finite set, then a countable set, then an uncountable set such as $]0, t]$.

  - Initially, let the process(es) X (and/or Z) be very easy versions of random variation, with only a finite number of possible sample values.

  - Similarly let the integrand g be an easily calculated function, such as a constant function or a step function.

  - Gradually increase the level of "sophistication", up to the level of recognizable stochastic integrals.

---

[6]There are other important stochastic integrals, such as $\int_\Omega (dX)^2$.

- This progression helps to develop a more robust intuition for this area of random variation. On that basis, the concept of "stochastic sums" is introduced. These are more flexible and more far reaching than stochastic integrals; and, unlike the latter, they are not over-burdened with issues involving weak convergence.

Chapter 2

Random Variation

2.1   What is Random Variation?

The previous chapter makes reference to random variables as functions which are measurable with respect to some probability domain. This conception of random variation is quite technical, and the aim of this chapter is to illuminate it by focussing on some fundamental features.

In broad practical terms, random variation is present when unpredictable outcomes can, in advance of actual occurrence, be estimated to within some margin of error. For instance, if a coin is tossed we can usually predict that heads is an outcome which is no more or no less likely than tails. So if an experiment consists of ten throws of the coin, it is no surprise if the coin falls heads-up on, let us say, between four and six occasions. This is an estimated outcome of the experiment, with estimated margin of error.

In fact, with a little knowledge of binomial probability distributions, we can predict that there is approximately 40 per cent chance that heads will be thrown on four, five or six occasions out of the ten throws. So if a ten-throw trial is repeated one hundred times, the outcome should be four, five, or six heads for approximately four hundred of the one thousand throws.

Such knowledge enables us to estimate good betting odds for placing a wager that a toss of the coin will produce this outcome. This is the "naive or realistic" view.

Can this fairly easily understandable scenario be expressed in the technical language of probability theory, as in Chapter 1 above? What is the probability space $(\Omega, \mathcal{A}, P)$? What is the P-measurable function which represents the random variable corresponding to a single toss of a coin?

The following remarks are intended to provide a link between the "naive or realistic" view, and the "sophisticated or mathematical" interpretation of this underlying reality.

The possible outcomes of an experiment consisting of a single throw of the coin are H (for heads) and T (for tails). Suppose a sample space Ω for this

Gauge Integral Structures for Stochastic Calculus and Quantum Electrodynamics, First Edition. Patrick Muldowney.
© 2021 John Wiley & Sons, Inc. Published 2021 by John Wiley & Sons, Inc.

experiment consists of the pair of numbers, 0 and 1. Let $\mathcal{A}$ be the family of all subsets of Ω:

$$\Omega = \{0,1\}, \qquad \mathcal{A} = \{\varnothing, \{0\}, \{1\}, \{0,1\}\};$$

and define a probability measure P by

$$P(\varnothing) = 0, \qquad P(\{0\}) = \frac{1}{2}, \qquad P(\{1\}) = \frac{1}{2}, \qquad P(\{0,1\}) = P(\Omega) = 1.$$

Then, trivially, $\mathcal{A}$ is a σ-algebra of subsets of Ω, and P is, trivially, countably additive[1] on $\mathcal{A}$, so $(\Omega, \mathcal{A}, P)$ is a probability measure space.

The set of outcomes of a single throw of a coin is the set $V = \{H, T\}$, and the family of subsets of V is

$$\mathcal{V} = \{\varnothing, \{H\}, \{T\}, \{H, T\}\};$$

and $(V, \mathcal{V})$ is a measurable space. Define the following function to represent the coin tossing experiment:

$$\mathcal{X}: \Omega \mapsto V, \qquad \mathcal{X}(0) = H, \qquad \mathcal{X}(1) = T.$$

Then, for $S \in \mathcal{V}$, $\mathcal{X}^{-1}(S) \in \mathcal{A}$ so $\mathcal{X}$ is a P-measurable function.

There are many different ways of defining the probability space. It is natural to use real-number-valued functions, so the outcomes V can be denoted 0 and 1 instead of H and T, and the measurable function $\mathcal{X}$ can then be the identity function. (It is usual that the values of a random variable are represented as real numbers[2]; with expected—mean or average—value, variance, and so on; which are also real numbers.)

But no matter what way this construction is done, the classical, rigorous mathematical representation by measurable function is evidently more complicated than the naive or natural view of the coin tossing experiment. In contrast, the purpose of this book is to provide a rigorous theory of stochastic integration/summation which (like [MTRV]) bypasses the "measurable function" view, and which is closer to the "naive realistic" view.

Example 3 *Throw a pair of dice and, whenever the sum of the numbers observed exceeds 10, pay out a wager equal to the sum of the two numbers thrown, and otherwise receive a payment equal to the smaller of the two numbers observed. If the two are the same number (with sum not exceeding 10) then the payout is that number.*

In Example 3 take sample space

$$\Omega = \{1,2,3,4,5,6\} \times \{1,2,3,4,5,6\}, \qquad \text{with}$$
$$\mathcal{A} = 2^{\Omega} \quad \text{(the set of all subsets of } \Omega\text{)},$$
$$P(A) = \frac{|A|}{|\Omega|} \quad \text{(the cardinality of } A \text{ divided by the cardinality of } \Omega\text{)}.$$

---

[1] In fact $\mathcal{A}$ is a finite family so there are only finitely many cases to check for additivity.
[2] [MTRV] also deals with complex-valued random variables. Also, in the classical definition of the Itô integral (**I1, I2, I3, I4** above) the integrands are measurable functions whose values are random variables (so the values are functions rather than numbers).

Observation of a throw of the pair of dice can be represented by a listing of the possible joint outcomes $\mathcal{X} = \{(x_1, x_2)\}$, $x_1 = 1, 2, \ldots, 6$, $x_2 = 1, 2, \ldots, 6$. Define a random variable $g(\mathcal{X}) : \Omega \mapsto \mathbf{R}$ by

$$g(x_1, x_2) = x_1 + x_2$$

for each $(x_1, x_2) \in \Omega$. Then, as in the previous example where the domain and range of $\mathcal{X}$ are finite sets, $g(\mathcal{X})$ is $(\Omega, \mathcal{A}, P)$-measurable and qualifies as a random variable, with expected value

$$E(g(\mathcal{X})) = \int_\Omega g(\mathcal{X}) \, dP = 7.$$

The integral in this case reduces to the sum of a finite number of terms.

The payoff from the wager in Example 3 is a randomly variable amount given by

$$\mathcal{Z} = f(\mathcal{X}), \qquad \mathcal{Z}(x_1 + x_2) = \begin{cases} -(x_1 + x_2) & \text{if} \quad x_1 + x_2 \geq 10, \\ x_1 & \text{if} \quad x_1 \leq x_2, \\ x_2 & \text{if} \quad x_2 \leq x_1. \end{cases}$$

In this case, $\mathcal{Z}$ is a composite of the deterministic function f with the random variable $\mathcal{X}$; and, just like $g(\mathcal{X})$, $\mathcal{Z}$ is (trivially) $(\Omega, \mathcal{A}, P)$-measurable, and is a random variable, with

$$E(\mathcal{Z}) = E(f(\mathcal{X})) = \int_\Omega \mathcal{Z}(\omega) \, dP = \frac{41}{36};$$

where, again, the Lebesgue integral reduces (trivially) to a finite sum of terms.

There are many alternative ways of representing mathematically the unpredictable payout of this wager, as the following illustration shows. The outcome of the wager is the value y, where

$$y = \begin{cases} -(x_1 + x_2) & \text{if} \quad x_1 + x_2 \geq 10, \\ x_1 & \text{if} \quad x_1 \leq x_2, \\ x_2 & \text{if} \quad x_2 \leq x_1. \end{cases}$$

Examining each of the 36 pairs in $\{(x_1, x_2)\}$ $(1 \leq x_1 \leq 6, 1 \leq x_2 \leq 6)$ in turn, the outcomes or payouts y are listed in Table 2.1.

For instance, of the 36 possible pairs of throws (x_1, x_2), a loss of 11 is incurred twice, with throws of (5,6) and (6,5).

Accordingly, let the sample space for the wager be

$$\Omega' = \{1, 2, 3, 4, 5, -11, -12\},$$

let the measurable sets $\mathcal{A}'$ consist of $2^{\Omega'}$, the family of all subsets of Ω', and, for $y \in \Omega'$ let $P(\{y\})$ be as set out in Table 2.1; so, for $A \in \mathcal{A}'$, $P(A)$ can be found by adding up the relevant probabilities in Table 2.1.

Payout	Probability
1	11/36
2	9/36
3	7/36
4	5/36
5	1/36
-11	2/36
-12	1/36

Table 2.1: Distribution of payouts.

Now define random variable $\mathcal{Y} : \Omega' \mapsto \mathbf{R}$ by the identity mapping $\mathcal{Y}(\omega') = y$ for $\omega' = y \in \Omega'$. Trivially, $\mathcal{Y}$ is $(\Omega', \mathcal{A}', P)$-measurable, and

$$\mathrm{E}(\mathcal{Y}) \;=\; \int_{\Omega'} \mathcal{Y}(\omega')\,dP \;=\; \frac{41}{36},$$

or slightly more than 1 euro.

The random variables $f(\mathcal{X})$ and $\mathcal{Y}$ are two equivalent ways of mathematically representing the wager. In [MTRV], $f(\mathcal{X})$ is described as a *contingent* form of the random variable, while $\mathcal{Y}$ is an *elementary* form.

Measurability ensures that the two forms are related. To illustrate, consider $A' = \{5, -11\}$, a subset in the range of the random variable $\mathcal{Y}$ (or $f(\mathcal{X})$). Then

$$f^{-1}(A') \;=\; \{(5,5),(5,6),(6,5)\}$$

which is a subset $A \in \mathcal{A}$ of the sample space Ω. Both A and A' are measurable sets (trivially), and f is a measurable function, with

$$P(A') \;=\; \tfrac{1}{36} + \tfrac{2}{36},$$
$$P\big(f^{-1}(A')\big) \;=\; \tfrac{1}{36} + \tfrac{1}{36} + \tfrac{1}{36},$$
$$P(A') \;=\; P\big(f^{-1}(A')\big) \;=\; P(A).$$

This kind of relationship is generally valid for contingent and elementary forms of random variables.

2.2   Probability and Riemann Sums

Elementary statistical calculation is often learned by performing exercises such as the following.

Example 4 *A sample of 100 individuals is selected, their individual weights are measured, and the results are summarized in Table 2.2.* **Estimate the mean weight and standard deviation of the weights in the sample.**

Weights (kg)	Proportion of sample
0 – 20	0.2
20 – 40	0.3
40 – 60	0.2
60 – 80	0.2
80 – 100	0.1

Table 2.2: Relative frequency table of distribution of weights.

I	$F(I)$	x	$f(x)$	$xF(I)$	$f(x)F(I)$
0 – 20	0.2	10	100	2	20
20 – 40	0.3	30	900	9	270
40 – 60	0.2	50	2500	10	500
60 – 80	0.2	70	4900	14	980
80 – 100	0.1	90	8100	9	810

Table 2.3: Calculation of mean and standard deviation.

Sometimes calculation of the mean and standard deviation is done by setting out the workings as in Table 2.3. The observed weights of the sample members are grouped or classified in intervals I, and the proportion of weights in each interval I is denoted by $F(I)$. A representative weight x is chosen from each interval I. The function $f(x)$ is x^2 since, in this case, these values are needed in order to estimate the variance. Completing the calculation, the estimate of the arithmetic mean weight in the sample is

$$\sum xF(I) = 44 \text{ kg,}$$

while the variance of the weights is approximately

$$\sum x^2 F(I) - (44)^2 = 2580 - 1936 = 644 \text{ kg}^2.$$

The latter calculation, involving $\sum x^2 F(I)$, has the form $\sum f(x)F(I)$ with $f(x) = x^2$. The expressions $\sum xF(I)$ and $\sum f(x)F(I)$ have the form of Riemann sums, in which the interval of real numbers $[0, 100]$ is partitioned by the intervals I, and where each x is a representative data-value in the corresponding interval I. Thus the sums

$$\sum xF(I) \quad \text{and} \quad \sum f(x)F(I)$$

are approximations to the Stieltjes (or Riemann–Stieltjes) integrals

$$\int_J x\, dF \quad \text{and} \quad \int_J f(x)\, dF, \quad \text{respectively;}$$

the domain of integration $[0,100]$ being denoted by J.

In Section 2.1 the variables are discrete. But the outcomes there can be expressed as discrete elements of a continuous domain provided the probabilities are formulated as atomic functions on the domain.

In contrast, the variables in Tables 2.2 and 2.3 are continuous, and their continuous domain is partitioned for Riemann sums in a natural way. Then Riemann sums can be formed as in Table 2.3.

2.3   A Basic Stochastic Integral

The following is similar to Example 2.

Example 5 *Suppose $s = 1, 2, 3, \ldots$ is time, measured in days. Suppose a share, or unit of stock, has value $x(s)$ on day s; suppose $z(s)$ is the number of shares held on day s; and suppose $c(s)$ is the change in the value of the shareholding on day s as a result of the change in share value from the previous day so $c(s) = z(s-1)(x(s) - x(s-1))$. Let $w(s)$ be the cumulative change in shareholding value at end of day s, so $w(s) = w(s-1) + c(s)$.* **If share value $x(s)$ and stockholding $z(s)$ are subject to random variability, how is the gain (or loss) from the stockholding to be estimated?**

Take initial value (at time $s = 0$) of the share to be $x(0)$ (or x_0), take the initial shareholding or number of shares owned to be $z(0)$ (or z_0). Then, at end of day 1 ($s = 1$),

$$c(1) \;=\; z(0) \times (x(1) - x(0)), \qquad w(1) \;=\; w(0) + c(1) \;=\; c(1). \quad (2.1)$$

At end of day s,

$$c(s) \;=\; z(s-1) \times (x(s) - x(s-1)), \qquad w(s) \;=\; w(s-1) + c(s). \quad (2.2)$$

After t days,

$$w(t) \;=\; \sum_{s=1}^{t} z(s-1)(x(s) - x(s-1)). \quad (2.3)$$

If the time increments are reduced to arbitrarily small size (so s represents number of "time ticks"—fractions of a second, say), with the meaning of the other variables adjusted accordingly, then

$$w(t) \;=\; \sum_{j=1}^{n} z(s_{j-1})(x(s_j) - x(s_{j-1})), \quad \text{or} \quad w(t) \;=\; \sum z(s) \Delta x(s). \quad (2.4)$$

The latter expressions are Riemann sum estimates of $\int_0^t z(s)\, dx(s)$ (a Stieltjes-type integral) whenever the latter exists.

Each of the expressions in (2.4) is sample value of a random variable

$$W(t) \;=\; \sum_{j=1}^{n} Z(s_{j-1})(X(s_j) - X(s_{j-1})) \quad \text{or} \quad \int_0^t Z(s)\, dX(s) \quad (2.5)$$

constructed from the random variables X, Z, and W. These notations symbolize—in a "naive" or "realistic" way—the *stochastic integral* of the process Z with respect to the process X. In chapter 8 of [MTRV], symbols s, or S, or $\mathcal{S}$ are used (in place of the symbol $\int$) for various kinds of stochastic integral. In the context described here, S would be the appropriate notation. (See (5.28) below.)

To illustrate the details of this basic stochastic integration, suppose the time increment is 1 day, so $s = 0, 1, 2, 3, 4$ tracks the process over four days. Suppose the initial value of the share at the start of day 1 is $x(0) = 10$. Suppose on each day the value of the share can change by +1 or −1. That is, an "up" increment (U) or "down" increment (D). (Although the probabilities involved will not be used at this stage, in order to keep random variability in mind suppose that, at the end of each day, U occurs with probability 0.5 and suppose D occurs with probability 0.5.)

Suppose initial stockholding at start of day 1 is $z(0) = 1$, or 1 share, and suppose the shareholder buys an extra share whenever the share value increases (U), and otherwise keeps the same number of shares. So there are no circumstances in which shareholding is decreased. (It is easy to imagine that the investor would apply a less optimistic and more prudent share purchasing strategy. But for purpose of illustration some particular strategy must be chosen, and this one is easy to describe.)

The up (U) or down (D) changes in share price over four days are listed in Table 2.4. There are 2^4, = 16, possible processes or histories, corresponding to the 16 possible permutations-with-repetition of the 2 symbols U and D, taken four at a time.

With $x(0) = 10$ and $z(0) = 1$, the histories or processes of interest are prices $X, = (X(1), X(2), X(3), X(4))$; holdings Z; and total gains W; represented by

$$\{x(1), x(2), x(3), x(4)\}, \quad \{z(1), z(2), z(3), z(4)\}, \quad \{w(1), w(2), w(3), w(4)\},$$

with $\{c(1), c(2), c(3), c(4)\}$ representing the daily changes in the value of the stockholding. For $s = 1, 2, 3, 4$, the 'tuple

$$x = (x(s))_{s=1,2,3,4} = (x(1), x(2), x(3), x(4))$$

is a *sample path* for the processes X. Similarly for 'tuples z, w and processes Z, W, respectively. One of the 16 underlying random transition sequences is no. 7 in Table 2.4:

$$U \quad D \quad D \quad U$$

With $x(0) = 10$, $z(0) = 1$, the corresponding sample paths

$$(x(s)), \quad (z(s)), \quad (w(s))$$

can be calculated for this particular sequence of share price U-D transitions, using

$$c(s) = z(s-1)(x(s) - x(s-1)) \quad \text{and} \quad w(s) = w(s-1) + c(s)$$

for $s = 1, 2, 3, 4$, as in Table 2.5.

	$s = 1$	$s = 2$	$s = 3$	$s = 4$	$w(4)$
1.	U	U	U	U	10
2.	D	U	U	U	5
3.	U	D	U	U	4
4.	D	D	U	U	1
5.	U	U	D	U	3
6.	D	U	D	U	0
7.	U	D	D	U	-1
8.	D	D	D	U	-2
9.	U	U	U	D	2
10.	D	U	U	D	-1
11.	U	D	U	D	-2
12.	D	D	U	D	-3
13.	U	U	D	D	-3
14.	D	U	D	D	-4
15.	U	D	D	D	-5
16.	D	D	D	D	-4

Table 2.4: UD sample paths for processes X, Z, W

s	0	1	2	3	4
		U	D	D	U
$x(s)$	10	11	10	9	10
$z(s)$	1	2	2	2	3
$c(s)$		$1(11-10)=1$	$2(10-11)=-2$	$2(9-10)=-2$	$2(10-9)=2$
$w(s)$	0	$0+1=1$	$1-2=-1$	$-1-2=-3$	$-3+2=-1$
		D	U	U	U
$x(s)$	10	9	10	11	12
$z(s)$	1	1	2	3	4
$c(s)$		$1(9-10)=-1$	$1(10-9)=1$	$2(11-10)=2$	$3(12-11)=3$
$w(s)$	0	$0-1=-1$	$-1+1=0$	$0+2=2$	$2+3=5$

Table 2.5: Calculations for two UD sample paths for processes X, Z, W

For transition sample path number 7, UDDU, the overall gain in shareholding value is

$$w(4) \;=\; \sum_{s=1}^{4} z(s-1)\,(x(s)-x(s-1)) \;=\; -1,$$

where $w(4) = -1$ is a "negative gain" or net loss. With $t = 4$, this can be interpreted as the Stieltjes integral[3]

$$w(t) \;=\; \int_{0}^{t} z(s)\,dx(s).$$

Observe that the number $z(s)$ of shares held at any time s depends on whether the share price $x(s)$ has moved up or down. So $z(s)$, $= z(x(s))$, is a deterministic function of $x(s)$; and the value of $z(s)$ varies randomly because $x(s)$ varies randomly.

The same applies to the values of $w(s)$, including the terminal value $w(4)$, or $w(t)$ with $t = 4$. Table 2.5 gives the respective process sample paths for processes, X, Z, W where the underlying share price process X follows sequence DUUU (sample number 2 in Table 2.4).

Regarding notation, the symbols X, Z, W, $X(s), Z(s), W(s)$ (and so on) are used here, in contrast to symbols $\mathcal{X}, \mathcal{Z}$ etc. which were used in discussion of stochastic calculus in Chapter 1. In the latter, the emphasis was on the classical rigorous theory in which random variables are measurable functions, and this is signalled by using $\mathcal{X}$ instead of X, etc.

Where X (rather than $\mathcal{X}$ etc.) is used, the purpose is to indicate the "naive" or "natural" outlook which sees random variability, not in terms of abstract mathematical measurable sets and functions, but in terms of actual occurrences such as tossing a coin, or such as the unpredictable rise and fall of prices.

A mathematically rigorous approach to random variation can be squarely based on the latter view, and in due course this will provide mathematical justification for notation X, Z, W etc.

Table 2.5 describes two out of a possible total of sixteen outcomes, or sample paths, for each of the processes involved. But the tables do not examine the probabilities of the various outcomes. So Table 2.4, for instance, does not really shed much light on how the investment policy of the portfolio holder (shareholder) is capable of performing. The alternative outcomes of the policy are displayed in Table 2.4, but on its own the list of outcomes does not say whether a gain of wealth is more likely than an overall loss.

What if, for instance, we wish to determine the expected overall gain in the value of the shareholding at the end of four days? With $t = 4$, what is the value[4] of $E[W(4)]$?

This can be answered directly as follows.

---

[3] A Stieltjes integral is "the integral of a point function $f(s)$ with respect to a point function $g(s)$", $\int f(s)\,dg$. See section 1.4 of [MTRV], pages 7–14.

[4] In accordance with the presentation in [MTRV], notation $E[\cdot]$ (square brackets) is used with random variables X (conceived "naively" or "realistically" as potential data, along with their probabilities), while round brackets $E(\cdot)$ are used when random variables are interpreted as P-measurable functions $\mathcal{X}$.

- Suppose the different possible amounts of total or net shareholding gain are known. Two of these, −1 and 5, are calculated above. There are 16 possible sample paths for the underlying process $(x(s))$ corresponding to the 16 permutations of U, D. So, allowing for duplication of values, there are at most 14 other values for total shareholding gains.

- The probability of each of the 16 values of $w(4)$ is the same as the probability of the corresponding underlying sample path $(x(s))$ (or $\{x(s) : s = 1, 2, 3, 4\}$). It is assumed that the probability of a U or D transition is 0.5 in each case. If it is further assumed that the transitions are independent, then the probability of each of the 16 sample paths is 0.5^4, = 0.0625, or one sixteenth. This is then the probability of each of 16 outcomes for total shareholding gain, including duplicated values.

The 16 values for $w(4)$ can be easily calculated, as in Table 2.5 above. In fact, the 16 outcomes for net wealth (shareholding value) gain are

$$-5, \ -4, \ -4, \ -3, \ -3, \ -2, \ -2, \ -1, \ -1, \ 0, \ 1, \ 2, \ 3, \ 4, \ 5, \ 10.$$

Since each of the $2^4 = 16$ transition sequences

$$(U,U,U,U), \ \ldots, \ (U,D,D,U), \ \ldots, \ (D,D,D,D)$$

has equal probability $\frac{1}{2} \times \frac{1}{2} \times \frac{1}{2} \times \frac{1}{2} = \frac{1}{16}$, each of the 16 values (including duplicates) for $w(4)$ has probability .0625, or one sixteenth (due to the assumption of independence). Therefore, when all the details are fully calculated out,

$$E[W(4)] \ = \ \sum w(4) \times \frac{1}{16} \ = \ 0, \tag{2.6}$$

the sum being taken over all 16 values (including duplicate values) of total gain $w(4)$.

When duplicate values are combined, there are 12 distinct outcomes for $w(4)$. Each of the duplicated outcomes $-1, -2, -3, -4$ has probability $\frac{1}{8}$, while each of the other 8 distinct outcomes has probability $\frac{1}{16}$.

To find the expected value of $W(4)$ (or $\mathcal{W}_4$) in accordance with the classical, rigorous mathematical theory of probability, it should be formulated in terms of a probability space $(\Omega, P, \mathcal{A})$, so

$$E(\mathcal{W}(4)), \ = \ E(\mathcal{W}_4) \ = \ \int_\Omega \mathcal{W}_4(\omega) \, dP. \tag{2.7}$$

There are many ways in which a sample space Ω can be constructed. One way is to let Ω be the set of numbers consisting of the *different* values of $w(4)$ (i.e. without duplicate values), of which there are 12, and let P be the appropriate atomic probability measure on these 12 values. Letting W_4 be the identity function on Ω, W_4 (or $\mathcal{W}_4$) is measurable (trivially), and

$$E(\mathcal{W}_4) \ = \ E[W(4)] \ = \ 0,$$

because the integral in (2.7) reduces to the sum in (2.6).

Now suppose that, at times $s = 1, 2, 3, 4$, the probability of an Up transition in $x(s)$ is $\frac{2}{3}$, while the probability of a Down transition in $x(s)$ is $\frac{1}{3}$:

$$P(U) = \frac{2}{3}, \qquad P(D) = \frac{1}{3},$$

and suppose, as before, that Up or Down transitions are independent of each other; so, for instance, the joint transition sequence U-D-D-U (and the corresponding $w(4) = -1$) has probability

$$P(UDDU) = \frac{2}{3} \times \frac{1}{3} \times \frac{1}{3} \times \frac{2}{3} = \frac{4}{81};$$

with similar probability calculations for each of the other 15 transition paths and their corresponding $w(4)$ values (including duplicates, such as D-U-U-U which also gives $w(4) = -1$).

The 16 outcomes (including replicated outcomes) for accumulated gain $w(4)$ are the same as before, but because the probabilities are different, the expected net gain is now $\qquad E(\mathcal{W}_4) = \int_\Omega W(\omega)\, dP = \frac{8}{3},\qquad$ or

$$E[W(4)] = \sum w(4)P(w(4)) = \frac{8}{3}. \qquad (2.8)$$

Both calculations reduce to the same finite sum of terms. It is seen here that the new probability distribution, favouring Up transitions, produces an overall net gain in wealth through the policy of acquiring shares on an up-tick, while not shedding shares on a down-tick—the "optimistic" policy, in other words.

If the joint transition probabilities

$$P(U) = P(D) = \frac{1}{2}, \qquad \text{or} \quad P(U) = \frac{2}{3}, \qquad P(D) = \frac{1}{3},$$

are **not** independent, then, provided the dependencies between the various transitions and events are known, it is still possible to calculate all the relevant joint probabilities. But generally this is not so simple as the rule (of multiplying the component probabilities) that obtains when the joint occurrences are independent of each other.

A key step in the analysis is the construction of the probabilities for the values $w(4)$ of the random variable $W(4)$ (or $\mathcal{W}_4$). The framework for this is as follows. Consider any subset A of the sample space

$$\Omega = \{-5,\ -4,\ -3,\ -2,\ -1,\ 0,\ 1,\ 2,\ 3,\ 4,\ 5,\ 10\} \qquad (2.9)$$

whose elements are the different values which can be taken by the variable $w(4)$. For instance, $A = \{-1, 0\}$, which is a member of the family $\mathcal{A}$ of all subsets of Ω.

Following through the logic of the classical theory, probability P is defined on the family $\mathcal{A}$ of measurable subsets of Ω. A random variable W is a real-number-valued, and $(P, \mathcal{A})$-measurable, function

$$\mathcal{W}_4 : \ \Omega \mapsto \mathbf{R}, \qquad \mathcal{W}_4(\omega) \to w(4)$$

in this case, where the potential values $w(4) \in \mathbf{R}$ are the numbers in the right-most column of Table 2.4. The latter set is finite; and every finite subset, such as $\{-1,0\}$, is measurable. In fact, with sample space Ω chosen in this way, $\mathcal{W}_4$ is the identity function, since we have chosen Ω so that its elements are the distinct values $w(4)$.

To find the probability of a set S of $w(4)$-outcomes, such as $S = \{-1,0\}$, the classical theory requires that the corresponding set $\mathcal{W}_4^{-1}(A) \in \mathcal{A}$ be found so that

$$P(S) = P\left(\mathcal{W}_4^{-1}(A)\right)$$

gives the probability of outcomes as the corresponding probability in the sample space. Conveniently, in this case Ω is chosen as simply the set of outcomes $\{w(4)\}$; $\mathcal{W}_4$ is the identity function; and

$$S = A = \mathcal{W}_4^{-1}(A), \quad \text{so} \quad P(S) = P\left(\mathcal{W}_4^{-1}(A)\right)$$

trivially. In effect, the random-variable-as-measurable-function approach of classical theory reduces to the "naive" or "realistic" method, in which the probabilities pertain to outcomes $w(4)$, and are not primarily inherited from some abstract measurable space Ω.

Alternatively, let the sample space be $\mathbf{R}$ and let $\mathcal{A}$ be the class of Borel subsets of $\mathbf{R}$ (so $\mathcal{A}$ includes the singletons $\{w\}$ for each $w \in \Omega$). Define P on $\mathcal{A}$ by $P(\mathbf{R} \smallsetminus \Omega) = 0$ and

$$P(\{w\}) = \begin{cases} \frac{1}{16} & \text{if } w = -5, 0, 1, 2, 3, 4, 5, \text{ or } 10, \\ \frac{2}{16} & \text{if } w = -1, -2, -3, \text{ or } -4, \\ 0 & \text{otherwise;} \end{cases}$$

so P is atomic. As before, with $S = \{-1,0\}$,

$$P(S) = P\left(\mathcal{W}_4^{-1}(A)\right) = \frac{3}{16}.$$

Classical probability involves a quite heavy burden of sophisticated and complicated measure theory. There are good historical reasons for this, and it is unwise to gloss over it. In practice, however, the sample space Ω, in which probability measure P is specified, is often chosen—as above—in such a way that measure-theoretic abstractions and complexities melt away, so that the "natural" or untutored approach, involving just outcomes and their probabilities, is applicable.

[MTRV] shows how to formulate an effective theory of probability which follows naturally from the naive or realistic approach described above, and which does not require the theory of measure as its foundation. The following pages are intended to convey the basic ideas of this approach.

Before moving on to this, here is an elaboration of a technical point of a financial character, which appeared in Example 5 above and in the ensuing discussion, and which is relevant in stochastic integration.

Example 6 *Expression (2.5) above gives two representations of a stochastic integral,*

$$W(t) = \sum_{j=1}^{n} Z(s_{j-1}) (X(s_j) - X(s_{j-1})), \qquad \int_0^t Z(s) \, dX(s),$$

based on sample value calculations (2.4):

$$w(t) = \sum_{s=1}^{t} z(s-1) (x(s) - x(s-1)), \qquad (2.10)$$

derived from (2.2) and (2.3):

$$c(s) = z(s-1) \times (x(s) - x(s-1)), \qquad w(s) = w(s-1) + c(s).$$

If $x(s)$ and $z(s)$ are to be treated as functions of a continuous variable s for $0 \le s \le t$, this suggests calculations or estimates on the lines of

$$w(t) = \sum_{j=1}^{n} z(s_{j-1}) (x(s_j) - x(s_{j-1})), \qquad (2.11)$$

where $0 = s_0 < s_1 < \cdots < s_n = t$ is a partition of $[0, t]$.

  For Example 5 the sample calculation (2.4) of total portfolio value leads unproblematically to the random variable representation (2.5), $W(t) = \int_0^t Z(s) \, dX(s)$. Though we have not yet settled on a meaning for stochastic integral, the discrete expression

$$W(t) = \sum_{j=1}^{n} Z(s_{j-1}) (X(s_j) - X(s_{j-1})),$$

points towards $\int_0^t Z(s) \, dX(s)$ as a continuous variable form of stochastic integral. It seems that the sample value form of the latter should be the Riemann-Stieltjes integral $\int_0^t z(s) \, dx(s)$, for which a Riemann sum estimate is

$$w(t) = \sum_{j=1}^{n} z(s_j') (x(s_j) - x(s_{j-1})) \qquad (2.12)$$

where $s_{j-1} \le s_j' \le s_j$ for $1 \le j \le n$.

  But (2.11) has $z(s_{j-1})$, not the $z(s_j')$ of (2.12). The logic of Example 5 indicates that only the left hand value $z(s_{j-1})$ is permitted in the Riemann sum estimates of the stochastic integral $\int_0^t Z(s) \, dX(s)$. Why is this?

  The issue is to choose between two forms of Riemann sum:

$$w(t) = \sum_{j=1}^{n} z(s_j') (x(s_j) - x(s_{j-1})), \qquad w(t) = \sum_{j=1}^{n} z(s_{j-1}) (x(s_j) - x(s_{j-1})).$$

The latter corresponds to the calculation

$$w(t) = \sum_{s=1}^{n} z(s-1) (x(s) - x(s-1))$$

of Example 5, where $z(s-1)$ is used, but not $z(s)$ or any value intermediate between $z(s-1)$ and $z(s)$.

The reasoning is as follows. At time $s-1$ the investor makes a policy decision to purchase a quantity $z(s-1)$ of shares whose value from time $s-1$ up to (but not including) time s is $x(s-1)$. This number of shares (the portfolio) is retained up to time s. At that instant of time s the decision cycle is repeated, and the investor adjusts the portfolio by taking a position of holding $z(s)$ number of shares, each of which has the new value $x(s)$.

In the time period $s-1$ to s, the gain in value of the portfolio level chosen at time $s-1$ is

$$z(s-1)x(s) - z(s-1)x(s-1), \quad = \quad z(s-1)\,(x(s) - x(s-1)),$$

not *$z(s)x(s) - z(s-1)x(s-1)$, since the portfolio quantity $z(s)$ operates in the time period s to $s+1$ (not $s-1$ to s). Reverting to continuous form, this translates to Riemann sum terms of the form*

$$z(s_{j-1})\,(x(s_j) - x(s_{j-1})), \quad not \quad z(s_j')\,(x(s_j) - x(s_{j-1})).$$

2.4   Choosing a Sample Space

It was mentioned earlier that there are many alternative ways of producing a sample space Ω (along with the linked probability measure P and family $\mathcal{A}$ of measurable subsets of Ω). The set of numbers

$$\{-5, \ -4, \ -3, \ -2, \ -1, \ 0, \ 1, \ 2, \ 3, \ 4, \ 5, \ 10\}$$

was used as sample space for the random variability in the preceding example of stochastic integration. The measurable space $\mathcal{A}$ was the family of all subsets of Ω, and the example was illustrated by means of two distinct probability measures P, one of which was based on Up and Down transitions being equally likely, where for the other measure an Up transition was twice as likely as a Down.

An alternative sample space for this example of random variability is

$$\Omega \ = \ \Omega_1 \times \Omega_2 \times \Omega_3 \times \Omega_4, \tag{2.13}$$

where $\Omega_j = \{U, D\}$ for $j = 1, 2, 3, 4$; so the elements ω of Ω consist of sixteen 4-tuples of the form

$$\omega \ = \ (\cdot, \cdot, \cdot, \cdot), \quad \text{such as } \omega = (U, D, D, U) \quad \text{for example.}$$

Let the measurable space $\mathcal{A}$ be the family of all subsets A of Ω; so $\mathcal{A}$ contains 2^{16} members, one of which (for example) is

$$A \ = \ \{(D,U,U,D), (U,D,D,U), (D,U,D,U), (U,U,U,U), (D,D,D,D)\},$$

with A consisting of five individual four-tuples. Assume that Up transitions and Down transitions are equally likely, and that they are independent events. Then, as before,

$$P(\{\omega\}) = \frac{1}{16}$$

for each $\omega \in \Omega$. For A above, $P(A) = \frac{5}{16}$.

To relate this probability structure to the shareholding example, let $\mathbf{R}^4 = \mathbf{R} \times \mathbf{R} \times \mathbf{R} \times \mathbf{R}$, and let

$$f : \Omega \mapsto \mathbf{R}^4, \qquad f(\omega) = ((x(1), x(2), x(3), x(4)), \tag{2.14}$$

using Table 2.4; so, for instance,

$$f(\omega) = f((U, D, D, U)) = (11, 10, 9, 10) = (x(1), x(2), x(3), x(4)),$$

and so on. Next, let $\mathbf{S}$ denote the stochastic integrals of the preceding section, so for $x = (x(1), x(2), x(3), x(4)) \in \mathbf{R}^4$,

$$\mathbf{S}(x) = \int_0^4 z(s) \, dx(s) = \sum_{s=1}^4 z(s-1)(x(s) - x(s-1)),$$

so $\mathbf{S}(x)$ gives the values $w(4)$ of Table 2.4. As described in Section 2.3, the rationale for deducing the probabilities of outcomes $\mathbf{S}(x)$, $= w(4)$, from the probabilities on Ω is the relationship

$$P(w(4)) = P(f^{-1}(\mathbf{S}^{-1}(w(4)))).$$

For this calculation to work in general, the functions involved (f and $\mathbf{S}$) must be measurable. But that is no problem in this case since all the sets involved are finite. To illustrate the calculation, take $w(4) = -2$. Then, referring to Table 2.4 whenever necessary,

$$\mathbf{S}^{-1}(w(4)) = \mathbf{S}^{-1}(-2) = \{(9, 8, 7, 8), \ (11, 10, 11, 10)\},$$
$$f^{-1}(\mathbf{S}^{-1}(w(4))) = f^{-1}(\mathbf{S}^{-1}(-2)) = \{(D, D, D, U), \ (U, D, U, D)\},$$
$$P(w(4)) = P(f^{-1}(\mathbf{S}^{-1}(-2))) = \tfrac{2}{16}.$$

As a further illustration, suppose we now take

$$\Omega = \mathbf{R}^4, \tag{2.15}$$

letting $\mathcal{A}$ be the sigma-algebra of Borel subsets of $\mathbf{R}^4$. For each $A \in \mathcal{A}$ define $P(A)$ as follows. Suppose $0 < \alpha_u < 1$ and $\alpha_d = 1 - \alpha_u$. Taking $x(0) = 10$, let $B \in \mathcal{A}$ be the set consisting of the 16 elements

$$x = (x(1), x(2), x(3), x(4)), \qquad x(s) = x(s-1) \pm 1 \quad \text{for } s = 1, 2, 3, 4;$$

and let P be an atomic measure with $P(\{y\}) = 0$ if $y \notin B$, and, for $x \in B$,

$$P(\{x\}) = \beta_1 \beta_2 \beta_3 \beta_4 \quad \text{where} \quad \beta_s = \begin{cases} \alpha_u & \text{if } x(s) = x(s-1) + 1, \\ \alpha_d & \text{if } x(s) = x(s-1) - 1 \end{cases} \tag{2.16}$$

for $s = 1, 2, 3$ or 4; with $P(A) = 0$ for all other $A \in \mathcal{A}$. Thus, if Up and Down transitions are independent and equally likely, then $\alpha_u = \alpha_d = \frac{1}{2}$ and

$$P(\{x\}) \;=\; \begin{cases} \frac{1}{16} & \text{for} \quad x \in B, \\ 0 & \text{for} \quad x \notin B. \end{cases} \tag{2.17}$$

As in previous versions of the sample space Ω, this construction imposes probabilities on the outcomes $w(4)$ or $\mathbf{S}(x)$, by means of the relation

$$\mathbf{S}^{-1}(S) \;=\; A \;\in\; \mathcal{A}, \qquad\qquad P(S) \;=\; P\big(\mathbf{S}^{-1}(S)\big) \;=\; P(A)$$

for measurable sets S in the range of $\mathbf{S}$, since $\mathbf{S}$ is clearly a measurable function on the domain Ω. Thus, for outcomes $w(4) = -1$ or -2, $S = \{-1, -2\}$ and

$$\begin{aligned} \mathbf{S}^{-1}(S) \;=\; A \;&=\; \{(x(1), x(2), x(3), x(4))\} \;\in\; \mathcal{A} \\ &=\; \{(11, 10, 9, 10),\ (9, 8, 7, 8),\ (9, 10, 11, 10),\ (11, 10, 11, 10)\}, \\ P(S) \;=\; P(A) \;&=\; \tfrac{4}{16} \ \ (\text{referring back to Table 2.4}) . \end{aligned}$$

2.5   More on Basic Stochastic Integral

The constructions in Sections 2.3 and 2.4 purported to be about stochastic integration. While a case can be made that (2.6) and (2.7) are actually stochastic integrals, such simple examples are not really what the standard or classical theory of Chapter 1 is all about. The examples and illustrations in Sections 2.3 and 2.4 may not really be much help in coming to grips with the standard theory of stochastic integrals outlined in Chapter 1.

This is because Chapter 1, on the definition and meaning of classical stochastic integration, involves subtle passages to a limit, whereas (2.6) and (2.7) involve only finite sums and some elementary probability calculations.

From the latter point of view, introducing probability measure spaces and random-variables-as-measurable-functions seems to be an unnecessary complication. So, from such a straightforward starting point, why does the theory become so challenging and "messy", as portrayed in Chapter 1?

As in Example 2, the illustration in Section 2.3 involves dividing up the time period (4 days) into 4 sections; leading to sample space $\Omega = \mathbf{R}^4$ in (2.15). Why not simply continue in this vein, and subdivide the time into 40, or 400, or 4 million steps instead of just 4; using sample spaces $\mathbf{R}^{40}$, or $\mathbf{R}^{400}$, or $\mathbf{R}^{4000000}$, respectively? The computations may become lengthier, but no new principle is involved; each of the variables changes in discrete steps at discrete points in time.[5]

Other simplifications can be similarly adopted. For instance, only two kinds of changes are contemplated in Section 2.3: increase (Up) or decrease (Down).

---

[5]R. Feynman proposed something on those lines to deal with an analogous problem in the path integral theory of quantum mechanics. See his comments on "subdivision and limiting processes" quoted in page 17 above.

But that is merely a slight technical limitation. Just as the number of discrete times can be increased indefinitely, so can the number of distinct, discrete values which can be potentially taken at any instant.

There is a plausible argument for this essentially discrete approach, at least in the case of financial shareholding. Actual stock market values register changes at discrete intervals of time (time ticks), and the amount of change that can occur is measured in discrete divisions (or basis points) of the currency.

So why does the mathematical model for such processes (as described in Chapter 1, for instance) require passages to a limit involving infinite divisibility of both the time domain, and the value range?

In fact there are sound mathematical reasons for this seemingly complicated approach. For one thing, instead of choosing one of many possible finite division points of time and values, passage to a limit—if that is possible—replaces a multiplicity of rather arbitrary choices by a single definite procedure, which may actually be easier to compute.

Furthermore, Brownian motion provides a good mathematical model for many random processes, and Brownian motion is based on continuous time and continuous values, not discrete.

The finite sum calculation $\sum Z(s-1)(X(s) - X(s-1))$ is a good enough starting point for stochastic integration. But it is only a starting point. The question is, by successive sub-division of time and/or values, can a single limiting result be found—a single random variable to replace all those obtained, as in Section 2.3, by computation with discrete times/values? In other words, is there convergence to a limit?

The account in **I1, I2, I3, I4** of Chapter 1 suggests that stochastic integrals converge, not in the strict sense of convergence, but only in some loose or weak manner. Our aim is to shed more light on the nature of convergence that appears in these constructions.

There is a more fundamental and compelling reason for introducing sophisticated measure theory into probability. Accessing the full power of probability, beyond the elementary calculations of Sections 2.3 and 2.4, requires an understanding of the operation, scope and limitations of results such as the Laws of Large Numbers, and the Central Limit Theorems of probability theory. This increased power was eventually achieved in the early twentieth century by A.N. Kolmogorov [93] and others, who formulated the theory of probability in terms of probability measure spaces; so random variables were understood, not as potential outcome data in association with their linked probabilities (as in Sections 2.3 and 2.4), but as measurable functions mapping an abstract probability space or sample space into a set of potential outcome data; thereby imposing a probability structure on the actual data.

The new power which Kolmogorov's innovation added to our understanding of probability springs from his interpretation of probability and expected value in terms of, respectively, measure theory and Lebesgue integral. The specific feature of the Lebesgue integral which enables this improvement is its enhanced convergence properties.

Earlier versions of integration—examples being the "integral as anti-derivative", and the Riemann integral (see Chapter 10)—did not provide adequate rationale for integrability of the limit of a convergent sequence of integrable functions. Lebesgue integration, on the other hand, has a dominated convergence theorem: if the absolute value of each member of a convergent sequence of integrable functions is less than some integrable function, then the limit function of the sequence of functions is integrable.

Introducing Lebesgue integration (and its dominated convergence theorem) into probability theory brought about the twentieth century's great advances in our understanding of random variation, including Brownian motion and stochastic integration. The price paid for this included the somewhat counter-intuitive and challenging notion that a random variable, such as a stochastic integral, is to be thought of as a measurable function.

But the unchallenging and intuitive finite sums of (2.6) look remarkably like Riemann sums. (This point is elaborated in [MTRV] pages 15–17.) What if Riemann integration can be adjusted so that it incorporates (for instance) a dominated convergence theorem?

A theory of probability on these lines is presented in [MTRV]. The purpose of Part I of this book is to examine some features of this Riemannian version of probability theory, particularly in relation to stochastic integration.

Similarly Part II investigates the Riemann approach to problems of path integration in quantum mechanics.

Chapter 3

Integration and Probability

3.1 -Complete Integration

Unlike algebra or calculus for instance, historians of the theory of probability often claim that, while it has a pre-history in gambling practice, this subject is a relative newcomer in mathematical terms.

Ideas of random variability and probability were put on a firmer mathematical basis in the course of the nineteenth century, and the modern form of the theory was well established by the mid-twentieth century.

An elementary link between statistics and probability is demonstrated in Sections 2.1 and 2.2 above, and the Riemann sum calculations of Example 4 indicate the central role of mathematical integration in analysis of random variation.

Twentieth century developments in probability and random variation are closely linked to developments in the theory of measure and integration culminating in Lebesgue's theory of the integral [100]. A.N. Kolmogorov [93] made this the foundation of probability theory by identifying–

- the probability of an event as the measure of a set,

- a random variable as a measurable function, and

- the expected value of a random variable as the integral of a measurable function with respect to a probability measure.

One of the standard ways of defining the Lebesgue integral of a μ-measurable function $f(\omega)$ $(\omega \in \Omega)$ is to form finite sums

$$s_j \;=\; \sum_{r=1}^{n} \phi_j^{(r)}(\omega)\,\mu\left(A_j^{(r)}\right)$$

of *simple functions* $\phi_j(\omega)$ which converge to f as $j \to \infty$, and then define the Lebesgue integral of f on Ω by

$$\int_{\Omega} f(\omega)\,d\mu \;=\; \lim_{j \to \infty} s_j. \tag{3.1}$$

Gauge Integral Structures for Stochastic Calculus and Quantum Electrodynamics, First Edition. Patrick Muldowney.
© 2021 John Wiley & Sons, Inc. Published 2021 by John Wiley & Sons, Inc.

The dominated convergence theorem emerges from this: *Suppose Lebesgue int-egrable functions $f_j(\omega)$ converge almost everywhere to $f(\omega)$, with $|f_j(\omega)| \le g(\omega)$ almost everywhere, g also being Lebesgue integrable. Then f is Lebesgue int-egrable, and $\int_\Omega f_j$ converges to $\int_\Omega f$ as $j \to \infty$.*

Depending on the measurable integrand f, the measurable sets $A_j^{(r)}$ in Defin-ition 3.1 can be intervals, open sets, closed sets, isolated points, and/or various countable combinations of these and other even more complicated sets. One could say that the means used to access the integral of f are themselves some-what arcane and inaccessible. The "cure" (finding appropriate measurable sets) could be worse than the "disease" (finding the integral).[1]

The entrance to measure theory and Lebesgue integration is guarded by such fearsome sets and functions as the Cantor set, and the Devil's Staircase or Cantor function [102]. But while it is unwise to enter the house of Lebesgue without keeping an eye out for monsters[2], in the simple examples of essentially finite domains of preceding chapters we managed to negotiate our way fairly painlessly through the relevant measurable sets/functions. (Would we be so lucky if the domains were infinite, or the functions a bit more complicated?)

These monsters will never completely go away. But perhaps it would be better to **not** have to wrestle with them as a pre-condition of gaining entry to the house. It would be nice if the monsters were kept locked up in the cellar, not on guard at the front door. Is there any other way to deal with probability which provides full mathematical power and rigour? Is there another house, one that is more easily accessible, and closer to the "naive" or realistic view of random variability, as outlined in Chapter 2 above, and in [MTRV] pages 15–17?

The answer lies in a version of Riemann integration, as implied by Example 4 which demonstrates that elementary statistical calculations take the form of Riemann sums.

The Riemann integral of a bounded, continuous function f on a real interval $[a, b]$ is sometimes defined by the Darboux method. Partitions $\mathcal{P}$ of $[a, b]$ are formed,

$$\mathcal{P} = \{]t_{j-1}, t_j]\}, \qquad a = t_0 < t_1 < \cdots < t_{n-1} < t_n = b,$$

from which upper and lower integrals of f are constructed, whose values converge to the Riemann integral of f on $[a, b]$, $\int_a^b f(s), ds$, as

$$|\mathcal{P}|, := \max(t_j - t_j - 1), \to 0.$$

This definition has features in common with the definition (3.1) of the Lebesgue integral above. Firstly, it provides a way of constructing or (in principle) calc-

---

[1] In practice, integrals are not always evaluated, but may be used for instance as intermed-iate constructions in the course of some larger analysis. In that case it is important that the integral have some robust meaning such as Lebesgue's. And whenever some integral must be evaluated there are various well-known devices which can be used; definition 3.1 is rarely used in practice. Chapter 10 below shows that definition 3.1 is equivalent to a Riemann-Stieltjes integration. But, even though that is a more basic and simpler kind of integral than Lebesgue's, in this case it does not simplify things very much.

[2] Hermite and Poincaré described them as "monsters". See Chapter 10.

ulating the value of the integral as a limit of a sequence of values. Secondly, the
assumption of continuity of the integrand (measurability in the case of (3.1))
ensures existence of the limit.

Here is an alternative definition of the Riemann integral.

Definition 1 *A function f is Riemann integrable on* $[a,b]$*, with integral*

$$\int_a^b f(s)\,ds = \alpha,$$

if, given $\varepsilon > 0$*, there exists* $\delta > 0$ *such that, for each partition* $\mathcal{P}$ *with* $|\mathcal{P}| < \delta$*, the
Riemann sum of f satisfies*

$$\left| \alpha - \sum_{j=1}^n f(s_j)(t_j - t_{j-1}) \right| < \varepsilon.$$

Note that $t_{r-1} \le s_r \le t_r$ for each r, or $s_r \in I^r = [t_{r-1}, t_r]$. If the Riemann sum for
partition $\mathcal{P}$ is denoted $(\mathcal{P}) \sum f(s)|I|$ then the Riemann integrability inequality
is

$$\left| \alpha - (\mathcal{P}) \sum f(s)|I| \right| < \varepsilon.$$

This definition imposes **no pre-conditions** (such as continuity or measurability) on the real- or complex-valued integrand f. But if, for instance, f
is continuous then the integrability inequality is easily shown to be satisfied.

Definition 1 uses intervals rather than measurable sets. But it does not
provide a dominated convergence theorem, so it is ultimately unsuitable for more
advanced analysis in probability. However a small adaptation of the definition
gives what is needed for probability. (So maybe, after all, the "cure" is **not**
worse than the "disease".)

A partition $\mathcal{P}$ consists of non-overlapping intervals $\{I\}$. In fact, by taking
half-open intervals $I^r =]t_{r-1}, t_r]$, we can say that the intervals in $\mathcal{P}$ are disjoint.
If $I =]t',t]$ and $t' \le s \le t$, we say that the couple (s,I) are an *associated
point-interval pair* (or that the interval I has *tag point* s. A finite collection
$\mathcal{D} = \{(s,I)\}$ of tagged intervals is a *division* of $]a,b]$ if the intervals I form a
partition[3] of $]a,b]$.

Given a positive function $\delta(s) > 0$ for $a \le s \le b$, the tagged interval $(s,I), =
(s,]t',t])$, is δ-*fine* if $|I| = t - t' < \delta(s)$. A division $\mathcal{D} = \{(s,I)\}$ of $[a,b]$ is δ-fine
if (s,I) is δ-fine for each $(s,I) \in \mathcal{D}$. Such a function $\delta(s) > 0$ is called a *gauge*.

Definition 2 *A function f is Riemann-complete integrable on* $[a,b]$*, with in-
tegral* $\int_a^b f(s)\,ds = \alpha$ *(or* $\int_{[a,b]} f(s)|I| = \alpha$*), if, given* $\varepsilon > 0$*, there exists a gauge*
$\delta(s) > 0$ *such that, for each* δ*-fine division* $\mathcal{D} = \{(s_j,]t_{j-1}, t_j])\}$ *of* $[a,b]$*, the
corresponding Riemann sum of f satisfies*

$$\left| \alpha - \sum_{j=1}^n f(s_j)(t_j - t_{j-1}) \right| < \varepsilon, \qquad (or\ \left| \alpha - (\mathcal{D}) \sum f(s)|I| \right| < \varepsilon).$$

[3]If $[a,b]$ is closed on the left, take the first partitioning interval to be $[t_0,t_1]$ or $[a,t_1]$.

The Riemann-complete integral (rc-integral, for short) has a dominated convergence theorem: *Suppose Riemann-complete integrable functions $f_j(s)$ converge almost everywhere[4] to $f(s)$ ($s \in [a,b]$), with $|f_j(s)| \le g(s)$ almost everywhere, g also being Riemann-complete integrable. Then f is Riemann-complete integrable, and $\int_\Omega f_j$ converges to $\int_{[a,b]} f$ as $j \to \infty$.*

This breakthrough for Riemann-type integration allows probability theory to be based, not on measurable sets/functions, but on the relatively more straightforward finite sums of Section 2.3. For full details see [MTRV] pages 183–256. The breakthrough was achieved in the 1950s by Ralph Henstock and (independently) Jaroslav Kurzweil.

This approach goes by various names in the literature, such as Henstock-Kurzweil integral, gauge integral, and generalized Riemann integral. In this book, and in [MTRV], it is referred to by its original name—Riemann-complete or rc-integral.

Historically, there are the Newton-Leibniz integral (anti-derivative or calculus integral), the Cauchy and Riemann integrals, the Stieltjes or Riemann-Stieltjes integral, the Lebesgue and Lebesgue-Stieltjes integrals, along with less familiar ones such as the Burkill, Radon, Perron and Denjoy integrals.[5] Each of these kinds of integral now has a "-complete" counterpart (or extension) on the lines of Definition 2. By adding the appendage "-complete", an important distinction is preserved between different kinds of integral which play different roles in different areas of mathematics.

In particular, a theme of this book is the stochastic or Itô integral, and in applying the approach of Definition 2, it will be evident that Stieltjes- and Burkill-type integration have a major role. A suitable terminology for these, in line with Definition 2, is Stieltjes-complete and Burkill-complete.

Comparing (3.1) and Definition 2, the latter uses a real interval domain such as $[a,b]$, while the Lebesgue integral (3.1) is defined on domain Ω, an abstract measurable space which **can be**—but is **not necessarily**—a real number domain. It is sometimes asserted that this is a defect in the Riemann-complete approach to integration, in that, unlike Lebesgue integration, the rc-integral is restricted to real number domains.

This is mistaken. In [123] (pages 31–33), it is shown that (3.1) can be expressed in Riemann-complete terms; and every Lebesgue integrable function, whether defined on a real domain or on a general measurable space Ω, is also Riemann-Stieltjes integrable. In particular, a Lebesgue-complete integral is available. See also (10.10), page 316 of Chapter 10 below.

What is Stieltjes-complete integration, and how does it differ from the Riemann-complete integral? The basic Riemann-Stieltjes integral is fairly well known. The Riemann integral involves a point function integrand $f(s)$ weighted by a distance or length function $|I| = |]t',t]| = t - t'$, with $t' \le s \le t$, giving Riemann sum $\sum f(s)|I|$.

---

[4]If rc-integration is not based on measurable sets and functions, what does it mean to say S is *null*, or a set of measure zero? The answer is as follows. Take integrand $f(s)$ to be $\mathbf{1}_S(s)$, the indicator function of S. Then S is null if $\int_a^b \mathbf{1}_S(s)\,ds = 0$.

[5]Many of these integrals are described in [MTRV]; see also [61].

The *Riemann-Stieltjes* integral is the integral of a point function $f(s)$ with respect to another point function $g(s)$. In this case, the length function $|I| = t - t'$ of Riemann integration is replaced by a weight function $g(I) := g(t) - g(t')$, giving Riemann sums $\sum f(s)g(I), = \sum f(s)(g(t) - g(t'))$. The following is a constructive definition for the Riemann-Stieltjes integral $\int_a^b f(s)\,dg$.

The definition states what existence of the integral means, but it does not provide conditions on f and g which are in themselves sufficient to ensure existence of $\int_a^b f(s)\,dg$ in accordance with Definition 3. Conditions of the latter sort are "f is continuous, g has bounded variation". These conditions will be revisited in the context of Itô stochastic integration.

Definition 3 *A point function $f(s)$ is Riemann-Stieltjes integrable with respect to a point function $g(s)$ on $[a,b]$, with integral $\int_a^b f(s)\,dg(s) = \alpha$, if, given $\varepsilon > 0$, there exists $\delta > 0$ such that, for each partition $\mathcal{P}$ with $|\mathcal{P}| < \delta$, the Riemann sum satisfies*

$$\left| \alpha - \sum_{j=1}^n f(s_j)(g(t_j) - g(t_{j-1})) \right| < \varepsilon.$$

In fact, starting with the latter definition, Definition 1 is implied by it if g is taken to be the identity function $g(s) = s$. Likewise, with g as identity function, Definition 4 of the Stieltjes-complete integral (sc-integral) reduces to Definition 2 of the rc-integral:

Definition 4 *A point function $f(s)$ is Stieltjes-complete integrable with respect to a point function $g(s)$ on $[a,b]$, with integral $\int_a^b f(s)\,dg = \alpha$, or*

$$\int_{[a,b]} f(s)\,dg(s), \quad = \quad \int_{[a,b]} f(s)g(I), \quad = \quad \alpha,$$

if, given $\varepsilon > 0$, there exists a gauge $\delta(s) > 0$ such that, for each δ-fine division $\mathcal{D} = \{(s_j, \,]t_{j-1}, t_j])\}$ of $[a,b]$, the corresponding Riemann sum satisfies

$$\left| \alpha - \sum_{j=1}^n f(s_j)(g(t_j) - g(t_{j-1})) \right| < \varepsilon, \qquad (or \quad \left| \alpha - (\mathcal{D})\sum f(s)g(I) \right| < \varepsilon).$$

How is Stieltjes integration connected to the Itô integral of stochastic calculus? In Chapter 1 the focus of interest is presented as

$$\int_\Omega \mathcal{Z}(s)\,d\mathcal{X}(s), \quad or \quad \int_\Omega f(\mathcal{X}(s))\,d\mathcal{X}(s),$$

and **I1** seeks to give some meaning to such formulations. In order to motivate further discussion, Chapter 1 posed questions as to the meaning of various aspects of the classical Itô integral. In particular, (1.1) brought up the standard treatment of the most basic form of $\int_\Omega f(\mathcal{X})\,d\mathcal{X}, = \int_\Omega d\mathcal{X}$, in which f is the simplest thing it can be—the function $f(\mathcal{X}(\omega)) = 1$ for all $\omega \in \Omega$.

As demonstrated in Section 2.3, the sample paths such as $(x(s))$ $(0 \le s \le t)$ form a point of contact between the so-called realistic view of stochastic integrals and the classical treatment outlined in Chapter 1. The sample path versions of

$$\int_\Omega d\mathcal{X}, \qquad \int_\Omega f(\mathcal{X}) \, d\mathcal{X},$$

are

$$\int_{s=0}^{s=t} dx(s), \qquad \int_{s=0}^{s=t} f(x(s)) \, dx(s),$$

because $\mathcal{X}(\omega) = (x(s))$, $0 \le s \le t$. If $\int_0^t dx(s)$ exists it is a real number, and then it can be interpreted as a "sample result" of $\int_\Omega d\mathcal{X}$; in the sense that $\mathcal{Y} = \int_\Omega d\mathcal{X}$ is a random variable, and, for $\omega \in \Omega$,

$$\mathcal{X}(\omega) = (x(s))_{0 \le s \le t}, \qquad\qquad \mathcal{Y}(\omega) = \int_0^t dx(s).$$

Likewise for $\int_\Omega f(\mathcal{X}) \, d\mathcal{X}$ and $\int_{s=0}^{s=t} f(x(s)) \, dx(s)$. The question remains, when does $\int_0^t dx(s)$ (or $\int_0^t f(x(s)) dx(s)$) exist? And how can the theory be understood if the integrals $\int_0^t dx(s)$ etc. do **not** exist?

The single most important instance of stochastic integration is when $\mathcal{X}$ is a Brownian motion. But for the moment let $\mathcal{X}$ be an arbitrary stochastic process. Given a function $x(s)$ for $0 \le s \le t$, the integral $\int_0^t dx(s)$, if it is defined and if it exists, is a Stieltjes-type integral of an (implicit) point integrand $f(s) \equiv 1$ with respect to the point function $x(s)$. Definition 3 deals with **definition**, not existence of the integral. It lays down no conditions on the point integrator function $g(s)$. It is permissible, therefore, to try out the sample function $x(s)$ in the integral.

In other words, "$\int_0^t dx(s)$" is admissible for consideration. To see whether this integral exists or not, let $\mathcal{P}$ be any partition $[0,t]$. For $I = \,]s',s]$ write $x(I) = x(s) - x(s')$. Then the Riemann sum is

$$(\mathcal{P}) \sum x(I) = \sum_{j=1}^n (x(t_j) - x(t_{j-1})) = x(t) - x(0)$$

by cancellation of terms. This holds for every partition $\mathcal{P}$. Therefore, taking $\alpha = x(t) - x(0)$ in Definition 3, the Riemann sum inequality holds, so, for every sample path $(x(s))$, $\int_0^t dx(s)$ exists and equals $x(t) - x(0)$.

On the strength of this it may be reasonable to conclude that the stochastic integral $\int_\Omega d\mathcal{X}$ exists, with value equal to the random variable $\mathcal{X}_t - \mathcal{X}_0$. This would then resolve the quibble expressed above in (1.1).

What about $\int_\Omega f(\mathcal{X}) \, d\mathcal{X}$? The sample path version of this is $\int_0^t f(x(s)) \, dx(s)$ which, again, is a Stieltjes-type integral which fits into the framework of Definition 3, and is therefore **defined** for all integrator functions x (or $x(I)$).

Does it **exist**? The answer is yes, if f is continuous, and if $x(s)$ has bounded variation and is continuous; so $f(x(s))$ is continuous. (This is a familiar property of Riemann-Stieltjes integration, in which the point integrand is continuous and the point integrator has bounded variation, or the other way round—point integrand has bounded variation and point integrator continuous.)

But there are problems with, not just $\int_0^t f(x(s)) \, dx(s)$, but also $\int_0^t dx(s)$. The most important case of stochastic integration is where $\mathcal{X}$ represents Brownian motion. In that case it can be arranged that all sample paths are continuous. But the set of sample paths $x(s)$ which have bounded variation is null. So, except for a null set of x, it is impossible to call upon the most familiar property[6] of Riemann-Stieltjes integration which ensures existence of the integral.

That is not the only difficulty, nor the most serious one. Probability theory, including stochastic integration, requires good convergence theorems, such as dominated convergence. Traditionally, these have been achieved by means of Lebesgue's theory, and this requires that the integrator function $x(I)$ (or $g(I) = g(s) - g(s')$ in Definition 3) be expressible as a measure function.

But, except for a null set of x, Brownian paths $x(s)$ have unbounded variation on every interval of time s. This seems to be the fundamental reason for the various devices employed by **I1, I2, I2, I4** of Chapter 1 to produce a meaningful stochastic integral.

However, as pointed out earlier, the -complete integral theory (including Stieltjes-complete or sc-integrals) not only delivers good convergence theorems, but is constructed by means of Riemann sums, using intervals I. So there is reason to attempt to re-formulate stochastic integration using -complete rather than Lebesgue integration.

Integrator functions such as the interval length function $|I|$ and the point integrator increment $g(I)$ are additive, in the following way. Given

$$I' = \;]s', s], \qquad I'' = \;]s, s''], \qquad I = \;]s', s''] = I' \cup I'',$$

then $|I| = |I'| + |I''|$, and

$$g(I) = g(s'') - g(s') = (g(s'') - g(s)) + (g(s) - g(s')) = g(I') + g(I'').$$

(In Lebesgue integration the integrator or measure function $\mu(S)$ is required to be countably additive on countable unions of disjoint measurable sets S.)

Thus, any point function integrator $g(s)$ corresponds to an interval function $k(I)$ which is additive on adjoining intervals, $k(I) = k(I') + k(I'')$; simply take $k(I) = g(I) = g(s'') - g(s')$. Conversely, given an additive interval function $k(I)$, there corresponds a point function $g(s)$:

$$g(s) := k(] - \infty, s]).$$

Accordingly, the Stieltjes-complete integral of Definition 4 can be expressed in terms of finitely additive integrator functions $k(I)$:

Definition 5 *A point function $f(s)$ is Stieltjes-complete integrable with respect to a finitely additive interval function $k(I)$ on $[a, b]$, with integral $\int_a^b f(s) \, k(I)$ $= \alpha$, or*

$$\int_a^b f(s) \, k(I), \;=\; \int_{[a,b]} f(s) k(I), \;=\; \alpha,$$

---

[6]Integrand $f(s)$ continuous, integrator $g(s)$ with bounded variation.

if, given $\varepsilon > 0$, there exists a gauge $\delta(s) > 0$ such that, for each δ-fine division $\mathcal{D} = \{(s_j, \,]t_{j-1}, t_j])\}$ of $[a, b]$, the corresponding Riemann sum satisfies

$$\left| \alpha - \sum_{j=1}^{n} f(s_j)\, k(]t_{j-1}, t_j]) \right| \; < \; \varepsilon, \qquad (or \; \left| \alpha - (\mathcal{D}) \sum f(s)k(I) \right| \; < \; \varepsilon).$$

3.2   Burkill-complete Stochastic Integral

In contrast, the *Burkill integral* ([13], [14], [68]) involves integrator functions $h(I)$ which are **not additive**—it is not required that $h(I) = h(I') + h(I'')$. (Of course, it is **not forbidden** either!)

It turns out that a version of the Burkill integral is very useful in a reformulated theory of stochastic integration, and in the Feynman integral theory of quantum mechanics. In [MTRV], in addition to dependence on cells I, an *extended Burkill integrand* $h(s, I)$ is allowed to depend also on tag points s of cells I; and from this is developed a *Burkill-complete* form of integration. (A Burkill-complete integrand h is not additive in respect of its dependence on cells I—if it **is** additive it receives a different designation.)

Definition 6 below fits into the -complete structure of definitions. It deals with integrands $h(s, I)$ which are functions of tagged intervals $(\bar{s}, I)$ (or associated point-interval pairs $(\bar{s}, I)$) ; for instance, with $I =]s', s'']$, $\bar{s} = s'$ or s'',

$$h(s, I) \; = \; \sqrt{\bar{s}}(s'' - s');$$

and a partition $\mathcal{P} = \{0 = s_0, s_1, \ldots, s_{n-1}, s_n = 1\}$ of $]0, 1]$ is a finite sample of points of the domain. Then $h(s, I) = \sqrt{\bar{s}_j}(s_j - s_{j-1})$ with $\bar{s}_j = s_j$ or s_{j-1}; and a Riemann sum $(\mathcal{P}) \sum h(s, I)$ is a functional of samples of points:

$$(\mathcal{P}) \sum h(x, I) \; = \; \sum_{j=1}^{n} \sqrt{\bar{s}_j (s_j - s_{j-1})} \tag{3.2}$$

This formulation changes the perspective of -complete integration from point-cell pairs to finite samples of points. Nevertheless, as in chapter 4 of [MTRV], the underlying structures can be readily conveyed in terms of relationships between cells or intervals I of the domain.

In effect, adjacent pairs of points (s_j, s_{j-1}) from the finite sample $\mathcal{P}$ must satisfy conditions corresponding to Axioms DS1 to DS8 in chapter 4 (pages 111–113 of [MTRV]). Of course, in simple domains such as $]0, 1]$ it is natural to visualize pairs of points (s_j, s_{j-1}) as intervals I_j. But in the more complicated and structured domains used in quantum field theory (Chapters 8 and 9 below), the alternative "samples of points" perspective may be helpful.

Lebesgue integration uses functions $\mu(A)$ of measurable subsets of a domain. In contrast, -complete integration uses functions $\mu(I)$ of subintervals of the domain. The latter can be replaced by $\mu(s', s'')$ (where $I =]s', s'']$). But measurable sets A can consist of infinitely many intervals and discrete points, ruling out the "finite sample of points" approach.

The Itô or stochastic integral $\int Z(s)\,dX(s)$ can be thought of as a Stieltjes integral, with sample form $\int z(s)\,dx(s)$—the integral of a point function $z(s)$ with respect to a point function $x(s)$.

It might be expected that, when converted into -complete integration, this becomes Stieltjes-complete. But this is not necessarily the case. It has been pointed out earlier that, in sampling or Riemann sum form, the integral estimate is $\sum z(s_{j-1})(x(s_j) - x(s_{j-1}))$. The variable s_{j-1} in $z(s_{j-1})$ is the left hand end point of the cell $]s_{j-1},\, s_j]$, and is not necessarily the tag point $\bar{s}$, which can be s_{j-1} **or** s_j.

Therefore the integrand $z(s_{j-1})(x(s_j) - x(s_{j-1}))$ should be thought of as a function $h(I)$ of cells $I =]s_{j-1},\, s_j]$; so h is a cell function which is not additive on pairs of adjoining cells. Therefore, in the -complete integration system, an integrand of the form $z(s')(x(s'') - x(s'))$ should be regarded as a Burkill integrand, not Stieltjes.

Note also that the dependence of h on tag points $\bar{s}$ is not necessarily explicit. It is used implicitly to construct δ-fine Riemann sums of integrand h over δ-fine divisions $(\bar{s}, I)$ of the domain of integration. Similarly, Example 7 below has a Burkill-complete integrand which depends on cells I of $(\bar{s}, I)$, but only in an unfamiliar way.

3.3 The Henstock Integral

The origins of the ideas in chapter 4 of [MTRV] are as follows. Starting with his 1948 PhD thesis, Ralph Henstock (1923–2007) worked in non-absolute integration, including the Riemann-complete or gauge integral which, independently, Jaroslav Kurzweil also discovered in the 1950s. As a Cambridge undergraduate (1941–1943) Henstock took a course of lectures, given by J.C. Burkill, on the integration of non-additive interval functions. Later, under the supervision of Paul Dienes in Birkbeck College, London, he undertook research into the ideas of Burkill (interval function integrands) and of Dienes (Stieltjes integrands); and he presented this thesis in December 1948.

In terms of overall approach and methods of proof, the thesis contains the germ of Henstock's later work as summarized in chapter 4 of [MTRV]. For example, a notable innovation is a set of axioms for constructing any particular system of integration. This approach highlights the features held in common by various systems, so that a particular property or theorem can, by a single, common proof, be shown to hold for various kinds of integration. These ideas are the basis of the theory in chapter 4 of [MTRV].

Within this approach, Henstock's thesis places particular emphasis on various alternative ways of selecting Riemann sums, as constituting the primary distinguishing feature of different systems of integration. This was central to his subsequent work and achievement. Accordingly, the theory in chapter 4 of [MTRV] is designated there as the *Henstock integral*, from which almost all systems of integration can be deduced.

Robert Bartle's book ([5], page 15) has a discussion of titles for this kind of integral—variously called Kurzweil-Henstock, gauge, or generalized Riemann. Bartle suggests that it could equally be called "the Denjoy-Perron-Kurzweil-Henstock integral". Evading this litany, Bartle settles for "generalized Riemann", or simply "the integral".

The first worked-out version of this kind of integration was in Henstock's *Theory of Integration* [70], published in 1962 and re-published in 1963, in which the integral was designated "Riemann-complete". In support of the "-complete" appendage, Henstock's presentation has theorems which justify the integration of limits of integrable functions, differentiation under the integral sign, Fubini's theorem, along with a theory of variation corresponding to measure theory.

In practice, different problems and different areas of application require different systems of integration. For instance, twentieth century probability theory is founded on the theory of measure and Lebesgue's integral. The theory of differential equations makes extensive use of the fundamental theorem of calculus—integrability of derivatives[7], differentiability of indefinite integrals.

In [MTRV], reference is made to the integration of Stieltjes and Burkill, and the different roles played by these in mathematical analysis. There are gauge or generalized Riemann versions of these kinds of integrals. But the need to distinguish between them remains. It would be confusing and ineffectual to designate them all indiscriminately as simply gauge (or generalized Riemann or Henstock-Kurzweil) integrals.

Accordingly, [MTRV] set up a nomenclature which, because of historic origins and distinct specialized usages, employed the original name as root, and added the appendage -complete in order to specify the gauge version of the original integral: Riemann-complete, Stieltjes-complete, and Burkill-complete are found to be the most useful in applications. (A Lebesgue-complete integral is defined in [123], definition 8.2, page 34.)

Here is the definition of the *bc-integral* (Burkill-complete integral):

Definition 6 *A point-interval function $h(s, I)$ is Burkill-complete integrable on $[a, b]$, with bc-integral $\int_a^b h(s, I) = \alpha$, or $\int_{[a,b]} h(s, I) = \alpha$, if, given $\varepsilon > 0$, there exists a gauge $\delta(s) > 0$ such that, for each δ-fine division $\mathcal{D} = \{(s_j, \,]t_{j-1}, t_j])\}$ of $[a, b]$, the corresponding Riemann sum satisfies*

$$\left| \alpha - \sum_{j=1}^n h(s_j, \,]t_{j-1}, t_j]) \right| < \varepsilon, \qquad (or \ \ |\alpha - (\mathcal{D}) \sum h(s, I)| < \varepsilon).$$

Definition 6 includes Definitions 2, 4, and 5, because the more general function format $h(s, I)$ can be taken to be the more specific $f(s)|I|$ or $f(s)g(I)$. Surprisingly perhaps, the Burkill-complete integral has all the useful properties one expects of a "good" integral, including Fubini's theorem and a dominated

---

[7]The latter idea has origins in the seventeenth century calculus of Newton and Leibniz. This is the form of integration implanted firmly in students' minds at an early age; and for the most part other conceptions of integration remain marginal or unknown, despite the fact that (for instance) quadrature, recognizably a version of Riemann summation, has a much more ancient pedigree.

convergence theorem: *If bc-integrable functions h_j converge to a function h, and if, for each j, $|h_j(s,I)| \le g(s,I)$ where g is bc-integrable, then h is bc-integrable and $\int h_j \to \int h$ as $j \to \infty$.* For details see chapter 4 of [MTRV].

Examples of bc-integrable functions[8] are given in [MTRV]. The function $h(s,I)$ in (3.2) above includes an expression $|I| = s'' - s'$. But since this factor has the form $\sqrt{|I|}$, the Riemann sums in (3.2) diverge to $+\infty$ as $|I| \to 0$, and h is not bc-integrable. Note that the inequality in Definition 6 can be used, not just to define the bc-integral, but also as a test for bc-integrability.

It has been mentioned earlier that the -complete system of integration is based essentially on Riemann sums formed from finite samples of points $\{s_j\}$ in the domain of integration; and that this can be accomplished without reference to subsets of the domain such as intervals I. To illustrate this further, here are some Burkill-type integrals.

Example 7 *A Burkill integrand in domain $]0,1]$ is $h(I)$, a function (not generally additive) of sub-intervals I of $]0,1]$. In accordance with [MTRV], an extended Burkill integrand is $h(s,I)$ where, in addition to dependence on intervals $I =]u,v]$, h is allowed to depend also on either of the tag points $s = u$ or $s = v$ of the associated cell $I =]u,v]$. The Burkill-complete integral of $h(s,I)$ is then given by Definition 6 above.*

The Dirichlet function is defined as having constant value 1 at the rational numbers in $[0,1]$, with constant value 0 at irrational numbers. Such a function can be useful in demonstrating integrability and non-integrability. The Dirichlet function q is not Riemann integrable, but is Riemann-complete integrable—see [MTRV], page 51. An incremental or Stieltjes version[9] of this function is non-integrable (see [MTRV] section 4.14, pages 178–181).

Let Q denote the rational numbers in the domain $]0,1[$, enumerated as

$$Q = \{r_1, r_2, r_3, \ldots\} \quad \text{and write} \quad Q_n = \{r_1, r_2, \ldots, r_{n-1}\}, \quad n = 1,2,3,\ldots.$$

For $I =]s', s''] \subset]0,1]$, and associated pairs (s,I) with $s = s'$ or $s = s''$, define

$$h_n(s,I) := \begin{cases} \frac{1}{2^j} & \text{if } s'' = r_j, \ 1 \le j \le n, \\ 0 & \text{otherwise.} \end{cases}$$

Note that $h_n(0,I) = h_n(1,I) = 0$. Also h_n does not contain an integrator factor such as $|I|$ or $\mu(A)$ (commonly denoted by notation such as ds or $d\mu$), a factor which generally contributes to convergence to an integral value. Nevertheless, though lacking such a factor, h_n is Burkill-complete integrable on $]0,1]$, with

$$\int_{]0,1]} h_n(s,I) = \sum_{j=1}^{n} \frac{1}{2^j}.$$

---

[8]In https://sites.google.com/site/stieltjescomplete/typo-s or [website], it is pointed out that example 3 on page 11 of [MTRV] has a typo in second last line, in which η should be $|\eta|$.

[9]This function is notable for being the first example of -complete integration [69].

To prove this, for given n let $P_n = Q_n \cup \{0, 1\}$,

$$P_n = \{0 = r_0, r_1, \ldots, r_{n-1}, r_n = 1\}, = \{0 = p_0, p_1, \ldots, p_{n-1}, p_n = 1\}$$

where $0 = p_0 < p_1 < \cdots < p_n = 1$. Define a gauge $\delta(s) > 0$ $(0 \le s \le 1)$ which conforms[10] to the rational numbers in P_n, as follows. For each k and each s with $p_k < s < p_{k+1}$, let

$$\delta(s) < \min\{s - p_k, \ p_{k+1} - s\}.$$

Let $\delta(0) < p_1$, $\delta(1) < 1 - p_{n-1}$, and otherwise let $\delta(p_k) > 0$ be arbitrary. Then every δ-fine division $\mathcal{D}$ of $]0, 1]$ contains terms $(s,]s', r_j])$ for each r_j of Q_n, and

$$(\mathcal{D}) \sum h_n(s, I) = \sum_{j=1}^{n} \frac{1}{2^j}, = \int_{]0,1]} h_n(s, I)$$

by Definition 6. (Pages 57–57 of [MTRV] provide more detail of how the conformance technique works in -complete integration.) Next, define

$$h(s, I) := \begin{cases} \frac{1}{2^j} & \text{if } s'' = r_j, \ j = 1, 2, 3, \ldots, \\ 0 & \text{otherwise.} \end{cases}$$

Then $h(s, I) = \lim_{n \to \infty} h_n(s, I)$, with $h_n(s, I) < h_{n+1}(s, I)$ for $n = 1, 2, 3, \ldots$. Now use theorem 57 of [MTRV] (monotone convergence theorem, pages 169–170) to deduce integrability of the limit function h from integrability of h_n for each n. Write $\int_{]0,1]} h_n = H_n$, so $H_n = \sum_{j=1}^{n} 2^{-j}$. The conditions of theorem 57 require that the numbers H_n be bounded above, with supremum H, as $n \to \infty$. In this case $H = \sum_{j=1}^{\infty} 2^{-j} = 1$. For each s there exists n such that $h(s, I) - h_n(s, I) = 0$, so the other requirements of theorem 57 are satisfied. Thus the bc-integral $\int_{]0,1]} h$ exists, with

$$\int_{]0,1]} h(s, I) = \lim_{n \to \infty} \int_{]0,1]} h_n(s, I) = \lim_{n \to \infty} \sum_{j=1}^{n} \frac{1}{2^j} = \sum_{j=1}^{\infty} \frac{1}{2^j} = 1.$$

To summarize, the -complete system of integration described in chapter 4 of [MTRV] provides new avenues for dealing with problems arising in stochastic processes, and in related problems in the path integral version of quantum mechanics. Common to both is the idea of finite samples of points as a basis for constructing Riemann sum approximations, which can be thought of as functionals of such point samples. In the course of this book, these functionals are referred to as *stochastic sums*, and *sampling sums*.

At root, stochastic integrals are Stieltjes integrals. In the -complete system of integration they fail to be Stieltjes and must be treated as Burkill integrals. But all of them are handled by means of gauges $\delta(\cdot)$ along with δ-fine divisions of the domain of integration.

---

[10]See [MTRV] pages 57–59.

The Henstock integral, as described in chapter 4 of [MTRV] (following [71, 72]), is not defined by means of gauges $\delta(\cdot)$. Therefore, according to the preceding interpretation, the Henstock integral is not itself a -complete integral. Instead, it is a framework of integration theory in which almost all systems of integration—including the -complete kind—can be located.

3.4 Riemann Approach to Random Variation

This section draws together various aspects of the preceding discussion. Broadly speaking, the mathematical theory of random variation is concerned with making numerical estimates of unpredictable events (or outcomes, or measurements). Any estimate is an approximation, the difference between the estimated value and the true value being the so-called "error".

The theory uses probabilities to establish exact numerical boundaries for the error. Though the exact value of the outcome is unknown, under various assumptions the exact range of limits of the extent of error can be stated exactly.

The logic of this can be reversed. If outcomes are known, perhaps by means of repeated experimentation, probabilities for the measurement involved can be deduced—or at least estimated—from the data. This knowledge can then be used in other contexts, bearing in mind that accurate probability values are often aspired to, rather than achieved. Likewise, an assumption of independence underlies many parts of the theory, and this too is often unjustified in practice.

In classical theory, data values are represented as the values taken by a measurable function (a random variable $\mathcal{X}$) defined on a P-measure space Ω (the sample space), and the probabilities of data values are deduced from P-values in Ω by means of the inverse function $\mathcal{X}^{-1}$.

But in all of the examples presented in Section 2.4, it was found that the random variable (as a function) could be taken to be the identity function, and sample space Ω could be taken to consist, essentially, of the potential data values themselves.

This corresponds to the naive or realistic view of random variation, which simply omits the measurable function part of the standard theory. If a dice is rolled, the outcomes or possible data values are the numbers 1, 2, 3, 4, 5, 6. Usually, for practical or intuitive reasons, probability $\frac{1}{6}$ is attached to each of these potential data values, and the set of data values $\Omega = \{1, \ldots, 6\}$ is taken to be the sample space.

In effect, this particular random variable consists of the set of potential data values $\{1, 2, \ldots, 6\}$ in association with their respective probabilities $\frac{1}{6}$:

$$X \simeq \left(\{1, 2, \ldots, 6\}, \left\{ \frac{1}{6}, \frac{1}{6}, \ldots, \frac{1}{6} \right\} \right).$$

The terms on the right capture the essential elements of what constitutes the experiment or random variable; while the symbol X on the left (which is **not** the measurable function notation $\mathcal{X}$) is the commonly used random variable notation. The symbol "$\simeq$" implies that the random variation elements on the right are denoted by the single symbol X on the left. The expected value of a single throw of the dice is

$$E[X] \;=\; \sum_{x=1}^{6} x\, P(x) \;=\; 1 \times \frac{1}{6} + 2 \times \frac{1}{6} + \cdots + 6 \times \frac{1}{6} \;=\; 3.5.$$

It was demonstrated in Section 2.4 that there is some freedom in choosing a sample space. In the case of the dice throwing experiment X above, instead of $\{1, \ldots, 6\}$ the sample space Ω can be taken to be the real line $\mathbf{R}$, the outcomes being intervals $I \subset \mathbf{R}$, and probabilities $P(I) = 1/6$ if I contains one of $j/6$, $j = 1, \ldots, 6$, and $P(I) = 0$ if I contains none of the values $j/6$. Then the random variable X is represented by $(\mathbf{R}, P(I))$. In [MTRV] the notation used is

$$X \;\simeq\; x[\mathbf{R}, P(I)],$$

with x symbolising the data value resulting from a throw of the dice, as distinct from some other data which may arise and which may be symbolised by some other letter, such as y or z. In this representation,

$$E[X] \;=\; \int_{\mathbf{R}} x\, P(I) \;=\; 3.5$$

because P is atomic and the integral reduces to the finite sum already calculated. The integral can be taken to be the Burkill-complete integral of Definition 6. Square brackets are used in $E[X]$ to distinguish it from the measurable function/Lebesgue integral approach of $E(\mathcal{X})$.

Suppose a wager is based on dice-throw, and suppose the wager pays out, not on the number x thrown, but on the square x^2. We may then be interested in the expected value of $f(X) = X^2$, the random variable now being

$$f(X) \;\simeq\; x^2[\mathbf{R}, P(I)].$$

This arises in the variance calculation

$$\mathrm{Var}[X] \;=\; E[X^2] - E[X]^2.$$

Alternatively, X^2 can be represented as

$$Y \;\simeq\; y[\mathbf{R}, Q(I)]$$

where $Q(I)$ is atomic on values $1^2, 2^2, \ldots, 6^2$. Without specifying any particular probabilities, [MTRV] sets out alternative representations $Y = f(X)$,

$$f(X) \;\simeq\; f(x)[\mathbf{R}, F_X(I)], \qquad Y \;\simeq\; y[\mathbf{R}, F_Y(I)],$$

the first being the *contingent* form and the second the *elementary* form. When the *expected value* (or *expectation*)

$$E[f(X)] \;=\; \int_{\mathbf{R}} f(x) F_X(I) \tag{3.3}$$

exists (as bc-integral) the *contingent observable* $f(X)$ is declared to be a *random variable*. (Likewise, the *elementary observable* Y is a random variable if $E[Y]$ = $\int_{\mathbf{R}} y F_Y(I)$ exists. In either case, $E[Y] = E[f(X)]$.)

By contrast, in the standard probability theory of Kolmogorov, $f(\mathcal{X})$ is a random variable if the composition of f with $\mathcal{X}$ is measurable. This is a more strict and demanding requirement than mere existence of expected value in the sense of bc-integrability as expressed above.

The bc-integrability formulation, replacing measurable function $\mathcal{X}$ by data values X, can, if desired, retain the Kolmogorov structure. In [123] it is established that any Lebesgue integral based on probability space $(\Omega, \mathcal{A}, P)$ can be expressed in -complete terms. But retaining abstract measurable space Ω as sample space adds a layer of complexity for which there is no necessity and no benefit, since the bc-integral on sample space $\mathbf{R}$ (—the potential data values) has a dominated convergence theorem, along with the other good properties of Lebesgue integration.

Expected value (3.3) uses the bc-integral on domain $\mathbf{R}$, as in Definition 7 below.[11] Partitions of $\mathbf{R}$ consist of a finite number of disjoint intervals, so in addition to bounded intervals $]s', s'']$, unbounded intervals $]-\infty, v]$, $]u, \infty[$ are included. Tag points (or associated points) for intervals of the unbounded kind are the formal "points" $-\infty$ and $+\infty$, respectively. A gauge on $\mathbf{R}$ is a function $\delta(s) > 0$ defined for each $s \in \mathbf{R}$, along with formal values $\delta(-\infty) > 0$ and $\delta(\infty) > 0$. Elements $(-\infty,]-\infty, v])$ and $(\infty,]u, \infty[)$ are δ-fine if, respectively,

$$v < -\frac{1}{\delta(-\infty)}, \qquad u > \frac{1}{\delta(\infty)}. \tag{3.4}$$

A division $\mathcal{D}$ of $\mathbf{R}$ is δ-fine if each element $(s, I) \in \mathcal{D}$ is δ-fine. A Riemann sum of integrand $h(s, I)$ over a division $\mathcal{D}$ of $\mathbf{R}$ is

$$\sum \{h(s, I) : (s, I) \in \mathcal{D}\} \qquad \text{or} \qquad (\mathcal{D}) \sum h(s, I).$$

In many cases $h(s, I)$ is undefined at $s = \pm\infty$, and, depending on the context, $h(s, I)$ may be taken to be zero for such cases. In general, $\delta(\infty)$ and $\delta(-\infty)$ are positive numbers, just like $\delta(s)$ for finite s. But the numbers $\delta(\pm\infty)$ act as cut-off points for terms admitted in Riemann sums on domain $\mathbf{R} =]-\infty, \infty[$.

With these additions, the definition of the -complete integral on $\mathbf{R}$ is the same as before:

Definition 7 *A function h, $= h(s, I)$, is Burkill-complete integrable on $\mathbf{R}$, with bc-integral $\int_{-\infty}^{\infty} h(s, I) = \alpha$, or $\int_{\mathbf{R}} h(s, I) = \alpha$, if, given $\varepsilon > 0$, there exists a gauge $\delta(s) > 0$ such that, for each δ-fine division $\mathcal{D} = \{(\bar{s},]s', s''])\}$ of $\mathbf{R}$, the corresponding Riemann sum satisfies*

$$\left| \alpha - (\mathcal{D}) \sum h(\bar{s}, \{s', s''\}) \right| < \varepsilon.$$

---

[11] Definition 6 extends only to compact domains $[a, b]$. Fortunately this definition can easily be adapted.

Since the bc-integral includes both the sc-integral and the rc-integral (Stieltjes-complete and Riemann-complete, respectively), Definition 7 can be taken to serve the other cases as well, just by substituting integrands $f(s)g(I)$ and $f(s)|I|$.

3.5   Riemann Approach to Stochastic Integrals

Aside from the specific examples above, can stochastic integrals in general be expressed in a -complete integral framework, and is there any advantage in doing so?

The sample space $\mathbf{R}^4$ in (2.15) was used to demonstrate how a simple "stochastic integral" can be modelled. Unlike most of the other illustrations above, the sample space is $\mathbf{R}^4$ instead of $\mathbf{R}$.

The reason for this is that the values $w(4)$ of the random variable $\mathbf{S}$ are determined, not by a single random event (—throw of dice/toss of coin, the idea is the same) but by four successive random events or tosses of a coin. The random variability in the set of possible outcomes $\mathbf{S}$ (or the set of possible values of $w(4)$) is generated by multiple repetitions of a single underlying (or basic) generator of randomness.

In Section 2.3 the number of basic repetitions was 4, delivering 4 basic outcomes $x = (x(1), x(2), x(3), x(4)) \in \mathbf{R}^4$ whose values $x(j) \in \mathbf{R}$ were used to calculate the value $w(4)$ of the stochastic integral $\mathbf{S}(x)$.

Unlike this example, a true stochastic process has infinitely many "repetitions" of joint sources $x(s)$ of random variability. The objective here is to develop a Burkill-complete model for this scenario, as alternative to the measurable function method.

By the preceding logic, an infinite Cartesian product $\mathbf{R}^\infty$ of $\mathbf{R}$ by itself should provide an appropriate sample space in which to locate the random variation which we are trying to model. The outcome of infinitely many successive repetitions of the basic experiment is then a compound element

$$x = x_{\mathbf{T}} = (x(s) : s \in \mathbf{T}),$$

where $\mathbf{T}$ is some infinite set such as $\{1, 2, 3, 4, \ldots\}$, or $\mathbf{T} =]0, t]$ where t is some positive value. In the latter case the sample space is $\mathbf{R}^{\mathbf{T}}$, an uncountably infinite Cartesian product of $\mathbf{R}$ by itself.

By the same logic as before, given an observable (such as accumulated gain $w(4)$, or $w(t)$, from stockholding), estimated or expected value may be found by calculating a -complete integral.

Up to this point, such integrals have been expressed as functions of values in $\mathbf{R}$. Our aim here is to define them, not on one-dimensional domain $\mathbf{R}$, but in multi-dimensional domains $\mathbf{R}^n$ or $\mathbf{R}^\infty$. For this it is necessary to set up a structure of tagged intervals (or associated point-interval pairs), along with gauges and Riemann sums.

To start, take a finite-dimensional Cartesian product $\mathbf{R}^n$ where n is any positive integer. Let $\bar{\mathbf{R}}$ denote $\mathbf{R}$ with formal points $-\infty, \infty$ adjoined; let x

denote $(x_1, \ldots, x_n)$ where any of the x_j can be $\pm\infty$ so $x \in \bar{\mathbf{R}}^n$. An interval of $\mathbf{R}^n$ is a product $I = I_1 \times \cdots \times I_n$ where each I_j is an interval of $\mathbf{R}$,

$$I_j = \;]-\infty, a], \quad \text{or} \quad]u, v], \quad \text{or} \quad]b, \infty[.$$

A point $x = (x_1, \ldots, x_n) \in \bar{\mathbf{R}}^n$ is a tag point (associated point) of I if, respectively,

$$x_j = -\infty, \quad \text{or} \quad x_j = u \text{ or } v, \quad \text{or} \quad x_j = \infty$$

for $1 \le j \le n$. And, given a gauge $\delta = \{\delta_j(x_j) > 0\}$ defined for $x_j \in \bar{\mathbf{R}}$ ($j = 1, 2, \ldots, n$), the associated point-interval pair (x, I) is δ-fine if, for each j,

$$a < -\frac{1}{\delta_j(x_j)}, \quad \text{or} \quad v - u < \delta_j(x_j), \quad \text{or} \quad b > \frac{1}{\delta_j(x_j)}, \quad \text{respectively.}$$

Let $\mathcal{I}, = \mathcal{I}(\mathbf{R}^n)$, denote the set of cells in the domain $\mathbf{R}^n$. A partition $\mathcal{P}$ of $\mathbf{R}^n$ is a finite collection of disjoint intervals $I \in \mathcal{I}(\mathbf{R}^n)$ whose union is $\mathbf{R}^n$. A division $\mathcal{D}$ of $\mathbf{R}^n$ is a finite collection of associated point-interval pairs (x, I) such that the intervals I form a partition of $\mathbf{R}^n$. Given a gauge δ defined for $x \in \bar{\mathbf{R}}^n$, a division $\mathcal{D}$ of $\mathbf{R}^n$ is δ-fine if each $(x, I) \in \mathcal{D}$ is δ-fine (so, for $j = 1, \ldots, n$, (x_j, I_j) is δ_j-fine).

The definition of the -complete integral on $\mathbf{R}^n$ of an integrand function $h(x, I)$ ($x \in \mathbf{R}^n$, $I \subset \mathbf{R}^n$) follows the same pattern as before:

Definition 8 *A function h, $=h(x,I)$, is Burkill-complete integrable on $\mathbf{R}^n$, with bc-integral $\int_{\mathbf{R}^n} h(x, I) = \alpha$ if, given $\varepsilon > 0$, there exists a gauge δ such that, for each δ-fine division $\mathcal{D}$ of $\mathbf{R}^n$, the corresponding Riemann sum satisfies*

$$\left| \alpha - (\mathcal{D}) \sum h(x, I) \right| < \varepsilon.$$

If $x = (x_1, \ldots, x_n)$ is a "point at infinity" (so $x_j = \pm\infty$ for one or more j ($1 \le j \le n$) and if the corresponding value of $h(x, I)$ is undefined, then it is usually convenient to take $h(x, I) = 0$ for such x. For definition of sc- and rc-complete integrands $f(x)g(I)$, $f(x)|I|$, respectively, just replace $h(x, I)$ in Definition 8 in order to get the appropriate definition for $\mathbf{R}^n$. The domain $\mathbf{R}^n$ can be replaced as domain in Definition 8 by any cell $J \in \mathcal{I}(\mathbf{R}^n)$.

Step function integrands depend take only a finite number of different values. Looked at from a slightly different angle, a *step function* f can be thought of as depending on only a finite number of elements x in the domain, with each other domain element y yielding a function value $f(y)$ equal to one of the finite number of values $f(x)$.

If the domain is multi-dimensional, such as $\mathbf{R} \times \mathbf{R} \times \cdots$ with elements $x = (x_1, x_2, \ldots)$, an integrand $f(x)$ may depend only on particular components x_j and not on others. For instance, $f(x_2, x_2, \ldots)$ may be the same as $f(y_1, x_2, \ldots)$ for all y_1. If the number of dimensions is infinite, and if f depends on only a finite number of dimensions, then f is *cylindrical* (or a *cylinder function*).

The following example illustrates Definition 8, and is a guide to the integration of cylinder functions.

Example 8 *Suppose $I = I_1 \times I_2$ is a representative interval in $\mathbf{R}^2, = \mathbf{R} \times \mathbf{R}$, so $I \in \mathcal{I}(\mathbf{R}^2)$. Let*

$$F_j(I_j) := \frac{1}{\sqrt{\pi}} \int_{I_j} e^{-x_j^2} dx_j \quad (j = 1, 2), \qquad F(I) := F_1(I_1) F_2(I_2).$$

Evaluation of Gaussian and Fresnel integrals by the -complete method is demonstrated in [MTRV] section 6.2 (pages 261–265). The following -complete integrals exist:

$$\int_{\mathbf{R}} F_1(I_1) = \int_{\mathbf{R}} F_2(I_2) = 1, \qquad \int_{\mathbf{R}} x_1 F_1(I_1) = \int_{\mathbf{R}} x_2 F_2(I_2) = 0.$$

Suppose $f_1(x_1) F_1(I_1)$ is integrable on $\mathbf{R}$, with $\int_{\mathbf{R}} f_1(x_1) F_1(I_1) = \alpha$. For $x = (x_1, x_2) \in \mathbf{R}^2$, let $f(x) := f_1(x_1)$. Then $\int_{\mathbf{R}^2} f(x) F(I) = \alpha$. In other words,

$$\int_{\mathbf{R} \times \mathbf{R}} f(x_1, x_2) F(I_1 \times I_2) = \int_{\mathbf{R}} f_1(x_1) F_1(I_1).$$

Proof. *The latter equation follows from Fubini's theorem ([MTRV] pages 159–165). But it is helpful to prove it by examining specific Riemann sums which arise in this case. With $\varepsilon > 0$ given, choose a gauge $\delta_1(x_1) > 0$ so that, for each δ_1-fine division $\mathcal{D}_1$ of $\mathbf{R}$,*

$$\left| (\mathcal{D}_1) \sum f_1(x_1) F_1(I_1) - \alpha \right| < \varepsilon.$$

Let $\delta_2(x_2) > 0$ be an arbitrary gauge for $x_2 \in \bar{\mathbf{R}}$, and let $\delta = \{\delta_1(x_1), \delta_2(x_2)\}$ be a gauge for $\mathbf{R}^2$. Let $\mathcal{D}$ be a δ-fine division of $\mathbf{R}^2$. The horizontal line $x_2 = 0$ in $\mathbf{R}^2$ intersects with some of the $I \in \mathcal{D}$. By drawing vertical lines in $\mathbf{R}^2$ through these points of intersection, as in figure 3.4 (page 91 of [MTRV]), a refinement of $\mathcal{D}$ is produced which forms a partially regular partition $\mathcal{P}$ of $\mathbf{R}^2$. Denote the intervals of $\mathcal{P}$ by J. Then each $I = I_1 \times I_2$ of $(x, I) \in \mathcal{D}$ is the union of a finite number of $J = J_1 \times I_2 \in \mathcal{P}$ and, by additivity of F_1,

$$f(x) F(I), = f(x_1, x_2) F(I_1 \times I_2) = f_1(x_1) \left(\sum_{J \subset I_1} F_1(J) \right) F_2(I_2); \quad with$$

$$(\mathcal{D}) \sum f(x) F(I) = (\mathcal{D}_1) \sum f_1(x_1) F_1(I_1)$$

since F_2 is additive with integral 1 on $\mathbf{R}$.   ○

As another illustration of Definition 8, -complete integration can be applied to the shareholding gain process of Section 2.3, in order to calculate $E[W(4)]$, the expected net gain at the end of 4 days trading.

This easily follows from Table 2.4 using the basic arithmetical method of taking the sample values $w(4)$, which are on the right hand margin of Table 2.4, and multiplying them by the appropriate probability values.

If up-ticks and down-ticks of the underlying share price process are equally likely, then (2.6) gives expected net gain to be 0. If up-ticks are twice as likely as down-ticks, then by (2.8) the expected net gain is 8/3.

These calculations can be formulated as -complete integrals. The random variable $W[4]$ can be understood in various ways. First consider it as an elementary random variable $Y \simeq y[\mathbf{R}, F_Y]$ where, for intervals I in $\mathbf{R}$,

$$F_Y(I) = \begin{cases} \frac{1}{8} & \text{if } I \text{ contains just one of } -1, -2, -3, -4, \\ \frac{1}{16} & \text{if } I \text{ contains just one of } -5, 0, 1, 2, 3, 4, 5, 10, \\ 0 & \text{if } I \text{ contains none of these numbers;} \end{cases} \quad (3.5)$$

the twelve numbers listed being the **distinct** possible values of $w(4)$ in Table 2.4. (For other intervals I, $F_Y(I)$ can be defined by addition.)

If $x \in \bar{\mathbf{R}}$ let $\delta(x) = 0.5$. This choice of $\delta(x)$ ensures[12] that, for every δ-fine division $\mathcal{D}$ of $\mathbf{R}$, if $(x, I) \in \mathcal{D}$ then $F_Y(I) = 0$ or $1/8$ or $1/16$, and

$$(\mathcal{D}) \sum y \, F_Y(I) = 0$$

since the Riemann sum reduces to the calculation in (2.6). Thus the bc-complete integral satisfies

$$\int_{\mathbf{R}} y \, F_Y(I) = \mathrm{E}[W(4)] = 0.$$

If the net gain calculation is regarded as a stochastic integral $\mathbf{S}((f(X))$, or $\int_{\mathbf{R}} Z \, dX$ (with $Z = f(X)$), then

$$\int_{\mathbf{R}} y \, F_Y(I) = \mathrm{E}[\mathbf{S}(f(X))] = 0.$$

The above interpretation of the random variability in net gain is the *elementary random variable* view, in which the potential values of net gain $w(t)$, and their probabilities, are the focus.

This is to be contrasted with the *contingent random variable view*, in which the random variability of the underlying share prices (which generates the random variability of net gain) is made explicit in the calculation of expected net gain.

In this case the processes are represented as operating on sample space (2.15); that is, $\mathbf{R} \times \mathbf{R} \times \mathbf{R} \times \mathbf{R}$. The basic outcome is a particular share price history $(x_1, x_2, x_3, x_4) \in \mathbf{R}^4$, where $x_j, = x(s)$ is the share price observed at end of day j (or day s). Then, with

$$x = (x_1, x_2, x_3, x_4) \in \mathbf{R}^4,$$

the share price process $X = (X(s)), = (X_j) \; (s = j = 1, 2, 3, 4,)$, is represented as the *joint-basic observable*[13]

$$X \simeq x[\mathbf{R}^4, F_X].$$

---

[12] The more delicate approach of *conformance* should be applied—see example 15 in page 57 of [MTRV]. This means that, if v is one of the 12 values listed in (3.5), and if $x < v$, then $\delta(x)$ should be less than $v - x$.

[13] The rationale for terminology *observable, random variable, basic, joint-basic, elementary, contingent* is given in [MTRV], chapter 5 pages 183–186.

(Probability distribution function notation $F(I)$ or $F_X(I)$ is used instead of probability measure notation P.) The potential price histories $x = (x_1, x_2, x_3, x_4)$ constitute a proper subset of $\mathbf{R}^4$; there are only 16 relevant points x (not the uncountably infinitely many x which constitute the whole of $\mathbf{R}^4$). These 16 points are:

$$
\begin{array}{llll}
(11, 12, 13, 14) & (9, 10, 11, 12) & (11, 10, 11, 12) & (9, 8, 9, 10) \\
(11, 12, 11, 12) & (9, 10, 9, 10) & (11, 10, 9, 10) & (9, 8, 7, 8) \\
(11, 12, 13, 12) & (9, 10, 11, 10) & (11, 10, 11, 10) & (9, 8, 9, 8) \\
(11, 12, 11, 10) & (9, 10, 9, 8) & (11, 10, 9, 8) & (9, 8, 7, 6).
\end{array}
\tag{3.6}
$$

Under the assumptions expressed in Section 2.3 there are only 16 possible price histories x, corresponding to the 16 possible combinations of up-ticks and down-ticks in Table 2.4.

Notation $X \simeq x[\mathbf{R}^4, F_X]$ indicates that probabilities (or probability distribution values) $F_X(I)$ are specified or known for intervals $I = I_1 \times I_2 \times I_3 \times I_4$ in $\mathbf{R}^4$.

As in Section 2.3, it is required that F_X be atomic, with $F_X(I) \neq 0$ only if $I = I_1 \times I_2 \times I_3 \times I_4$ contains one or more of the 16 points $x = (x_1, x_2, x_3, x_4)$ of (3.6).

If up-ticks and down-ticks are equally likely, then each of the 16 point-outcomes is equally likely, with probability $1/16$. So if I contains one, and only one, of the 16 points of (3.6), then $F_X(I) = 1/16$. The other values of $F_X(I)$ are deducible from this.

With $t = 4$ the "stochastic integral" is

$$
\int_0^t Z(s)\, dX(s) \;=\; \sum_{j=1}^{4} Z_{j-1}(X_j - X_{j-1})
$$

where Z_j is the number of shares held at the end of day j, equal to Z_{j-1} or $Z_{j-1}+1$ depending on whether, by the end of day j, the share price has decreased or increased. Corresponding to the possible histories x in (3.6), the 16 possible values for the "stochastic integral" $\sum_{j=1}^{4} Z_{j-1}(x_j - x_{j-1})$ are the values $w^{(r)}$, $r = 1, 2, \ldots, 16$, in the following list:

$$
\begin{array}{rrrr}
10 & 5 & 4 & 1 \\
3 & 0 & -1 & -2 \\
2 & -1 & -2 & -3 \\
-3 & -4 & -5 & -4.
\end{array}
\tag{3.7}
$$

The expected value of net gain has been calculated earlier. The result is 0. But the present exercise is to re-calculate it as $E[S(X)]$ based on the contingent representation

$$
\mathbf{S}(X) \;\simeq\; \mathbf{S}(x)[\mathbf{R}^4, F_X],
$$

where $\mathbf{S}(x)$ is the "stochastic integral"

$$\int_0^t z(s)\,dx(s) \;=\; \sum_{j=1}^{4} z_{j-1}\,(x_j - x_{j-1}),$$

and

$$E[\mathbf{S}(X)] \;=\; \int_{\mathbf{R}^4} \mathbf{S}\,(x_1, x_2, x_3, x_4)\, F_X\,(I_1 \times I_2 \times I_3 \times I_4),$$

the latter being interpreted as the bc-integral on domain $\mathbf{R}^4$. To evaluate this 4-dimensional integral in accordance with Definition 8, a suitable gauge δ, $= \{\delta_j\}$ ($j = 1, \ldots, 4$), must be defined to check whether the Riemann sum condition of Definition 8 holds for this integrand.

The gauge δ must be defined in such a way that any δ-fine division $\mathcal{D}$ of $\mathbf{R}^4$ includes 16 intervals $I = I_1 \times I_2 \times I_3 \times I_4$, each of which contains one and only one of the points listed in (3.6). The method by which this is achieved is used fairly extensively in -complete integration, and is explained in example 15, page 57 of [MTRV], where it is called *conformance*.

Broadly speaking, conformance is manifested as follows. If $v = (v_1, v_2, v_3, v_4)$ is, in turn, any one of the 16 points of (3.6), and if $(x_1, x_2, x_3, x_4), (y_1, y_2, y_3, y_4) \in \mathbf{R}^4$ with $x_j < v_j < y_j$, $1 \le j \le 4$, then let $\delta_j(x_j)$ be less than $\min\{v_j - x_j,\ y_j - x_j\}$ for $j = 1, 2, 3, 4$.

This device ensures that, in any δ-fine division $\mathcal{D} = \{(x, I)\}$ of domain $\mathbf{R}^4$, the elements (x, I) of the division include 16 elements

$$(x, I) \;=\; ((x_1, x_2, x_3, x_4),\ ]u_1, v_1] \times]u_2, v_2] \times]u_3, v_3] \times]u_4, v_4])$$

satisfying $x_j = v_j$ for $j = 1, 2, 3, 4$ for each of the 16 histories x of (3.6). Thus, for any δ-fine Riemann sum,

$$(\mathcal{D}) \sum \mathbf{S}(x) F_X(I) \;=\; \sum_{r=1}^{16} w^{(r)} \frac{1}{16} \;=\; 0;$$

so, with $\alpha = 0$, the condition

$$\left| (\mathcal{D}) \sum \mathbf{S}(x) F_X(I) - \alpha \right| \;<\; \varepsilon$$

of Definition 8 is satisfied, and $E[\mathbf{S}(X)] = \int_{\mathbf{R}^4} \mathbf{S}(x) F_X(I) = 0$. This confirms that the joint-basic contingent representation and elementary representation,

$$\mathbf{S}(X) \;\simeq\; \mathbf{S}(x)[\mathbf{R}^4, F_X], \qquad Y \;\simeq\; y[\mathbf{R}, F_Y],$$

respectively, yield the same expected value.

Here is another finite-dimensional illustration of expected value calculated by means of a -complete integral.

Example 9 *Suppose, on n successive throws of a dice, a wager pays out an amount 2^{-n} if 6 is thrown, and an amount 0 otherwise. What is the expected payout for this wager?*

Take $n = 3$; so the answer can be easily found by elementary methods. A little calculation gives expected payout

$$\frac{1}{6}\left(\frac{1}{2} + \frac{1}{2^2} + \frac{1}{2^3}\right) = \frac{7}{48}.$$

To see how this works by -complete integration in $\mathbf{R}^3$, the wager can be represented as joint basic observable $X, = (X_j), = (X_1, X_2, X_3)$ and $x = (x_j) = (x_1, x_2, x_3)$, with

$$X \simeq x\left[\mathbf{R}^3, F_X\right] = (x_1, x_2, x_3)\left[\mathbf{R} \times \mathbf{R} \times \mathbf{R}, F_{(X_1, X_2, X_3)}\right].$$

Assuming each throw of the dice is independent, then

$$F_X(I) = F_{(X_1, X_2, X_3)}(I_1 \times I_2 \times I_3) = F_{X_1}(I_1)F_{X_2}(I_2)F_{X_3}(I_3),$$

where, for $j = 1, 2, 3$,

$$F_{X_j}(I_j) = \begin{cases} \frac{1}{6} & \text{if } \frac{1}{2^j} \in I_j \text{ and } 0 \notin I_j, \\ \frac{5}{6} & \text{if } \frac{1}{2^j} \notin I_j \text{ and } 0 \in I_j, \\ 1 & \text{if } \frac{1}{2^j} \in I_j \text{ and } 0 \in I_j, \\ 0 & \text{if } \frac{1}{2^j} \notin I_j \text{ and } 0 \notin I_j. \end{cases}$$

With $f(x) = x_1 + x_2 + x_3$ define a contingent observable

$$f(X) \simeq f(x)\left[\mathbf{R}^3, F_X\right].$$

Then $\mathrm{E}[f(X)] = \int_{\mathbf{R}^3} f(x)F_X(I) =$

$$= \int_{\mathbf{R} \times \mathbf{R} \times \mathbf{R}} (x_1 + x_2 + x_3)F_{X_1}(I_1)F_{X_2}(I_2)F_{X_3}(I_3) = \sum_{j=1}^{3} x_j F_{X_j}(I_j)$$

if the integral exists. The gauge-conformance/Riemann sum method of Example 8 above can be used here. Let $y = (y_1, y_2, y_3)$ denote any of the 8 points for which $y_j = 0$ or 2^{-j} for $j = 1, 2, 3$. Following Definition 8, define a gauge $\delta = \{\delta_j\}$ for $x = (x_j) \in \mathbf{R}^3$ $(j = 1, 2, 3)$ as follows.

For each $x = (x_1, x_2, x_3)$ let $0 < \delta_j(x_j) < 2^{-3}$, and if $x_j \neq 2^{-j}$ and $x_j \neq 0$ for one or more of $j = 1, 2, 3$, let

$$\delta_j(x_j) < \min\left\{|x_j|, |x_j - 2^{-j}|\right\}, \quad j = 1, 2, 3.$$

This ensures that if associated (x, I) is δ-fine in $\mathbf{R}^3$ and if $x = (x_1, x_2, x_3)$ has $x_j \neq 2^{-j}$ and $x_j \neq 0$ for one or more of $j = 1, 2, 3$, then I does not contain any of the 8 points (y_1, y_2, y_3). Therefore, if $\mathcal{D}$ is a δ-fine division of $\mathbf{R}^3$, $\mathcal{D}$ contains 8 pairs (y, I) with $y_j = 0$ or 2^{-j} for $j = 1, 2, 3$. So for any such division $\mathcal{D}$ the Riemann sum satisfies

$$(\mathcal{D}) \sum f(x)F_X(I) = \left(\frac{1}{2} \times \frac{1}{6}\right) + \left(\frac{1}{2^2} \times \frac{1}{6}\right) + \left(\frac{1}{2^3} \times \frac{1}{6}\right) = \frac{7}{48},$$

so $E[f(X)] = \int_{\mathbf{R}^3} f(x)F_X(I)$ exists and equals 7/48.

The next step is to expand this approach in order to deal with actual stochastic processes involving infinitely many joint random variables. So -complete integration needs to be defined in an infinite-dimensional Cartesian product domain $\mathbf{R}^{\infty}$.

Chapter 4

Stochastic Processes

Section 3.5 provides some elementary illustrations of how to apply Riemann methods to stochastic processes. This chapter seeks to provide a bridge to a Riemann-type theory of Brownian motion in Chapter 5, expanding on the presentation in chapters 7 and 8 of [MTRV].

4.1 From $\mathbf{R}^n$ to $\mathbf{R}^\infty$

The net gain process (W_s) of Section 2.3 was based on a price process (X_s) whose data x_s took only a finite number of Up and Down values at times $s = 1, 2, 3, 4$. But, by locating these values in $\mathbf{R}^4$, the possibility was left open to allow each x_s to take a continuum of values; and this can be accomplished in the -complete system just by using a suitable distribution function F in $f(X) \simeq f(x)[\mathbf{R}^n, F_X]$.

But time values s also form a continuum, just like data values x_s. So, instead of a finite number of x_j $(j = 1, 2, \ldots, n)$, there arises the possibility of infinitely many x_s (uncountably infinitely many possibilities in case $0 < s < t$).

Before tackling an uncountable infinity of observables X_s, consider the following example, which introduces a countable infinity.

Example 10 *Suppose dice are thrown successively at times $s = 1 - p^{-1}$ ($p = 2, 3, 4, \ldots$), and a wager involves paying out an amount*

$$x_p = \frac{1}{2^p}$$

whenever a 6 is thrown at time $s = 1 - p^{-1}$; and amount $x_p = 0$ whenever $1, 2, 3, 4$ or 5 are thrown. What is the overall expected payout from the wager?

Again, this can be calculated[1] by elementary traditional methods. But the objective here is to formulate and answer the question in the contingent observable format.

---

[1] Expected payout is $\frac{1}{6}$.

Gauge Integral Structures for Stochastic Calculus and Quantum Electrodynamics, First Edition. Patrick Muldowney.
© 2021 John Wiley & Sons, Inc. Published 2021 by John Wiley & Sons, Inc.

The total payout is $f(X)$, where, for any particular joint outcome $x = (x_2, x_3, \ldots)$,

$$f(x) = \sum_{p=2}^{\infty} x_p.$$

Denote the payout process by $X = (X_p)$, $x = (x_p)$, $(p = 2, 3, \ldots)$. It will be possible to represent this process as a joint basic observable

$$X \simeq x[\mathbf{R} \times \mathbf{R} \times \mathbf{R} \times \cdots, \; F_X], \qquad \text{or} \quad X \simeq x[\mathbf{R}^\infty, F_X]$$

provided the probability distribution function $F_X(I)$ can be defined for intervals $I \subset \mathbf{R}^\infty$. Then the expected total gain will be found by the -complete integral

$$\mathrm{E}[f(X)] = \int_{\mathbf{R}^\infty} f(x) F_X(I)$$

—provided the -complete integral on $\mathbf{R}^\infty$ can be defined. That is the next step.

First, the factor spaces $\mathbf{R}$ of the Cartesian product domain (or sample space) $\mathbf{R}^\infty$ need to be labelled. The labels are the elements

$$s = 1 - \frac{1}{p} \text{ of } T = \left\{ 1 - \frac{1}{p} : p = 2, 3, 4, \ldots \right\}; \quad s_1 = 1 - \frac{1}{2} = \frac{1}{2}, \; s_2 = 1 - \frac{1}{3} = \frac{2}{3}, \ldots;$$

so, instead of notation $\mathbf{R}^\infty$, denote this domain by

$$\mathbf{R}^T, \; = \prod_{s=1}^{\infty} \mathbf{R}_s, \quad \text{where} \quad \mathbf{R}_s = \mathbf{R} \quad \text{for each} \quad p. \qquad (4.1)$$

(For purpose of motivation, in the illustration above the elements s of T are discrete and countable. But, without any change of notation or wording, the following terms and definitions apply equally to a set $\mathbf{T}$ in which the elements s are dense and uncountable; for instance, the elements s of $\mathbf{T} = \,]0, 1]$.)

Let $\mathcal{N}, = \mathcal{N}(T)$, denote the family of all finite subsets N of T. Given

$$N = \{s_1, s_2, \ldots, s_n\} \in \mathcal{N}, \quad \text{and} \quad x \in \mathbf{R}^T,$$

let $x(N)$ denote $(x_{s_1}, x_{s_2}, \ldots, x_{s_n})$, an element of the finite-dimensional $\mathbf{R}^N$.

For given $N \in \mathcal{N}$ and for each $s_j \in N$ let I_{s_j} (or I_j for short), be an interval of $\mathbf{R}, = \mathbf{R}_{s_j}$; so

$$I_{s_j}, \; = I_j, \; = \;]-\infty, b] \quad \text{or} \quad]u, v] \quad \text{or} \quad]a, \infty[,$$

(unbounded below, bounded above and below, and unbounded above, respectively); and

$$I(N) := \prod_{j=1}^{n} I_{s_j} = \prod_{j=1}^{n} I_j$$

is an interval of the finite-dimensional Cartesian product domain

$$\mathbf{R}^N, \; = \prod_{j=1}^{n} \mathbf{R}_{s_j} = \mathbf{R}^n.$$

For given $N \in \mathcal{N}$, an interval $I, = I[N]$, of the infinite-dimensional Cartesian product domain $\mathbf{R}^T$ is

$$I[N] := I(N) \times \prod \{\mathbf{R}_s : s \in T \smallsetminus N\}.$$

Thus $I[N]$ is restricted in dimensions $s \in N$, and it is unrestricted in dimensions $s \in T \smallsetminus N$. Let $\mathcal{I}, = \mathcal{I}(\mathbf{R}^T)$, denote the family of intervals $I[N]$ (all $N \in \mathcal{N}$) in $\mathbf{R}^T$.

A point $x = (x_s) \in \mathbf{R}^T$ and an interval $I[N] \in \mathcal{I}(\mathbf{R}^T)$ form an associated point-interval pair—in other words, x is a tag point of $I[N]$—if $(x(N), I(N))$ are associated in $\mathbf{R}^N$ (so $x(N)$ is a tag point of $I(N)$). A partition $\mathcal{P}$ of $\mathbf{R}^T$ is a finite collection of disjoint intervals $I, = I[N]$, whose union is $\mathbf{R}^T$, so

$$\mathcal{P} = \{I^r[N^r] : r = 1, 2, \dots q\}, \quad \bigcup_{r=1}^q I^r[N^r] = \mathbf{R}^T, \quad I^r \cap I^{r'} = \varnothing \ (r \neq r').$$

A division $\mathcal{D}$ of $\mathbf{R}^T$ is a finite collection of associated point-interval pairs $(x, I[N])$ such that the intervals $I[N]$ form a partition of $\mathbf{R}^T$.

In [MTRV] a gauge in $\mathbf{R}^T$ was defined as a pair of mappings (L, δ) given by

$$\bar{\mathbf{R}}^T \overset{L}{\mapsto} \mathcal{N}, \qquad \bar{\mathbf{R}}^T \times \mathcal{N} \overset{\delta}{\mapsto}]0, \infty[,$$
$$x \to L(x), \qquad (x, N) \to \delta(x, N).$$

But the problems in this book involve structured Cartesian products of $\mathbf{R}$. Because of this, a different approach is indicated, as follows:

$$\bar{\mathbf{R}}^T \overset{L}{\mapsto} \mathcal{N}, \qquad \bar{\mathbf{R}}^N \overset{\delta_N}{\mapsto}]0, \infty[^N, \ (N \in \mathcal{N}),$$
$$x \to L(x), \qquad x(N) \to \{\delta_N(x_t) > 0 : t \in N\}, \ (N \in \mathcal{N}).$$

(4.2)

Thus[2] the mapping L assigns to each $x \in \bar{\mathbf{R}}^T$ a finite set of dimension labels $s \in T$; and the mapping δ_N assigns[3] a positive number $\delta_N(x_t)$ to each $x_t \in \bar{\mathbf{R}}$ for $t \in N$. A gauge γ in $\mathbf{R}^T$ is

$$\gamma = (L, \{\delta_N\}_{N \in \mathcal{N}}) = (L, \delta_{\mathcal{N}}); \quad \text{or simply} \quad (L, \delta).$$

Given γ, an associated point-interval[4] pair $(x, I[N])$ is γ-fine if

$$N \supseteq L(x) \quad \text{and} \quad I(N) \text{ is } \delta\text{-fine in } \mathbf{R}^N.$$

(4.3)

---

[2] For further discussion see Chapter 7.

[3] Instead of δ_N (which may be different for different $N \in \mathcal{N}$), a mapping $\delta, = \{\delta_t\}_{t \in T}$, could be used. This may deliver integrability for the simpler kinds of integrands. But δ_N gives more discrimination in selection of divisions. Most integrands in this book have the form $f(x(N))$, requiring gauge δ_N; a different gauge $\delta_{N'}$ may be needed for $f(x(N'))$, $N' \neq N$; and even if $t \in N \cap N'$, the gauge value $\delta_N(x(t))$ need not be the same as $\delta_{N'}(x(t))$. (See Example 14 below.) This particular construction of the product space gauge γ is practically the same as the corresponding definition in [73] (Henstock, 1973). The version in [MTRV] is an innovation.

[4] See comment below, in which $(x, I[N])$ is replaced by triple $(x, N, I[N])$.

A division $\mathcal{D}$ of $\mathbf{R}^T$ is γ-fine if each $(x, I) \in \mathcal{D}$ is γ-fine. Finally, a function $h(x, I)$ is bc-integrable on $\mathbf{R}^T$ with $\int_{\mathbf{R}^T} h = \alpha$ if, given any $\varepsilon > 0$, there exists a gauge γ such that, for any γ-fine division $\mathcal{D}$ of $\mathbf{R}^T$, the corresponding Riemann sum satisfies

$$\left| (\mathcal{D}) \sum h(x, I) - \alpha \right| < \varepsilon. \tag{4.4}$$

In fact, integrands in $\mathbf{R}^T$ often take the form $h(x, N, I[N])$ in which h depends explicitly on elements $s_j \in N \in \mathcal{N}$, in addition to dependence on variables such as x_j (relating to points x of $\mathbf{R}^T$) and $v_j - u_j$ (relating to lengths of the edges of intervals $I = I[N] \in \mathcal{I}$). The variable N is already implicitly included in the pair-formulation $(x, I), = (x, I[N])$, since the only intervals allowed in $\mathbf{R}^T$ are $I[N]$, with variable $N \in \mathcal{N}$. But because, in practice, N appears explicitly in integrands h, it is useful also to make N explicit in the association relationship $(x, N, I[N])$. One way to think of the association relationship is as follows:

- Choose any point $x = x_T \in \bar{\mathbf{R}}^T$.

- Choose any finite set of dimensions (or "times") $N \in \mathcal{N} = \mathcal{N}(\mathbf{T})$.

- Given the chosen x and N, choose an interval $I = I[N] \in \mathcal{I}(\mathbf{R}^T)$ which is associated with the given elements x and N. Association of I with N means:

    - I is restricted in dimensions N,
    - I is unrestricted in dimensions $T \smallsetminus N$,
    - so $I = I[N]$.

- Now consider elements $x(N)$ and $I(N)$ in the finite-dimensional space $\mathbf{R}^N$; namely, points $x(N) = (x_1, \ldots, x_n) \in \bar{\mathbf{R}}^N$ and intervals $I(N) = I_1 \times \cdots \times I_n \in \mathcal{I}(\mathbf{R}^N)$.

- Association of $x(N)$ and $I(N)$ means that each component x_j of $x(N)$ is an associated point (or tag point) of the corresponding component I_j of $I(N)$ in the finite-dimensional space $\mathbf{R}^N$.

- Given a gauge $\gamma = (L, \delta_N)$, an associated pair $(X, I[N])$ in $\mathbf{R}^T$ is γ-fine if $N \supset L(x)$ and $(x(N), I(N))$ is δ_N-fine in $\mathbf{R}^N$ (so, for each $s_j \in N$, (x_j, I_j) is δ_N-fine in $\mathbf{R}$).

Thus, because of the additional and new kind of variability of N in the integrand h, the idea of association and γ-fineness is extended from pairs (x, I) to triples $(x, N, I[N])$. The only change to (4.3) and (4.4) is to replace associated pairs $(x, I[N])$ by associated triples $(x, N, I[N])$.

The elements s of T perform the role of labels for "dimensions" in a multi-dimensional type of integration, just like $1, 2, 3$, or x, y, z in three-dimensional integrals. In broad terms, the appearance and utilisation of variable "dimension sets" N of finite order (or cardinality) n, suggest that integration of functions in the infinite-dimensional $\mathbf{R}^T$ is based on the more familiar integration in the n-dimensional domain $\mathbf{R}^N$.

To some extent, this is an impression worth holding on to, even though on second thoughts it may be somewhat contradictory. After all, integration in the plane $\mathbf{R}^2$ is not at all reducible to integration on the line $\mathbf{R}$. (Though Tonelli's theorem may sometimes allow us to evaluate a double integral by iterated single integrals.)

It is the first part of (4.3) that forms a kind of link between $\mathbf{R}^N$ and $\mathbf{R}^T$. In fact, if followed through in detail, it forms a link between lower-dimensional spaces and higher-dimensional (even infinite-dimensional) space. The set $L(x) \in \mathcal{N}$ is a "lower bound" for variable sets N in $\mathcal{N}$ which can, though remaining finite, be as large as we please. Since this lower bound $L(x)$ is selected in the course of selecting gauge γ, in effect the order of (or number of dimension labels s in) $L(x)$ becomes successively larger—just as, in one dimension, the value of $\delta(x) > 0$ becomes successively smaller by successive choices of gauge.

These introductory comments are expanded in [MTRV], pages 83–106; part-icularly the diagrams on pages 81, 87, 91, 102. There is further discussion of infinite-dimensional integration in [website].[5]

It is now possible to set out the -complete sample space framework for Example 10. Let

$$T = \left\{ 1 - \frac{1}{j} : j = 2, 3, 4, \dots \right\},$$

and let sample space Ω be $\mathbf{R}^T$. For basic observable $X, = X_T, = (X_j)$, take

$$(X_j) = X \simeq x\left[\mathbf{R}^T, F_X\right] \quad \text{where} \quad x = (x_j) = (x_1, x_2, \dots, x_j, x_{j+1}, \dots) \in \mathbf{R}^T.$$

For this to be meaningful the probability distribution function $F_X(I[N])$ must be defined for all $I[N] \in \mathcal{I}(\mathbf{R}^T)$ (the family of intervals or cells $I = I[N]$ in $\mathbf{R}^T$). For $j = 1, 2, 3, \dots$ and $I_j \in \mathcal{I}(\mathbf{R})$, define $F_{X_j}(I_j)$ as in Example 9. That is

$$F_{X_j}(I_j) = \begin{cases} \frac{1}{6} & \text{if } \frac{1}{2^j} \in I_j \text{ and } 0 \notin I_j, \\ \frac{5}{6} & \text{if } \frac{1}{2^j} \notin I_j \text{ and } 0 \in I_j, \\ 1 & \text{if } \frac{1}{2^j} \in I_j \text{ and } 0 \in I_j, \\ 0 & \text{if } \frac{1}{2^j} \notin I_j \text{ and } 0 \notin I_j. \end{cases}$$

Independence of X_j is assumed (so the outcome of any one throw of dice does not affect the outcome of any other throw). Thus[6] for any $N = \{s_1, \dots, s_n\} \in \mathcal{N}$,

$$F_{X_N}(I(N)) = F_{(X_{s_1}, X_{s_2}, \dots, X_{s_n})}(I_{s_1} \times I_{s_2} \times \cdots \times I_{s_n}) = \prod_{j=1}^{n} F_{X_{s_j}}(I_{s_j}),$$

which is a probability distribution function in the finite-dimensional domain $\mathbf{R}^N$. For $X = X_T = (X_1, X_2, X_3, \dots)$ and for each $N \in \mathcal{N}$ and each $I = I[N] \in \mathcal{I}(\mathbf{R}^T)$,

---

[5] The gauge version of integration in infinite-dimensional product spaces originated in Henstock's 1973 paper [73]. It is touched upon in [71] (Henstock, 1968). The version in this book is similar to [73]. The general idea goes back to Jessen [83] and Daniell [19].

[6] Classically, *consistency* arises at this point. This is addressed in [MTRV] pages 184, 213. It can also be applied to the definition of $F_{X(N)}(I(N))$.

define $F_X(I[N]), = F_{X_T}(I[N])$, by

$$F_X(I[N]) := F_{X_N}(I(N)).$$

The payout from the wager of Example 10 is $f(x_T) = x_1+x_2+x_3+\cdots$, where $x_j = 0$ or 2^{-j} for each j. If (x_j) is regarded as an element of $\mathbf{R}^T$ then $x_1 + x_2 + x_3 + \cdots$ is generally divergent. Therefore, define $f : \mathbf{R}^T \mapsto \mathbf{R}$ by

$$f(x_T) := \begin{cases} \sum_{j=1}^{\infty} x_j & \text{if } 0 \le x_j \le \frac{1}{2^j}, \quad (j = 1,2,3,\ldots), \\ 0 & \text{otherwise,} \end{cases}$$

so f is well defined. Now define a contingent observable

$$f(X) \simeq f(x)\left[\mathbf{R}^3, F_X\right],$$

representing the payout from the wager of Example 10. Then

$$\mathrm{E}[f(X_T)] = \int_{\mathbf{R}^T} f(x_T)F_{X_T}(I[N]) = \int_{\mathbf{R}\times\mathbf{R}\times\mathbf{R}\times\cdots} \left(\sum_{j=1}^{\infty} x_j\right) F_{X_T}(I[N])$$

(4.5)

if the integral exists. One might be tempted, on intuitive grounds, to immediately express the latter as

$$\lim_{j\to\infty} \int_{\mathbf{R}^n} \left(\sum_{j=1}^{n} x_j\right) \prod_{j=1}^{n} F_{X_j}(I_j) = \lim_{j\to\infty} \left(\sum_{j=1}^{n} \int_{\mathbf{R}} x_j F_{X_j}(I_j)\right) \qquad (4.6)$$

which reduces the infinite-dimensional integral to successive one-dimensional integrals (and which, of course, gives the right answer to Example 10). But the objective here is to apply Definition 4.4 to evaluate the infinite-dimensional integral $\int_{\mathbf{R}^T} f(x_T)F_{X_T}(I[N])$ in (4.5), in order to fill in the intermediate steps between (4.5) and (4.6).

The evaluation of (4.5) is similar to that of Example 9. Let $\varepsilon > 0$ be given. Define a gauge $\gamma = (L, \delta)$ in $\mathbf{R}^T$ as follows. For each $x \in \mathbf{R}^T$ let

$$L(x) = \left\{1 - \frac{1}{2}, \quad 1 - \frac{1}{3}, \quad \ldots \quad , \quad 1 - \frac{1}{p}\right\},$$

with p to be determined later. Suppose $N \in \mathcal{N}$ is chosen, with $N = L(x) \cup M$ where $M \in \mathcal{N}$ (including $M = \varnothing$), so N contains $L(x)$ as laid down in Definition 4.4. Any $x \in \bar{\mathbf{R}}^T$ can be associated with any $N \in \mathcal{N}$, but with x given, the relation of γ-fineness between x_T and N can be valid only for those N which contain $L(x)$. So the finite set N can be arbitrarily large, but is "bounded below" by $L(x)$.

For the purpose of evaluating expectation $\mathrm{E}[f(X)], = \int_{\mathbf{R}^T} f(x)F_X(I[N])$, the function $\delta_j(x(s_j)) > 0$ must be defined for each $s_j \in N$. Each $N \supset L(x)$ consists of a finite number of elements of T,

$$N = \{s_1, s_2, \ldots, s_n\}$$

in increasing order, with $s_j = 1 - 1/j$ for $1 \le j \le p$. Suppose $s_n = 1 - \frac{1}{q}$, where $q \ge p$ is the largest denominator of the list of N. If, for some j $(1 \le j \le q)$, the jth co-ordinate x_j of $x(N)$ is not equal to 0 or 2^{-j}, let

$$0 < \delta_N(x_j) < \min\{|x_j|, |x_j - 2^{-j}|\}$$

for each such j. This ensures that if associated $(x, N, I[N])$ is γ-fine in $\mathbf{R}^T$ and if $x(N) = (x_{s_1}, x_{s_2}, \dots, x_{s_q})$ has $x_{s_j} \ne 2^{-j}$ and $x_{s_j} \ne 0$ for one or more of $j = 1, 2, \dots, q$, then $I(N)$ (an interval of the finite-dimensional space $\mathbf{R}^N$) does not contain[7] any of the 2^q points $(y_{s_1}, y_{s_2}, \dots, y_{s_q}) \in \mathbf{R}^N$ for which $y_{s_j} = 0$ or 2^{-j}. Therefore if $\mathcal{D}$ is a γ-fine division of $\mathbf{R}^T$, $\mathcal{D}$ contains 2^q triples $(y_T, N, I[N])$ with $y_{s_j} = 0$ or 2^{-j} for $j = 1, 2, \dots, q$. By definition of f and $F_X(I[N])$, and for any p and $N \supset L(x) = \{1 - \frac{1}{2}, \dots, 1 - \frac{1}{p}\}$,

$$f(x_T) = x_1 + x_2 + \cdots + x_p + \sum_{j=p+1}^{\infty} x_{s_p},$$

$$F_{X_T}(I[N]) = \left(\prod_{j=1}^{p} F_{X_j}(I_j)\right) \times \left(\prod_{j=p+1}^{q} F_{X_{s_j}}(I_{s_j})\right).$$

In the first product, the labels $j = 1, \dots, p$ are the same for each element $(x, N, I[N])$ of a γ-fine division $\mathcal{D}$, while in the second product the labels s_j (and the number $q - p$ of those labels) may differ in different terms of a γ-fine division $\mathcal{D}$.

Thus, for any γ-fine division $\mathcal{D}$ of $\mathbf{R}^T$, the Riemann sum has the form

$$(\mathcal{D})\sum f(x_T)F_X(I[N]) = \sum\left(\left(\sum_{j=1}^{p} x_j F_X(I[N])\right) + \left(\sum_{j=p+1}^{q} x_{s_j} F_X(I[N])\right)\right),$$

and, using the partial regularization argument of Example 8 above, the Riemann sum $(\mathcal{D})\sum f(x_T)F(I[N])$ satisfies

$$\sum_{j=1}^{p}\left(\frac{1}{2^j}\right)\frac{1}{6} = \sum_{j=1}^{p}\left(\int_{\mathbf{R}} x_j F_{X_j}(I_j)\right) \le (\mathcal{D})\sum f(x_T)F(I[N]) < \sum_{j=1}^{\infty}\left(\frac{1}{2^j}\right)\frac{1}{6} = \frac{1}{6}.$$

Choosing p so that

$$\left|\sum_{j=1}^{p}\frac{1}{2^j} - 1\right| < \varepsilon, \qquad \left|\sum_{j=1}^{p}\left(\frac{1}{2^j}\right)\frac{1}{6} - \frac{1}{6}\right| < \varepsilon,$$

this establishes that, for every γ-fine division $\mathcal{D}$ of $\mathbf{R}^T$,

$$\left|(\mathcal{D})\sum f(x_T)F(I[N]) - \frac{1}{6}\right| < \varepsilon,$$

---

[7]See, for instance section 2.14 (pages 71–73) of [MTRV] for further illustration of this frequently used argument.

so, by Definition 4.4, $f(x_T)F(I[N])$ is bc-integrable on $\mathbf{R}^T$, and

$$E[f(X)] \; = \; \int_{\mathbf{R}^T} f(x_T)F_X(I[N]) \; = \; \frac{1}{6}.$$

The following example shows that, at the most basic level of infinite-dimensional domains, cells combine in a fairly "reasonable" way in Riemann sums.

Example 11 *Let $S = \{s_1, s_2, s_3, \ldots\}$ be a countable set, and let*

$$\Omega \; = \;]0,1]^S \; = \; \prod_{s \in S}]0,1] \; = \;]0,1] \times]0,1] \times]0,1] \times \cdots.$$

For cells $I = I[N] \in \mathcal{I}(\Omega)$, let $|I[N]| := |I(N)| = \prod_{s \in N} |I(s)|$, the product of the one-dimensional interval lengths. The volume function $|\cdots|$ in Ω behaves as might be expected; that is, if $\mathcal{P} = \{I[N]\}$ is any partition of Ω, then

$$(\mathcal{P}) \sum |I[N]| \; = \; 1, \qquad so \qquad \int_\Omega |I[N]| \; = \; 1.$$

But in the infinite-dimensional domain Ω it requires some proof. It would be obviously true if the cells $I[N]$ in $\mathcal{P}$ were each restricted in the same set of dimensions, so there is a finite set $M \subset S$ with $N = M$ for all $I[N] \in \mathcal{P}$. In that case $(\mathcal{P}) \sum |I[N]|$ reduces to $\sum |I(M)|$ where $\{I(M)\}$ is a partition of the finite-dimensional product $]0,1]^M$, and then

$$(\mathcal{P}) \sum |I[N]| \; = \; \sum |I(M)| \; = \; 1.$$

The problem is that two cells $I[N]$ and $I'[N']$ in $\mathcal{P}$ do not necessarily have $N = N'$. (See figures 3.2 and 3.5 of [MTRV], pages 87 and 102.) To get round this, let $M = \bigcup\{N : I[N] \in \mathcal{P}\}$, so, for given $\mathcal{P}$, $M \in \mathcal{N}(S)$ is a fixed finite subset of S. Now construct a regularization of $\mathcal{P}$ (see Example 8 above; and also Figures 3.3 and 3.4, page 91 of [MTRV]), so each $I[N] \in \mathcal{P}$ is itself partitioned,

$$I[N] \; = \; \bigcup\{J[M]\}, \qquad I(N) \; = \; \bigcup\{J(M)\}, \qquad |I(N)| \; = \; \sum |J(M)|,$$

with M the same for all $I[N] \in \mathcal{P}$, giving

$$(\mathcal{P}) \sum |I[N]| \; = \; (\mathcal{P}) \sum \left(\sum_{J[M] \subset I[N]} |J[M]| \right) \; = \; \sum_{I \in \mathcal{P}} |J(M)| \; = \; 1.$$

This way of calculating Riemann sums in infinite-dimensional domains is demonstrated further in theorem 157, page 276 of [MTRV], and is discussed in more detail in Supplement of [website].

Countability of T made no contribution to the argument, which works equally well for uncountable T. On the other hand the factor domain $]0,1]$ has unit length. For $\Omega = [a,b]^T$, the Riemann sum $(\mathcal{P}) \sum |I[N]|$ evaluates as $(b-a)^m$ where m is the cardinality of M, so $\int_\Omega |I[N]| = 0$ if $b - a < 1$, and the integral diverges to $+\infty$ if $b - a > 1$. On the other hand, $\int_{\mathbf{R}^T} k(I) = 1$ if, for instance, $k(I[N])$ is Gaussian, $k(I[N]) = \prod_{s \in N} k(I(s))$, with

$$k(I(s)) = \int_{I(s)} \frac{e^{-\frac{1}{2}y^2}}{\sqrt{2\pi}} dy$$

for one-dimensional cells $I(s)$. Similarly,

$$\int_{[a,b]^T} k(I[N]) = 1 \quad if \quad k(I(s)) = \frac{|I(s)|}{b - a}, \quad k(I[N]) = \prod_{s \in N} k(I(s)).$$

4.2   Sample Space $\mathbf{R}^{\mathbf{T}}$ with T Uncountable

The labelling set (or dimension set) T in the domain $\mathbf{R}^T$ of (4.1) is a countable set of dimensions or labels. But the labels can be taken to be an uncountable set $\mathbf{T}$, such as the continuum $]0, 1]$, and then the domain is

$$\mathbf{R}^{\mathbf{T}} := \prod_{s \in \mathbf{T}} \mathbf{R}_s, \qquad \mathbf{R}_s = \mathbf{R} \quad \text{for each } s \in \mathbf{T}. \qquad (4.7)$$

With this change in meaning of $\mathbf{R}^{\mathbf{T}}$, the concepts and notation introduced for the definition (4.4) of the integral of a function h in $\mathbf{R}^{\mathbf{T}}$ then carry over unchanged in the new context of uncountable $\mathbf{T}$.

The following example illustrates the use of $\int_{\mathbf{R}^{\mathbf{T}}} h$, with uncountable $\mathbf{T}$, by means of a calculation of broadly stochastic integral type.

Example 12 *Suppose, at each instant s of the time interval $]0, \tau]$, a share takes random value $x(s)$ (or x_s). Suppose $x(0) = x_0 = 1$ (with probability 1), and suppose, at each time s ($0 < s \leq \tau$), the share price takes value $x(s)$, $0 \leq x(s) \leq 1$, with uniform probability on $[0, 1]$. Suppose the value at time s is independent of the value taken at any other time. Suppose an investor takes an initial shareholding $z(0)$, or z_0, of 1 share (so $z_0 = 1$), and suppose the shareholding or number of shares $z(s)$ varies randomly at m fixed times τ_k between initial time 0 and terminal time τ,*

$$0 = \tau_0 < \tau_1 < \tau_2 < \cdots < \tau_m = \tau; \qquad M := \{\tau_1, \tau_2, \ldots, \tau_m\}.$$

Thus $]0, \tau]$ can be a period of m days, with shareholding changing randomly at the end of each day. Suppose the random value z_{τ_k} (or simply z_k) of shares at time τ_k is independent of the value $z_{k'}$ at any other time $\tau_{k'}$, and independent of the value $x(s)$ of the share at any time s. To keep things uncomplicated suppose that, at any time τ_k, the shareholding z_k is 1 with probability 0.5, and $z_k = 0$ with probability 0.5. **What is the expected payout at terminal time t from this shareholding?**

The intention is to apply a stochastic integral calculation. But the probability distributions are deliberately chosen so that the expected payout is fairly obvious on intuitive grounds. Then it can be seen whether the stochastic integral calculation confirms what common sense indicates.

The share price varies uniformly (and continuously) between 0 and 1, so the average share price over the period is 0.5. The shareholding (or number of shares held) varies between 0 and 1, with average shareholding equal to 0.5. Therefore the average value of the portfolio throughout the period $]0, \tau]$ is 0.5×0.5, $= 0.25$. At initial time 0 the shareholding consists of 1 share, purchased at a cost of 1, so initial portfolio value is 1. Therefore the **net** portfolio value at any time t $(0 < t \leq \tau)$ is $1 - 0.25$, which is -0.75, an overall net loss.

The result might be different if the distribution of share price values were non-uniform, and/or if the random variation of the shareholding were not simply as the toss of a coin. It could be postulated, for instance, that shareholding varies continuously (not just 0 and 1), and/or that the shareholding depends in some manner on the random values of the share.

The values of the portfolio at times s, s' $(s < s')$ are $z(s)x(s)$ and $z(s')x(s')$, and the increment can be taken to be

$$z(s)x(s) - z(s')x(s').$$

Writing $\beta(s) = z(s)x(s)$, the shareholding value increment is then $\beta(s') - \beta(s)$, and the overall increment in the period $]0, \tau]$ is

$$\sum (\beta(s) - \beta(s')), \quad = \quad \beta(\tau) - \beta(0) \quad \text{or} \quad \int_{]0,\tau]} d\beta(s) \quad = \quad \beta(\tau) - \beta(0),$$

by cancellation or "telescoping". This calculation is unhelpful insofar as it omits the role of the investor or trader, who makes investment decisions at times s for $0 < s < \tau$. What is really needed is to find a way to put a value on investment strategy or policy.

Thus, for purposes of financial calculation, it is appropriate to consider the quantity $z(s)$ to be a "look forward" value, a "gambling" amount whose payoff depends on the value $x(s)$ at time s and on potential future share values $x(t)$ at later times $t > s$. (In Example 12, the investor strategy[8] is to toss a coin at times τ_j, and at times $s \neq \tau_j$ to keep the shareholding constant.)

In other words, at time s the investor "goes long in" (or purchases) a number $z(s)$ of shares to be held until the later instant $s' > s$; at which point a new investment decision is made. So the growth in shareholding value is

$$\cdots \; + \; z(s)(x(s') - x(s)) \; + \; z(s')(x(s'') - x(s')) \; + \; \cdots.$$

This gives an overall valuation for the investment decisions or strategy in the period $]0, \tau]$.

These are the ideas underlying a net gain calculation for shareholding value as outlined in (2.1), (2.2), (2.3), (2.4), (2.5) of Section 2.3.

Taking this point of view the aggregate of value increments is

$$\sum z(s)(x(s') - x(s)), \quad \text{or} \quad W_\tau = \int_0^\tau Z_s(X_{s'} - X_s), \; = \int_{]0,\tau]} Z_s \, dX_s.$$

---

[8]In financial theory a typical financial strategy is *elimination of arbitrage*. In [MTRV] this is described in pages 436–440.

In other words it appears to be a calculation of the stochastic integral kind, aimed at finding a random variable expression W_τ for the shareholding value at time τ. But the question posed is to find the **expected value** $E[W_\tau]$ of the latter random variable. If it is the case that, as a random variable, the stochastic integral W_τ can be expressed as

$$W_\tau \simeq w_\tau [\mathbf{R}, F_{W_\tau}], \qquad (4.8)$$

with distribution function F_{W_τ} to be determined somehow, then

$$E[W_\tau], \ = \ \int_{\mathbf{R}} w_\tau \, dF_{W_\tau},$$

is the solution to the question posed in Example 12—provided the latter integral expression is meaningful.

As presented above, it seems that shareholding value depends jointly on:

- the random values taken by share value $x(s)$ at each time s, $0 < s \le \tau$; and

- the random values taken by shareholding (number of shares in the portfolio) at times $\tau_k \in M$.

Consider the following formulation of sample space Ω:

$$\Omega \ = \ \mathbf{R}^M \times \mathbf{R^T}. \qquad (4.9)$$

The factor $\mathbf{R}^M$ accommodates the random variability of $(z_s)_{s \in M}$ while the factor $\mathbf{R^T}$ accommodates the random variability of $(x_s)_{s \in \mathbf{T}}$.

This formulation seems to neglect a key point. The random variability that occurs at times $\tau_k \in M$ is linked to the times s at which the random variation occurs in $\mathbf{R^T}$. In other words a shareholding amount $z(\tau_k)$ applies to the share price values $x(s)$ for $\tau_k < s \le \tau_{k+1}$. Because of this complication, postpone for a moment the final designation of the sample space Ω, until the details become clearer.

For $0 < s \le \tau$, define $[s]$ by

$$[s] \ := \ \max\{\tau_k \in M \ : \ \tau_k \le s\}.$$

Choose $N = \{s_1, \ldots, s_n\} \in \mathcal{N}(\mathbf{T})$ with $N \supseteq M$. For $s_j \in N$, $[s_j]$ is $\tau_k \in M$ where τ_k is the largest $\tau_k \in M$ for which $\tau_k \le s_j$.

For joint variables $(x_{\mathbf{T}}, z_{\mathbf{T}})$ and $M \subset N \in \mathcal{N}(\mathbf{T})$, let

$$f\left((x_{\mathbf{T}}, z_{\mathbf{T}}), N\right) \ := \ \sum_{j=1}^{n} z_{[s_{j-1}]}\left(x(s_j) - x(s_{j-1})\right);$$

and, since intermediate terms in $\sum \left(x(s_j) - x(s_{j-1})\right)$ cancel,

$$f\left((x_{\mathbf{T}}, z_{\mathbf{T}}), N\right) \ = \ \sum_{k=1}^{m} z(\tau_{k-1})\left(x(\tau_k) - x(\tau_{k-1})\right).$$

Note that f does not depend **explicitly** on N or M in this case; so it is not unreasonable to express it as $f(x_{\mathbf{T}}, z_{\mathbf{T}})$. But, writing $z(\tau_k) = z_k$, $x(\tau_k) = x_k$, the cancellation of terms above shows that the contingent joint observable

$$
\begin{aligned}
f((x_{\mathbf{T}}, z_{\mathbf{T}}), N) &= f(x_M, z_M) \\
&= \sum_{k=1}^{m} z(\tau_{k-1})\,(x(\tau_k) - x(\tau_{k-1})) \\
&= -z_0 x_0 + \sum_{k=1}^{m} x_k\,(z_k - z_{k-1}), \qquad (4.10)
\end{aligned}
$$

reducing the random variation to the variability expressed by the finite-dimensional domain $\mathbf{R}^M$. But the randomness manifests itself in **two** distinct variables, $Z_{\mathbf{T}}$ and $X_{\mathbf{T}}$; and the joint occurrences are at times τ_k. Therefore the sample space Ω should be formulated as $\mathbf{R}^M \times \mathbf{R}^M$, or, to be more precise,[9]

$$
\Omega = (\mathbf{R}_\tau \times \mathbf{R}_\tau)_{\tau \in M} = (\mathbf{R}^2)^M.
$$

For each $s \in \mathbf{T}$, and for $I_s \in \mathcal{I}(\mathbf{R})$, $I_s = \,]u, v]$, write $J_s = I_s \cap [0,1]$, and take $F_{X_s}(I_s) = |J_s|$, where $|J_s| := 0$ if $J_s = \varnothing$. Let $F_{Z_s}(I_s)$ be atomic on 0 and 1. Thus the distribution functions $F_{X_s}(I_s)$ and $F_{Z_s}(I_s)$ for the (independent) random variables X_s and Z_s are

$$
F_{X_s}(I_s) = \begin{cases}
v - u & \text{if } 0 \le u < v \le 1, \\
1 - u & \text{if } 0 \le u < 1 < v, \\
v - 0 & \text{if } u < 0 < v \le 1, \\
1 - 0 & \text{if } u \le 0 < 1 \le v, \\
0 & \text{if } I_s \cap [0,1] \text{ is empty or is not an interval.}
\end{cases} \qquad (4.11)
$$

And

$$
F_{Z_s}(I_s) = \begin{cases}
\frac{1}{2} & \text{if } 0 \in I_s,\ 1 \notin I_s, \\
\frac{1}{2} & \text{if } 0 \notin I_s,\ 1 \in I_s, \\
1 & \text{if } 0 \in I_s,\ 1 \in I_s, \\
0 & \text{if } 0 \notin I_s,\ 1 \notin I_s,
\end{cases} \qquad (4.12)
$$

if $s = \tau_j$, $j = 1, 2, \ldots, m$. Thus the probability distribution of the share price $x(s)$ is uniform on $[0,1]$; while the shareholding value $z(s)$ is distributed equally, and atomically, on 0 and 1, for $s = \tau_j$, $1 \le j \le m$. For $s \neq \tau_j$,

$$
F_{Z_s}(I_s) = \begin{cases}
1 & \text{if } z_{[s]} \in I_s, \\
0 & \text{if } z_{[s]} \notin I_s;
\end{cases}
$$

---

[9] Domain $(\mathbf{R}^M)^2$ consists of pairs of M-tuples, while $(\mathbf{R}^2)^M$ consists of M-tuples of pairs.

so the random variables Z_s $(0 < s \leq \tau)$ (which are independent of the random variables X_s) are not independent of each other.

For purposes of experimentation and investigation, these distribution values can be varied and tested. Likewise for other parameters. Note also that the sample space Ω can have various forms, to be chosen as a matter of convenience; such as

$$\Omega = \{0,1\}^m \times [0,1]^T, \quad \text{or} \quad \Omega = \mathbf{R}^M \times \mathbf{R}^T. \tag{4.13}$$

The latter form is implicitly chosen by (4.11) and (4.12).

The next step is to define a joint distribution function $F_{XZ}, = F_{(X_\mathbf{T}, X_\mathbf{T})}$, for the pair of joint processes $X_\mathbf{T}$ and $Z_\mathbf{T}$. Suppose

$$N = \{t_1, \ldots, t_n\} \in \mathcal{N}(\mathbf{T}), \quad 0 = t_0 < t_1 < \cdots < t_n = \tau,$$

$$M = \{\tau_1, \ldots, \tau_n\}, \quad 0 = \tau_0 < \tau_1 < \cdots < \tau_n = \tau,$$

so N is variable and M is fixed. Given $I[N] \in \mathcal{I}(\mathbf{R^T})$ and $J = J(M) \in \mathcal{I}(\mathbf{R}^M)$, the former infinite-dimensional, the latter finite-dimensional,

$$F_{X_\mathbf{T}}(I[N]) = \prod_{j=1}^{n} F_{X_{t_j}}(I(t_j)),$$

$$F_{Z_\mathbf{T}}(J(M)) = \prod_{k=1}^{m} F_{Z_{\tau_k}}(I(\tau_k)).$$

For $I_\tau, J_\tau \in \mathcal{I}(\mathbf{R})$ write $K_\tau = I_\tau \times J_\tau \in \mathcal{I}(\mathbf{R} \times \mathbf{R})$, so

$$K(M) = \prod_{\tau \in M} K_\tau = \prod_{\tau \in M} I_\tau \times J_\tau,$$

$$F_{X_\mathbf{T} Z_\mathbf{T}}(K(M)) = F_{X_M Z_M}(K(M)) = \prod_{\tau \in M} F_{X_\tau}(I_\tau) F_{Z_\tau}(J_\tau).$$

Because of cancellation of random variables described above in (4.10), the contingent joint observable for gain in shareholding value

$$f(X_M, Z_M) \simeq f(x_M, z_M)\left[(\mathbf{R} \times \mathbf{R})^M, F_{X_M Z_M}\right],$$

and expected gain is $\mathrm{E}[f(X_M, Z_M)] =$

$$= \int_{(\mathbf{R} \times \mathbf{R})^M} \left(\sum_{k=1}^{m} z(\tau_{k-1})\left(x(\tau_k) - x(\tau_{k-1})\right)\right) F_{X_M Z_M}(K(M)).$$

It is almost immediately obvious that this integral exists, with value -0.75, and it is hardly necessary to conduct a formal analysis. To make it slightly more obvious, write the atomic integrator F_{Z_s} as $d\mu_{z_s}$, and the uniform integrator F_{X_s} as dx_s, and write the point-integrand $\sum_{k=1}^{m} z(\tau_{k-1})\left(x(\tau_k) - x(\tau_{k-1})\right)$ as

$$\sum_{k=1}^{m} z_{k-1}(x_k - x_{k-1}).$$

Then, in more traditional format, $\mathrm{E}[f(X_M, Z_M)] =$

$$\int_{z_1=0}^{1} \int_{x_1=0}^{1} \cdots \int_{z_m=0}^{1} \int_{x_m=0}^{1} \left(\sum_{k=1}^{m} z_{k-1}(x_k - x_{k-1})\right) d\mu_{z_1} dx_1 \ldots d\mu_{z_m} dx_m. \tag{4.14}$$

If required, this can be set out in gauge-Riemann sum format, using the "conformance" technique as described in previous examples. For $m = 1$,

$$E[f(X_M, Z_M)] \;=\; \int_{z_1=0}^{1} \int_{x_1=0}^{1} z_0(x_1 - x_0) d\mu_{z_1} dx_1 \;=\; -\frac{1}{2},$$

as predicted earlier by means of elementary probability considerations. For $m = 2$ the integral is

$$\int_{z_1=0}^{1} \int_{x_1=0}^{1} \int_{z_2=0}^{1} \int_{x_2=0}^{1} \left(z_0(x_1 - x_0) + z_1(x_2 - x_1) \right) d\mu_{z_1} dx_1 d\mu_{z_2} dx_2$$

with $z_0 = x_0 = 1$. This multiple integral can be evaluated by elementary methods, and the result is again -0.75, as predicted. Similarly for any positive integer m. Alternatively some formal integration theory can be used, such as Tonelli's theorem; again giving

$$E[f(X_M, Z_M)] \;=\; -0.75. \tag{4.15}$$

This is the result which was predicted on elementary intuitive or common sense grounds. So the preceding analysis is like using a sledge-hammer to crack a nut. On the other hand, it is best to learn how to swing a sledge-hammer just in case we ever have to crack a rock.

Also, the above analysis does not depend on creating a stochastic integral

$$\sum z(s)(x(s') - x(s)) \quad \text{or} \quad \int_0^\tau z(s)\, dx(s).$$

Such a formulation appears momentarily in (4.10), only to disappear quickly from the ensuing calculations. Stochastic integrals for Example 12 are considered in Section 4.3 below.

Note that, as flagged by the cancellations in (4.10), uncountably infinite random variability reduced to finite-dimensional random variation, with sample space $\Omega = (\mathbf{R} \times \mathbf{R})^M$. What if, in the above analysis, we have some reason to keep track of the random variation in $x(s)$ for $s \in \mathbf{T} \setminus M$, even though we know that this random variability cannot affect the value of $E[f(X_\mathbf{T}, Z_M)]$, which is the stated objective in this case?

Example 15 below explores this scenario, and is a further examination of uncountably infinite joint random variability.

4.3   Stochastic Integrals for Example 12

Example 12 above has two independent stochastic processes,

$$Z_\mathbf{T} \;\simeq\; z_\mathbf{T}\left[\mathbf{R}^\mathbf{T}, F_{Z_\mathbf{T}} \right], \qquad X_\mathbf{T} \;\simeq\; x_\mathbf{T}\left[\mathbf{R}^\mathbf{T}, F_{X_\mathbf{T}} \right],$$

with joint process $(Z_\mathbf{T}, X_\mathbf{T})$ expressed by

$$(Z_\mathbf{T}, X_\mathbf{T}) \;\simeq\; (z_\mathbf{T}, x_\mathbf{T})\left[(\mathbf{R} \times \mathbf{R})^\mathbf{T}, F_{(Z_\mathbf{T}, X_\mathbf{T})} \right]; \tag{4.16}$$

where each sample path $z_{\mathbf{T}} = (z(s) : 0 < s \leq \tau)$ is constant for $\tau_{j-1} \leq s < \tau_j$, $(1 \leq j \leq m)$.

The calculation in (4.10) enabled us to disregard the random variation in $x(s)$ for $\tau_{j-1} < s < \tau_j$, $(1 \leq j \leq m)$, so the joint processes can be expressed as

$$(Z_M, X_M) \simeq (z_M, x_M) \big[(\mathbf{R} \times \mathbf{R})^M , F_{(Z_M, X_M)} \big]; \qquad (4.17)$$

and the latter formulation enabled us to perform a calculation for the expected gain in portfolio value (or shareholding value).

Stochastic integrals $\int_0^\tau \cdots$ on domain $\mathbf{T} = \,]0, \tau]$ can be formulated from version (4.16). The objective is to express the gains (or losses) in portfolio value $w(t)$ $(0 < t \leq \tau)$ in terms of joint sample paths $(z(s), x(s))$ $(0 < s \leq \tau)$ of the joint process $(Z_{\mathbf{T}}, X_{\mathbf{T}})$,

$$w(t) \; = \; \sum_{0 \leq s < s' \leq t} z(s)\,(x(s') - x(s)), \quad \text{or} \qquad (4.18)$$

$$w(t) \; = \; \int_0^t z(s)\,dx(s), \quad (0 < t \leq \tau). \qquad (4.19)$$

Thus $w(t)$ depends on the joint outcomes $((z(s), x(s)) : 0 < s \leq t)$, or $w(t) = h(z_{\mathbf{T}}, x_{\mathbf{T}})$ where $\mathbf{T} = \,]0, t]$ and h is the deterministic function given by the Stieltjes integral $\int_0^t z(s)\,dx(s)$—if and when the latter integrals exist. These integrals are sample path versions of a stochastic integral $\int_0^t Z(s)\,dX(s)$, and can be examined further, in terms of particular sample paths, in order to try to understand whether or not they exist, and what other kinds of issues can arise with them.

Before undertaking this task, the random variability in the outcomes $(w(t) : 0 < t \leq \tau)$ can be examined. Write $U_{\mathbf{T}}$ for the joint processes $(Z_{\mathbf{T}}, X_{\mathbf{T}})$, so a sample path for $U_{\mathbf{T}}$ is $u_{\mathbf{T}} = (z_{\mathbf{T}}, x_{\mathbf{T}})$. Then (4.16) gives

$$W_{\mathbf{T}} \; = \; h(U_{\mathbf{T}}) \simeq h(u_{\mathbf{T}}) \big[(\mathbf{R} \times \mathbf{R})^{\mathbf{T}} , F_{U_{\mathbf{T}}} \big], \qquad (4.20)$$

where $F_{U_{\mathbf{T}}}$ is the joint distribution function $F_{(Z_{\mathbf{T}}, X_{\mathbf{T}})}$ mentioned in (4.16), and $h(U_{\mathbf{T}})$ is the stochastic integral $\int_0^t Z(s)\,dX(s)$. Thus $W_{\mathbf{T}}$ is a contingent process depending on the joint values of the processes $Z_{\mathbf{T}}, X_{\mathbf{T}}$.

The details of $F_{(Z_M, X_M)}$ were described above, but not those of $F_{(Z_{\mathbf{T}}, X_{\mathbf{T}})}$. These are provided in Section 4.4 below.

Another issue is the function h, which is determined by integrals $\int_0^t z(s)\,dx(s)$. The existence of these integrals has not been established. They are now to be examined in more detail by looking at particular sample paths.

According to Example 12, a sample path for the process $Z_{\mathbf{T}}$ is

$$z(s) = \alpha_j \text{ for } \tau_j \leq s < \tau_{j+1}, \; j = 0, 1, 2, \ldots, m - 1; \text{ with } z(\tau) = z(\tau_m) = \alpha_m,$$

where each α_j is 0 or 1.

Now consider some sample paths for $X_{\mathbf{T}}$:

1. $x(s)$ *constant for* $0 \le s \le \tau$.

   Then the Stieltjes integral $\int_0^t z(s)\,dx(s)$ exists, with value zero, so $w(t) = 0$ for $0 < t \le \tau$.

2. $x(s) = s$ *for* $0 \le s \le \tau$.

   Let k be the largest integer j for which $\tau_j \le t$. The Stieltjes integral $\int_0^t z(s)\,dx(s)$ exists, with value

   $$w(t) \;=\; \sum_{j=1}^{k-1} \alpha_{j-1}\,(\tau_j - \tau_{j-1}) \;+\; \alpha_k\,(t - \tau_k).$$

3. *Let the sample function* $x(s)$ *be a step function on* $[0, \tau]$.

   Existence of $w(t) = \int_0^t z(s)\,dx(s)$ is easily deduced.

4. *Suppose the sample path* $x(s)$ *is continuous for* $0 \le s \le \tau$.

   From theory of Riemann-Stieltjes integration, $w(t) = \int_0^t z(s)\,dx(s)$ exists for each t, $0 < t \le \tau$.

5. *Similarly if the sample path* $x(s)$ *has bounded variation in* $]0, \tau]$.

Continuing in this vein, if it were the case that $w(t) = h(z_{\mathbf{T}}, x_{\mathbf{T}}) = \int_0^t z(s)\,dx(s)$ exists for **every** pair of sample paths $(z_{\mathbf{T}}, x_{\mathbf{T}})$, it would be reasonable to declare that the stochastic integral process exists, with

$$W(t) \;=\; h(Z_{\mathbf{T}}, X_{\mathbf{T}}) \;=\; \int_0^t Z(s)\,dX(s),$$

$$W(t) \;\simeq\; h(z_{\mathbf{T}}, x_{\mathbf{T}}) \left[(\mathbf{R} \times \mathbf{R})^T, F_{(Z_{\mathbf{T}}, X_{\mathbf{T}})} \right]. \qquad (4.21)$$

But the set of sample paths $x_{\mathbf{T}}$ in $\mathbf{R}^{\mathbf{T}}$ which are "well-behaved" (continuous, bounded variation etc.) is merely a subset of all potential sample paths $\mathbf{R}^{\mathbf{T}}$. Example 13 below is a counter-example.

6. *Suppose* $x(s)$ *is the Dirichlet function of Example 13.*

   That is, $x(s) = 1$ if s is a rational number, and $x(s) = 0$ otherwise. The argument in Example 13 below shows that $w(t) = \int_0^t z(s)\,dx(s)$ **exists if and only if the step function** $z(s)$ **is constant for** $0 \le s \le t$.

This counter-example, though rather extreme and unusual, demonstrates that construction of stochastic integrals can be problematic even in a simple case such as Example 12. Nevertheless, calculation of $E[W(\tau)]$ for Example 12 was accomplished in (4.15) **without** constructing stochastic integrals.

Example 13 *Consider the Dirichlet function* $d(s)$ *for* $0 \le s \le 1$; $d(s) = 1$ *if* s *is a rational number, and* $d(s) = 0$ *if* s *is not rational. For intervals* $J =]s', s] \subset$

[0, 1] *define the Dirichlet interval function (or incremental Dirichlet function)* D *by*

$$D(J) := d(s) - d(s').$$

The interval function D *is the Stieltjes version of the point function* d. *Accordingly* D *is finitely additive on disjoint, adjoining intervals, and* D *is integrable on* $]0, 1]$. *(If* $\mathcal{P}$ *is given by* $0 = s_0 < s_1 < s_2 < \cdots < s_{n-1} < s_n = 1$, *then*

$$(\mathcal{P}) \sum D(I) = \sum_{j=1}^{n} (d(s_j) - d(s_{j-1})) = d(s_n) - d(s_0) = 1 - 1 = 0,$$

and this holds for all partitions, so $\int_0^1 D(I) = 0$.) *But, according to theorem 67, page 180 of [MTRV], the expression* $f(s)D(I)$ *is integrable on* $]0, 1]$ *if and only if* $f(s)$ *is constant for* $0 \le s \le 1$.

Proof. If $f(s)$ has constant value κ, say, then, for any partition $\mathcal{P}$, and $s_{j-1} \le s'_j \le s_j$ $(1 \le j \le n)$,

$$(\mathcal{P}) \sum f(s)D(I) = \sum_{j=1}^{n} f(s'_j)(d(s_j) - d(s_{j-1})) = \kappa \sum_{j=1}^{n} (d(s_j) - d(s_{j-1})) = 0;$$

so $f(s)D(I)$ is integrable in this case. Now suppose there exists c, $0 < c \le 1$, such that $f(0) \neq f(c)$; so f is not constant. We can establish that $f(s)D(I)$ is not integrable (in the -complete sense) on $]0, c]$, and hence is not integrable on $]0, 1]$. For simplicity take $c = 1$ if c is a rational number. Define a gauge $\delta(s) > 0$ in $[0, 1]$ as follows. Let $\delta(0) < 1$ and $\delta(1) < 1$. For each s in $0 < s < 1$ let

$$\delta(s) < \min\{s, 1 - s\}. \tag{4.22}$$

Then every δ-fine division $\mathcal{D}$ of $]0, 1]$ consists of associated pairs $\{(s, I)\}$:

$$(s'_1,]s_0, s_1]), \quad \ldots, \quad (s'_j,]s_{j-1}, s_j]), \quad \ldots, \quad (s'_n,]s_{n-1}, s_n]),$$

where $s'_1 = s_0 = 1$, $s'_j = s_{j-1}$ or s_j, and $s'_n = s_n = 1$; and where $s_j - s_{j-1} < \delta(s'_j)$ for $1 \le j \le n$. The points 0 and 1 have to be tag points of their respective associated intervals because of the "conformance" condition (4.22) on $\delta(s)$ for $s \neq 0, 1$. Then the Riemann sum for $\mathcal{D}$ is

$$(\mathcal{D}) \sum f(s)D(I) = \sum_{j=1}^{n} f(s'_j)(d(s_j) - d(s_{j-1})).$$

Each of the partition points s_j $(1 \le j \le n - 1)$ is either a rational or irrational number. Consider any two adjoining pairs

$$(s'_j,]s_{j-1}, s_j]), \quad (s'_{j+1},]s_j, s_{j+1}]). \tag{4.23}$$

If s_j is rational $(1 \le j \le n - 1)$, replace it by an irrational s''_j as follows. We have

$$s_j - s_{j-1} < \delta(s'_j), \qquad s_{j+1} - s_j < \delta(s'_{j+1}),$$

and since the irrationals are dense there exists an irrational number s_j'' which can replace s_j without breaching the gauge inequalities (4.23). Make the replacements, so the δ-fine division $\mathcal{D}$ now has irrational partition points s_j for $1 \le j \le n-1$, and

$$(\mathcal{D}) \sum f(x)D(I) \;=\; f(0) - f(1).$$

Using the density argument again, form another δ-fine partition $\mathcal{D}'$ so that the partition points s_j are all rational numbers. Then $(\mathcal{D}') \sum f(x)D(I) = 0$ and

$$\left|(\mathcal{D}) \sum f(x)D(I) - (\mathcal{D}') \sum f(x)D(I)\right| \;=\; |f(0) - f(1)| \;>\; 0.$$

If $\varepsilon > 0$ is chosen so that $\varepsilon < |f(0) - f(1)|$ then there is **no** gauge $\delta(s) > 0$ in $[0,1]$ capable of satisfying the Riemann sum inequality of Definition 6 of the -complete integral, and $f(s)D(I)$ is not integrable. If c is an irrational number, then a modification of the above argument gives the same result. If c is irrational, then $0 < c < 1$, and a gauge $\delta(s) > 0$ can be defined so that every δ-fine division of $]0,1]$ must include terms

$$(c,\ ]s_{m-1}, c]) \quad \text{and} \quad (c,\ ]c, s_{m+1}])$$

where $]0,1]$ is partitioned by points

$$0 \;=\; s_0 < s_1 < \cdots < s_{m-1} < s_m \;=\; c < s_{m+1} < \cdots < s_{n-1} < s_n \;=\; 1.$$

This is *gauge-conformance to the point* c and, as before, is ensured as follows. For each $a < c$ $(0 \le a < c)$, take $\delta(a) < c - a$; and for each $b > c$ $(c < b \le 1)$, take $\delta(b) < b - c$. Choose a δ-fine division $\mathcal{D}$ of $]0,1]$ with partition points $\{0 = s_0, s_1, \ldots, s_m = c, \ldots, s_n = 1\}$. As before make very small adjustments to these partition points to ensure that s_j is rational for $1 \le j \le m-1$, and s_j is irrational for $m+1 \le j \le n-1$. The rest of the argument follows as before. ◯

Gauges $\gamma = (L, \delta_N)$ for infinite product spaces are defined in (4.2), and it is noted that, for $N \ne N'$, and for $t \in N \cap N'$, the value $\delta_N(x(t))$ need not be the same as $\delta_{N'}(x(t))$. The following example illustrates this point.

Example 14 *For domain* $\Omega = [0,1]^{\mathbf{T}}$, *samples* $N = \{t_1, \ldots, t_n\} \in \mathcal{N}(\mathbf{T})$, *and cells* $I[N] \in \mathcal{I}(\Omega)$, *define a cell function* $h(I[N])$ *by*

$$h(I[N]) \;:=\; \prod_{j=1}^{n-1} |I_{t_j}| \, D(I_{t_n})$$

where D *is the incremental Dirichlet function of Example 13.   Then, provided* $\delta_N(x(t_n)) > 0$ *is chosen according to the method of Example 13 (with* $\delta_N(x(t_j)) > 0$ *arbitrary for* $j < n$*),* h *is integrable on* Ω *with* $\int_\Omega h(I[N]) = 0$. *Consider* $N' = \{s_1, \ldots, s_m\}$ *with* $t_n = s_j$ *where* $j < m$. *Then* $t = t_n \in N \cap N'$. *But, in the Riemann sums for* $\int_\Omega h$*, the gauge value* $\delta_{N'}(x(t)) > 0$ *is arbitrary, while* $\delta_{N'}(x(s_m))$ *has to be chosen as in Example 13. Therefore* $\delta_N(x(t))$ *is not generally the same as* $\delta_{N'}(x(t))$.

4.4   Example 12

Example 13 demonstrates that stochastic integrals can fail to exist even in the relatively simple case of Example 12. It has been pointed out that (4.15) delivers $E[W(\tau)]$ **without** constructing stochastic integrals. The joint distribution function used in that calculation is $F_{U_M}, = F_{(Z_M, X_M)}$ as in (4.17).

Compare this with (4.8), where the sample space for the *elementary* form of the random variable $W(\tau)$ is $\mathbf{R}$, with distribution function F_{W_τ}. The difference between the elementary and joint basic representations of joint random variability was described in Section 3.5. In the *elementary* format (4.8), the expected value of $W(\tau)$ is obtained by integration on $\mathbf{R}$ with respect to F_{W_τ}. Likewise for any contingent observable $f(W(\tau))$ that might arise.

In contrast, (4.20) and (4.16) employ $(\mathbf{R} \times \mathbf{R})^{\mathbf{T}}$ as sample space for W_τ; and the joint distribution function is $F_{U_{\mathbf{T}}}, = F_{(Z_{\mathbf{T}}, X_{\mathbf{T}})}$ (or $F_{Z_{\mathbf{T}} X_{\mathbf{T}}}$); but this distribution function was left unspecified. At this point it is useful to pursue this approach to $E[W(\tau)]$ a bit further. Here is a summary of what is involved:

- With $W(\tau) \simeq w(\tau)[\mathbf{R}, F_{W(\tau)}]$, expected gain in portfolio value at time τ can be calculated as

$$E[W(\tau)] \ = \ \int_{\mathbf{R}} w(\tau) F_{W(\tau)}(I_\tau).$$

- With $W(\tau) = f(X_M, Z_M) \simeq f(x_M, z_M)\left[(\mathbf{R} \times \mathbf{R})^M, F_{X_M Z_M}\right]$, expected gain in portfolio value at time τ can be calculated as

$$\begin{aligned} E[W(\tau)] \ &= \ E[f(X_M, Z_M)] \\ &= \ \int_{(\mathbf{R} \times \mathbf{R})^M} f(x_M, z_M) F_{X_M Z_M}\left(I_{X_M}(M) \times I_{Z_M}(M)\right). \end{aligned}$$

Interval notation I_{X_M}, I_{Z_M}, refers to events (or sets of potential occurrences) of random variables X_M, Z_M, respectively.

- The two preceding models have been examined already, and it was demonstrated that the expected gain (actually a loss of -0.75) corresponded with the value indicated by simple, intuitive probabilistic ideas. What remains is a model in which the sample space for the random variation is $(\mathbf{R} \times \mathbf{R})^{\mathbf{T}}$:
  With $W(\tau) = h(U_{\mathbf{T}}) = h(X_{\mathbf{T}}, Z_{\mathbf{T}}) \simeq h(x_{\mathbf{T}}, z_{\mathbf{T}})\left[(\mathbf{R} \times \mathbf{R})^{\mathbf{T}}, F_{(X_{\mathbf{T}}, Z_{\mathbf{T}})}\right]$, *expected gain in portfolio value at time τ can be calculated as*

$$\begin{aligned} E[W(\tau)] \ &= \ E[h(U_{\mathbf{T}})] \ = \ E[h(X_{\mathbf{T}}, Z_{\mathbf{T}})], \\ E[h(U_{\mathbf{T}})] \ &= \ \int_{(\mathbf{R} \times \mathbf{R})^{\mathbf{T}}} h(u_{\mathbf{T}}) F_{U_{\mathbf{T}}}\left(I_{U_{\mathbf{T}}}[N]\right), \qquad (4.24) \\ E[h(X_{\mathbf{T}}, Z_{\mathbf{T}})] \ &= \ \int_{(\mathbf{R} \times \mathbf{R})^{\mathbf{T}}} h(x_{\mathbf{T}}, z_{\mathbf{T}}) F_{X_{\mathbf{T}} Z_{\mathbf{T}}}\left(I_X[N] \times I_Z[N]\right). \end{aligned}$$

The function h corresponds to a stochastic integral construction as in (4.20) and (4.21). (We make no distinction between $(x_{\mathbf{T}}, z_{\mathbf{T}})$ and $(z_{\mathbf{T}}, x_{\mathbf{T}})$,

etc. The order in which these two variables are placed does not affect
the result.) But Example 13 shows that the stochastic integral used in
(4.21) does not always exist as a Stieltjes-type integral; so in general the
construction $\int_0^\tau Z(s)dX(s)$ of (4.19) is not fully meaningful as it stands.
But the construction

$$w(t) = h(x_{\mathbf{T}}, z_{\mathbf{T}}) = \sum_{0 \le s < s' \le t} z(s)\,(x(s') - x(s))$$

of (4.18) is simply a finite Riemann sum; and, essentially, this is what is
proposed here as the integrand for (4.24).

For (4.24) to be valid, it must be established that the integral on the right exists,
with value -0.75.

To start, consider the distribution function for random variables Z_s, $s \in \mathbf{T}$.
For $s \in M$ (so $s = \tau_j$ for some j, $1 \le j \le m$), $Z(s) = 0$ or 1 with equal likelihood.
For other values of s, $z(s)$ has value $z(s_j)$ where

$$[s] = \tau_j \quad \text{if} \quad \tau_j \le s < \tau_{j+1}.$$

For each $s \in M$, $F_{Z_s}(I_s) = \frac{1}{2}$ if I_s contains just one of the numbers 0 or 1; and
$= 0$ if I_s contains neither 0 nor 1; and $= 1$ if I_s contains both 0 and 1. Thus
$F_{Z_s}(I_s)$ is atomic. For $s \notin M$, $F_{Z_s}(I_s) = 1$ if $[s] \in I_s$, and $= 0$ otherwise.
For $s \in M$, the random variables Z_s are independent, so for $i \ne j$

$$F_{Z(\tau_i, \tau_j)}\,(I(\tau_i) \times I(\tau_j)) = F_{Z(\tau_i)}(I(\tau_i))F_{Z(\tau_j)}(I(\tau_j)).$$

But the random variables Z_s are not independent: given $z([s]) = 0$, then $z(s) = 0$
for $[s] = \tau_j < s < \tau_{j+1}$. Likewise if $z([s]) = 1$. Thus, if $[s] = \tau_j < s < \tau_{j+1}$, and
if $0 \in I([s]), 1 \notin I([s])$ and $0 \in I(s), 1 \notin I(s)$, then $F_{Z([s])}\,(I([s])) = \frac{1}{2}$ and
$F_{Z(s)}\,(I(s)) = \frac{1}{2}$, but

$$\frac{1}{4} = \frac{1}{2} \times \frac{1}{2} = \left(F_{Z([s])}\,(I([s]))\right)\left(F_{Z(s)}\,(I(s))\right)$$

$$\ne F_{(Z([s]),Z(s))}\,(I([s]) \times I(s)) = \frac{1}{2}.$$

The joint distribution functions for $Z_{\mathbf{T}}$ are as follows. For $N \in \mathcal{N}(\mathbf{T})$ let

$$N \supset M = \{\tau_1, \ldots, \tau_m\},$$

and resolve N into disjoint subsets $N_1 \cup \cdots \cup N_m$ where $s \in N_k$ implies $\tau_{k-1} \le s < \tau_k$
for $k = 1, 2, \ldots, m$. Then, for $I(N) = \prod_{s \in N} I(s)$ and $I[N] = \{x_{\mathbf{T}} \in \mathbf{R}^{\mathbf{T}} : x(N) \in I(N)\}$

$$F_{Z_N}(I(N)) = F_{Z_M}\left(\prod_{k=1}^m I(\tau_k)\right) = \prod_{k=1}^m F_{Z(\tau_k)}I(\tau_k),$$

$$F_{Z_{\mathbf{T}}}(I[N]) = F_{Z_N}(I(N)) = \prod_{k=1}^m F_{Z(\tau_k)}I(\tau_k).$$

The distribution function for the process $X_{\mathbf{T}}$ is as follows. For $0 < s \le \tau$ and $I(s) \in \mathcal{I}(\mathbf{R})$, $F_{X(s)}(I(s)) = |J|$ where $J = I(s) \cap [0,1]$ is an interval. (If the intersection $I(s) \cap [0,1]$ is empty, or not an interval, then $F_{X(s)}(I(s)) = 0$.) The random variables $X(s)$ are independent so

$$F_{X(s,s')}(I(s) \times I(s')) = F_{X(s)}(I(s)) F_{X(s')}(I(s')).$$

For each finite set $N \subset \mathbf{T}$,

$$F_{X_N}(I(N)) = \prod_{s \in N} I(s), \qquad F_{X_{\mathbf{T}}}(I[N]) = F_{X_N}(I(N)).$$

For the joint processes $U_{\mathbf{T}} = (Z_{\mathbf{T}}, X_{\mathbf{T}})$ of Example 12, the outcomes are

$$u_{\mathbf{T}} = (z_{\mathbf{T}}, x_{\mathbf{T}}) \in K[N] = (I_Z[N] \times I_X[N]) \subset (\mathbf{R} \times \mathbf{R})^{\mathbf{T}},$$

and the joint distribution function is

$$F_{U_{\mathbf{T}}}(K[N]) = F_{(Z_{\mathbf{T}}, X_{\mathbf{T}})}(I_Z[N] \times I_X[N]) = F_{Z_{\mathbf{T}}}(I_Z[N]) F_{X_{\mathbf{T}}}(I_X[N]).$$

If (4.21) is a valid representation of the portfolio value observable in this case, then the expected value (if it exists) of $W(t) \simeq h(z_{\mathbf{T}}, x_{\mathbf{T}})\left[(\mathbf{R} \times \mathbf{R})^{\mathbf{T}}, F_{(Z_{\mathbf{T}}, X_{\mathbf{T}})}\right]$ would be given by

$$E[W(t)] = E[h(Z_{\mathbf{T}}, X_{\mathbf{T}})] = \int_{(\mathbf{R} \times \mathbf{R})^{\mathbf{T}}} h(z_{\mathbf{T}}, x_{\mathbf{T}}) F_{(Z_{\mathbf{T}}, X_{\mathbf{T}})}(I_Z[N] \times I_X[N]).$$

The latter calculation has $(\mathbf{R} \times \mathbf{R})^{\mathbf{T}}$ as domain, and instead of proceeding further on these lines we try a few simpler illustrations in preparation.

Example 15 *Suppose, at each instant s of the time interval $\mathbf{T} = {]}0, t]$, a share takes random value $x(s)$ (or x_s). Suppose $x(0) = x_0 = 1$ (with probability 1), and suppose, at each time s $(0 < s \le t)$, the share price takes value $x(s)$, $0 \le x(s) \le 1$ with uniform probability on $[0,1]$; and suppose the value at time s is independent of the value taken at any other time. Suppose an investor takes an initial shareholding $g(0)$, or g_0, of 1 share, so $g(0) = g_0 = 1$; and suppose the shareholding varies non-randomly at each instant s of time between initial time 0 and terminal time t, with shareholding at time s given by $g(s)$. What is the expected payout at terminal time t from this shareholding?*

In this case the investor policy[10] is, at each instant s $(0 < s \le t)$ to "go long in" (or hold) a variable number $g(s)$ of shares—regardless of the values $x(s)$ of a share—where g is some deterministic function of s.

For $N = \{t_1, \ldots, t_n = t\} \in \mathcal{N}(\mathbf{T})$ the payout from the shareholding can be estimated as

$$f(g, x_{\mathbf{T}}, N) = \sum_{j=1}^{n} g(t_{j-1}) (x(t_j) - x(t_{j-1})); \qquad (4.25)$$

---

[10]The fact that an investment policy such as this makes no practical financial sense is irrelevant for the moment.

This has the form $\sum g(s)\Delta x(s)$, a Riemann sum estimate of $\int_0^t g(s)\,dx(s)$; the latter expression being a sample element of the stochastic integral $\int_0^t g(s)\,dX(s)$. In Example 15, the random function $Z(s)$ of Example 12 is replaced by a deterministic function $g(s)$.

No commitment is being made here as to whether this supposed stochastic integral $\int_0^t g(s)\,dX(s)$ actually exists, or as to what it means if it does exist. But if $f(g, x_{\mathbf{T}}, N)$ is taken as a "typical" or representative history of the random variability involved in Example 15, then it is not unreasonable to attempt to estimate the expected payout as

$$\mathrm{E}[f(g, X_{\mathbf{T}}, N)] \;=\; \int_{\mathbf{R}^{\mathbf{T}}} f(g, x_{\mathbf{T}}, N) F_{X_{\mathbf{T}}}(I[N]) \qquad (4.26)$$

—if the latter integral exists. This approach evades the issue of whether the stochastic integral $\int_0^t f(s)\,dX(s)$ (whatever it is) exists, since no such expression or formulation appears explicitly in (4.26).

On the other hand it can be said that, while this manoeuvre evades or postpones the $\int_0^t \cdots dX_{\mathbf{T}}(s)$ issue, in doing so it is confronted (in (4.26)) by the $\int_{\mathbf{R}^{\mathbf{T}}} \cdots F(I[N])$ issue—which, for all we know at this stage, may be worse for us. Evaluating or analysing such an expression may turn out to be a very formidable task.

Against that, even if the stochastic integral $\int_0^t f(s)\,dX(s)$ is found, the problem of finding the expected value of a stochastic integral still remains. What is the appropriate sample space Ω for the random variability of a stochastic integral? What is the appropriate probability function for it? Whichever direction we take, such questions may prove challenging.

At any rate, instead of attempting stochastic integrals at this point, we can now explore the $\int_{\mathbf{R}^{\mathbf{T}}}$ approach of (4.26) in the hope of, at least, gaining more insight into problems of joint random variability. And any such insight may turn out to be of some assistance in analysing stochastic integrals. This is the motivation in Example 15.

To sum up, (4.25) suggests two alternative directions—towards the stochastic integral, or towards the multi-dimensional analysis (4.26) of joint variability. For now we take the latter direction.

In (4.25) the stochastic integrand $g(s)$ does not depend on the random process (X_s); nor is it random in any other fashion. Looking at examples and illustrations of stochastic integrals in [MTRV] (pages 391–392), it resembles stochastic integrand g_5, or $s(x(s') - x(s))$ (where $s' > s$). Due to typographical error described in [website] the stochastic integral $\int_0^t s\,dX(s)$ is stated (page 397 of [MTRV]) to be equal to $\int_0^t X_s ds$, where the correct form[11] should be

$$\int_0^t s\,dX(s) \;=\; tX_t - \int_0^t X_s ds.$$

The latter integral $\int_0^t X_s ds$ may or may not exist, depending on the meaning

---

[11]See Section 6.8 below.

of (X_s). (Its sample value $\int_0^t x_s \, ds$, though random, is, strictly speaking, **not** a sample of a stochastic integral, since the integrator is ds, not $dx(s)$.)

More generally, for some unspecified deterministic function $g(s)$ (such as s^2 or e^{-s}), the sample path versions $\int_0^t g(s) \, dx(s)$ of a supposed stochastic integral $\int_0^t g(s) \, dX(s)$ may or may not exist. In [MTRV] example 60 (page 395–396) it is claimed that the stochastic integral $\int_0^t f(s) \, dX(s)$ exists if $f(s)$ is a step function. This is actually true if $X_\mathbf{T}$ is a relatively "well-behaved" process such as Brownian motion. But it is **not** generally true, as demonstrated Example 13 (page 94 above), and in *Clarifications and Corrections* of the [website] for [MTRV] where theorem 174 (integrability of step functions) is discussed.

In the present discussion, beyond referring to Riemann sums such as (4.25), it has not been declared what is meant by "stochastic integral". In fact, the task set in the above examples is not to produce stochastic integral representation of the payout observable. Rather, the task set is to establish an expected value for the payout from a shareholding, bypassing the issue of stochastic integration.

One might suspect that, corresponding to Riemann sums like (4.25), there may be an actual integral (*stochastic integral*) representation of payout. But up to this point, the *expected payout* has generally been sought, not from a stochastic integral, but by modelling the multi-factor random variability in sample spaces $\mathbf{R}^n$, $\mathbf{R}^M$, $\mathbf{R}^\mathbf{T}$, and $\mathbf{R}^\mathbf{T} \times \mathbf{R}^\mathbf{T}$.

So for Example 15 we continue in the latter vein, leaving consideration of the stochastic integral approach until later. Accordingly, the expected payout can be calculated as

$$\mathrm{E}[f(g, X_\mathbf{T}, N)] \;=\; \int_{\mathbf{R}^\mathbf{T}} f(g, x_\mathbf{T}, N) F_{X_\mathbf{T}}(I[N]) \tag{4.27}$$

$$=\; \int_{\mathbf{R}^\mathbf{T}} \left(\sum_{j=1}^n g(t_{j-1}) \, (x(t_j) - x(t_{j-1})) \right) F_{X_\mathbf{T}}(I[N]) \tag{4.28}$$

$$=\; \int_{\mathbf{R}^\mathbf{T}} \left(\sum_{j=1}^n g(t_{j-1}) \, (x(t_j) - x(t_{j-1})) \right) \prod_{j=1}^n F_{X_{t_j}}(I(t_j)).$$

With $x(t_j) = x_j$, and variables $x_1, \ldots, x_n$, the integrand in (4.28) might suggest that the more familiar integration process in $\mathbf{R}^n$ can be attempted here. But this would be a misinterpretation. The form $f(\cdots, N)$ of the integrand function in (4.27) implies that, in addition to the integrand variable $x_\mathbf{T}$, the element N is also an integrand variable, so the number n of elements x_n of the variable $x_\mathbf{T}$ is not fixed. The expression (4.28) shows that the $x_\mathbf{T}$-variable appears in the integrand in the form $(x(t_1), \ldots, x(t_n))$ where the integer n, and indeed the dimensions (indices or labels) $t_1, \ldots, t_n$, are variable. (There is variability, too, in that the set N, and its cardinality n, increase without limit in the Riemann sum process of integration.)

Thus f is an example of a *sampling function* integrand, which, after step functions and cylinder functions, are the next level of integrand in infinite-dimensional integration.

Previously, some -complete integral evaluations were achieved by forming

Riemann sums. Formation of Riemann sums depends on partitioning the domain of integration.

It is fairly easy to visualise partitions of domains $\mathbf{R}$ or $\mathbf{R}^n$ for finite n. Such partitions are illustrated diagrammatically in [MTRV] pages 81 and 91. Partitions for infinite-dimensional domain $\mathbf{R}^\mathbf{T}$ are illustrated in page 87 (and page 91), and discussed in section 3.2 (pages 88–89 of [MTRV]). The latter discussion is revisited and tidied up in *Clarifications and Corrections* of [website] where further diagrams are presented in order to get a sense of how an infinite-dimensional Cartesian product domain $\mathbf{R}^\mathbf{T}$ can be partitioned for -complete integration. These partitions can be more difficult to visualise than corresponding partitions of $\mathbf{R}^n$. So, instead of going directly to Riemann sums for (4.27), the integration in Examples 12 and 15 can be simplified by doing it in easier stages.

For that purpose, amend Example 15 as follows.

Example 16 *In Example 15, assume g is continuous on* $[0,\tau]$.

As in (4.26), the aim is to estimate expected payout as

$$\mathrm{E}[f(g, X_\mathbf{T}, N)] \;=\; \int_{\mathbf{R}^\mathbf{T}} f(g, x_\mathbf{T}, N) F_{X_\mathbf{T}}(I[N]); \qquad (4.29)$$

and, as before, this depends on proving that the latter integral exists. It might seem that the new assumption, that the deterministic function $g(s)$ is continuous in s, will enable a proof. Using the integration theory of [MTRV], an attempted proof might run as follows.

1. If $g(s)$ has constant value α for $0 \le s \le \tau$, then for any $N = \{t_1,\ldots,t_n\}$,

$$f(g, x_\mathbf{T}, N) \;=\; \sum_{j=1}^{n} \alpha\left(x(t_j) - x(t_{j-1})\right) \;=\; \alpha\left(x(\tau) - x(0)\right),$$

by cancellation; and in this case (with g constant),

$$\begin{aligned}\mathrm{E}[f(g, X_\mathbf{T}, N) &= \int_{\mathbf{R}^\mathbf{T}} f(g, x_\mathbf{T}, N) F_{X_\mathbf{T}}(I[N]) \\ &= \int_{\mathbf{R}^\mathbf{T}} \alpha\left(x(\tau) - x(0)\right) F_{X_\mathbf{T}}(I[N]) \\ &= \alpha\left(\mathrm{E}[X_\tau] - \mathrm{E}[X_0]\right) \;=\; \alpha\left(\frac{1}{2} - 1\right) \;=\; -\frac{1}{2}\alpha.\end{aligned}$$

This result depends on proving that $\sum F_{X_\mathbf{T}}(I[N]) = 1$, the sum being taken over any partition of the domain $\mathbf{R}^\mathbf{T}$. This is not as obvious as it might seem at first glance. Sections 3.2 and 3.9 of [MTRV] (pages 88–110) cover some aspects of the issues involved.

2. By continuity, there exists a real number κ for which $g(s) < \kappa$ for $0 \le s \le \tau$. For $\mathbf{T} =]0, \tau]$, a given integer $m > 0$, and $k = 1, 2, \ldots, 2^m$, define $\tau_k, = \tau_k^{(m)}$, and $g_m(s)$, by

$$\tau_k := \frac{k\tau}{2^m}, \qquad g_m(s) := \min_{\tau_{k-1} \le s \le \tau_k} g(s) \quad \text{for } \tau_{k-1} < s \le \tau_k;$$

so $g_m(s)$ has constant value α_{km} for $\tau_{k-1} < s \le \tau_k$; and define $g_m(\tau_0)$, = $g_m(0)$, to be $g(0)$.

3. For $0 < s \le \tau$, we have $g_m(s) \le g(s)$ with $g_m(s) \to g(s)$ as $s \to \infty$; so, for each s, $g_m(s)$ is monotone increasing as $m \to \infty$.

4. Then the function $f(g_m, x_{\mathbf{T}}, N)$ has the following properties:

   (a) For each $x_{\mathbf{T}}$ and each N, $f(g_m, x_{\mathbf{T}}, N)$ converges to $f(g, x_{\mathbf{T}}, N)$ as $m \to \infty$;

   (b) for each m, $f(g_m, x_{\mathbf{T}}, N) \le f(g_{m+1}, x_{\mathbf{T}}, N)$, so $f(g_m, x_{\mathbf{T}}, N)$ is monotone increasing for each $x_{\mathbf{T}}$ and each N;

   (c) $f(g_m, x_{\mathbf{T}}, N)$ is $F_{X_{\mathbf{T}}}$-integrable on $\mathbf{R}^{\mathbf{T}}$ for each m; and

   (d) for all m, $f(g_m, x_{\mathbf{T}}, N)$ has upper bound which is $F_{X_{\mathbf{T}}}$-integrable on $\mathbf{R}^T$.

5. By [MTRV] theorem 57 (Levi monotone convergence theorem, pages 169–170), the limit function $f(g, x_{\mathbf{T}}, N)$ is $F_{x_{\mathbf{T}}}$-integrable on $\mathbf{R}^{\mathbf{T}}$, and

$$
\begin{aligned}
\mathrm{E}[f(g_m, X_{\mathbf{T}}, N)] &= \int_{\mathbf{R}^{\mathbf{T}}} f(g_m, x_{\mathbf{T}}, N) F_{X_{\mathbf{T}}}(I[N]) \\
&\to \int_{\mathbf{R}^{\mathbf{T}}} f(g, x_{\mathbf{T}}, N) F_{X_{\mathbf{T}}}(I[N]) \\
&= \mathrm{E}[f(g, X_{\mathbf{T}}, N)]
\end{aligned}
$$

as $m \to \infty$.

If the preceding outline is valid then, for continuous g, arguments which are familiar from standard integration theory give existence of expected values $\mathrm{E}[f(g_m, X_{\mathbf{T}}, N)]$ and $\mathrm{E}[f(g, X_{\mathbf{T}}, N)]$, with

$$
\lim_{m\to\infty} \mathrm{E}[f(g_m, X_{\mathbf{T}}, N)] = \mathrm{E}[f(g, X_{\mathbf{T}}, N)].
$$

Also, as pointed out below, for any given m the integrand $f(g_m, x_{\mathbf{T}}, N)$ is cylindrical, and can be replaced by an integrand $f(g_m, x(M))$ in $\mathbf{R}^m$. Thus the value of $\mathrm{E}[f(g_m, X_{\mathbf{T}}, N)]$ can in principle be calculated as a finite-dimensional integral (a *cylindrical integral*) $\int_{\mathbf{R}^m} \cdots$ rather than the infinite-dimensional integral $\int_{\mathbf{R}^{\mathbf{T}}} \cdots$.

Unfortunately the outline proof above is invalid, as item 4(b) above is not generally true. First, consider whether or not $f(g_m, x_{\mathbf{T}}, N)$ converges to $f(g, x_{\mathbf{T}}, N)$ as m tends to infinity. For $N = \{t_1, \ldots, t_n\}$,

$$
f(g_m, x_{\mathbf{T}}, N) = \sum_{j=1}^{n} g_m(t_{j-1})\left(x(t_j) - x(t_{j-1})\right),
$$

$$
f(g, x_{\mathbf{T}}, N) = \sum_{j=1}^{n} g(t_{j-1})\left(x(t_j) - x(t_{j-1})\right);
$$

so, for fixed N and fixed $x_{\mathbf{T}}$, $f(g_m, x_{\mathbf{T}}, N) \to f(g, x_{\mathbf{T}}, N)$ as $m \to \infty$, so 4(a) is true. But the convergence is not monotone, since, for any given $x_{\mathbf{T}}$ and N, each factor $x(t_j) - x(t_{j-1})$ can be positive, negative, or zero. Therefore 4(b) is false.

Regarding 4(d), for any given $x_{\mathbf{T}}$ and N,

$$\sum_{j=1}^{n} g_m(t_{j-1}) \left(x(t_j) - x(t_{j-1}) \right) \; < \; \kappa \sum_{j=1}^{n} \left(x(t_j) - x(t_{j-1}) \right) \; = \; \kappa(x(\tau) - x(0)),$$

and the latter is $F_{X_{\mathbf{T}}}$-integrable on $\mathbf{R^T}$.

What about 4(c)? Denote the set $\{\tau_k^{(m)} : 1 \le k \le 2^m\}$ by M. Write

$$f(g_m, x(M)) \; := \; \sum_{k=1}^{m} g_m \left(\tau_{k-1}^{(m)} \right) \left(x \left(\tau_k^{(m)} \right) - x \left(\tau_{k-1}^{(m)} \right) \right).$$

If $N \supseteq M$, then, by cancellation of terms, $f(g_m, x_{\mathbf{T}}, N) = f(g_m, x_{\mathbf{T}})$, so the integrand in $\mathbf{R^T}$ is cylindrical, and, because of the form of the integrator function $F_{X_{\mathbf{T}}}$, it can be replaced by the finite-dimensional integrand $f(g_m, x(M))$ with integrator F_{X_M} in finite-dimensional domain $\mathbf{R}^M$, $= \mathbf{R}^m$.

The replacement is permissible because, in the integration in $\mathbf{R^T}$, gauges $\gamma = (L, \delta_N)$ can be chosen for which $L(x_{\mathbf{T}}) \supset M$ for each $x_{\mathbf{T}}$, so in the resulting γ-fine Riemann sums, each appearance of N satisfies $N \supseteq M$.

4.5   Review of Integrability Issues

In the course of the preceding discussion, while some challenging features were encountered, there were occasions when integrability was fairly easily established.

Starting with some of the more troublesome issues, the sample paths $(x(s))$ of the price process $X_{\mathbf{T}}$ in Examples 15 and 16 include paths whose extreme oscillation mirrors that of the Dirichlet function $d(s)$ of Example 13.

Here are some further issues:

- Suppose a domain Ω can be partitioned into sub-domains Ω_j, and suppose an integrand g is a step function taking constant values κ_j in domain Ω_j for each j. Then, even if κ_j is integrable on Ω_j for each j, it is not necessarily the case that f is integrable on Ω.

- If a sequence of such step functions converges pointwise to a function f it is not necessarily the case that f is integrable.

- Dirichlet-type oscillation can occur in the sample functions $x_{\mathbf{T}}$ of a stochastic process $X_{\mathbf{T}}$. This phenomenon presents integrability problems.

On the other hand, integrals on infinite-dimensional domains sometimes reduce to more familiar finite-dimensional integrals. Some aspects of this phenomenon can be summarized as follows. Suppose $\mathbf{T}$ is an infinite labelling set such as $]0, t]$, and suppose

- $x_{\mathbf{T}} \in \mathbf{R}^{\mathbf{T}}$;

- $f(x_{\mathbf{T}})$ is an integrand in $\mathbf{R}^{\mathbf{T}}$;

- $F(I)$ is an integrator function defined on the cells I of $\mathbf{R}^{\mathbf{T}}$.

The integral on $\mathbf{R}^{\mathbf{T}}$ of $f(x_{\mathbf{T}})$ with respect to $F(I)$ (if it exists) is $\int_{\mathbf{R}^{\mathbf{T}}} f(x_{\mathbf{T}})F(I)$, which for present purposes can be denoted as $\int_{\mathbf{R}^\infty} f(x)\, dF$.

When $\mathbf{T}$ is infinite (that is, when $\mathbf{T}$ has infinite cardinality) the cells I are cylindrical, as indicated in the notation $I = I[N]$ for finite subsets $N = \{t_1, \ldots, t_n\}$ of $\mathbf{T}$. Accordingly, some aspects of the finite Cartesian product

$$\mathbf{R}^n, \;=\; \mathbf{R} \times \cdots \times \mathbf{R} \;=\; \mathbf{R}_{t_1} \times \cdots \times \mathbf{R}_{t_n},$$

already make an appearance in the integration.

In general, for positive integers $m > n$, the domains $\mathbf{R}^m$ and $\mathbf{R}^n$ are entirely different. A plane is quite different from a line. So $\int_{\mathbf{R}^m}$ is not the same as $\int_{\mathbf{R}^n}$. But is it possible, because of the parameters N in $I[N] \subseteq \mathbf{R}^\infty$, that $\int_{\mathbf{R}^\infty} f(x_{\mathbf{T}})\, dF$ reduces to $\int_{\mathbf{R}^n} f(x(N))\, dF$? That would be very convenient!

In fact this kind of reduction is sometimes possible when the integrand f depends, for instance, on only a fixed, finite number of the variables x_s for $s \in \mathbf{T}$. The basic idea is as follows.

Suppose a function f_1 is defined for $x = (x_1, x_2)$, $0 \le x_j \le 1$, $j = 1, 2$; and suppose the values taken by f_1 depend only on the variable x_1, and are independent of the variable x_2. Thus there is a function $f(x_1)$, defined for $0 \le x_1 \le 1$ such that

$$f_1(x_1, x_2) \;=\; f(x_1), \qquad 0 \le x_j \le 1, \quad j = 1, 2.$$

Consider the integrals

$$\int_0^1 \int_0^1 f_1(x_1, x_2)\, dx_1 dx_2, \qquad\qquad \int_0^1 f(x_1)\, dx_1.$$

The following questions can then be posed.

- *If $f_1(x_1, x_2)$ is integrable on $[0, 1] \times [0, 1]$, is $f(x_1)$ integrable on $[0, 1]$, and if so, are the two integrals equal?*

- *If $f(x_1)$ is integrable on $[0, 1]$, is $f_1(x_1, x_2)$ integrable on $[0, 1] \times [0, 1]$, and if so, are the two integrals equal?*

Illustrations are provided in [MTRV], section 3.7 (pages 98–101), section 3.9 (pages 106–110), and also pages 279–282. Example 55 (page 279) and theorems 159 and 160 (page 280) are discussed further in [website]. To sum up, there is a progression of kinds of integrands,

1. Step functions,

2. Cylinder functions,

3. Sampling functions,

which can be exploited when integrating in an infinite-dimensional domain.

Chapter 5

Brownian Motion

5.1 Introduction to Brownian Motion

Section 5.6 below provides a summary of various different kinds of stochastic integral, whose intuitive meaning can be obtained from the elementary examples and illustrations in preceding chapters, and which are presented here in terms of the theory provided in [MTRV].

The stochastic processes of the previous sections are somewhat artificial and selective. They were chosen because they are fairly easily intelligible and relatively straightforward.

Nevertheless, their simpler and more easily formulated scenarios are not necessarily the most manageable in mathematical terms—because, for instance, of Dirichlet oscillation as illustrated in Example 13. Also, the preceding examples, though they may help to provide a feel for the subject, are not the kind of processes which are important in practice.

One of the most important stochastic processes is Brownian motion. It is not so easy to formulate; it reflects some of the complexity of real random phenomena. Nonetheless it is relatively amenable to some well-established mathematical techniques.

Before actually defining Brownian motion, Example 17 below is a version of it which demonstrates how the Dirichlet-type oscillation of Examples 12, 15, 16 may be evaded. It is intended to be a bridge joining those examples to the standard Brownian motion to be discussed in this chapter.

The preceding examples are located—like Brownian motion—in domains of the form $\mathbf{R}^T$. But this was somewhat artificial. In reality their random variability extended only to $[-1, 1]$, not to $\mathbf{R} =]-\infty, \infty[$. To emphasize this point, the following example uses domain $\Omega =]-1, 1[^T$, not $\mathbf{R}^T$. Also, the notation and arguments of [MTRV] and [website] are given more prominence.

Example 17 *With $\mathbf{T} =]0, \tau]$ suppose an asset price process $X_{\mathbf{T}}$ is represented as*

$$X_{\mathbf{T}} \simeq x_{\mathbf{T}}[\Omega, F_{X_{\mathbf{T}}}] \quad where \quad \Omega = (]-1, 1[)^{\mathbf{T}}.$$

Gauge Integral Structures for Stochastic Calculus and Quantum Electrodynamics, First Edition. Patrick Muldowney.
© 2021 John Wiley & Sons, Inc. Published 2021 by John Wiley & Sons, Inc.

Thus, for $0 < s \le \tau$, the asset price $x(s)$ can take a value between -1 and $+1$,

$$-1 < x(s) < 1, \qquad 0 < s \le \tau.$$

Suppose the probability distribution function $F_{X_\mathbf{T}}$ satisfies the following conditions:

[S1] For $0 < s \le \tau$, $F_{X_s}(\,]-1,1[) = 1$.

[S2] For any s $(0 < s \le \tau)$, $\mathrm{E}[X_s], = \int_{-1}^{1} x_s\, F_{X_s}(I_s), = \mu$, a constant for all $s \in \mathbf{T}$. (For instance, the distribution functions F_{X_s} can be the same for all s.)

Nominally, the objective is to estimate in advance the earnings of the asset portfolio. Since the mean value (or expected value) of the portfolio at any time t is μ, the growth in value of the portfolio is obviously $\mu - \mu'$, where μ' is the initial cost of setting up the portfolio. However, even though the final result is obvious, the real purpose of this exercise is to use the ideas of stochastic integration to track the random variability of the portfolio value; and in doing so, to introduce some of the concepts which are used in the theory of Brownian motion. In order to make the ideas of stochastic integration workable in this Example, here are some further conditions:

[S3] If $N, N' \in \mathcal{N}(\mathbf{T})$ with $N' = N \cup \{t'\}$, and if $I_{t'} =]-1,1[$, then

$$F_{X_\mathbf{T}}(I[N']) \;=\; F_{X_\mathbf{T}}(I[N]).$$

This is the consistency *property.*

*[S4] Let C denote the set of $x_\mathbf{T} \in \Omega$ such that $x(s)$ is uniformly continuous for each $s \in \mathbf{T}$; and let $D = \Omega \smallsetminus C$. We **postulate** that $F_{X_\mathbf{T}}$ should satisfy*

$$\int_J 1_D(x_\mathbf{T}) F_{X_\mathbf{T}}(I[N]) \;=\; 0$$

for all $J \in \mathcal{I}(\Omega)$, including $J = \Omega$.

For each s $(0 < s \le t)$, the asset portfolio (number of units of the asset held by the investor) is given by the deterministic function $g(s)$. Suppose initial asset value $x(0)$ is 1, and initial asset holding $g(0) = 1$ (so $\mu' = 1$).

Firstly, does such a distribution function $F_{X_\mathbf{T}}$ actually exist? It cannot be definitely asserted here that it does, but something similar will be demonstrated later. So we proceed with Example 17 on that basis.

A familiar first step in many investigations involving integrals is to consider a step function approximation to the integrand. A property of *measurability* may then imply that properties established for the step function are inherited by the original integrand.

In case of Example 17, the integrand function g is assumed continuous and, as in Example 15, for each positive integer m we form step function g_m,

$$\tau_k := \frac{\tau k}{2^m}, \quad (0 \le k \le 2^m); \qquad g_m(s) := \inf\{g(s) : \tau_{k-1} \le s \le \tau_k\}$$

for $1 \le k \le 2^m$, with $g_m(s) := g(0)$ for $0 = \tau_0 \le s \le \tau_1$; so $g_m(s) \to g(s)$ for each s as $m \to \infty$. Let $N \in \mathcal{N}(\mathbf{T})$,

$$N = \{t_1, \ldots, t_n\}, \qquad 0 = t_0 < t_1 < \cdots < t_n = \tau,$$

With $M = \{\tau_1, \tau_2, \ldots, \tau_m\}$, write

$$f(g, x_\mathbf{T}, N) = \sum_{j=1}^{n} g(t_{j-1}) \left(x_{t_j} - x_{t_{j-1}} \right),$$

$$f_m(g_m, x(M)) := \sum_{k=1}^{2^m} g_m(\tau_{k-1}) \left(x_{\tau_k} - x_{\tau_{k-1}} \right),$$

$$f(g_m, x_\mathbf{T}, N) := \sum_{j=1}^{2^m} g_m(t_{j-1}) \left(x_{t_j} - x_{t_{j-1}} \right), \qquad (5.1)$$

where the latter sum reduces to the preceding one by cancellation whenever $N \supseteq M$. For $N \supseteq M$ define

$$f_M(g, x_\mathbf{T}, N) := f(g_m, x_\mathbf{T}, N) = f_m(g_m, x(M)).$$

The difference between f_M and f_m is that f_M is a cylinder function in infinite-dimensional domain $]-1, 1[^\mathbf{T}$, while f_m has finite-dimensional domain $]-1, 1[^M$. The latter domain is simply

$$]-1,1[\times]-1,1[\times \cdots \times]-1,1[=]-1,1[^M -]-1,1[^m.$$

Recall that N is a variable of integration in the expression f; while, for the function f_m, the set $M = \{\tau_k\}$ is a fixed finite set. Both f and f_m can be interpreted as sample estimates of a randomly variable payout (at terminal time τ) of the asset portfolio.

As in Example 15, and because of the Riemann sum format of f and f_m, a further question arises. That is, whether a formulation $h(g, x_\mathbf{T}) = \int_0^\tau g(s) \, dx(s)$, or (in random variable format) $h(g, X_\mathbf{T}) = \mathbf{S}^g(X_\mathbf{T})$, can be meaningfully considered in this context. In other words, a stochastic integral $\mathbf{S}$ instead of a Riemann sum (5.1).

Thus there are three alternative formulations of random variability to be considered. The three proposed "random variables" are:

$$f(g, X_\mathbf{T}, N), \quad f_M(g, X_\mathbf{T}, N), \quad \int_0^\tau g(s) \, dX(s). \qquad (5.2)$$

Remember, though, that a random variable—whether expressed in Kolmogorov measurable function terms or in -complete terms—is in itself just a mathematical device or abstraction. It is an imaginative construction required by the mathematical theory in order to deliver "real-world" estimates or "solutions" in actual measurements of unpredictably varying quantities.

These "solutions" might be correlations, or expectations, or error estimates in approximation. But the "random variable" element in this is just a means to an end. From this point of view, the existence of expected value—a "prediction"—is what makes the "random variable" concept meaningful. (In [MTRV], each of the expressions f, f_M and $\mathbf{S}^g(X_{\mathbf{T}})$ of (5.2) would be designated as *observables*, and would receive the title *random variables* only **after** existence of their *expected values* had been established.) Therefore it is necessary to address the expected value of each of the expressions in (5.2):

$$f(g, X_{\mathbf{T}}, N), \qquad f_M(g, X_{\mathbf{T}}, N), \qquad \mathbf{S}^g(X_{\mathbf{T}}).$$

The easiest evaluation is $\mathrm{E}[f_M(g, x_{\mathbf{T}}, N)], = \int_\Omega f_M(g, x_{\mathbf{T}}, N) F(I[N])$. First, note that [S3] above implies that, for a given fixed s, [S2] gives

$$\mathrm{E}[X_s] \;=\; \int_\Omega x_s F_{X_{\mathbf{T}}}(I[N]) \;=\; \int_0^1 x_s F_{X_s}(I_s) \;=\; \mu, \qquad (5.3)$$

where $\Omega = \,]-1, 1[^{\mathbf{T}}$. As before, for all $N \supseteq M$, $\;f_M(g, x_{\mathbf{T}}, N) \;=$

$$= \; \sum_{k=1}^{2^m} g_m(\tau_{k-1}) \left(\sum \{x(t_k) - x(t_{k-1}) \,:\, \tau_{k-1} \le t_{k-1} < t_k \le \tau_{k-1}\} \right)$$

$$= \; \sum_{k=1}^{2^m} g_m(\tau_{k-1}) \left(x(\tau_k) - x(\tau_{k-1}) \right) \;=\; f_m(g_m, x(M))$$

$$= \; -g_m(\tau_0)x(\tau_0) \,+\, g_m(\tau_{m-1})x(\tau) \,+\, \sum_{k=1}^{2^m-1} \left(g_m(\tau_k) - g_m(\tau_{k-1}) \right) x(\tau_k);$$

$$= a + b + c \qquad\qquad\qquad\qquad\qquad\qquad\qquad\qquad\qquad\qquad (5.4)$$

where

$$a = -g_m(\tau_0)x(\tau_0), \quad b = g_m(\tau_{m-1})x(\tau), \quad c = \sum_{k=1}^{2^m-1} \left(g_m(\tau_k) - g_m(\tau_{k-1}) \right) x(\tau_k);$$

and the expectation $E[f_M(g, X_{\mathbf{T}}, N)]$ is

$$\int_{]-1,1[^{\mathbf{T}}} f_M(g, x_{\mathbf{T}}, N) F_{X_{\mathbf{T}}}(I[N]), \;=\; \int_{]-1,1[^M} f_m(g_m, x(M)) F_{X_M}(I(M)).$$

This equals

$$\int_{]-1,1[^M} \left(a \,+\, g_m(\tau_{m-1})x(\tau) \,+\, \sum_{k=1}^{2^m-1} \left(g_m(\tau_k) - g_m(\tau_{k-1}) \right) x_{\tau_k} \right) F_{X_M}(I(M)).$$

and since $\mathrm{E}[X_{\tau_k}] = \mu$ for each k, (5.3) implies that $E[f_M(g, X_{\mathbf{T}}, N)]$ reduces to

$$-g_m(\tau_0)x(\tau_0) \,+\, \mu g_m(\tau_{2^m-1}) \,+\, \mu \sum_{k=1}^{2^m-1} \left(g_m(\tau_k) - g_m(\tau_{k-1}) \right)$$

$$= \; -g_m(\tau_0)x(\tau_0) \,+\, \mu g_m(\tau_{m-1}) \,-\, \mu g_m(\tau_{m-1}) \,+\, \mu g_m(\tau_0)$$

$$= \; g(0)\left(\mu - x(0) \right).$$

If $g(0)$ and $x(0)$ are both 1, then the latter is $\mu - 1$; and is the same for all m.

Turning to the observable $\mathbf{S}^g(X_\mathbf{T})$, note that, by Postulate [S4] above, the set D of $x_\mathbf{T} \in \Omega$ which fail to be uniformly continuous on $[0, \tau]$ form an $F_{X_\mathbf{T}}$-null set in Ω.

Now add another condition; this time on the integrand $g(s)$:

[S5] *Suppose g has bounded variation on $[0, \tau]$.*

This condition ensures, for instance, that g does not oscillate excessively. Also, the following standard result of elementary integration can be invoked: *Suppose h_1, h_2 are point functions $h_i(s)$ defined for $a \leq s \leq b$, $i = 1, 2$; and suppose h_1 is continuous and h_2 has bounded variation on $[a, b]$. Then the Riemann-Stieltjes integral $\int_a^b h_2(s)\, dh_1(s)$ of h_2 with respect to h_1 exists; so that, if $\varepsilon > 0$ is given there exists a number $\delta > 0$ and if $\mathcal{P}, = \{t_j\}$, is any partition of $[a, b]$ with $|\mathcal{P}| = \max_j(t_j - t_{j-1}) < \delta$, and $s_{j-1} \leq s_j' \leq s_j$, then*

$$\left| \int_a^b h_2(s)\, dh_1(s) - (\mathcal{P}) \sum h_2(s_j')\, (h_1(s_j) - h_1(s_{j-1})) \right| < \varepsilon.$$

Thus, for each sample path $x_\mathbf{T} \notin D$, the Riemann-Stieltjes integral of $g(s)$ with respect to $x(s)$ exists, and, as $m \to \infty$,

$$f_m(g_m, x_M) = \sum_{k=1}^{2^m} g_m(\tau_{k-1})\, (x(\tau_k) - x(\tau_{k-1})) \to \int_0^\tau g(s)\, dx(s) = \mathbf{S}^g(x_\mathbf{T}).$$

Integrating with respect to $F_{X_\mathbf{T}}(I[N])$ in Ω, $\quad \int_\Omega f_M(g, x_\mathbf{T}, N) F_{X_\mathbf{T}}(I[N]) =$

$$= \int_\Omega \left(\sum_{k=1}^{2^m} g_m(\tau_{k-1})\, (x(\tau_k) - x(\tau_{k-1})) \right) F_{X_\mathbf{T}}(I[N]) \;-\; \mu - 1.$$

Can we deduce from this that

$$E[\mathbf{S}^g(X_\mathbf{T})], = \int_\Omega \mathbf{S}^g(x_\mathbf{T}) F_{X_\mathbf{T}}(I[N]),$$

exists, and that it equals $\lim_{m \to \infty} \int_\Omega f_M(g, x_\mathbf{T}, N) F_{X_\mathbf{T}}(I[N]) = \mu - 1$ (where, conveniently, the sequence in m has the same value for every m)?

In fact the function $g(s)$, continuous on $[0, \tau]$, is bounded on $[a, b]$, with $|g(s)| < \kappa_1$, say. Also the uniformly continuous $x(s)$ is bounded above by 1, and g has bounded variation κ_2, say. Thus m_0 can be chosen so that $m > m_0$ implies

$$\left| \sum_{k=1}^{2^m - 1} (g_m(\tau_k) - g_m(\tau_{k-1}))\, x(\tau_k) \right| \leq \sum_{k=1}^{2^m - 1} |(g_m(\tau_k) - g_m(\tau_{k-1}))| < 2\kappa_2.$$

Thus, using (5.4), for all $m > m_0$,

$$|f_M(g, x_\mathbf{T}, N) F_{X_\mathbf{T}}(I[N])| < (2\kappa_1 + 2\kappa_2)\, F_{X_\mathbf{T}}(I[N]),$$

and the latter is integrable on Ω with integral $2\kappa_1 + 2\kappa_2$. Therefore the dominated convergence theorem ([MTRV], theorem 61, page 173) applies, and $\mathbf{S}^g(x_{\mathbf{T}})$ is integrable on Ω with respect to the integrator function $F_{X_{\mathbf{T}}}(I[N])$, with integral equal to $\mu - 1$.

It is therefore reasonable to say that $\int_0^\tau g(s)\,dX(s)$ (the stochastic integral of $g(s)$ with respect to $X(s)$) exists; and that it is a joint-basic contingent random variable $\mathbf{S}^g(X_{\mathbf{T}})$, dependent on the process $X_{\mathbf{T}}$, whose expected value is $\mu - 1$.

Also, $f_M(g, x_{\mathbf{T}}, N)$ has $F_{X_{\mathbf{T}}}$-expectation $\mu - 1$ in Ω, so $f_M(g, X_{\mathbf{T}}, N)$ is a joint-contingent random variable, and, as $m \to \infty$,

$$f_M(g, X_{\mathbf{T}}, N) \;\to\; \mathbf{S}^g(X_{\mathbf{T}}),$$

where, conveniently, all terms have the same expected value $\mu - 1$.

Finally, what about $\mathrm{E}[f(g, X_{\mathbf{T}}, N)]$, $= \int_\Omega f(g, x_{\mathbf{T}}, N) F_{X_{\mathbf{T}}}(I[N])$? Does this integral exist, and if so, does its value—the expected value of the random variable $f(g, X_{\mathbf{T}}, N)$—equal $\mu - 1$ like the other forms of this random variable? In some ways, this question provides more interesting insights than the preceding discussion of the other two expressions in (5.2).

The argument can be broken down into successive steps. The general idea is to construct Riemann sums which establish integrability on Ω in accordance with the Riemann sum inequality of (4.4).

1. The "fact" (assumption or postulate, really) that the set D of discontinuous $x_{\mathbf{T}}$ in Ω is null (with respect to probabilities $F_{X_{\mathbf{T}}}$) means that $\int_\Omega \mathbf{1}_D(x) F_{X_{\mathbf{T}}}(I[N]) = 0$. With $\varepsilon > 0$ given, choose a gauge $\gamma^{(1)}$, $= \left(L^{(1)}, \delta_N^{(1)}\right)$, in Ω so that, for every $\gamma^{(1)}$-fine division $\mathcal{D}_1$ of Ω,

$$\left|(\mathcal{D}_1)\sum \mathbf{1}_D(x) F_{X_{\mathbf{T}}}(I[N])\right|, \;=\; (\mathcal{D}_1)\sum \mathbf{1}_D(x) F_{X_{\mathbf{T}}}(I[N]), \;<\; \varepsilon. \quad (5.5)$$

2. Given any continuous $x_{\mathbf{T}} \in \Omega$, postulate [S5] implies that the Riemann-Stieltjes integral $\int_0^\tau g(s)\,dx(s)$ $(= \alpha$, say) exists. So choose a positive number η so that, for any partition $\mathcal{P}$ of $\mathbf{T} = [0, \tau]$ with $|\mathcal{P}| < \eta$,

$$\left|(\mathcal{P})\sum g(s'')\,(x(s) - x(s')) - \alpha\right| \;<\; \varepsilon, \qquad (5.6)$$

where s' and s are successive partition points of $\mathcal{P}$ and $s' \leq s'' \leq s$ for each term. The number η depends on the particular $x_{\mathbf{T}}$, so write $\eta = \eta(x_{\mathbf{T}})$.

3. Now define a gauge $\gamma^{(2)} < \gamma^{(1)}$ in Ω, $\gamma^{(2)} = \left(L^{(2)}, \delta_N^{(2)}\right)$, as follows.

   - For each continuous $x_{\mathbf{T}}$ let $L^{(2)}(x_{\mathbf{T}}) \supseteq L^{(1)}(x_{\mathbf{T}})$ also satisfy $s - s' < \eta(x_{\mathbf{T}})$ for each successive s', s in $L^{(2)}(x_{\mathbf{T}})$; so (5.6) is satisfied.

   - For each $x_{\mathbf{T}} \in \Omega$ and each $N \in \mathcal{N}(\mathbf{T})$ let $\delta_N^{(2)}(x_t) \leq \delta_N^{(1)}(x_t)$ for $t \in N$.

   - Then every $\gamma^{(2)}$-fine division $\mathcal{D}^{(2)}$ of Ω is also $\gamma^{(1)}$-fine.

4. From preceding discussion, the stochastic integral $\mathbf{S}^g(x_\mathbf{T}), = \int_0^\tau g(s)\,dx(s)$, considered as an integrand in Ω, is itself integrable with respect to $F_{X_\mathbf{T}}$ on Ω, with integral value $\mu - 1$. Accordingly, define a gauge $\gamma^{(3)}, = \left(L^{(3)}, \delta_\mathcal{N}^{(3)}\right)$ so that for each $x_\mathbf{T}$, and for each N,

$$L^{(3)}(x_\mathbf{T}) \supseteq L^{(2)}(x_\mathbf{T}), \qquad \delta_\mathcal{N}^{(3)} \leq \delta_\mathcal{N}^{(2)},$$

so (5.5) and (5.6) are satisfied, and

$$\left|(\mathcal{D}^{(3)}) \sum \mathbf{S}^g(x_\mathbf{T}) F_{X_\mathbf{T}}(I[N]) - (\mu - 1)\right| < \varepsilon. \tag{5.7}$$

The objective is to show that $f(g, x_\mathbf{T}, N)$ is $F_{X_\mathbf{T}}$-integrable on Ω, with integral value $\mu - 1$. Choose a γ_3-fine division $\mathcal{D}^{(3)}$ of Ω. Then

$$\left|(\mathcal{D}^{(3)}) \sum f(g, x_\mathbf{T}, N) F_{X_\mathbf{T}}(I[N]) - (\mu - 1)\right|$$
$$\leq \left|(\mathcal{D}^{(3)}) \sum f(g, x_\mathbf{T}, N) F_{X_\mathbf{T}}(I[N]) - (\mathcal{D}^{(3)}) \sum \mathbf{S}^g(x_\mathbf{T}) F_{X_\mathbf{T}}(I[N])\right| +$$
$$+ \left|(\mathcal{D}^{(3)}) \sum \mathbf{S}^g(x_\mathbf{T}) F_{X_\mathbf{T}}(I[N]) - (\mu - 1)\right|$$
$$\leq (\mathcal{D}^{(3)}) \sum |f(g, x_\mathbf{T}, N) - \mathbf{S}^g(x_\mathbf{T}| F_{X_\mathbf{T}}(I[N]) +$$
$$+ \left|(\mathcal{D}^{(3)}) \sum \mathbf{S}^g(x_\mathbf{T}) F_{X_\mathbf{T}}(I[N]) - (\mu - 1)\right|$$
$$< (\mathcal{D}^{(3)}) \sum \varepsilon F_{X_\mathbf{T}}(I[N]) + \varepsilon = 2\varepsilon.$$

Since this holds for every γ_3-fine $\mathcal{D}_3$ and arbitrary $\varepsilon > 0$, the function $f(g, x_\mathbf{T}, N)$ is $F_{x_\mathbf{T}}$-integrable, with

$$\mathrm{E}[f(g, X_\mathbf{T}, N)] = \int_\Omega f(g, x_\mathbf{T}, N) F_{X_\mathbf{T}}(I[N]) = \mu - 1;$$

which equals, in turn,

$$\mathrm{E}[f(g_m, X_\mathbf{T}, N)], \qquad \lim_{m \to \infty} \mathrm{E}[f(g_m, X_\mathbf{T}, N)], \qquad \mathrm{E}[\mathbf{S}^g(X_\mathbf{T})].$$

This resolves an issue raised by the three representations in (5.2). Each of them yields the same expected terminal portfolio value for Example 17.

The formulation of Example 17 above postulates some properties which are usually observed in Brownian motion (and also, crucially, some properties which are not).

These postulated properties enabled us to establish a connection between three different formulations (5.2) of a stochastic integral with respect to the process X of Example 17. In the following sections similar connections will be sought for stochastic integrals with respect to Brownian motion.

A -complete (or Riemann sum) version of Brownian motion is presented in [MTRV], pages 305–381. The basic ideas are examined in the next section.

5.2   Brownian Motion Preliminaries

Chapter 7 of [MTRV] contains a mathematical account of Brownian motion as a random variation phenomenon, from the -complete standpoint rather than the classical Itô/Kolmogorov/Lebesgue standpoint. Without repeating all the technicalities, some aspects can be reviewed here with the stochastic integral issue in view.

Small but visible particles suspended in some medium such as gas or water are seen to undergo rapid, irregular motion. Successive impacts on such a particle by invisible molecular-scale particles of the medium produce successive spatial transitions of the visible particle. Under molecular particle impact, the visible particle follows a straight line trajectory or transition until the next molecular impact produces a new trajectory or transition. The successive transitions are small, but whenever observable by sight they are seen to follow a zig-zag course made up of continuous straight line segments, or polygonal-type paths through space.

- The length of any one transition does not depend on the length of the immediately preceding transition or, indeed, on any of the preceding transitions.

- The lengths of individual line segments or transitions are mostly small, but longer segments or transitions occur less frequently.

- It is observed that that the square of net distance traversed by a visible particle from some initial starting point is, on average, proportional to the time elapsed.

Comparable behaviour was observed in the changes or movements of share prices in stock markets over any given time period:

- Price changes, like Brownian particle transitions, are uncertain or unpredictable.

- Over any given time period the range or spread of possible price change tends on average to correlate with the time elapsed.

- There tend to be many small price changes, with larger price changes being rarer.

Some of the examples and illustrations in the preceding sections show that, for a system involving only a finite number of transitions, or even a countable number of discrete transitions (i.e. discrete times), a mathematical representation is not too difficult to find.

But for processes (X_s) with continuous s, such as $s \in \mathbf{T} = {]0, \tau]}$, it is not immediately obvious how to construct a mathematical system which conveys the above list of properties (and other properties not listed) without falling into contradiction and inconsistency. Also, though actual physical phenomena may appear to be discrete, and may not manifest the "infinite divisibility" of a continuum, it was found that mathematical representation of the Brownian phenomenon was best accomplished in continuous time and continuous values.

For -complete integration, the steps involved in the mathematical construction of processes $(X(s))$ representing Brownian motion are presented in [MTRV], chapters 6 and 7; including the question of continuity of sample paths $x(s)$ for $0 \le s \le \tau$ (pages 315–320).

Some of the key points are as follows. Consider a basic case in which each of the particle transitions is normally distributed with mean zero and variance equal to the lapse of time in a representative time interval $]s_{j-1}, s_j]$, where $0 = s_0 < s_1 < \cdots < s_n = \tau$. Let $x(s)$ denote any one of the space co-ordinates of the particle at time s; and suppose that at given time s_{j-1} the particle is located at given position $x_{j-1} = x(s_{j-1})$.

The corresponding co-ordinate $x_j = x(s_j)$ at a given subsequent time s_j is unpredictable, but from the description above it is not unreasonable to suppose that x_j is normally distributed, with mean value x_{j-1} and with a spread of values (or variance) proportional to (in fact, equal to in this case) the lapse of time $s_j - s_{j-1}$. So the particle position co-ordinate x_j is equally likely to be less than or greater than the starting value x_{j-1}, and the greater the lapse of time the greater the difference (transition) in co-ordinate values is likely to be.

Thus, for an interval I_j such as $I_j =]u_j, v_j]$, the probability, given x_{j-1}, that x_j is contained in the interval I_j (the *transition probability*) is

$$P\left(x_j \in I_j \mid x_{j-1}\right) = \sqrt{\frac{1}{2\pi(s_j - s_{j-1})}} \int_{I_j} \exp\left(-\frac{(x_j - x_{j-1})^2}{2(s_j - s_{j-1})}\right) dx_j.$$

With $x_0 = x(s_0) = x(0) = 0$ given, independence of successive normally distributed transitions implies that the probability that $x_j \in I_j$ for $j = 1, 2, \ldots, n$ (the joint transition probability) is

$$\prod_{j=1}^{n} \left(\sqrt{\frac{1}{2\pi(s_j - s_{j-1})}} \int_{I_j} \exp\left(-\frac{(x_j - x_{j-1})^2}{2(s_j - s_{j-1})}\right) dx_j\right) =$$

$$= \frac{\int_{I(N)} \exp\left(-\sum_{j=1}^{n} \frac{(x_j - x_{j-1})^2}{2(s_j - s_{j-1})}\right) dx(N)}{\prod_{j=1}^{n} \sqrt{2\pi(s_j - s_{j-1})}} \tag{5.8}$$

where $N = \{s_1, \ldots, s_n\}$, $I(N) = I_1 \times \cdots \times I_n$ and $dx(N) = dx_1 \times \cdots \times dx_n$. (The "=" in (5.8) requires use of Fubini's theorem, [MTRV] page 60, theorem 54.)

In a *drifting* Brownian motion, with *drift rate* μ and *variance rate* σ^2, for given x_{j-1} the mean value of X_j (the various possible values of x_j) is $x_{j-1} + \mu(s_j - s_{j-1})$ and the variance of X_j is $\sigma^2(s_j - s_{j-1})$. Then the expression (5.8) becomes

$$\frac{\int_{I(N)} \exp\left(-\sum_{j=1}^{n} \frac{(x_j - (x_{j-1} + \mu(s_j - s_{j-1})))^2}{2\sigma^2(s_j - s_{j-1})}\right) dx(N)}{\prod_{j=1}^{n} \sqrt{2\pi\sigma^2(s_j - s_{j-1})}} \tag{5.9}$$

For $I[N] \in \mathcal{I}(\mathbf{R}^{\mathbf{T}})$ (an interval of the infinite-dimensional sample space $\Omega = \mathbf{R}^{\mathbf{T}}$), denote (5.9) by $G^{\mu\sigma}(I[N])$. Then, with $\mu = 0$ and $\sigma = 1$, (5.8) is

$$G(I[N]) := G^{0,1}(I[N]); \tag{5.10}$$

and it is established in [MTRV] (chapter 7) that $G(I[N])$ is the joint probability distribution function[1] for standard Brownian motion. Included in chapter 7 is a demonstration of the *continuous modification*, so that the *continuity* postulate for Brownian motion is satisfied. This means that if D is the set of sample paths $x_{\mathbf{T}} \in \mathbf{R}^{\mathbf{T}}$ for which $x(s)$ fails to be uniformly continuous at any s $(0 \le s \le \tau)$, then

$$\int_{\mathbf{R}^{\mathbf{T}}} 1_D(x_{\mathbf{T}})G(I[N]) = 0, \tag{5.11}$$

provided the meaning of "$\int_{\mathbf{R}^{\mathbf{T}}}$" is modified in a particular way. This technical modification of $\int$ is applied below wherever it is required by the context.[2]

There is a proportionality between typical spatial increments $x_j - x_{j-1}$, on the one hand, and, on the other hand, the corresponding $\sqrt{s_j - s_{j-1}}$ in the time increments. In consequence of this, the "typical" sample paths $x(s)$ are everywhere continuous and nowhere differentiable.

This mathematical property is indicated by the "jagged path" diagrams often used to illustrate Brownian motion. These are polygonal paths or joined up line segments; so they are continuous. But since such graphs are differentiable almost everywhere, they are closer to the physical or observed form of Brownian motion than they are to its mathematical representation—no matter how useful and effective the latter is found to be in theory.

In the case of share price transitions $z_j - z_{j-1}$ (or $z(s_j) - z(s_{j-1})$) in stock markets, the following model has been used. Writing $y(s) = \ln z(s)$, the process $(Y(s))$ is presumed to be Brownian motion with variance rate σ^2 and drift rate $\rho - \frac{1}{2}\sigma^2$. Then, as explained[3] in [MTRV] section 7.3 pages 308–314 and section 8.16 pages 440–442, the probability distribution for the price transitions represented by $z(N) \in J[N] \in \mathcal{I}(\mathbf{R}_+^{\mathbf{T}})$ is given by $\mathcal{G}(J[N]) =$

$$= \int_{J(N)} \left(\prod_{j=1}^{n} \frac{\exp\left(\dfrac{-\left((\ln z_j - \ln z_{j-1}) - \rho(s_j - s_{j-1}) + \frac{\sigma^2}{2}(s_j - s_{j-1}) \right)^2}{2\sigma^2(s_j - s_{j-1})} \right)}{z_j \sigma \sqrt{2\pi(s_j - s_{j-1})}} \right) dz(N) \tag{5.12}$$

where $\mathbf{R}_+ =]0, \infty[$, the positive real numbers. (Share prices take only positive numerical values.) The share price process $(Z_s)_{s \in \mathbf{T}}$,

$$Z_{\mathbf{T}} \simeq z_{\mathbf{T}}\left[\mathbf{R}_+^{\mathbf{T}}, \mathcal{G} \right],$$

is then called a *geometric* Brownian motion with *growth rate* ρ and *volatility* σ.

In the -complete approach to Brownian motion, the formidable subtleties of continuous-time, continuous-valued processes, as referred to earlier, are absorbed into the content and meaning of the -complete integrals involved. In

---

[1] There are some related typos in [MTRV]. Page 306, line 5: π should be πt (twice). Page 306, line 10: π should be $\pi(t-s)$ (twice). Page 308, line 4: $t_j - t_{j-1}$ should be $\sqrt{t_j - t_{j-1}}$. Pages 313, 314 and 442 have a + and − mixed up; more details of this can be found in [**website**].

[2] In Part II of this book, cell function $G(I[N])$ is complex-valued.

[3] Typos are noted and corrected in [**website**].

effect, existence of

$$E[f(X_{\mathbf{T}})] = \int_{\mathbf{R^T}} f(x_{\mathbf{T}})G(I[N]), \qquad E[k(Z_{\mathbf{T}})] = \int_{\mathbf{R_+^T}} k(z_{\mathbf{T}})\mathcal{G}(J[N])$$

$$(5.13)$$

is the foundation of this interpretation of the mathematical conception of geometric/Brownian motion.

On the face of it, the -complete integrals (5.13), involving as they do the rather fearsome looking integrator expressions (5.8) and (5.12), might appear to add a new challenge to the already formidable mathematics of Brownian motion.

But, formidable or not, the substance of both of these expressions is already inherent in every interpretation of the subject, and cannot be evaded. Each of the above expressions is defined on cells $I(N)$ of an n-dimensional Cartesian product space $\mathbf{R}^n$ (or $\mathbf{R}_+^n$) for each positive integer n. Such intervals are themselves fairly easy to grasp, as are the corresponding cylindrical intervals $I[N]$ in the infinite-dimensional domain $\mathbf{R^T}$ (or $\mathbf{R_+^T}$).

Also, the analytical forms of (5.8) and (5.12) are fairly familiar Gaussian expressions; though it is easy to trip over details of the algebraic combinations of the terms which compose them. Getting the +'s and −'s right can be a bit of a chore.

In fact, calculating Riemann sums of (5.8) and (5.12) is not so difficult as might appear. Essentially, the deeper subtleties of the subject are contained, not in (5.8) and (5.12), but in the specific form and details of the construction of the -complete integral. Much of that construction is already familiar from basic Riemann integration; and, with the experience gained from routine practice and experimentation, it is not too hard to achieve competence in the -complete version of the well-known Riemann theory—even when (5.8) and (5.12) are used.

How does this compare with the standard theories of Brownian motion? There is no escaping the expressions (5.8) and (5.12). Essentially, the second formula is obtained from the first by changes of variable, as demonstrated in [MTRV], pages 308–314. In the classical approach, the first of these expressions is extended to a probability measure on the Borel sets in $\mathbf{R^T}$ (or $\mathbf{R_+^T}$) generated by the cylindrical intervals $I[N]$. In the standard theory of Brownian motion these constructions are used to give meaning to integral, expected value, etc., by means of limits of the so-called *simple functions* of Lebesgue integration theory.

The same approach can be used for the second expression (5.12). More commonly, the Brownian probability measure can be transformed into the geometric Brownian by change of variable. In any event, (5.8) and (5.12) are central to the mathematical representation of Brownian motion.

5.3   Review of Brownian Probability

About half of [MTRV] is taken up with (5.8), (5.12), and related expressions. When motivational and explanatory material is included, these expressions, along with their properties and implications, constitute almost all of the present book whenever the Feynman quantum mechanical expression (8.29) of Section 8.6 below is included.

[MTRV] uses the symbol $\mathcal{G}$ for geometric Brownian distribution function (5.12). while the symbol $G_c(I[N])$ is used to denote the joint distribution function for c-Brownian motion. So $G_{-\frac{1}{2}}$ gives standard or classical Brownian motion (5.8); while $G_{\frac{\iota}{2}}$ is used in [MTRV] for the Feynman theory of (8.29).

But in this book the symbol G is used without subscript, allowing the context (Brownian motion or quantum mechanics) to show which meaning is intended.

To sum up, the theory of (5.8), (5.12), and (8.29) is covered in chapters 6, 7, and 8 of [MTRV]. It is not proposed to rehearse this theory here; but, instead, to highlight some aspects of it which are particularly relevant to the topics of this book.

In [MTRV], Brownian motion, geometric Brownian motion, and Feynman path integration (for single particle mechanical phenomena) are united in a single theory based on a version of Fresnel's integral using a parameter $c = a + \iota b$, where $\iota = \sqrt{-1}$, $a \leq 0$, and $c \neq 0$ (so a and b are not both zero.)

The case $c = -\frac{1}{2}$ (real, negative) leads to Brownian and geometric Brownian motion. The case $c = \frac{\sqrt{-1}}{2}$ (pure imaginary) gives Feynman path integrals. The designation c-Brownian motion is intended to cover all cases, including those which are "intermediate" between real negative and pure imaginary.

The Fresnel evaluation is in theorem 133 (pages 261–262 of [MTRV]). For $c < 0$ (real, negative) lemmas 12 and 13 (page 262) show that finite compositions (or addition) of normal distributions are normal, so that these distributions are additive in some sense. And provided the real part of c is non-positive (with $c \neq 0$), lemmas 12 and 13 are valid for complex-valued c.

These results are crucial in going from finite compositions of distributions to infinite compositions, giving a theory of infinitely many (and "infinitely divisible" or continuum) of normal distributions (or c-normal distributions), leading to the theory of c-Brownian motion.

Joint probability distributions are defined on domains $\mathbf{R}^{\mathbf{T}}$ where $\mathbf{T}$ is typically a real interval open on the left and closed on the right, such as

$$]0, t], \quad]0, \tau], \quad]\tau', \tau].$$

A joint distribution such as (5.8), (5.12), or (8.29), is constructed for samples

$$N = \{t_1, t_2, \ldots, t_{n-1}, t_n\} \subset \mathbf{T},$$

with t_0 taken to be the left hand boundary of interval $\mathbf{T}$, and t_n the right hand boundary point.

Samples N are arbitrary choices of finite subsets of $\mathbf{T}$; so they need not necessarily contain the right hand boundary point. But the analysis is generally more straightforward when the right hand boundary point of $\mathbf{T}$ is included in the sampling. This can be ensured by appropriate choice of gauge $\gamma = (L, \delta_N)$; just include the right hand boundary point in $L(x_{\mathbf{T}})$ for each $x_{\mathbf{T}}$. Then each sample $N \supseteq L(x_{\mathbf{T}})$ must include t_n.

The domain for analysis of Brownian motion is typically $\mathbf{R^T}$. But for quantum mechanics the domain is $\mathbf{R}^T$, where interval T is open on the left and also open on the right:

$$]0, t[, \quad]0, \tau[, \quad]\tau', \tau[;$$

and sampling times are

$$N = \{t_1, t_2, \ldots, t_{n-1}\} \subset T,$$

with left and right hand boundaries of T denoted by t_0 and t_n, respectively. Even though n is variable, the point t_n is the fixed right hand boundary of T.

For both $\mathbf{R^T}$ and $\mathbf{R}^T$, cell functions (5.8), (5.12), and (8.29) are denoted in this book by $G(I[N])$, $\mathcal{G}(I[N])$, and $G(I[N])$, respectively; the distinction between the $\mathbf{T}$-form and T-form of the cell function G being determined by the context. But whenever it is necessary to deal with $\mathbf{R^T}$ and $\mathbf{R}^T$ together, both contexts are present. In that case the following convention is applied:

$$N = \{t_1, \ldots, t_{n-1}\} \quad \subset \mathbf{T}, \qquad N^+ = \{t_1, \ldots, t_{n-1}, t_n\} \subset \mathbf{T}, \quad \text{or}$$
$$N = \{t_1, \ldots, t_{n-1}, t_n\} \subset \mathbf{T}, \qquad N^- = \{t_1, \ldots, t_{n-1}\} \qquad \subset T.$$

where t_n denotes the fixed right hand boundary point of both T and $\mathbf{T}$. Then $G(I[N])$ or $G(I[N^-])$ is the Brownian (or quantum mechanical) cell function for domain $\mathbf{R}^T$, while $G(I[N^+])$ or $G(I[N])$, respectively, is the Brownian (or quantum mechanical) cell function for domain $\mathbf{R^T}$; depending on which representation is more convenient. The following relationship holds for $I[N^+] = I[N] \times I_{t_n}$:

$$G(I[N^+]) = \int_{I_{t_n}} G(I[N]) \, dx_{t_n}. \tag{5.14}$$

(This follows from Fubini's theorem.)

In $\mathbf{R^T}$, the cell function G (or $G(I[N^+])$) is a probability distribution function (in the sense of [MTRV] pages 183 and 305), since it is finitely additive on adjoining, disjoint cells ($I[N^+]$), with $\int_{\mathbf{R^T}} G(I[N^+]) = 1$. Then $G(I[N])$ can be thought of as a function of

$$I[N] \times \{x_{t_n}\}, \ = \ I_{t_1} \times \cdots \times I_{t_{n-1}} \times \{x_{t_n}\}$$

in the domain $\mathbf{R^T}$, so, by (5.14), $G(I[N])$ is a probability density function in $\mathbf{R^T}$; and, in accordance with chapter 7 of [MTRV], integrals

$$\int_{\mathbf{R^T}} f(x_{\mathbf{T}}) G(I[N^+]), \qquad \int_{\mathbf{R^T}} f(x_{\mathbf{T}}) G(I[N])$$

are, respectively, expectation values $\mathrm{E}[f(X_{\mathbf{T}})]$ and wave functions $\psi(\xi)$ where $\xi = x(t_n)$ is fixed in the latter case; with

$$\mathrm{E}[f(X_{\mathbf{T}})] = \int_{\mathbf{R}} \psi(\xi) \, d\xi = \int_{\mathbf{R}} \psi(x_{t_n}) \, dx_{t_n}$$

when the integrals exist. Thus wave function $\psi(\xi)$ can be thought of as a *marginal density of expectation*. See page 320, chapter 7 of [MTRV].

The set $\mathbf{T} \setminus T$ consists of a single point, the right hand boundary point of both $\mathbf{T}$ and T. In some situations this single point difference has only slight or no effect in analysis. For instance, a function $f(s)$ can be integrated over domains $\mathbf{T}$ and T using the Riemann-complete, or Lebesgue, or some other method:

$$\int_{\mathbf{T}} f(s)\,ds, \qquad \int_{T} f(s)\,ds.$$

If f is "well-behaved" (for instance, a differentiable function), then we can expect

$$\int_{\mathbf{T}} f(s)\,ds = \int_{T} f(s)\,ds. \tag{5.15}$$

But the difference between $\mathbf{R}^{\mathbf{T}}$ and $\mathbf{R}^{T}$ is far from insignificant; it is a difference in dimensionality, like $\mathbf{R}^2$ (a plane) and $\mathbf{R}$ (a line). Thus the pair

$$\int_{\mathbf{R}^T} h_T(x_T)G(I[N]), \qquad \int_{\mathbf{R}^{\mathbf{T}}} h_{\mathbf{T}}(x_{\mathbf{T}})G(I[N^+]), \tag{5.16}$$

are as profoundly different as

$$\int_{\mathbf{R}} h_1(x_1)\,dx_1, \qquad \int_{\mathbf{R}\times\mathbf{R}} h_2(x_1,x_2)\,dx_1 dx_2. \tag{5.17}$$

Figures 3.1 and 3.2 on page 87 of [MTRV] should be borne in mind in order to keep these differences in perspective. In figure 3.1, x_T is represented as a graph of dependent values $x(t)$ plotted against independent values t. In Figure 3.2, the elements t are labels for uncountably many "mutually perpendicular" axes, for which x_T is a single point.

Such issues were the backdrop for a popular nineteenth century book *Flatland, A Romance of Many Dimensions* by Edwin Abbott Abbott [1].

5.4   Brownian Stochastic Integration

A stochastic integral with respect to Brownian processes can have forms

$$\int_0^t g(s)dX(s), \qquad \int_0^t f(X(s))dX(s), \qquad \int_0^t Z(s)dX(s).$$

Each random variable $X(s)$ ($0 < s \le t$) is normally distributed, so individually they are not too difficult.

But $\int_0^t$ involves a continuum of such normal distributions. Section 4.4 mentions step functions, cylinder functions, and sampling functions as a progression of stages in dealing with this problem. This section applies a cylinder function approach to Brownian stochastic integrals. The idea is to replace the continuum $]0,t]$ by discrete times $0 = \tau_0 < \tau_1 < \cdots < \tau_n = t$. (In Part II, R. Feynman's path integrals of cylinder functions use a countable infinity of discrete times τ_j.)

For $0 < t \le \tau$ let $\mathbf{T}$ denote $]0,t]$, closed at boundary t; and let T denote $]0,t[$, open at boundary t. Suppose an asset price process $X_{\mathbf{T}}$ can be represented as

$$X_{\mathbf{T}} \simeq x_{\mathbf{T}}\big[\mathbf{R}^{\mathbf{T}}, G\big]$$

where $x(0) = 0$ for all sample paths, and G is the joint probability distribution function $G(I[N])$ of (5.8) for standard Brownian motion. Assume the asset is some portfolio which (unlike shares) can take unbounded positive and negative values. In other words the "asset" (or portfolio) can also be a liability.

A distinction can be made between the *value of the portfolio* at any time t, and the *earnings of the portfolio* at time t.

The latter is intended to denote the stake of the investor or holder of the portfolio, taking account of the initial expenditure (denoted below by β) paid out by the investor in order to acquire possession of the portfolio to begin with.

The former represents a third party view of the portfolio, disregarding any cost of acquisition. If $w(t)$ is the value of the portfolio then earnings equal $w(t) - \beta$ for all t; where β denotes the upfront cost to the investor of acquiring the portfolio at time $t = 0$.

The value of the portfolio at any time s depends on the size $\nu(s)$ of (or number of units of the assets/liabilities in) the portfolio. Then the value of the portfolio at time s is $\nu(s)x(s)$. For the purpose of investigating stochastic integrals, the number $\nu(s)$ can have various interpretations, such as

$$g(s), \quad Z(s), \quad f(X(s));$$

where, for $0 \le s \le t$, $g(s)$ is a deterministic[4] function, $(Z(s))$ is a random process independent of the Brownian motion $(X(s))$, and $(f(X(s)))$ is a process which depends on $(X(s))$.

Each of these scenarios has been encountered in the preceding examples. Stochastic integrals have appeared in some of the examples, not in the form of the **value** $\nu(s)x(s)$ of the portfolio, but in the form of its **growth in value**. For a time increment $]s, s']$, the increment unit value is $x(s') - x(s)$, and the corresponding increment in portfolio value is

$$\nu(s')x(s') - \nu(s)x(s). \tag{5.18}$$

But, as in (2.1)–(2.5), for the finance reasons described in Example 6, the value increment is expressed as

$$\nu(s)\left(x(s') - x(s)\right), \tag{5.19}$$

and (5.18) does not apply.[5] For both formulations (5.18) and (5.19), the aggregated increase in total portfolio value can then be estimated in Riemann sum format as

$$\sum_{j=1}^{n} \left(\nu(s_j)x(s_j) - \nu(s_{j-1})x(s_{j-1})\right), \qquad \sum_{j=1}^{n} \nu(s_{j-1})\left(x(s_j) - x(s_{j-1})\right),$$

respectively. By cancellation, the former reduces to

$$\nu(s_n)x(s_n) - \nu(s_0)x(s_0) \quad = \quad \nu(t)x(t) - \nu(0)x(0),$$

---

[4]Deterministic functions s^{-1} or e^{-s} $(0 < s \le t)$ are examples.

[5]Since the function ν is evaluated at the left hand boundary s' of cell $]s, s']$, (5.19) is a Riemann sum term of a Burkill-complete integrand. A Stieltjes-complete term would have the form $\nu(\bar{s})\left(x(s') - x(s)\right)$ where $\bar{s}$ is tag point (s' **or** s) of cell $]s, s']$.

which is merely the difference between the terminal and initial portfolio values; while the form based on (5.19) can be regarded as a Riemann sum estimate of the integral $\int_0^t \nu(s) \, dx(s)$; or, in stochastic integral form,

$$\int_0^t \nu(s) \, dX(s).$$

Using the rationalization of Example 6, the latter is the construction which should be applied—usually with $\nu(s) = f(X(s))$, a random variable depending on Brownian motion $(X(s))$.

Example 18 *For $0 \leq s \leq t$ let $g(s)$ denote the quantity of assets/liabilities in the portfolio (measured as number of units) at time s.* **The aim is to find mathematical representation for portfolio value $w(t)$.**

For present purposes the asset values are taken to be Brownian.[6] The function g can have many forms. It can be taken to be random or deterministic; for instance, it can be constant, a step function, continuous, and/or have bounded variation. Some of these scenarios are explored in the following paragraphs and in further Examples.

Suppose g is a deterministic function of bounded variation. Then, for any continuous sample path $x_\mathbf{T} = (x(s))_{s \in \mathbf{T}} \in \mathbf{R}^\mathbf{T}$, the Riemann-Stieltjes integral $w(t) = w(x_\mathbf{T}) = \int_0^t g(s) \, dx(s)$ exists because of a well-known property of Riemann-Stieltjes integration. The set of C of continuous $x_\mathbf{T}$ is of full G-measure in $\mathbf{R}^\mathbf{T}$ (in the modified sense of theorem 188, [MTRV] page 300), so the stochastic integral $W(t) = w(X_T) = \int_0^t g(s) \, dX(s)$ is an observable in domain $\Omega = \mathbf{R}^\mathbf{T}$,

$$w(X_\mathbf{T}) \simeq w(x_\mathbf{T}) \left[\mathbf{R}^\mathbf{T}, G \right],$$

where $w(x_\mathbf{T}) = \int_0^t g(s) \, dx(s)$ and $w(X_\mathbf{T}) = \int_0^t g(s) \, dX(s)$.

The observable $\int_0^t g(s) \, dX(s)$ is a random variable with respect to probability function G if $\mathrm{E}\left[\int_0^t g(s) \, dX(s)\right] = \int_{\mathbf{R}^\mathbf{T}} \left(\int_0^t g(s) \, dx(s)\right) G(I[N])$ exists. If the unpredictable real numbers $w = w(t) \in \mathbf{R}$ possess a probability distribution function $F_W(J)$ for cells J in $\mathbf{R}$ then $W \simeq w[\mathbf{R}, F_W]$ is an observable, and it is a random variable if its expectation $E[W] = \int_\mathbf{R} w \, F(J)$ exists, with

$$\mathrm{E}\left[\int_0^t g(s) \, dX(s)\right] = \mathrm{E}[W];$$

and then W is the elementary form of the contingent joint-basic $\int_0^t g(s) \, dX(s)$.

Some simple cases can be easily checked. A function g has bounded variation if it does not oscillate too much; for instance, if it is constant, or a step function, or is monotone increasing. The Dirichlet function $d(s)$, whose value is 1 if s is rational and 0 otherwise, is discontinuous everywhere and has unbounded

---

[6]More usually, asset values are said to be geometric Brownian, though that assumption is also unreliable. See [MTRV], section 9.9 pages 479–485.

variation. A "typical jagged path" $x(s)$ of Brownian motion is continuous with $x(0) = 0$, is nowhere differentiable, and has unbounded variation.

Take $g(s) = \alpha$ (constant) for $0 \leq s \leq t$. Then

$$w(t) = \int_0^t g(s)\, dx(s) = \alpha\, (x(t) - x(0)) = \alpha x(t),$$

and

$$
\begin{aligned}
E[W(t)] &= \int_{\mathbf{R^T}} w(t) G(I[N]) = \alpha \int_{\mathbf{R^T}} x(t) G(I[N]) \\
&= \alpha E[X(t)] = \alpha \times 0 = 0 \qquad (5.20)
\end{aligned}
$$

since $E[X(t)] = 0$ for all $t > 0$.

The value of the portfolio is equally likely to increase or decrease. So, intuitively, the average value of the portfolio at any time $t > 0$ will be equal to the initial value (time $t = 0$). Note also that, for $0 < t \leq \tau$ (supposing τ fixed and t variable) and $\mathbf{T} =]0,t]$, $W_{\mathbf{T}}$ is a Brownian process, since $W(t) = \alpha X(t)$, and (X_t) is Brownian.

If g is a step function, with $g(s) = \alpha_k$ for

$$0 = s_0 < s_1 < \cdots < s_{k-1} \leq s < s_k < \cdots < s_n = t$$

and $1 \leq k \leq n$, then g has bounded variation as in Example 18. The preceding argument may then suggest that

$$\int_0^t g(s)\, dX(s) = \sum_{k=1}^m \alpha_k \left(X\left(\frac{kt}{m}\right) - X\left(\frac{(k-1)t}{m}\right) \right), \qquad (5.21)$$

on the supposed grounds that

$$\int_0^t g(s)\, dx(s) = \sum_{k=1}^m \alpha_k \left(x\left(\frac{kt}{m}\right) - x\left(\frac{(k-1)t}{m}\right) \right) \qquad (5.22)$$

for every sample function $(x(s))$ of the Brownian motion. But this argument is not complete. See example 60, page 395 of [MTRV] which contains an error of this kind. The correction is in [website].

Example 13 shows that (5.22) fails for certain Brownian motion paths; for instance the sample path $x(s)$ which is zero for rational values of s, and 1 otherwise (a Dirichlet "path"). But all is not lost, since Dirichlet is discontinuous, and, with the "continuous modification" described in chapter 6 of [MTRV], the set of Brownian sample paths which are not uniformly continuous form a null set in the sample space $\Omega = \mathbf{R^T}$.

In fact (5.21) is true, even though (5.22) is not always true. This is because (5.22) is true **for continuous sample paths** $x(s)$, which form a set of *full measure* (with respect to $G(I[N])$) in $\mathbf{R^T}$.

A correct argument can be formulated on the following lines. For $k = 0, 1, \ldots, m$ write $\tau_k = kt/m$. Choose a gauge $\delta(s) > 0$ $(0 \leq s \leq t)$ so that every δ-fine partition $\mathcal{P} = \{s_j\}$ of $[0,t]$ conforms to $[\tau_k, \tau_{k+1}]$. Choose one such partition

$$0 = \tau_0 = s_0 < s_1 < \cdots < s_k' < \tau_k < s_k'' < \cdots < s_n = \tau_m = t,$$

with Riemann sum

$$\beta \;=\; \sum_{j=1}^{n} g(s_{j-1})\,(x(s_j) - x(s_{j-1})),$$

and observe that the Riemann sum includes terms

$$g(\tau_j)\,(x(\tau_k) - x(s_k')) \;+\; g(\tau_j)\,(x(s_k') - x(\tau_k)) \;=\; +\, a_k + b_k + c_k$$

where

$$
\begin{aligned}
a_k &= g(\tau_{j-1})\,(x(\tau_k) - x(s_k')), \\
b_k &= (g(\tau_j) - g(\tau_{j-1}))\,(x(\tau_k) - x(s_k')), \\
c_k &= g(\tau_j)\,(x(s_k') - x(\tau_k)).
\end{aligned}
$$

By inserting correction term b_k for each k, every increment $x(s_j) - x(s_{j-1})$ for which

$$\tau_{k-1} \;\leq\; s_{j-1} \;<\; s_j \;\leq\; \tau_k$$

can be given coefficient $g(\tau_{j-1}) = \alpha_{j-1}$, so $\beta = \alpha + \eta$ where

$$\alpha \;=\; \sum_{k=1}^{m} \alpha_{k-1}\,(x(\tau_k) - x(\tau_{k-1})), \qquad \eta \;=\; \sum_{k=1}^{m} (\alpha_k - \alpha_{k-1})\,(x(\tau_k) - x(\tau_{k-1})).$$

The gauge δ can be chosen so that, for each k and each **continuous** sample path x,

$$|x(\tau_k) - x(s_k')| \;<\; \frac{\varepsilon}{m\,|\alpha_j - \alpha_{j-1}|}; \tag{5.23}$$

giving $|\beta - \alpha| = |\eta| < \varepsilon$. Therefore for every x outside a G-null set of discontinuous sample paths, $\int_0^t g(s)\,dx(s)$ exists, and

$$\int_0^t g(s)\,dx(s) \;=\; \sum_{k=1}^{m} \alpha_{k-1}\,(x(\tau_k) - x(\tau_{k-1})), \tag{5.24}$$

so the stochastic integral $\mathbf{S}_{\mathbf{T}}^g(X)$ exists and

$$\mathbf{S}_{\mathbf{T}}^g(X) \;:=\; \int_0^t g(s)\,dX(s) \;=\; \sum_{k=1}^{m} \alpha_{k-1}\,(X(\tau_k) - X(\tau_{k-1})). \tag{5.25}$$

As this is a finite linear combination of Brownian motions, the stochastic integral representing accumulated incremental portfolio value has normal distribution, and is a random variable—in the sense that its expected value $E[W(t)]$ exists. The latter can be calculated as in (5.20).

In Example 18 the integrand function g is supposedly of bounded variation. Then

- the Stieltjes integral $\int_0^t g(s)\,dx(s)$ exists for G-almost all x (in the modified sense of [MTRV] pages 316–320); and

- the stochastic integral $\mathbf{S}^g(X_\mathbf{T}), = \int_0^t g(s) \, dX(s)$, exists and is an observable:

$$\mathbf{S}^g(X_\mathbf{T}) \simeq \mathbf{S}^g(x_\mathbf{T})[\mathbf{R}^\mathbf{T}, G].$$

It has been established above that if g is constant then $\mathbf{S}^g(X_\mathbf{T})$ is a random variable in the sense that its expected value $\int_{\mathbf{R}^\mathbf{T}} \mathbf{S}_\mathbf{T}^g(x_\mathbf{T}) G(I[N])$ exists, with

$$E[\mathbf{S}_\mathbf{T}^g(X_\mathbf{T})] = \int_{\mathbf{R}^\mathbf{T}} \mathbf{S}_\mathbf{T}^g(x_\mathbf{T}) G(I[N]) = \int_{\mathbf{R}^\mathbf{T}} \left(\int_\mathbf{T} g(s) \, dx(s) \right) G(I[N]) = 0.$$

The following Example extends this to step function g.

Example 19 *If $g(s)$ is a step function then the stochastic integral*

$$\int_\mathbf{T} g(s) \, dX(s), = \mathbf{S}^g(X_\mathbf{T}),$$

is a random variable with expected value $E[\mathbf{S}^g(X_\mathbf{T})] = 0$. *This can be deduced from (5.24) and (5.25) as follows.*

$$
\begin{aligned}
E[\mathbf{S}_\mathbf{T}^g(X_\mathbf{T})] &= \int_{\mathbf{R}^\mathbf{T}} \left(\int_0^t g(s) \, dx(s) \right) G(I[N]) \\
&= \int_{\mathbf{R}^\mathbf{T}} \left(\sum_{k=1}^m \alpha_{k-1} \left(x(\tau_k) - x(\tau_{k-1}) \right) \right) G(I[N]) \\
&= \sum_{k=1}^m \alpha_{k-1} \left(\int_{\mathbf{R}^\mathbf{T}} \left(x(\tau_k) - x(\tau_{k-1}) \right) G(I[N]) \right) \\
&= \sum_{k=1}^m \alpha_{k-1} E[X_\mathbf{T}(\tau_k) - X_\mathbf{T}(\tau_{k-1})] = 0
\end{aligned}
$$

since the process $(X_\mathbf{T})$ is Brownian, and $X_\mathbf{T}(\tau_k) - X_\mathbf{T}(\tau_{k-1})$ is normally distributed with mean value zero for each k. (Use theorem 160, page 280 of [MTRV].)

This converts $\int_{\mathbf{R}^\mathbf{T}}$ into $\int_{\mathbf{R}^m}$ using the cylinder function method.[7] Only the most basic properties of integrals are used in Examples 18 and 19. The following goes a bit further.

Example 20 *In Example 19, g is a deterministic step function. Suppose, for $0 \le s \le t$, the function $g(s)$ is a continuous limit of such step functions. Does the corresponding stochastic integral $\mathbf{S}^g(X_T)$ (or $\int_T g(s) \, dX(s)$) exist in that case? Continuity of $g(s)$ for $0 \le s \le t$ implies g is bounded, with upper bound $\kappa > 0$, say. Suppose it is also assumed that g has bounded variation. Bounded variation implies the stochastic integral $\mathbf{S}^g(X_T)$ (or $\int_T g(s) \, dX(s)$) exists, as in Example 18. Continuity also implies that, for $m = 2, 3, \ldots$, there are step functions*

$$g_m(s) = \alpha_k, \qquad \frac{(k-1)t}{2^m} \le s < \frac{kt}{2^m}, \qquad 1 \le k \le 2^m,$$

---

[7] Is it possible to go from $\int_{\mathbf{R}^m}$ to $\int_{\mathbf{R}^\mathbf{T}}$? This was the method used by R. Feynman in [39] (or [F1]) to define path integrals for quantum mechanics.

such that $g_m(s) \leq \kappa$ for each m and for $0 \leq s \leq t$; and $g_m(s) \to g(s)$ for each s ($0 \leq s \leq t$) as $m \to \infty$. With all these helpful assumptions, is it the case that the observable $\int_T g(s) \, dX(s)$ is a random variable, with expected value 0?

If the answer to this question is "yes", then the first step in a proof might involve establishing that, as $m \to \infty$,

$$\int_T g_m(s) \, dx(s) \quad \to \quad \int_T g(s) \, dx(s),$$

for each continuous sample function $x(s)$; so $\int_T g_m(s) \, dX(s) \to \int_T g(s) \, dX(s)$.

To get a sense of what might be involved in this, some relevant features of sample functions of Brownian processes are considered in Section 5.5 below.

In Example 19 the step function integrand g is taken to be deterministic. What if randomness is allowed in this part of the integrand? Some random step functions are shown to be stochastically integrable in Examples 21 and 22.

Example 21 *In place of deterministic $g(s)$, introduce a random integrand $Z(s)$ where $Z(s)$ is the same random variable $Z_k(s)$ for $\tau_k \leq s < \tau_{k+1}$; so $Z_k \simeq z_k[\Omega_k, F_{Z_k}]$ is a given observable for each k, and*

$$Z(s) \;=\; Z_k \quad for \;\; \tau_k = \frac{t(k-1)}{m} \leq s < \tau_{k+1} = \frac{tk}{m}, \quad k = 1, 2, \dots, m.$$

Designate this as a step function process *$(Z_k(s))$. The proposed stochastic integral is $\int_0^t Z_k(s) \, dX(s)$, the integrand being a step function process $(Z_k(s))$, where each $Z_k \simeq z_k[\Omega_k, F_{X_k}]$, $k = 1, 2, \dots, m$. (Each sample space Ω_k could be $\mathbf{R}$, for instance. Assume, for convenience, that Z_0 is a degenerate random variable with constant value z_0.)*

Existence of the stochastic integral $\int_0^t Z_k(s) \, dX(s)$ follows as in Example 19. To see this, note that each sample path $z(s), = z_k(s)$, $(k = 1, \dots, m)$, is Stieltjes integrable with respect to each of the uniformly continuous Brownian sample paths $x(s)$ which form a set of G-full measure in sample space $\Omega = \mathbf{R}^T$; so $\int_0^t z(s) \, dx(s)$ exists for all relevant sample paths $z(s)$ and $x(s)$.

In Example 19 it was possible to assert that, for deterministic step function g and variable t, the stochastic integral

$$Y(t) \;=\; \mathbf{S}_T^g(X) \;=\; \int_0^t g(s) \, dX(s)$$

itself forms a Brownian motion $(Y(t))$. Now, with g random, and writing $y_k = z_{k-1}(x_k - x_{k-1})$, a sample value of $Y = \mathbf{S}_T^Z(X), = \int_0^t Z_k(s) \, dX(s)$, is

$$\sum_{k=1}^m z_{k-1}(x_k - x_{k-1}), \;=\; \sum_{k=1}^m y_k.$$

Then y_1 is a sample value of a normal distribution whose mean value is $z_0 x_0 = 0$ (since $x_0 = 0$); and, for $1 < k \leq n$, y_k is a sample value of $Z_{k-1}(X_k - X_{k-1})$. So if the process Z is independent of the Brownian motion X then, for each k,

$$Y_k \;\simeq\; y_k[\Omega_k \times \mathbf{R}, \; F_{Z_k}\Phi_k]$$

where Φ_k is the standard normal distribution function[8] which regulates the increments $X_k - X_{k-1}$ of the Brownian motion X.

What if the integrand process Z depends on the Brownian integrator X, in the sense that there is a deterministic step function g by which the number of units in the portfolio is calculated from the price process $(X(s))$? In that case $g(s) = g(X(s))$.

Example 22 *Suppose g is a step function such that, as random variables, each $g(X(s))$ is the random variable $g(X(\tau_{k-1}))$ for each s satisfying*

$$\tau_{k-1} = \frac{(k-1)t}{m} \leq s < \frac{kt}{m} = \tau_k, \qquad k = 1, 2, \ldots, m.$$

Then the stochastic integral $\int_0^t g(X(s)) \, dX(s)$ exists.

The sample path version of this stochastic integral is $\int_0^t g(x(s)) \, dx(s)$. Using (5.23) as before, for each of the G-full set of continuous sample paths $x \in \Omega = \mathbf{R}^T$,

$$\int_0^t g(x(s)) \, dx(s) = \sum_{k=1}^m g(x(\tau_{k-1})) \, (x(\tau_k) - x(\tau_{k-1})).$$

This holds for all sample paths outside a G-null set, so $\int_0^t g(X(s)) \, dX(s)$ exists and equals $\sum_{k=1}^m g(X(\tau_{k-1})) \, (X(\tau_k) - X(\tau_{k-1}))$. This stochastic integral can be regarded as a joint random variable $Y = (Y_1, \ldots, Y_m)$ where Y_k has sample space $\Omega_k = \mathbf{R}$ with underlying distribution function $F_{X_k} = \Phi$, a normal distribution.

Another way to represent this stochastic integral is as follows. For each continuous sample path x of the underlying Brownian motion, write

$$w(t) = \sum_{k=1}^m g(x(\tau_{k-1})) \, (x(\tau_k) - x(\tau_{k-1})).$$

Then $w(t)$, is a sample value of a random variable $W(t) \simeq w(t) [\mathbf{R}, F_W]$. If g happens to be a linear function, then $F_W = \Phi$, since a finite linear combination of normally distributed random variables $X(\tau_k)$ is itself normally distributed. In that case the independence properties of (X_k) are inherited by $W(t)$, so $W(t)$, as a function of t, is itself a Brownian motion.

The random variable $W(t)$ is the *elementary* form of the stochastic integral $\mathbf{S}^g(X) = \int_0^t g(X(s)) \, dX(s)$, the latter being a *contingent* form of random variable.[9]

5.5 Some Features of Brownian Motion

Example 19 above suggests there is a need to consider some extreme behaviour of Brownian paths.

---

[8]See [MTRV] page 309 for definition of Φ as normal distribution function in probability theory. (Notation $\mathbf{N}$ is used for normal distribution in [MTRV].)

[9]A contingent random variable W is a function of a random variable Y, so $W = f(Y)$; or, in terms of sample values, $w = f(y)$. These ideas are elaborated in [MTRV], pages 21–35.

Mathematical Brownian motion is very "bad". A stereotypical pictorial representation of a sample element of Brownian motion is a "jagged-path" graph consisting of straight line segments adjoining each other consecutively with sharp corners at the points where each one adjoins the next one.

Mathematically, however, a typical sample path is nowhere differentiable. This is much "worse" than the jagged-path graphical representation. Except for their end points, line segments are smooth, or differentiable. So the class of all such jagged paths are a G-null subset of the sample space $\Omega = \mathbf{R}^{\mathbf{T}}$.

The reason for this "badness" is that, typically, the increments or transitions $x(s') - x(s)$ vary as the square root of the time increment $s' - s$. Calculating a derivative for $x(s)$ at s involves

$$\frac{x(s') - x(s)}{s' - s} = \frac{1}{\sqrt{s' - s}}\left(\frac{x(s') - x(s)}{\sqrt{s' - s}}\right)$$

which diverges as $s' \to s$ for "typical" x of Brownian motion, since the final factor remains finite for such x.

From a different perspective, mathematical Brownian motion is very "good". This is because, typically[10], its sample paths are uniformly continuous. The reason for this "goodness" is that, typically, the increments or transitions $x(s') - x(s)$ vary as the square root of the time increment $s' - s$. So if $s' \to s$ then $\sqrt{s' - s} \to 0$ and hence $x(s') \to x(s)$.

These issues are discussed in detail in chapter 6 of [MTRV], and in many other presentations of the subject

Brownian motion includes sample paths which resemble the Dirichlet function of Example 13, and it includes straight lines, and it includes everything in between these two extremes.

In this book stochastic integrals have been presented as some kind of Stieltjes integral, involving integration of one point function $h_1(s)$ with respect to a different point function $h_2(s)$. (In Section 5.4 the integrator function $h_2(s') - h_2(s)$ was supposed to be a Brownian sample path increment $x(s') - x(s)$; while the integrand function $h_1(s)$ was generally designated as $g(s)$.)

A basic Riemann-Stieltjes integral has Riemann sum approximations of the form

$$\sum h_1(s'')\left(h_2(s') - h_2(s)\right)$$

where s'' satisfies $s \le s'' \le s'$; so s'' could be taken to be s for every term of the Riemann sum. This fits in with the usual form of the stochastic integral (notably in finance) where s'' is taken to be the initial s of the time increment $[s, s'[$.

In the -complete version of the Stieltjes integral (see [MTRV], page 43) the evaluation point s'' of the integrand h_1 is the same as the tag point (or associated point)$\bar{s}$ of the interval $[s, s'[$ in the associated point-interval pair $(\bar{s}, [s, s'[)$.

In [MTRV] (see page 44) the convention has been adopted of requiring that the tag point $\bar{s}$ be one or other of the end-points s or s' of the associated interval.

---

[10]This holds provided a modification of the integral on $\mathbf{R}^{\mathbf{T}}$ is applied. See [MTRV] pages 316–320.

(This makes no difference to the resulting integral if the integrator function is additive on adjoining intervals. See [MTRV], page 56.)

But, unlike "ordinary" Stieltjes integration, which permits **any** point s'' ($s \leq s'' \leq s'$) to be the evaluation point, in -complete integration it is not generally possible to compel the tag point or associated point $\bar{s}$ to be always the same as the left hand end-point s of the associated interval $[s, s'[$, which is the usual way in which stochastic integrals are formulated.

So how can stochastic integrals be accommodated in the -complete system? The answer is to use the Burkill-complete definition (Definition 6). The integrand

$$h(\bar{s}, [s, s'[) := h_1(s)(h_2(s') - h_2(s)), = g(s)(x(s') - x(s)),$$

depends on the elements s and s' of the interval $[s, s'[$. In other words, h is a function of the intervals $[s, s'[$. But h does **not** include a separate function $f(\bar{s})$ of the tag points $\bar{s}$ from which δ-fine partitions are constructed. This is the feature which distinguishes Burkill-complete[11] integration from, for instance, Riemann-complete integration (also known as Kurzweil-Henstock or generalized Riemann integration).

Thus, where stochastic integrals have been occasionally viewed in preceding chapters as Stieltjes-type integrals, once the -complete perspective is finally and fully adopted it is necessary to treat them as Burkill-type.

Inequality (5.23) is the crucial step in establishing integrability of step function g with respect to uniformly continuous Brownian sample path x. Even though tag points $\bar{s}$ do not appear explicitly anywhere, the -complete method of choosing $\delta(\bar{s})$-fine partitions is needed for (5.23), as well as for conformance of partitions with the sub-intervals $[\tau_{k-1}, \tau_k[$ on which step function g takes constant values.

Whether or not these elements appear explicitly in particular instances or applications, the underlying structure of tag points $\bar{s}$ linked to their associated point-pairs $\{s, s'\}$—cells or partitioning intervals I—constitutes the essential technical machinery which operates the -complete system of integration.

Now return to stochastic integration of integrand $g(s)$ with respect to Brownian motion $x(s)$. As stated earlier, mathematical Brownian motion is very "bad" in some senses, and very "good" in others. For instance, in every time interval $[s, s']$, no matter how small, a "typical" Brownian path x has unbounded variation, and therefore has infinite length.

Brownian motion includes sample paths which resemble the Dirichlet function of Example 13; and it includes straight lines; and it includes everything in between, such as the everywhere continuous/nowhere differentiable paths of unbounded variation which, mathematically, are the "typical" Brownian paths.

Example 13 shows that, if $x(s)$ is Dirichlet $d(s)$, the only function $g(s)$ which is integrable (in the -complete sense) with respect[12] to $x(s)$ on $]0, t]$ is a

---

[11]In addition to [13], [14], and [68], Burkill integration is described in [MTRV], page 24.

[12]That is, with respect to the incremental Dirichlet function D.

function $g(s)$ which is constant for $0 \leq s \leq t$. So if $g(s)$ is a non-constant step function, then g is not (Burkill-complete) integrable with respect to x.

A Dirichlet function is non-differentiable at every point; it is discontinuous at every point; and it has unbounded variation in every interval $[s, s']$. But sample paths $x(s)$ as "extreme" as this constitute a G-null set in $\mathbf{R}^{\mathbf{T}}$. A "typical" Brownian sample path x is non-differentiable at every point; has unbounded variation in every interval; but is uniformly continuous in every interval.

Section 5.4 establishes that any step function g is Burkill-complete integrable with respect to a "typical" Brownian path x; so, in effect, the stochastic integral $\int_0^t g(s)\, dX(ds)$ exists for step function $g(s)$ and Brownian $(X(s))$.

Here is a broad categorisation of stochastic integrands g and stochastic integrators x (or X):

- If g is "very good" (such as g constant), then x can be "very bad" (such as x Dirichlet).

- If g is "fairly good" (such as g a step function), then x can be "not very bad" (such as x Brownian).

- If g is "very bad" (such as g having unbounded variation), then x should be "very good" (such as x differentiable everywhere).

Such comments can be tested out in further exploration of practical examples. They can be compared to the well known result in basic Riemann-Stieltjes integration: $h_1(s)$ is Riemann-Stieltjes integrable with respect to $h_2(s)$ if

- h_1 is continuous and h_2 has bounded variation; or if

- h_2 is continuous and h_1 has bounded variation.

For Stieltjes integrability, either one of h_1, h_2 must be "fairly good", while the other one is "not too bad".

5.6   Varieties of Stochastic Integral

This section provides a summary and overview of stochastic integrals. The notation for stochastic integrals in a Riemann setting is set out in section 8.2 (pages 386–390) of [MTRV]. The family of cells or intervals in a domain Ω is denoted by $\mathcal{I}(\Omega)$. So if Ω is, respectively, a real interval such as $]0, t]$, a finite-dimensional domain $\mathbf{R}^M$, or an infinite-dimensional domain $\mathbf{R}^{]0,t]}$, then $\mathcal{I}(\Omega)$ consists of cells which are denoted, respectively, by

$$\imath, \quad I(M), \quad I[N],$$

where M and N are finite sets. For ease of reference, relevant content of section 8.2 of [MTRV] is repeated here.

Suppose $\mathbf{T} =]\tau', \tau]$ (closed on the right) and suppose $F_X, = F_{X_{\mathbf{T}}}$, is a distribution defined on $\mathcal{I}(\mathbf{R}^{\mathbf{T}})$; so $X \simeq x[\mathbf{R}^{\mathbf{T}}, F_X]$ is a joint-basic observable.

Suppose $\imath, = \imath_s, =]s, s'] \in \mathcal{I}(\mathbf{T})$ and $f(x, \{s, s'\}) = x(s') - x(s) = \mathbf{x}(\imath_s)$. We then have a contingent joint observable

$$f(X, \{s, s'\}) \simeq f(x, \{s, s'\})[\mathbf{R}^{\mathbf{T}}, F_X], \quad \text{or} \quad \mathbf{X}(\imath_s) \simeq \mathbf{x}(\imath_s)[\mathbf{R}^{\mathbf{T}}, F_X].$$

Suppose g is a function of the elements $\mathbf{x}(\imath)$ for $\imath \in \mathcal{I}(\mathbf{T})$. For instance, $g(\mathbf{x}(\imath))$ could be the function

$$g(\mathbf{x}(\imath)), = g(\mathbf{x}(\imath_s)), = \mathbf{x}(\imath_s)^2 = (x(s') - x(s))^2.$$

The family of finite subsets of $\mathbf{T}$ is denoted by $\mathcal{N}(\mathbf{T})$. If $N = \{t_1, t_2, \ldots, t_n\} \in \mathcal{N}(\mathbf{T})$ with $t_0 = \tau'$ and $t_n = \tau$, we can[13] write $\imath_j =]t_{j-1}, t_j]$. Thus the cells $\{\imath_j\}$ form a partition of the domain $\mathbf{T}$. For simplicity, let the symbol N denote:

- partition points $\{t_j\}$, or

- partition $\{\imath_j\}$, or

- division $\{(\bar{s},]t_{j-1}, t_j])\}$, with associated points (or tag points) $\bar{s} = t_{j-1}$ or $\bar{s} = t_j$.

The first of these is the primary meaning of N, but for simplicity we allow it, in appropriate contexts, to also represent the corresponding partition or division of $\mathbf{T}$. For the given function g and for some fixed x, the function $\imath \to g(\mathbf{x}(\imath))$ can be thought of as a function

$$\phi: \mathcal{I}(\mathbf{T}) \mapsto \mathbf{R}, \quad \phi(\imath) = g(\mathbf{x}(\imath)).$$

Thus, for any given x, g depends ultimately on cells $\imath \in \mathcal{I}(\mathbf{T})$; and, since it is not necessarily additive on adjoining disjoint cells $\imath$, it is generally a Burkill- rather than Stieltjes-type function. Denote Riemann sums of $\phi(\imath), = g(\mathbf{x}(\imath))$, by

$$(N) \sum g(\mathbf{x}(\imath)), = \sum_{j=1}^n g(\mathbf{x}(\imath_j)), = \sum_{j=1}^n g(x(t_j) - x(t_{j-1})).$$

For this particular x, if $\phi(\imath), = g(\mathbf{x}(\imath))$, is integrable in a Burkill sense, or in a Burkill-complete sense, then this Riemann sum is an approximation to, or estimate of, the integral $\int_{\mathbf{T}} g(\mathbf{x}(\imath)), = \int_{\mathbf{T}} \phi(\imath)$.

This integral (if it exists) depends on the given element $x_{\mathbf{T}}$; and if it exists for variable $x_{\mathbf{T}} \in \mathbf{R}^{\mathbf{T}}$ the integral can be regarded as a contingent datum of a contingent observable. That is, writing

$$f(x) = \int_{\mathbf{T}} \phi(\imath) = \int_{\mathbf{T}} g(\mathbf{x}(\imath)),$$

---

[13]There is some inconsistency in writing $\imath_s =]s, s']$ and $\imath_j =]t_{j-1}, t_j]$. The first notation uses the left-hand end point of the cell to label the cell, while the second notation makes reference to the j of the right-hand end point t_j. However, s generally belongs to the set $\mathbf{T}$, while j is an integer used to label elements t_j of $\mathbf{T}$, so there is a distinction between the usages.

$f(X)$ is a contingent observable

$$f(X) \;\simeq\; f(x)\big[\mathbf{R}^{\mathbf{T}}, F_X\big] \quad \text{or} \quad \int_{\mathbf{T}} g(\mathbf{X}(\imath)) \;\simeq\; \int_{\mathbf{T}} g(\mathbf{x}(\imath))\big[\mathbf{R}^{\mathbf{T}}, F_X\big].$$

Because the integrator function is $\mathbf{x}(\imath)$, which is dependent on the stochastic process X, this is a *stochastic integral*. In [MTRV] stochastic integrals are given dedicated notations $\mathbf{s}(X)$, $\mathbf{S}(X)$, and $\mathcal{S}(X)$ to replace the general purpose symbols f or $\int$.

In addition to dependence on the elements $\mathbf{x}(\imath_s)$, $= x(s') - x(s)$, the function g can be allowed to depend on other parameters such as $x(s)$, s, and $|\imath_s|$, $= s' - s$. In fact, what can be expected in a stochastic integrand g are elements s and x_s, and differences $s' - s$ and $x_{s'} - x_s$. In order to make these elements explicit and visible, write the integrand as

$$g(x_s, s, \mathbf{x}(\imath_s), |\imath_s|), \quad \text{or simply} \quad g(x_s, s, \mathbf{x}(\imath_s), \imath_s).$$

As an integrand, g has the integrand form laid down in chapter 4 of [MTRV], so the Henstock integral of g on domain $\mathbf{T}$ is well defined. In fact, for any given $x_{\mathbf{T}} \in \mathbf{R}^{\mathbf{T}}$, the integration of g on $\mathbf{T}$ is covered by chapter 4 of [MTRV], so g is a Burkill-complete integrand. If g happens to be an additive function of adjoining, disjoint cells $\imath$, it can be considered to be a Stieltjes-complete integrand.

What about integrability of g on $\mathbf{T}$? A Riemann sum for integrand g has the form[14]

$$
\begin{aligned}
(N) \sum g(x_s, s, \mathbf{x}(\imath_s), \imath_s), \;&=\; \sum_{j=1}^{n} g(x_{t_{j-1}}, t_{j-1}, \mathbf{x}(\imath_j), |\imath_j|), \\
&=\; \sum_{j=1}^{n} g(x_{t_{j-1}}, t_{j-1}, x(t_j) - x(t_{j-1}),\, t_j - t_{j-1}); \\
&=:\; \mathcal{R}_{\mathbf{T}}^{g}(x, N), \;=\; \oint_{\mathbf{T}}^{g}(x, N).
\end{aligned}
\tag{5.26}
$$

This formulation of integrand g (noted in superscript) and Riemann sum (composed of terms g) is sufficient to cover those cases that generally arise in applications. Notable features of the usual stochastic integrands g are:

- absence of dependence on associated points $\bar{s}$ of the point-cell pairs $(\bar{s}, \imath_j)$ of divisions of $\mathbf{T}$;

- dependence on $x_{t_{j-1}}$ where the t_{j-1} are the left-hand end points of the partitioning cells $\imath_j$.

In this chapter these features are demonstrated in evaluation of particular stochastic integrals.

---

[14]Notation $\mathcal{R}_{\mathbf{T}}^{g}(x, N)$ is used in [MTRV], with $\mathcal{R}$ intended to suggest "Riemann sum". In Chapter 6 these Riemann sums are given priority over stochastic integrals; and the symbol $\oint_{\mathbf{T}}^{g}(x, N)$ is used instead of $\mathcal{R}_{\mathbf{T}}^{g}(x, N)$. The subscript $\mathbf{T}$ can be omitted if the domain is apparent from the context.

The presence of $x_{t_{j-1}}$ as the *evaluation point* in the integrand g is intended to convey that the integrand depends on the value that x takes at the left-hand vertex of the cell $\imath = \,]t_{j-1}, t_j]$. This is a form of functional dependence, not on the associated point $\bar{s}$, but on the cell $\imath$.

In place of t_{j-1}, some versions of the theory include dependence on

$$x_{t_{j-1}} + \frac{x_{t_j} - x_{t_{j-1}}}{2}.$$

Some stochastic integrands have the form $g(x(\underline{s}, \ldots)$, where $\underline{s} = s + \lambda(s' - s)$ with $0 < \lambda < 1$ fixed. In terms of -complete integration, none of the commonly used variants involves dependence on associated points $\bar{s}$, or on $x(\bar{s})$. But each of these variants can be handled by the method presented in this section.

Using a formal notation (as in page 368 of [MTRV]), the integral is

$$\int_{\imath \in \mathcal{I}(\mathbf{T})}^{\bar{s} \in \imath} g(x_s, s, \mathbf{x}(\imath_s), \imath_s).$$

This notation indicates, on the one hand, dependence of the integrand g on x_s—with x_s depending, in turn, on $x_{\mathbf{T}}$ and on the partitioning points $\{s, s'\}$ (or cell $\imath_s = \,]s, s'])$—and, on the other hand, it indicates that g does *not* depend explicitly on the associated points $\bar{s}$ used to form Riemann sums. The role of $\bar{s}$ is to regulate (by means of gauge function $\delta(\bar{s})$) the terms of the Riemann sums of integrand g.

The $\mathbf{T}$-integrand g includes dependence on x ($x \in \mathbf{R}^{\mathbf{T}}$), which itself will be allowed to vary; so the $\mathbf{T}$-integral of g (the integral of g on $\mathbf{T}$) thereby becomes a function defined on $\mathbf{R}^{\mathbf{T}}$. At that point the $\mathbf{T}$-integral of g, if it exists, enters the theoretical framework of -complete integrands of associated elements in a domain $\mathbf{R}^{\mathbf{T}}$, as described in chapter 4 of [MTRV]. In other words, if, for each $x \in \mathbf{R}^{\mathbf{T}}$, the $\mathbf{T}$-integral of g is multiplied by a distribution or weighting function F_X, then the F_X-mean in $\mathbf{R}^{\mathbf{T}}$ of the $\mathbf{T}$-integral of g,

$$\int_{\mathbf{R}^{\mathbf{T}}} \left(\int_{\mathbf{T}} g \right) F_X(I[N]),$$

if it exists, is defined in [MTRV] (definition 17, page 112).

For some $\mathbf{T}$-integrands g, and for certain classes of elements x, $\int_{\mathbf{T}} g$ may **not** exist. This means that the $\mathbf{T}$-Riemann sums (Riemann sums on $\mathbf{T}$) of g do not converge.

Nonetheless, it is always possible to form Riemann sums $f_{\mathbf{T}}^g(x, N)$ of the $\mathbf{T}$-integrand g. And, for certain $x_{\mathbf{T}}$-dependent $\mathbf{T}$-integrands g, it is possible that the following integral exists:

$$\int_{\mathbf{R}^{\mathbf{T}}} \left(f_{\mathbf{T}}^g(x, N) \right) F_X(I[N]).$$

That is, the F_X-mean in $\mathbf{R}^{\mathbf{T}}$ of such (N-dependent) $\mathbf{T}$-Riemann sums $f_{\mathbf{T}}^g(x, N)$ may exist, in accordance with definition 17 of the -complete integral in $\mathbf{R}^{\mathbf{T}}$ (page 112 of [MTRV]).

In that case, $f_T^g(x, N)$ is F_X-observable as a Riemann-sum-type joint function of $x_T \in \mathbf{R}^\mathbf{T}$ and partitions $N \in \mathcal{N}(\mathbf{T})$. This makes it possible to investigate the F_X-random variability of such Riemann sum functions even when, for particular elements $x_T \in \mathbf{R}^\mathbf{T}$, these Riemann sums do not converge to an integral on $\mathbf{T}$.

In other words, the methods of [MTRV] make it possible to avoid altogether the issue of integrability of g on $\mathbf{T}$. The idea of random variability, in sample space $\mathbf{R}^\mathbf{T}$, of the integral on $\mathbf{T}$ of g, can be abandoned in favour of random variability, in sample space $\mathbf{R}^\mathbf{T}$, of the Riemann sum function $f_T^g(x, N)$ which depends jointly on x_T and N. This seems to be the best way of analyzing the F_X-random variability of integral-like functions "$\int_\mathbf{T} g$" of $x \in \mathbf{R}^\mathbf{T}$ when the functions g are not actually integrable[15] on $\mathbf{T}$.

We then have three distinct forms of the integral on $\mathbf{T}$ of g, as follows.

- If g depends on x but not on $\mathbf{x}(\imath)$, then, as x varies, denote the integral of g by $s_\mathbf{T}^g(x)$. In other words, letting $x \in \mathbf{R}^\mathbf{T}$ vary in $\mathbf{R}^\mathbf{T}$, the integral on $\mathbf{T}$ of the integrand $g(x_s, s, \imath_s)$ can be regarded as a function of $x \in \mathbf{R}^\mathbf{T}$. This interpretation is prompted by introducing the notation

$$s_\mathbf{T}^g(x) \;=\; \int_\mathbf{T} g(x_s, s, \imath_s); \qquad (5.27)$$

  which, writing $\phi(\imath) = g(x_s, s, \imath_s)$, is simply the integral $\int_\mathbf{T} \phi(\imath)$ of the cell function ϕ on $\mathbf{T}$.

- If g depends on $\mathbf{x}(\imath)$, then, as x varies, denote the integral of g by $\mathbf{S}^g(x)$,

$$\mathbf{S}_\mathbf{T}^g(x) \;=\; \int_\mathbf{T} g(x_s, s, \mathbf{x}(\imath_s), \imath_s). \qquad (5.28)$$

  We call this the **strong stochastic integral**. The traditional theory of stochastic integration does not include this concept.

- If g depends on $\mathbf{x}(\imath)$, and if the integral of g on $\mathbf{T}$ exists only in some weak sense as adverted to above, then, for variable $x \in \mathbf{R}^\mathbf{T}$, denote the weak integral of g by $\mathcal{S}_\mathbf{T}^g(x)$,

$$\mathcal{S}_\mathbf{T}^g(x) \;=\; \text{``}\int_\mathbf{T}\text{''} g(x_s, s, \mathbf{x}(\imath_s), \imath_s), \qquad (5.29)$$

  whose exact meaning is established in chapter 8 of [MTRV]. This corresponds to the traditional idea of the stochastic integral.

In a more familiar integration context, symbols $\int_0^t$, $\int_{]0,t]}$, or $\int_\mathbf{T}$ may be written as $\int$ if the domain $]0, t]$ is obvious from the context. Likewise $s_\mathbf{T}$, $\mathbf{S}_\mathbf{T}$, and $\mathcal{S}_\mathbf{T}$ may be written

$$\mathbf{s}, \quad \mathbf{S}, \quad \mathcal{S}$$

---

[15]Because of this, the more suggestive notation $f_T^g(x, N)$ is used instead of $\mathcal{R}_T^g(x, N)$ in Chapter 6 below; and the terminology "*stochastic sum*" is used without any suggestion of convergence to any form of stochastic integral.

with domain understood. With $\mathbf{T} =]0, t]$ and $\imath_s =]s, s'] \in \mathcal{I}(\mathbf{T})$, an example of a stochastic integrand g is

$$g(x_s, s, \mathbf{x}(\imath_s), \imath_s) \;=\; s^3 |\imath_s| + x_s^2 |\imath_s| + \mathbf{x}(\imath) + x_s \mathbf{x}(\imath_s). \qquad (5.30)$$

The final two terms of the right-hand expression are the ones that make the function g a stochastic integrand, since they contain the increment factor $\mathbf{x}(\imath)$.

Consider each of the four right hand terms of (5.30) separately. The first term contains no random element; and it is integrable on $\mathbf{T} =]0, t]$ with integral equal to $\frac{1}{4}t^4$. The second term has random component x_s; and it is integrable on $\mathbf{T}$ if, for instance, X is a Brownian motion. (If X is Brownian then $x(s)$ can be assumed to be a continuous function of s). The third and fourth terms include a random increment $\mathbf{x}(\imath)$, and therefore, if they are integrable in some sense, their integrals will be called stochastic integrals. We will see that the third term is integrable (in the Stieltjes-complete sense) on $\mathbf{T}$ for all x, with integral equal to $x(t)$. In theorem 235 of [MTRV] (page 403) it is shown that, if the process X happens to be standard Brownian motion, the fourth and final term is integrable in a weak sense.

With domain $\mathbf{T} =]0, t]$, denote the Riemann sums of an integrand g on $\mathbf{T}$ by

$$\oint_{\mathbf{T}}^{g}(x, N), \;=\; (N) \sum g(x_s, s, \mathbf{x}(\imath), \imath),$$

where $\oint_{\mathbf{T}}^{g}(x, N)$ can be written $\oint^{g}(x_{\mathbf{T}}, N)$. If, for each $x \in \mathbf{R}^{\mathbf{T}}$, the function g is integrable on $\mathbf{T}$, denote the integral by

$$\mathbf{S}^{g}(x_{\mathbf{T}}), \;=\; \int_{\mathbf{T}} g(x_s, s, \mathbf{x}(\imath), \imath).$$

Allowing x to vary in $\mathbf{R}^{\mathbf{T}}$, then (with $\mathbf{R}$ and $\mathbf{C}$ denoting real and complex numbers, respectively)

$$\oint_{\mathbf{T}}^{g}: \quad \mathbf{R}^{\mathbf{T}} \times \mathcal{N}(\mathbf{T}) \;\mapsto\; \mathbf{R} \text{ or } \mathbf{C}, \qquad \text{and} \qquad \mathbf{S}_{\mathbf{T}}^{g}: \quad \mathbf{R}^{\mathbf{T}} \;\mapsto\; \mathbf{R} \text{ or } \mathbf{C},$$

are functions defined on $\mathbf{R}^{\mathbf{T}}$. Therefore

$$\mathbf{S}_{\mathbf{T}}^{g}(X) \;\simeq\; \mathbf{S}_{\mathbf{T}}^{g}(x)\big[\mathbf{R}^{\mathbf{T}}, F_X\big]$$

is a joint-contingent observable with respect to the joint-basic observable

$$X \simeq x\big[\mathbf{R}^{\mathbf{T}}, F_X\big].$$

Similarly, for any given N, $\oint_{\mathbf{T}}^{g}(X, N) \simeq \oint_{\mathbf{T}}^{g}(x, N)\big[\mathbf{R}^{\mathbf{T}}, F_X\big]$ is a contingent observable.

The symbols $\int$, s, S, and $\mathcal{S}$ are each versions of the letter S, indicating summation. The first three symbols denote -complete integrals, but "s" and "S" have been introduced because "$\int$" does not usually carry the connotation of a function "$f(x)$" of a variable x. The symbols "s(x)", "S(x)", and "$\mathcal{S}(x)$"

are a reminder that any such expression is a functional of variable sample paths $x = x_{\mathbf{T}} \in \mathbf{R}^{\mathbf{T}}$, corresponding to contingent observables $s(X)$, $\mathbf{S}(X)$, and $\mathcal{S}(X)$. Likewise, $f(x, N)$ and $f(X, N)$.

For an integrand g that does not depend on $\mathbf{x}(\imath)$—in other words, does not contain a factor $dx(s)$—the expression $\mathbf{s}_{\mathbf{T}}^g(x)$ denotes $\int_{\mathbf{T}} g$.

The symbol "$\mathcal{R}$" used in [MTRV] further indicates "Riemann sum", which, in addition to dependence on variable x, is also dependent on variable N. Because these Riemann sums turn out to be more useful than any kind of stochastic integral (see Chapter 6 below), the term *stochastic sum* is applied to them, with the new and more suggestive notation $f_{\mathbf{T}}^g(x_{\mathbf{T}}, N)$—or $f^g(x, N)$ when domain $\mathbf{T}$ is clear from the context.

For ease of reference, here are some stochastic integrals[16] which appear in the literature of Brownian processes, and which are discussed in chapter 8 of [MTRV]—see (8.16), page 419. (In [MTRV], instead of $\mathbf{T}$, the symbol $\mathcal{T}$ is used to denote $]0, t]$.)

$$
\begin{aligned}
\int_{\mathbf{T}} dX_s &= X_t, & \int_{\mathbf{T}} (dX_s)^2 &= t, \\
\int_{\mathbf{T}} (dX_s)^3 &= 0, & \int_{\mathbf{T}} X_s dX_s &= \tfrac{1}{2}X_t^2 - \tfrac{1}{2}t, \\
\int_{\mathbf{T}} s\, dX_s &= tX_t - \int_{\mathbf{T}} X_s ds, & \int_{\mathbf{T}} X_s (dX_s)^2 &= \int_{\mathbf{T}} X_s ds, \\
\int_{\mathbf{T}} X_s^2 dX_s &= \tfrac{1}{3}X_t^3 - \int_{\mathbf{T}} X_s ds, & \int_{\mathbf{T}} d(X_s^3) &= X_t^3.
\end{aligned}
\tag{5.31}
$$

In [MTRV], these stochastic integrals $\int_{\mathbf{T}} \cdots$ are separated into strong stochastic integrals (denoted $\mathbf{S}_{\mathbf{T}}\cdots$), and weak stochastic integrals (denoted $\mathcal{S}_{\mathbf{T}}\cdots$).

The purpose of the next chapter is to replace stochastic integrals (whether $\int_{\mathbf{T}} \cdots$, $\mathbf{S}_{\mathbf{T}}\cdots$, or $\mathcal{S}_{\mathbf{T}}\cdots$) by stochastic sums $f_{\mathbf{T}} \cdots$. Again, for ease of reference, with sample times $N = \{s_1, \ldots, s_{n-1}, s_n\}$ and corresponding sample values $x_j, = x(s_j)$, $j = 1, \ldots, n$, the sample summands for each of the expressions in (5.31) is

$$
\begin{aligned}
g_1(\mathbf{x}(\imath_j)) &= x_j - x_{j-1}, & g_2(\mathbf{x}(\imath_j)) &= (x_j - x_{j-1})^2, \\
g_3(\mathbf{x}(\imath_j)) &= (x_j - x_{j-1})^3, & g_4(\mathbf{x}(\imath_j)) &= x_{j-1}(x_j - x_{j-1}), \\
g_5(\mathbf{x}(\imath_j)) &= s_{j-1}(x_j - x_{j-1}), & g_6(\mathbf{x}(\imath_j)) &= x_{j-1}(x_j - x_{j-1})^2, \\
g_7(\mathbf{x}(\imath_j)) &= x_{j-1}^2(x_j - x_{j-1}), & g_8(\mathbf{x}(\imath_j)) &= x_j^3 - x_{j-1}^3.
\end{aligned}
\tag{5.32}
$$

Thus, for stochastic integral $\int_{\mathbf{T}} X_s dX_s = \tfrac{1}{2}X_t^2 - \tfrac{1}{2}t$ (for instance), the next chapter will present stochastic sum evaluation

$$
f_{\mathbf{T}}^{g_4}(X_{\mathbf{T}}, \mathcal{N}) = \sum_{j=1}^{n} X_{j-1}(X_j - X_{j-1})^2 \overset{G}{=} \frac{1}{2}X_t^2 - \frac{1}{2}t;
$$

or, in sample times/sample value form,

$$
f_{\mathbf{T}}^{g_4}(x_{\mathbf{T}}, N) = \sum_{j=1}^{n} x_{j-1}(x_j - x_{j-1})^2 \overset{G}{=} \frac{1}{2}x_t^2 - \frac{1}{2}t;
$$

---

[16]The value of $\int_{\mathbf{T}} s\, dX_s$ is incorrectly given as $\int_{\mathbf{T}} X_s ds$ in [MTRV].

where notation $\overset{G}{=}$ implies

$$\int_{\mathbf{R}^{\mathbf{T}}} \left| \sum_{j=1}^{n} x_{j-1}\left(x_j - x_{j-1}\right)^2 \ - \ \left(\frac{1}{2}x_n^2 - \frac{1}{2}s_n\right) \right| G(I[N]) \ = \ 0, \quad \text{with} \ \ s_n = t, \ x_n = x_t.$$

Strong stochastic integration requires non-absolute convergence, and is therefore unavailable in the standard (Lebesgue/Itô) version of the theory.

Stochastic summation can be used in place of strong stochastic integration— but in practice there is no advantage or benefit in that; it is just a formality. The significant benefit of stochastic sums is to avoid having to use weak stochastic integrals, which are the only kind available in the standard or Lebesgue-Itô theory.

Each equality in the list (5.31) is formulated as

$$\int_{\mathbf{T}} g(X_{\mathbf{T}}, \cdots) \ = \ f(X_{\mathbf{T}}). \tag{5.33}$$

No matter how we choose to interpret the term "random variable/process", setting one side equal to the other in (5.33) is not simply a statement that one number equals another, or one function is the same as another. What is implied in (5.33) is

$$\int_{\mathbf{T}} g(X_{\mathbf{T}}, \cdots) \ \overset{F}{=} \ f(X_{\mathbf{T}}) \tag{5.34}$$

where F is a distribution function for the joint-basic variables or processes involved.

The meaning and implications of this are spelled out in theorem 83 (page 200 of [MTRV]), which confirms that, for reasonably well-behaved observables, each side of (5.33) has the same range of sample values and the same distribution function when interpreted as elementary random variable.

In other words, "=" or "$\overset{F}{=}$" in (5.33) or (5.34) means:

- Each side delivers the same numerical sample values, and

- Each side has the same distribution function when expressed in elementary form.

This applies also to evaluation of stochastic sums, and should be borne in mind in Chapter 6 below; in which step functions and cylinder functions are replaced by a sampling sum technique.

Chapter 6

Stochastic Sums

In Chapter 1 the classical or standard concept of stochastic integral, including Itô's integral, is outlined. The mathematical need or motive for some concept of stochastic integration has been illustrated in preceding chapters by means of various examples. It is illustrated in particular by the manifestation, in the form of a stochastic integral, of the value at any time t of a shareholding (or portfolio) of a quantity $g(s)$ of shares whose value at time s $(0 \leq s \leq t)$ is $x(s)$:

$$\int_0^t g(s)\, dx(s),$$

where $g(s)$ is a deterministic or random function of time s $(0 \leq s \leq t)$ and $x(s)$ $(0 < s \leq t)$ is a sample path of a process $X = (X_s)_{0<s\leq t}$.

For example, in financial terms it is natural to suppose that the stockholding function $g(s)$ may itself depend on the recent sample values taken by the stock price X_s, so $g(s) = g(X(s))$, and the stochastic integral is $\int_0^t g(X(s))\, dX(s)$. The literature of the subject includes, for instance,

$$\int_0^t X(s)\, dX(s), \qquad \int_0^t X(s)^2\, dX(s).$$

In earlier illustrations of these concepts, such as Example 12, there were elementary examples of the form $\int_0^t z(s)\, dx(s)$. In this case the integrand $g(s)$ is random but does not depend on the share price process $X(s)$; and $z(s)$ and $x(s)$ are, respectively, sample values at time s of processes

$$Z = (Z_s)_{0<s\leq t}, \qquad X = (X_s)_{0<s\leq t};$$

and $\int_0^t z(s)\, dx(s)$ is a sample value of the stochastic integral

$$\int_0^t Z(s)\, dX(s).$$

In section 8.2 of [MTRV], pages 386–390, various kinds of "stochastic integral" are put forward:

$$\mathcal{R}_{\mathrm{T}}^g(X, N), \qquad \mathbf{s}_{\mathrm{T}}^g(X), \qquad \mathbf{S}_{\mathrm{T}}^g(X), \qquad \mathcal{S}_{\mathrm{T}}^g(X).$$

Gauge Integral Structures for Stochastic Calculus and Quantum Electrodynamics, First Edition. Patrick Muldowney.

The latter expression $\mathcal{S}_{\mathbf{T}}^g(X)$ corresponds to the classical $\int_0^t g(s)\,dX(s)$ (which is the Itô integral if (X_s) is a Brownian motion). The other three expressions are innovations. They are introduced in MTRV, which includes discussion of $\mathbf{s}_{\mathbf{T}}^g(X)$ and $\mathbf{S}_{\mathbf{T}}^g(X)$, along with a brief outline of the first one, $\mathcal{R}_{\mathbf{T}}^g(X,N)$.

In mathematical discussion of integration, including stochastic integration, it is customary to use a notation with three components

$$\{\text{ integral symbol }\}\{\text{ point integrand }\}\{\text{ differential }\}\quad\text{or}\quad\left(\int\right)(f(y))\,(dy).$$

In the -complete integration of [MTRV], this is expressed as $\int f(y)\,k(I)$, where $k(I) = |I|$ corresponds to the traditional differential symbol dy.

Traditionally, a stochastic integral may take the form $\int f(X)\,dX$ where X is a stochastic process. But [MTRV] breaks with this notation. Instead of $\int$ we have symbols $\mathbf{s}$, $\mathbf{S}$, and $\mathcal{S}$; each used in particular contexts (see section 8.2, [MTRV] pages 386–390). Integrand elements such as $f(X)\,dX$ are denoted by some expression g which is attached to the relevant integration symbol as a superscript, giving

$$\mathbf{s}^g,\qquad \mathbf{S}^g,\qquad \mathcal{S}^g.$$

[MTRV] also introduces another such functional, $\mathcal{R}^g$, which is a Riemann sum rather than an integral, and which is now to be written $\oint^g$.

As well as integration, these procedures have a functional aspect, in the sense that the final result depends on the choice of sample path $x_{\mathbf{T}}$. The innovations in notation are intended to emphasize the functional rather than the integration aspect. So with the integrand g safely relegated to superscript position, the functional dependence on $x_{\mathbf{T}}$ is denoted by

$$\mathbf{s}^g(x_{\mathbf{T}}),\qquad \mathbf{S}^g(x_{\mathbf{T}}),\qquad \mathcal{S}^g(x_{\mathbf{T}});$$

and, whenever needed for clarity, the domain $\mathbf{T}$ is placed as subscript,

$$\mathbf{s}_{\mathbf{T}}^g(x_{\mathbf{T}}),\qquad \mathbf{S}_{\mathbf{T}}^g(x_{\mathbf{T}}),\qquad \mathcal{S}_{\mathbf{T}}^g(x_{\mathbf{T}}),$$

just as it is in $\int_T$. The same general idea is in the stochastic sum notation

$$\mathcal{R}_{\mathbf{T}}^g(x_{\mathbf{T}}),\qquad \oint_{\mathbf{T}}^g(x_{\mathbf{T}}).$$

These two symbols are equivalent, but the latter symbol is given precedence because of the suggestion it contains of "sum replacing integral".

The aim of this chapter is to amplify and extend the ideas behind $\mathcal{R}_{\mathbf{T}}^g(X)$— or $\mathcal{R}_{\mathbf{T}}^g(X,\mathcal{N})$, or $\oint_{\mathbf{T}}^g(X_{\mathbf{T}},\mathcal{N})$—as a simpler and more comprehensive way of dealing with stochastic integration; so that the single formulation $\oint_{\mathbf{T}}^g(X_{\mathbf{T}},\mathcal{N})$ replaces each of $\mathbf{s}_{\mathbf{T}}^g(X)$, $\mathbf{S}_{\mathbf{T}}^g(X)$, $\mathcal{S}_{\mathbf{T}}^g(X)$, and $\int_{\mathbf{T}} f(X)\,dX$.

6.1   Review of Random Variability

To set the scene for this, here is an overview of the -complete approach to random variability, with emphasis on those aspects which reinforce and validate the replacement of stochastic integrals by stochastic sums.

Random variability is associated with observation or measurement of some quantity whose precise value is not known, but for which estimated values x can be given. Suppose that, even though the precise or true value is not known, there is some method of assessing the accuracy of estimated values x. Suppose, in fact, that the degree or level of accuracy of the estimate x can itself be estimated by means of a distribution function F or F_X defined on intervals I of possible values of x in domain Ω (called the *sample space*). Then the term *observable* is applied to the notion of measurement or estimate X, with possible values x in sample space Ω, equipped with accuracy function (or *likelihood distribution function*) F_X,

$$X \simeq x[\Omega, F_X].$$

The measured value (or *occurrence*, or *datum*) x can be a number (usually real), so $\Omega = \mathbf{R}$. Or x can consist of jointly measured values $x = (x_s)$ where $s \in \mathbf{T}$; so if $\mathbf{T}$ is a finite set of cardinality n, then $\Omega = \mathbf{R}^n$ and $x = (x_1, \ldots, x_n)$. In that case, $X \simeq x[\mathbf{R}^n, F_X]$ is a *joint-basic* observable, with *distribution function* F_X defined on cells

$$I = I_1 \times \cdots \times I_n \subset \mathbf{R}^n, \qquad x_j \in I_j, \quad j = 1, \ldots n.$$

If the measurement is a real value $f(x)$ formed by means of a deterministic function f of the basic observable x, then $f(X)$ is a *contingent observable*

$$f(X) \simeq f(x)[\Omega, F_X].$$

An *event* is a set of occurrences. An observable $f(X)$ is a *random variable* (or is an F_X-random variable) if its expected value $\mathrm{E}[f(X)]$ exists:

$$\mathrm{E}[f(X)] = \int_\Omega f(x) F_X(I). \tag{6.1}$$

In this formulation a joint-basic observable $X = (X_s)$ ($s \in \mathbf{T}$) cannot be regarded as a random variable if $\mathbf{T}$ has cardinality greater than one, since "$\int_{\mathbf{R}^\mathbf{T}} x F_X(I)$" has no meaning as a -complete integral in that case. The -complete theory of random variability hinges on (6.1). Additionally, $f(X)$ is an *absolute random variable*, if $|f(X)|$ is a random variable with

$$\mathrm{E}[|f(X)|] = \int_\Omega |f(x)| F_X(I), \tag{6.2}$$

and then all the extra analytical powers of *measurability* are available.[1] It is often useful to express contingent observables $f(X)$ in *elementary* form Y:

$$f(X) \simeq f(x)[\Omega, F_X], \qquad Y \simeq y[\mathbf{R}, F_Y], \quad \text{with} \quad y = f(x) \quad \text{so} \quad Y = f(X).$$

Analysis of random variability sometimes involves deducing F_Y from F_X. And, generally speaking, problems of random variability are addressed in the first instance by using distribution functions such as F_Y and F_X from which the

---

[1] See [MTRV], appendix A, pages 491–500.

probability of an event may be deduced. The starting point for such analysis is existence of expected values in the form of integrals on the sample space.

The calculations (6.1) and (6.2), $\int_\Omega f(x)F_X(I)$ and $\int_\Omega |f(x)|F_X(I)$, are central. On the face of it, these calculations involve infinitely many values of x, such as $x \in [0,1]$ or $x \in \mathbf{R}$—a continuum.

However, x itself is an unpredictable occurrence or *outcome of a single measurement*. So how can the infinite continuum of x-values be dealt with, or understood, in calculations (6.1) and (6.2)?

In practice, multiple values of x can sometimes be obtained by means of finite samples or by means of repeated measurements. Calculations (6.1) and (6.2) are a mathematical rather than a practical problem. A mathematical solution is to use Riemann sums of the form

$$(\mathcal{D}) \sum f(x)F_X(I) \quad \text{or} \quad \sum_{k=1}^m f(x^{(k)})F_X(I^{(k)}), \qquad (6.3)$$

where the set of data values or occurrences $x^{(1)},\ldots,x^{(m)}$ can be regarded as a **finite event** (or finite set of possibilities) weighted by likelihood values $F_X(I^{(1)}), \ldots, F_X(I^{(m)})$. Or they can be regarded as a **finite sample** of measurement outcomes, if the measurement is repeated a finite number of times.

Thus the Riemann sum construction converts the infinite or continuum-type calculation $\int_\Omega \cdots$ into a discrete, finite addition $(\mathcal{D}) \sum \cdots$. We do not, after all, have to wrestle with infinitely many occurrences or sample values x.

This Riemann summation perspective can be extended to help with problems of stochastic integration and stochastic summation.

6.2 Riemann Sums for Stochastic Integrals

This section seeks to extend the Riemann sum stratagem described above in order to simplify and unify various conceptions of strong and weak stochastic integration; and to replace stochastic integrals by stochastic sums.

A stochastic process is a family of random variables $X = X_\mathbf{T} = (X_s)$, $s \in \mathbf{T}$, where $\mathbf{T}$ is an infinite set such as $]0,t]$. Stochastic integration is a device which constructs a random variable Z from a process $X_\mathbf{T}$; such as

$$Z = \int_0^t X_s\, dX_s.$$

Example 23 *Constructions of this kind have been given a variety of interpretations and meanings in chapter 8 (pages 383–446) of [MTRV], such as strong and weak stochastic integrals:*

$$\mathbf{S}_T^g(X_\mathbf{T}), \qquad\qquad \mathcal{S}_\mathbf{T}^g(X_\mathbf{T}),$$

where (in this case) the integrand g is $X_s(X_{s'} - X_s), = X_s\, dX_s$, $(0 \le s < s' \le t)$. In fact, provided $X_\mathbf{T}$ is standard Brownian motion, $\int_0^t X_s\, dX_s$ is $\mathcal{S}_\mathbf{T}^g(X_\mathbf{T})$, a *weak stochastic integral which evaluates as $\frac{1}{2}X_t^2 - \frac{1}{2}t$. (See example 63, pages 405–406 of [MTRV].)*

Expressed in terms of sample values x_s $(0 < s < t)$, or in terms of sample path x_T, this result states that, with $x_t = x(t)$ given,

$$S_T^g(x_T) \;=\; \int_0^t x_s \, dx_s \;=\; \int_0^t x(s) \, dx(s) \;=\; \frac{1}{2}x_t^2 - \frac{1}{2}t, \qquad (6.4)$$

in some weak sense; where $\int_0^t x(s) \, dx(s)$ is a Stieltjes-type integral of the point-function $x(s)$ with respect to (increments of) the point-function $x(s)$.

  (a) Equation (6.4) is a "weak" equation, which can only be valid in some sense of "average value" of one or other side, or both.

  (b) Furthermore, the left hand side of (6.4) references infinitely many values x_s, corresponding to the infinitely many time instants $0 < s < t$. As in (6.3), this suggests infinitely many sample measurements x_s. This is counter-intuitive as a method of calculation. It is not practically possible to sample every instant s of time.

In the discussion below, both of these issues are addressed by using (as in (6.4)) a Riemann sum method for the averaging required by (a), so each Riemann sum involves a finite sample consisting of only a finite number of times s.

To sum up, random variability consists, essentially, of measurements whose precise values cannot be predicted but whose accuracy can be assessed. The analysis of random variability commences with calculation of expected values for these measurements.

The random variability scenario outlined in Example 6.4 involves an infinite number of potential joint occurrences (of stochastic processes) x_T, in which each of the individual processes x_T involves a basic occurrence x_s at infinitely many times s for $0 < s \leq t$.

What is proposed here is to obtain expected values by means of the Riemann sum method of [MTRV]. This method uses (at each stage) only a finite sample of processes x_T, each of which uses only a finite sample of times $s_1, \ldots, s_{n-1}, s_n$. Remember, the observable we are dealing with is

$$\int_0^t g(X_s, X_{s'}), \qquad \text{or} \qquad S^g(X_T), \quad \mathcal{S}^g(X_T)$$

where exemplar function g is $X_s(X_s' - X_s)$, so these stochastic integral calculations can be represented in sample format as Stieltjes-type integral

$$\int_T x(s) \, dx(s).$$

Each of these formulations involves infinitely many elements s. To evade this problematic infinity of s, we choose a partition

$$N \;=\; \{s_1, \ldots, s_{n-1}, s_n\}, \qquad s_0 \;=\; 0, \qquad s_n \;=\; t,$$

and replace $\int_{\mathbf{T}} x(s)\,dx(s)$ by a Riemann sum expression

$$f(x_{\mathbf{T}}, N) \;=\; (N)\sum g(x_j, x_{j-1}) \;:=\; \sum_{j=1}^{n} x(s_{j-1})\,(x(s_j) - x(s_{j-1}))\,.$$

In a more suggestive notation[2], instead of $f(x_{\mathbf{T}}, N)$ denote the Riemann sum expression $(N)\sum(\cdots)$ by $\oint_{\mathbf{T}}^{g}(x_{\mathbf{T}}, N)$, so, in the case $g = X_s\,dX_s$,

$$\oint_{\mathbf{T}}^{g}(x_{\mathbf{T}}, N) \;=\; (N)\sum g(x_j), \quad\text{which equals}\quad \sum_{j=1}^{n} x(s_{j-1})\,(x(s_j) - x(s_{j-1}))\,.$$

The function g is the *stochastic integrand*; or, in the present[3] interpretation, the *stochastic summand*; and the functional $\oint_{\mathbf{T}}^{g}(x_{\mathbf{T}}, N)$ is the *stochastic sum*.

The expression $\sum_{j=1}^{n} x(s_{j-1})\,(x(s_j) - x(s_{j-1}))$ looks like a Riemann sum approximation to a Stieltjes-type integral $\int_{\mathbf{T}} x(s)\,dx(s)$. But we do **not** claim that

$$``\oint^{g}(x_{\mathbf{T}}, N) \quad\rightarrow\quad \int_{\mathbf{T}} x(s)\,dx(s)"$$

for suitable choices of partitioning points N. Instead, "$\oint_{\mathbf{T}}^{g}(X_{\mathbf{T}}, \mathcal{N})$" is **substituted** for "$\int_{\mathbf{T}} x(s)\,dx(s)$".

With a suitable distribution function $F(I[N])$, an expected value for the stochastic sum is defined by

$$\mathrm{E}\left[\oint_{\mathbf{T}}^{g}(X_{\mathbf{T}}, \mathcal{N})\right] \;=\; \int_{\mathbf{R}^{\mathbf{T}}} \left(\oint_{\mathbf{T}}^{g}(x_{\mathbf{T}}, N)\right) F(I[N])\,.$$

For example, if $X_{\mathbf{T}}$ is a standard Brownian process then F is the Gaussian distribution function $G(I[N])$. In that case the function $\oint_{\mathbf{T}}^{g}(X_{\mathbf{T}}, \mathcal{N})$ is candidate to be the substitute for the weak stochastic integral

$$\int_{0}^{t} X_s\,dX_s, \;=\; S_{\mathbf{T}}^{g}(X_{\mathbf{T}})\,,$$

whose value is given in example 63 (pages 405–406 of [MTRV]) as $\frac{1}{2}X_t^2 - \frac{1}{2}t$. So, if $\oint_{\mathbf{T}}^{g}(X_{\mathbf{T}}, \mathcal{N})$ is to succeed as replacement candidate for $\int_0^t X_s\,dX_s$ then, for instance, the following should hold:

$$\mathrm{E}\left[\oint_{\mathbf{T}}^{g}(x_{\mathbf{T}}, N)\right] \;=\; \mathrm{E}\left[X(t)^2 - \frac{1}{2}t\right], \quad\text{or}$$

$$\int_{\mathbf{R}^{\mathbf{T}}} \left(\oint_{\mathbf{T}}^{g}(x_{\mathbf{T}}, N)\right) G(I[N]) \;=\; \int_{\mathbf{R}}\left(\frac{1}{2}y^2 - \frac{1}{2}t\right)\frac{e^{-\frac{1}{2}y^2}}{\sqrt{2\pi}}\,dy;$$

the latter being the expectation calculation for contingent random variable $f(Y)$ with Y normally distributed. Details are given in Sections 6.8 and 6.12 below.

---

[2]Notation $\mathcal{R}$ was introduced in [MTRV] (as in **R**iemann sum); but $\oint$ provides a more visual prompt to the meaning.

[3]The summand $x_s\,dx_s$, or $x_s(x_{s'} - x_s)$, is denoted as g_4 in [MTRV] page 391.

6.3 Stochastic Sum as Observable

A new type of observable is required:

$$f(X_{\mathbf{T}}, \mathcal{N}) \;\simeq\; f(x_{\mathbf{T}}, N)\left[\mathbf{R}^{\mathbf{T}}, F_{X_{\mathbf{T}}}\right]$$

where $F_{X_{\mathbf{T}}}$ is a distribution function defined for $I[N] \in \mathcal{I}(\mathbf{R}^{\mathbf{T}})$ (the set of cells in $\mathbf{R}^{\mathbf{T}}$),

$$F_{X_{\mathbf{T}}} : \quad \mathcal{I} \;\mapsto\; [0,1], \qquad 0 \le F_{X_{\mathbf{T}}}(I[N]) \le 1.$$

In addition to dependence on joint occurrences $(x_s) = x_{\mathbf{T}}$, an observable f is permitted to depend explicitly on partitions $N = \{s_1, \ldots, s_{n-1}, s_n\}$ of $\mathbf{T}$. Likewise, a distribution function $F_{X_{\mathbf{T}}}$ depends on cells $I = I[N]$, and may depend explicitly on the partitions $N = \{s_1, \ldots, s_{n-1}, s_n\}$ of $\mathbf{T}$ which (with $s_n = t$) are the "cylinder labels", or dimension labels, of the cylindrical intervals $I[N]$ in $\mathbf{R}^{\mathbf{T}}$. For example, with $I_t =]u_t, v_t]$ $(t \in N)$, the incremental Gaussian distribution function G of (5.8) (see page 115 above) depends explicitly on the parameters u_t, v_t, and t, for $t \in N$.

A left hand limit (or vertex) u_t for a partitioning component cell $I_t =]u_t, v_t]$ $(t \in N)$ is a right hand limit or vertex of an adjoining cell I'_t. Thus, choice of a partition $\mathcal{P}$ of domain $\mathbf{R}^{\mathbf{T}}$ reduces to choice of finite samples N of times, along with choices $\{u_t\}$ of finite samples of vertices for $t \in N$.

As outlined in Section 6.2, the fundamental step is to define the expectation $\mathrm{E}[f(X_{\mathbf{T}}, \mathcal{N})]$; that is, to define the integral of $f(x_{\mathbf{T}}, N)$ with respect to distribution function $F(I[N])$. In particular, when $f(x_{\mathbf{T}}, N) = \mathcal{R}_{\mathbf{T}}^g(x_{\mathbf{T}}, N)$, (or $\mathcal{f}_{\mathbf{T}}^g(x_{\mathbf{T}}, N)$),

$$\mathrm{E}\left[\mathcal{R}_{\mathbf{T}}^g(X_T, \mathcal{N})\right], \quad = \quad \int_{\mathbf{R}^{\mathbf{T}}} \left(\mathcal{R}_{\mathbf{T}}^g(x_{\mathbf{T}}, N)\right) F(I[N]),$$

$$\text{or } \mathrm{E}\left[\mathcal{f}_{\mathbf{T}}^g(X_T, \mathcal{N})\right], \quad = \quad \int_{\mathbf{R}^{\mathbf{T}}} \left(\mathcal{f}_{\mathbf{T}}^g(x_{\mathbf{T}}, N)\right) F(I[N]);$$

so $\mathcal{f}_{\mathbf{T}}^g(X_{\mathbf{T}}, \mathcal{N})$ is a random variable. Chapter 4 (pages 111–182) of [MTRV] deals with the integration in $\mathbf{R}^S$ of integrands of the form $h(x_S, N, I[N])$, where S is any infinite set (such as intervals of time $\mathbf{T}$ or T), including integrands $f(x_S, N)F(I[N])$. Briefly, $f(x_S, N)F(I[N])$ is integrable on $\mathbf{R}^S$, with integral

$$\int_{\mathbf{R}^S} f(x_S, N)F(I[N]) \;=\; \alpha,$$

if, given $\varepsilon > 0$, there exists a gauge $\gamma = (L, \delta_N)$ such that, for every γ-fine division $\mathcal{D}$ of $\mathbf{R}^S$, the corresponding Riemann sums satisfy

$$\left| \alpha - (\mathcal{D}) \sum f(x_S, N)F(I[N]) \right| \;<\; \varepsilon.$$

Chapter 4 of [MTRV] provides a *theory of variation* for functions $h(x, N, I[N])$ which is applicable to functions $F(I[N])$ and $f(x, N)F(I[N])$. So, for instance, $F(I[N])$ (defined on cells $I[N]$) can be extended to an "outer measure" on arbitrary subsets A of $\mathbf{R}^S$. Chapter 4 also provides limit theorems for integrals (such as integrability of limits of integrable functions), and Fubini's theorem for integrands defined on product domains of the form $\mathbf{R}^{S'} \times \mathbf{R}^{S''}$.

Measurability of sets and functions is covered in chapter 5 of [MTRV], pages 186–197, and in appendix A, pages 491–500. Measurability permits approximation by step functions, and is important in many aspects of the theory of random variation. However chapter 5 deals mainly with integrands $f(x_\mathbf{T})$, and not with functions of the form $f(x_\mathbf{T}, N)$. For the latter, see section A.1 of [MTRV], pages 491–500. As in chapter 4, section A.1 of [MTRV] covers integrands $h(x_\mathbf{T}, N, I[N])$ which include as a special case the functions $f(x_\mathbf{T}, N) F(I[N])$, including when

$$ f(x_\mathbf{T}, N) \;=\; {\fint_\mathbf{T}}^{g}(x_\mathbf{T}, N) \;=\; (N)\sum g \;=\; \sum_{j=1}^{n} g(s_j) $$

where $N = \{s_1, \ldots, s_{n-1}, s_n\}$ is a partition of $\mathbf{T} = \,]0, t]$, with $s_n = t$.

For any given $x_\mathbf{T} \in \mathbf{R}^\mathbf{T}$, the *stochastic summand* g, or g_s, corresponds to an integrand $f(y)\,dy$ in an ordinary integral $\int_a^b f(y)\,dy$. In the notation of -complete integration, the latter integral is $\int_{[a,b]} f(y)|I|$ where I is a typical partitioning cell $]u, v]$ of domain $[a, b]$. Instead of $f(y)|I|$ the integrand can be a function $h(y, I)$, with integral represented as $\int_{[a,b]} h(y, I)$.

A summand g_s can be a function depending on any or all of:

- points s in $0 \le s \le t$, cells $\imath_s$, $= \,]s, s']$, in $]0, t]$, values $x(s)$; and

- values $\mathbf{x}(\imath_s)$, where $\mathbf{x}(\imath_s)$ depends on the pair of values $x(s)$ and $x(s')$, and is usually the increment $x(s') - x(s)$. Thus g_s could be $\mathbf{x}(\imath_s)^2, = (x(s') - x(s))^2$.

Specific examples of stochastic summands g or g_s are given in [MTRV], section 8.3 (pages 391–392. The underlying ideas and motivation are explained in section 8.2 (pages 386–390). The symbol used in [MTRV] for the stochastic summation procedure is $\mathcal{R}_\mathbf{T}^g(x, N)$, where $\mathcal{R}$ is intended to suggest "Riemann sum". In place of the latter notation the "sum-integral" symbol $\fint$ is introduced here for stochastic summation; so $\mathcal{R}_\mathbf{T}^g(x, N)$ is the same as $\fint_\mathbf{T}^g(x, N)$.

6.4   Stochastic Sum as Random Variable

This section follows through on the definitions of Section 6.3, using familiar examples to illustrate the theory of stochastic sums, as replacement for both strong and weak stochastic integrals in chapter 8 of [MTRV]. The examples are based on the functions g_1 to g_9 of pages 391–392; also listed in (5.31) and (5.32) at the end of Chapter 5 above.

The notation is as set out in section 8.2 of MTRV (pages 386–390); with

$\mathbf{T} =]0,t]$ replacing the symbol $\mathcal{T} =]0,t]$ of [MTRV]. For any given $x_{\mathbf{T}} \in \mathbf{R}^{\mathbf{T}}$,

$$
\begin{aligned}
\imath_s &=]s,s'], & 0 \le s < s' \le t, \\
\mathbf{x}(\imath_s) &= x(s') - x(s), & x = x_{\mathbf{T}} \in \mathbf{R}^{\mathbf{T}}, \\
g &= g(x_s, s, \mathbf{x}(\imath_s), \imath_s), & \text{a } \mathbf{stochastic\ summand}\ (\text{or } \mathbf{integrand}), \\
\mathbf{X}(\imath_s) &= X(s') - X(s), & X = X_{\mathbf{T}} \text{ an observable in sample space } \mathbf{R}^{\mathbf{T}}, \\
N &= \{s_1, \ldots, s_{n-1}, s_n\}, & \text{a partition of } \mathbf{T}, \text{ or finite subset of } \mathbf{T}, \\
\imath_j &=]s_{j-1}, s_j], & j = 1, \ldots, n, \quad s_0 = 0, \ s_n = t.
\end{aligned}
$$

For $g = g_1, \ldots, g_9$ of (8.16) in page 419 of [MTRV], evaluations of stochastic integrals (strong and weak) have been given in [MTRV]. The idea here is to illustrate stochastic summation by replacing[4] the stochastic integrals $\mathbf{S}_{\mathbf{T}}^{g_j}(X_{\mathbf{T}})$ or $\mathcal{S}_{\mathbf{T}}^{g_j}(X_{\mathbf{T}})$ with corresponding stochastic sums of the form

$$
\oint_{\mathbf{T}}^{g_j}(X_{\mathbf{T}}, \mathcal{N}) = \sum_{j=1}^{n} g_j(X_{s_j}, s_j, \mathbf{X}(\imath_{s_j}), \imath_{s_j}).
$$

The three sample versions of these are

$$
\mathbf{S}_{\mathbf{T}}^{g_j}(x_{\mathbf{T}}), \qquad \mathcal{S}_{\mathbf{T}}^{g_j}(x_{\mathbf{T}}), \qquad \oint_{\mathbf{T}}^{g}(x_{\mathbf{T}}, N) \ \text{ or } \ \sum_{j=1}^{n} g(x_{s_j}, s_j, \mathbf{x}(\imath_{s_j}), \imath_{s_j}),
$$

respectively; in which first two are functions of sample processes $x_{\mathbf{T}}$, and the third is a function of finite samples of times s and occurrences $x(s)$.

What is meant by one random variable "replacing" another random variable? The idea is that of definition 30 and theorem 45, page 149 of [MTRV]; also section 4.9 (pages 152–154). Given functions $f_1(x, N)$, $f_2(x, N)$, and $F(I[N])$, f_1 is F-equivalent to f_2, written

$$
f_1 \overset{F}{=} f_2,
$$

if $f_1(x, N)F(I[N])$ is variationally equivalent to $f_2(x, N)F(I[N])$. If $F \ge 0$ then, by theorems 35–38 (page 146 of [MTRV]),

$$
\int_{\mathbf{R}^{\mathbf{T}}} |f_1(x, N) - f_2(x, N)|\, F(I[N]) = 0 \quad \text{implies} \quad f_1 \overset{F}{=} f_2; \tag{6.5}
$$

giving, for instance,

$$
E[f_1(X, \mathcal{N})] = E[f_2(X, \mathcal{N})]
$$

if F is a probability distribution; and likewise for other parameters of random variability; so contingent F-observables $f_1(X, \mathcal{N})$ and $f_2(X, \mathcal{N})$ exhibit identical characteristics as random variables. This is based on theorem 83 (page 200 of [MTRV]), and is summarized in (5.33) and (5.34) above.

---

[4]In some cases there is no advantage in substituting or replacing strong stochastic integration $\mathbf{S}$ by stochastic summation $\oint$, as $\mathbf{S}$ is more direct and clear. But in general terms, replacement/substitution remain valid.

In order to justify replacement of stochastic integrals by stochastic sums, we now seek to show that, for $j = 1, 2, \ldots, 9$,

$$\mathbf{S}_{\mathbf{T}}^{g_j}(x_{\mathbf{T}}) \overset{G}{=} \oint_{\mathbf{T}}^{g}(x_{\mathbf{T}}, N) \qquad \text{or} \qquad \mathcal{S}_{\mathbf{T}}^{g_j}(x_{\mathbf{T}}) \overset{G}{=} \oint_{\mathbf{T}}^{g}(x_{\mathbf{T}}, N)$$

for the Brownian distribution function $G(I[N])$ of (5.8); and similarly for other distribution functions $F(I[N])$.

Example 24 *With $x(0) = 0$ and $x(t) \in \mathbf{R}$ given, let $\mathbf{R}^{\mathbf{T}} = (x(s) : 0 < s \le t)$ be the sample space. Let g_1 be $g_1(\mathbf{x}(\imath_s)) = \mathbf{x}(\imath_s)$ be the summand. For $N = \{s_1, \ldots, s_n = t\} \in \mathcal{N}(\mathbf{T})$, the stochastic sum for g_1 is the Riemann sum over partition (or time sample) N,*

$$
\begin{aligned}
\oint_{\mathbf{T}}^{g_1}(x_T, N) &= \sum_{j=1}^{n} g_1(\mathbf{x}(\imath_s)) \\
&= \sum_{j=1}^{n} (x(s_j) - x(s_{j-1})), \\
&= x(s_n) - x(s_0) \;=\; x_n \;=\; x(t)
\end{aligned}
$$

since $x(0) = 0$. This holds for every time sample or partition N of $\mathbf{T}$; so, as in example 59 (pages 392–393 of [MTRV]),

$$\oint_{\mathbf{T}}^{g_1}(x_T, N) = x_t = \mathbf{S}_{\mathbf{T}}^{g_1}(x_{\mathbf{T}}), \qquad\qquad \oint_{\mathbf{T}}^{g_1}(X_{\mathbf{T}}, \mathcal{N}) = X_t = \mathbf{S}_{\mathbf{T}}^{g_1}(X_{\mathbf{T}});$$

and

$$\int_{\mathbf{R}^{\mathbf{T}}} \left| \oint_{\mathbf{T}}^{g_1}(x_{\mathbf{T}}, N) - \mathbf{S}_{\mathbf{T}}^{g_1}(X_{\mathbf{T}}) \right| G(I[N]) \;=\; \int_{\mathbf{R}^{\mathbf{T}}} 0 \cdot G(I[N]) \;=\; 0,$$

giving the result, in accordance with (6.5). (Also the result holds generally for arbitrary non-negative distribution functions $F(I[N])$.)

Of course this argument is highly redundant, and it is unnecessary to invoke (6.5). Simple inspection of cancellation of terms in $\oint_{\mathbf{T}}^{g_1}(x_{\mathbf{T}}, N)$ gives the result, just as it does for $\mathbf{S}_{\mathbf{T}}^{g_1}(x_{\mathbf{T}})$; and there is no advantage in replacing the (strong) stochastic integral by the corresponding stochastic sum.

For summand g_1, the stochastic sum approach (denoted by $\oint$) to stochastic integration/summation reduces to the strong stochastic integral approach (denoted by $\mathbf{S}$) as described in [MTRV] (example 59, pages 392–393).

The traditional Itô calculus is based on Lebesgue integration, which requires absolute convergence. Regarded this way, sample values of random variable "$\int_0^t dX_s$" have the form of a Lebesgue-Stieltjes integral "$\int_0^t dx_s$" where (x_s) is a sample path of standard Brownian motion.

In a Lebesgue integral context, this is meaningful only if $|dx_s|$ is integrable. And this fails since a "typical" Brownian sample path has infinite variation. But, by definition, Brownian displacement x_t at time t is a normally distributed random variable which is the sum of incremental displacements $x_{s'} - x_s$ for $0 < s < s' \le t$. Thus, "by definition", the Itô integral "$\int_0^t dx_s$" must be x_t for all sample paths (x_s). In other words $\int_0^t dX_s = X_t$.

In effect the integral statement $\int_T dX_s = X_t$ is an axiom of traditional Itô calculus. In that context it is a statement about Brownian motion, not about mathematical integration. Example 24 above shows that it can be upgraded to a valid result of integration theory if non-absolute integration is used. This is discussed further in [MTRV] section 8.3, pages 391–398.

6.5   Introduction to $\int_T (dX_s)^2 = t$

The basis of traditional Itô calculus is the *isometry property* $\int_T (dX_s)^2 = t$. For this to be valid for Brownian motion $X = X_T = (X_s : 0 < s \le t)$, and if an appropriate meaning or interpretation can be given to the "integral" expression of the isometry property, then the statement $\int_T (dx_s)^2 = t$ must in some sense be valid for "typical" Brownian sample paths $x = x_T = (x_s)$.

Traditional Itô calculus provides such an interpretation. The following discussion aims to provide a sense of what is involved.

In Example 24, **every** sample path $(x(s))$ satisfies $\int_T dx_s = x_t$ provided the Stieltjes-complete definition of $\int_T$ is used. Examples in section 8.4 of [MTRV] (pages 398–399) show that this approach does not work for $\int_T (dx_s)^2$.

If $\int_T (dx_s)^2$ has some meaning as an integral then it is not unreasonable to seek to approximate it by means of some kind of Riemann sum expression of the form

$$(N) \sum (x_{s'} - x_s)^2, \; = \; \sum_{j=1}^{n} (x_j - x_{j-1})^2, \; \text{where } N = \{s_1, \ldots, s_{n-1}, s_n\}$$

is a partition of $\mathbf{T} =]0, t]$ with $s = s_{j-1} < s_j = s'$, $j = 1, \ldots, n$; $0 = s_0$, $s_n = t$.

Such "typical" sample paths (x_s) have unbounded variation (so the Lebesgue-style $\int_0^t |dx_s|$ typically diverges to $+\infty$). But "dx_s^2" is typically less than "$|dx_s|$", so an aggregation of the form $\int_0^t dx_s^2$ may turn out to have a finite value. The expression $\sum_{j=1}^{n} (x_j - x_{j-1})^2$ is a sample occurrence of a stochastic sum $\int_T^{g_2} (X_T, \mathcal{N})$ where the summand g_2 is

$$g_2(\imath_s) \; = \; (x(s') - x(s))^2 \; \text{for } \imath_s =]s, s'], \; \imath_j =]s_{j-1}, s_j], \; x_{s_j} = x(s_j) = x_j.$$

For all partitions N we have

$$X_t \; = \; \sum_{j=1}^{n} (X_j - X_{j-1}) \; = \; \int_T^{g_1} (X_T, \mathcal{N})$$

as in Example 24, and

$$E[X_t] \; = \; E\left[\int_T^{g_1} (X_T, \mathcal{N})\right] \; = \; E\left[\sum_{j=1}^{n} (X_j - X_{j-1})\right] \; = \; \sum_{j=1}^{n} E\left[(X_j - X_{j-1})\right] \; = \; 0$$

with $E[(X_j - X_{j-1})] = 0$ for each j since the increments of standard Brownian motion have mean 0. Again, according to the theory of Brownian motion the increments $X_j - X_k$ are independent for all choices of j, k, including $k = 0$ and $j = n$, with variance $t_j - t_k$ in each case. Recall that, for any random variable Y, the variance $\text{Var}[Y]$ is $E\left[(Y - E[Y])^2\right]$. Therefore

$$\text{Var}[X_j - X_{j-1}] \;=\; E\left[(X_j - X_{j-1})^2\right] \;=\; t_j - t_{j-1},$$

$$\text{Var}[X_t] \;=\; \text{Var}[X_t - X_0] \;=\; E\left[X_t^2\right] \;=\; t$$

since $X_0 = 0$ by definition. For each partition $N = \{s_1, \dots, s_{n-1}, s_n\}$ of $\mathbf{T} =]0, t]$ (with $s_0 = 0$, $s_n = t$),

$$t \;=\; \text{Var}[X_t] \;=\; \text{Var}\left[\sum_{j=1}^{n}(X_j - X_{j-1})\right] \;=\; \sum_{j=1}^{n}\text{Var}[(X_j - X_{j-1})] \qquad (6.6)$$

by independence of the Brownian increments (theorem 100 of [MTRV], pages 218–219). Since

$$\text{Var}[(X_j - X_{j-1})] \;=\; E\left[(X_j - X_{j-1})^2\right],$$

further confirmation comes from

$$\sum_{j=1}^{n}\text{Var}[(X_j - X_{j-1})] \;=\; \sum_{j=1}^{n}(s_j - s_{j-1}), \quad \text{or}$$

$$\sum_{j=1}^{n}E\left[(X_j - X_{j-1})^2\right] \;=\; t; \quad \text{giving}$$

$$E\left[\sum_{j=1}^{n}(X_j - X_{j-1})^2\right] \;=\; t \qquad (6.7)$$

because expectation E is finitely additive (theorems 88 and 97, pages 203 and 213 of [MTRV]). Suppose the number n of partition points s_j in N increases, with $s_j - s_{j-1} \to 0$ for all j. There is then some plausibility in writing $X_j - X_{j-1}$ as dX_j, so (6.7) might be written as

$$\int_0^t dX_s^2 \;=\; t$$

in some weak sense of mean or average denoted by $E[\cdots]$ in (6.7). Thus equation (6.7) provides some intuitive support for the isometry property of Itô calculus.

But (6.7) by itself has only limited weight as justification for isometry. Intuitively, $\int_0^t dX_s^2$ (if it means anything at all) depends, not on a finite number $X_1, \dots, X_{n-1}, X_n$ of random variables, but on **every** random variable X_s of the Brownian motion. There are uncountably infinitely many of these, corresponding to each s. But, even with increasing n, only a finite number of random variables X_j appears in equation (6.7). Only a finite number of values s_j $(j = 1, \dots, n-1)$ is selected out of the infinitely many possible times s $(0 < s \le t)$ at which the random variability of Brownian motion manifests itself.

It is suggested above that n could tend to infinity. But in that case some way would have to be found to make such a procedure meaningful and unambiguous, ensuring that **every** random variable X_s $(0 < s < t)$ contributes to the calculation—in addition to $X_n, = X(t)$.

Equation (6.7) provides some confirmation of consistency between the various properties that define Brownian motion. It also provides some basic intuition for the isometry property $\int_{\mathbf{T}} (dX_s)^2 = t$.

The calculation $E[\cdots]$ of the expression $\sum_{j=1}^{n} (X_j - X_{j-1})^2$ in equation (6.7) above is essentially as follows. Suppose M is a fixed, finite subset of $\mathbf{T}$ with $M = \{\tau_1, \ldots, \tau_m\}$ and $0 = \tau_0 < \tau_1 < \cdots < \tau_m = t$. (Notation M, m, τ, $\cdots$, is consistent with the cylindrical or finite-dimensional integration in [MTRV], pages 90–97 and elsewhere.) For each cell $I(M) \subset \mathbf{R}^M$ let $G(I(M))$ denote the joint Brownian transition probability that

$$x(0) = 0, \quad x(\tau_j) \in I(\tau_j), \quad j = 1, 2, \ldots, m-1, m.$$

In other words,

$$G(I(M)) = \int_{\mathbf{R}^M} \left(\prod_{j=1}^{m} \frac{\exp\left(\frac{-(y_j - y_{j-1})^2}{2(\tau_j - \tau_{j-1})} \right)}{\sqrt{2\pi(\tau_j - \tau_{j-1})}} \right) dy_1 dy_2 \cdots dy_{m-1} dy_m.$$

Then the evaluation in (6.7) is

$$E\left[\sum_{j=1}^{m} (X_j - X_{j-1})^2 \right] = \int_{\mathbf{R}^M} \left(\sum_{j=1}^{m} (x_j - x_{j-1})^2 \right) G(I(M))$$

$$= \sum_{j=1}^{m} \int_{\mathbf{R}^M} (x_j - x_{j-1})^2) G(I(M))$$

$$= \sum_{j=1}^{m} (\tau_j - \tau_{j-1}) = \tau_m - \tau_0 = \tau_m = t.$$

6.6   Isometry Preliminaries

Some properties of finite-dimensional Gaussian integrals can be used to establish a version of the isometry property of Brownian motion.

P1 Assume $c < 0$. Consider the one-dimensional integral $h(I) = \int_u^v y^2 e^{cy^2} dy$ with I a cell such as $]u, v]$. In [MTRV] (page 263) integration by parts is applied, giving

$$\int_u^v y^2 e^{cy^2} dy = \frac{1}{2c} \int_u^v y \left(e^{cy^2} 2cy \right) dy$$

$$= \frac{1}{2c} \left[y e^{cy^2} - \int e^{cy^2} dy \right]_u^v$$

$$= \frac{1}{2c} \left(v e^{cv^2} - u e^{cu^2} - \int_u^v e^{cy^2} dy \right).$$

P2 Suppose $x \in \mathbf{R}$, $f(x) = x^2$, and, for cells $I =]u, v]$ in $\mathbf{R}$, the cell function $g(I)$ is $\int_I e^{cy^2} dy$. For associated (x, I) in $\mathbf{R}$, consider integrand $f(x)g(I)$ in domain $\mathbf{R}$. It is easy to show that $f(x)g(I)$ is variationally equivalent[5] to $h_0(I) = \int_u^v y^2 e^{cy^2} dy$. Since the latter is an additive cell function, it is the indefinite integral[6] of the integrand $f(x)g(I)$; and, by the preceding calculation, the indefinite integral of $f(x)g(I)$ can be expressed as the additive cell function

$$h_0(I) = \frac{1}{2c}\left(v e^{cv^2} - u e^{cu^2} - \int_u^v e^{cy^2} dy\right). \tag{6.8}$$

The purpose of presenting the indefinite integral of integrand $x^2 \int_I e^{cy^2} dy$ in the form (6.8) is to establish the isometry property of Brownian motion.

P3 Next, consider finite-dimensional domain $\mathbf{R}^m$ with points and cells

$$x = (x_1, \ldots, x_{m-1}, x_m), \qquad I = I_1 \times \cdots \times I_{m-1} \times I_m,$$

respectively. Let $h_1(I) =$

$$= \int_I \left(y_1^2 + \cdots + y_{m-1}^2 + y_m^2\right) e^{\left(c_1 y_1^2 + \cdots + c_{m-1} y_{m-1}^2 + c_m y_m^2\right)} dy_1 \ldots dy_{m-1} dy_m$$

if the integral exists. Assume $c_j < 0$ for $j = 1 \ldots, m$. Regarding existence, for any k $(1 \le k \le m)$, with $I_k =]u_k, v_k]$, (6.8) implies

$$\int_I \left(y_k^2 e^{\sum_{j=1}^m c_j y_j^2}\right) dy_1 \ldots dy_m =$$

$$= \int_{I_k} y_k^2 e^{c_k y_k^2} dy_k \prod\left(\int_{I_j} e^{c_j y_j^2} dy_j : j = 1, 2, \ldots, m, \ j \ne k\right)$$

$$= \frac{1}{2c_k}\left(v_k e^{c_k v_k^2} - u_k e^{c_k u_k^2} - \int_{I_k} e^{c_k y^2} dy_k\right)\prod_{j \ne k}\left(\int_{I_j} e^{c_j y_j^2} dy_j\right), \tag{6.9}$$

so the first integral $\int_I \cdots$ exists. Thus $h_1(I) =$

$$= \sum_{k=1}^m \left(\frac{1}{2c_k}\left(v_k e^{c_k v_k^2} - u_k e^{c_k u_k^2} - \int_{I_k} e^{c_k y_k^2} dy_k\right)\prod_{j \ne k}\left(\int_{I_j} e^{c_j y_j^2} dy_j\right)\right),$$

and $h_1(I)$ is finitely additive on disjoint cells I. This ensures that $h_1(I)$ is integrable on $\mathbf{R}^m$. Now define $h(I) :=$

$$:= h_1(I) \prod_{j=1}^m \left(\frac{\pi}{-c_j}\right)^{-\frac{1}{2}} = \int_{I_1 \times \cdots \times I_m} \left(\sum_{j=1}^m y_j^2 e^{\sum_{j=1}^m c_j y_j^2}\right) \frac{dy_1}{\sqrt{\frac{\pi}{-c_1}}} \cdots \frac{dy_m}{\sqrt{\frac{\pi}{-c_m}}},$$

---

[5]For the meaning of this, see, for instance, page 148 of [MTRV].

[6]In traditional calculus the indefinite integral of a function $f(x)$ is a function $F(x) = \int_a^x f(y)\,dy$, while the definite integral is a number $\int_a^b f(y) dy$. Pages 131–132 of [MTRV] describe how these concepts of definite/indefinite integral are expressed in -complete integration.

(where $\int_{\mathbf{R}} e^{c_j y_j^2} \frac{dy_j}{\sqrt{\frac{\pi}{-c_j}}} = 1$ for each j by theorem 133, [MTRV] page 261).

Note that each of $v_k e^{c_k v_k^2}$ and $u_k e^{c_k u_k^2}$ tends to zero as $|v_k|, |u_k|$ tend to infinity. Therefore, using the -complete integral construction on $\mathbf{R}$ ([MTRV] pages 69–78, corresponding to improper Riemann integration),

$$\int_{\mathbf{R}^m} h(I) = h(\mathbf{R}^m) = \sum_{k=1}^{m} \frac{-1}{2c_k}.$$

P4 For $c_j < 0$ and associated (x, I) in $\mathbf{R}^m$ let $h(x, I)$ denote the integrand

$$\left(x_1^2 + \cdots + x_m^2\right) \int_I e^{(c_1 y_1^2 + \cdots + c_m y_m^2)} \frac{dy_1}{\sqrt{\frac{\pi}{-c_1}}} \cdots \frac{dy_m}{\sqrt{\frac{\pi}{-c_m}}}. \tag{6.10}$$

As in **P1**, $h(x, I)$ is variationally equivalent to the additive cell function

$$h(I) = \int_I \left(y_1^2 + \cdots + y_m^2\right) e^{(c_1 y_1^2 + \cdots + c_m y_m^2)} \frac{dy_1}{\sqrt{\frac{\pi}{-c_1}}} \cdots \frac{dy_m}{\sqrt{\frac{\pi}{-c_m}}}. \tag{6.11}$$

The latter is integrable, and is therefore the indefinite integral of $h(x, I)$; so, by **P3**,

$$\int_{\mathbf{R}^m} h(x, I) = \int_{\mathbf{R}^m} h(I) = h(\mathbf{R}^m) = \sum_{k=1}^{m} \frac{-1}{2c_k}.$$

P5 Suppose $0 = s_0 < s_1 < \cdots < s_{m-1} < s_m = t$, with $c_j = \frac{-1}{2(s_j - s_{j-1})}$ above for $j = 1, 2, \ldots, m$. Then $\int_{\mathbf{R}^m} h(x, I) = \int_{\mathbf{R}^m} h(I) =$

$$= h(\mathbf{R}^m) = \sum_{k=1}^{m} \frac{-1}{2c_k} = \sum_{k=1}^{m} (s_k - s_{k-1}) = s_m = t. \tag{6.12}$$

For $j = 1, \ldots, m$ replace x_j^2 and y_j^2 by $(x_j - x_{j-1})^2$ and $(y_j - y_{j-1})^2$ in (6.10) and (6.11), leaving dy_j unchanged, and denote the amended integrands by $p(x, I)$ and $p(I)$ respectively. Using appropriate changes of variables, the result is

$$\int_{\mathbf{R}^m} p(x, I) = \int_{\mathbf{R}^m} p(I) = s_m = t. \tag{6.13}$$

6.7 Isometry Property for Stochastic Sums

The second integrand/summand in the lists (5.31) and (5.32) is the function g_2. By adding more detail to Section 6.5 the formulation $\int_0^t dX_s^2 = t$ can now be brought into a framework of stochastic sums.

In (6.7) the partition points τ_j of M are taken to be fixed times for the purpose of calculating the expected value $E\left[\sum_{j=1}^{m} (X_j - X_{j-1})^2\right]$ in a finite-dimensional sample space $\mathbf{R}^M$, with $M = \{\tau_1, \ldots, \tau_{m-1}, \tau_m\}$. In contrast, Sections 6.3 and 6.4 have provided expressions such as $\sum_{j=1}^{n} (X_j - X_{j-1})^2$ in equation (6.7) with an enhanced meaning as a new kind of observable or random variable,

$$\oint_{\mathbf{T}}^{g_2}(X_{\mathbf{T}}, \mathcal{N}) \;=\; \sum_{j=1}^{n} (X_j - X_{j-1})^2. \tag{6.14}$$

Here, $\oint_{\mathbf{T}}^{g_2}(X_{\mathbf{T}}, \mathcal{N})$ is an observable in sample space $\mathbf{R}^{\mathbf{T}}$ with distribution function $G(I[N])$ for times $N \subset \mathbf{T}$:

$$\oint_{\mathbf{T}}^{g_2}(X_{\mathbf{T}}, \mathcal{N}) \;\simeq\; \oint_{\mathbf{T}}^{g_2}(x_{\mathbf{T}}, N)\,[\mathbf{R}^{\mathbf{T}}, G],$$

in which N is variable, so sample values $\oint_{\mathbf{T}}^{g_2}(x_{\mathbf{T}}, N)$ are constructed from samples of times $s_j \in N \subset \mathbf{T}$, with corresponding sample values $x_j, = x(s_j)$ of the random variables $X_j, = X(s_j)$, of the process $X_{\mathbf{T}}$.

If observable $\oint_{\mathbf{T}}^{g_2}(X_{\mathbf{T}}, \mathcal{N})$ has expected value it is a random variable. And, in that case, it is an **absolute** random variable (and therefore measurable) since its sample values are non-negative. (See theorems 76 and 250, [MTRV] pages 193 and 494.) Example 25 below confirms these properties, with

$$\mathrm{E}\left[\oint_{\mathbf{T}}^{g_2}(X_{\mathbf{T}}, \mathcal{N})\right] \;=\; \int_{\mathbf{R}^{\mathbf{T}}} \left(\oint_{\mathbf{T}}^{g_2}(x_{\mathbf{T}}, N)\right) G(I[N]) \;=\; t. \tag{6.15}$$

Thus

$$\oint_{\mathbf{T}}^{g_2}(X_{\mathbf{T}}, \mathcal{N}), \;=\; \sum_{s_j \in N} (X(s_j) - X(s_{j-1}))^2 \text{ with variable } N \in \mathcal{N}, \tag{6.16}$$

is the meaning we ascribe to $\int_{\mathbf{T}} dX_s^2$, validating the latter as a random variable contingent on the Brownian process $X_{\mathbf{T}}$, so

$$\mathrm{E}\left[\int_{\mathbf{T}} dX_s^2\right] \;=\; \int_{\mathbf{R}^{\mathbf{T}}} \left(\int_{\mathbf{T}} dx_s^2\right) G(I[N]) \;=\; t. \tag{6.17}$$

In this way, Example 25 supports the traditional Itô calculus interpretation of "$\int_{\mathbf{T}} dX_s^2$" as a weak integral which converges "in the mean" to value t.

In stochastic sum format, isometry means variational equivalence of the expressions

$$\oint_{\mathbf{T}}^{g_2}(x_{\mathbf{T}}, N)\,G(I[N]), \qquad t\,G(I[N]),$$

implying (6.15) above. Variational equivalence follows from (6.20) below.

Example 25 *The link between (6.7) and (6.16) is the sample values x_j which appear in both. Where (6.16) differs is in the variable parameters s_j, parameters which are fixed in (6.7); because, in addition to variable sample values x_j, (6.16) has variable sample times s_j. Notation $\oint_{\mathbf{T}}^{g_2}(X_{\mathbf{T}}, \mathcal{N})$ has contingent observable format $f(X)$ with a new kind of "sample times observable" N selected from $\mathcal{N}$, giving a formulation $f(X, \mathcal{N})$. The particular form of function f needed for stochastic sums is represented by the symbol $\oint$.*

The notation "$\int_{\mathbf{T}} dX_s^2$" purports to be a statement about all possible occurrences $x(s)$ at all possible times s $(0 < s < t)$. But analysis of random variability *starts with calculations based on observation of actual sample values. We are now extending this in order to include, in these calculations, observation of variable sample times s as well as variable sample values $x(s)$.*

To complete this construction (6.15) must be proved. The proof is similar to proofs already given in [MTRV]. For any joint occurrence (x_T, N) in (6.15), with $N = \{s_1, \ldots, s_n\}$,

$$\oint_T^{g_2}(x_T, N) = \sum_{j=1}^{n}(x(s_j) - x(s_{j-1}))^2 = \sum_{j=1}^{n}(x_j - x_{j-1})^2, \qquad (6.18)$$

and

$$\int_{R^T}\left(\oint_T^{g_2}(x_T, N)\right)G(I[N]) = \int_{R^T}\left(\sum_{j=1}^{n}(x_j - x_{j-1})^2\right)G(I[N])$$

$$= \sum_{j=1}^{n}\left(\int_{R^T}(x_j - x_{j-1})^2\,G(I[N])\right).$$

By theorem 171 ([MTRV] page 286), the integrand $\left(\sum_{j=1}^{n}(x_j - x_{j-1})^2\right)G(I[N])$ is variationally equivalent to $H_2(I[N]) =$

$$= \int_{I(N)}\left(\sum_{j=1}^{n}(y_j - y_{j-1})^2\prod_{j=1}^{n}\frac{\exp\frac{-(y_j-y_{j-1})^2}{2(s_j-s_{j-1})}}{\sqrt{2\pi(s_j - s_{j-1})}}\right)dy_1\cdots dy_n; \qquad (6.19)$$

with corresponding result for the separate factors $(x_j - x_{j-1})^2\,G(I[N])$, each j. For more details of this see theorems 162, 176, and 177 (pages 281 and 287 of [MTRV]). Thus

$$E\left[\oint_T^{g_2}(X_T, \mathcal{N})\right] = \int_{R^T}\left(\oint_T^{g_2}(x_T, N)\right)G(I[N]) = \int_{R^T}H_2(I[N]).$$

*Now it must be shown that the latter integral evaluates to t. A finite-dimensional version of this has been proved in (6.13) of **P5** above.*

To recapitulate the argument, for fixed $N = \{s_1, \ldots, s_n\}$ the finite-dimensional integral

$$\int_{R^N}H_2(I(N)) = \int_{R^N}\left(\sum_{j=1}^{n}(y_j - y_{j-1})^2\prod_{j=1}^{n}\frac{\exp\frac{-(y_j-y_{j-1})^2}{2(s_j-s_{j-1})}}{\sqrt{2\pi(s_j - s_{j-1})}}\right)dy_1\cdots dy_n$$

$$= \sum_{j=1}^{n}(s_j - s_{j-1}) = s_n = t \qquad (6.20)$$

by successive evaluation of Gaussian integrals based on lemma 15 ([MTRV] page 263).

Now consider $\int_{R^T}H_2(I[N])$. The method of theorem 168 ([MTRV] page 285) shows that this is integrable, with integral value t. Let $\mathcal{D}$ be a division of *the infinite-dimensional* $\mathbf{R}^{\mathbf{T}}$. *Form a regular partition* $\mathcal{P}$ *of* $\mathbf{R}^{\mathbf{T}}$ *by extending the faces of each* $I[N] \in \mathcal{D}$. *With* $\tau_0 = 0$, $\tau_m = t$, *let*

$$\{\tau_1, \ldots, \tau_m\} = M_0 = \bigcup \{N : (x, I[N]) \in \mathcal{D}\}, \qquad \mathcal{P}_{M_0} = \{I(M_0) : I[M_0] \in \mathcal{P}\};$$

so $\mathcal{P}_{M_0}$ *is a regular partition of the finite-dimensional product domain* $\mathbf{R}^{M_0}$. *Then, by additivity of the cell functions* $H_2(I[N])$ *and* $H_2(I(N))$,

$$
\begin{aligned}
(\mathcal{D}) \sum H_2(I[N]) &= (\mathcal{P}) \sum H_2(I[M_0]) \\
&= (\mathcal{P}_{M_0}) \sum H_2(I(M_0)) \\
&= \int_{\mathbf{R}^{M_0}} H_2(I(M_0)),
\end{aligned}
$$

and, by the earlier evaluation, the latter integral has value τ_m, $= t$. *This holds for all* $\mathcal{D}$, *so* $\int_{\mathbf{R}^{\mathbf{T}}} H_2(I[N]) = t$, *and therefore*

$$\mathrm{E}\left[\fint_{\mathbf{T}}^{g_2}(X_{\mathbf{T}}, \mathcal{N}) \right] = \int_{\mathbf{R}^{\mathbf{T}}} \left(\fint_{\mathbf{T}}^{g_2}(x_{\mathbf{T}}, N) \right) G(I[N]) = t.$$

A corollary is that the non-negative cell function

$$F(I[N]) := \frac{H_2(I[N])}{t}$$

is a probability distribution function on $\mathbf{R}^{\mathbf{T}}$. *(In other words,* F *is an additive cell function with integral 1.) To show that*

$$\fint_{\mathbf{T}}^{g_2}(X_{\mathbf{T}}, \mathcal{N}) \overset{G}{=} \mathcal{S}_{\mathbf{T}}^{g_2}(X_{\mathbf{T}}), \qquad or, \; in \; other \; words, \qquad \fint_{\mathbf{T}}^{g_2}(X_{\mathbf{T}}, \mathcal{N}) \overset{G}{=} t,$$

consider the following integrand function:

$$h(x, N, I[N]) := \fint_{\mathbf{T}}^{g_2}(x_{\mathbf{T}}, N)\, G(I[N]) - t\, G(I[N]).$$

The preceding analysis shows that h *is integrable on* $\mathbf{R}^{\mathbf{T}}$, *with* $\int_{\mathbf{R}^{\mathbf{T}}} h = 0$, *so*

$$\int_{\mathbf{R}^{\mathbf{T}}} \left(\fint_{\mathbf{T}}^{g_2}(x_{\mathbf{T}}, N) - t \right) G(I[N]) = 0.$$

Using theorem 18 of [MTRV] (Saks-Henstock lemma, page 132),

$$\int_{\mathbf{R}^{\mathbf{T}}} \left| \fint_{\mathbf{T}}^{g_2}(x_{\mathbf{T}}, N) - t \right| G(I[N]) = 0,$$

giving the required result (see theorem 39, [MTRV] page 147). $\bigcirc$

A stochastic sum $\fint_{\mathbf{T}}^{g_2}(X_{\mathbf{T}}, \mathcal{N})$ (or stochastic integral $\int_{\mathbf{T}}(dX_s)^2$) is a combination of random variables. Example 6.18 above says that each of these can be "replaced" by the quantity t,

$$\fint_{\mathbf{T}}^{g_2}(X_{\mathbf{T}}, \mathcal{N}) \overset{G}{=} t,$$

a non-random quantity. This is somewhat counter-intuitive. Though any particular sampling $\mathit{f}_{\mathbf{T}}^{g_2}(x_{\mathbf{T}}, N)$ (or $\int_{\mathbf{T}}(dX_s)^2$) will differ somewhat from the fixed quantity t, the differences are G-null in the sense described in Example 6.18. Since it is "replaceable" by a constant, the stochastic sum $\mathit{f}_{\mathbf{T}}^{g_2}(X_{\mathbf{T}}, \mathcal{N})$—or stochastic integral $\int_{\mathbf{T}}(dX_s)^2$—is a "degenerate" random variable.

6.8   Other Stochastic Sums

In section 8.3 of [MTRV], pages 391–392 provide examples of stochastic integrals with the following integrands, also listed in (5.31) and (5.32) above,

$$
\begin{array}{lllllll}
dX_s, & = & g_1; & (dX_s)^2, & = & g_2; & (dX_s)^3, & = & g_3; \\
X_s dX_s, & = & g_4; & s dX_s, & = & g_5; & X_s dX_s^2, & = & g_6; \\
X_s^2 dX_s, & = & g_7; & d(X_s^3), & = & g_8; & d(X_s^p), & = & g_9;
\end{array}
$$

whose integrals $\int_0^t \cdots$ are shown[7] to be, respectively,

$$
X_t, \ \ t, \ \ 0, \ \ \frac{1}{2}X_t^2 - \frac{1}{2}t, \ \ tX_t - \int_0^t X_s\,ds, \ \ \int_0^t X_s\,ds, \ \ \frac{1}{3}X_t^3 - \int_0^t X_s\,ds, \ \ X_t^3, \ \ X_t^p.
$$

These results (with [MTRV] page references) are as follows, where, in the notation of [MTRV], **S** denotes strong stochastic integration and $\mathcal{S}$ denotes weak stochastic integration:

$$
\begin{array}{llll}
\mathbf{S}_{\mathbf{T}}^{g_1}(X_{\mathbf{T}}) & = & X_t & \text{for any process } X_{\mathbf{T}}, \text{ page 392 of [MTRV]}; & (6.21) \\
\mathcal{S}_{\mathbf{T}}^{g_2}(X_{\mathbf{T}}) & = & t & \text{if } X_{\mathbf{T}} \text{ is Brownian, page 402}; & (6.22) \\
\mathcal{S}_{\mathbf{T}}^{g_3}(X_{\mathbf{T}}) & = & 0 & \text{if } X_{\mathbf{T}} \text{ is Brownian, page 413}; & (6.23) \\
\mathcal{S}_{\mathbf{T}}^{g_4}(X_T) & = & \dfrac{1}{2}X_t^2 - \dfrac{1}{2}t & \text{if } X_{\mathbf{T}} \text{ is Brownian, page 406}; & (6.24) \\
\mathbf{S}_{\mathbf{T}}^{g_5}(X_{\mathbf{T}}) & = & tX_t - \displaystyle\int_0^t X_s\,ds & \text{if } X_{\mathbf{T}} \text{ is Brownian, page 396}; & (6.25) \\
\mathcal{S}_{\mathbf{T}}^{g_6}(X_{\mathbf{T}}) & = & \displaystyle\int_0^t X_s\,ds & \text{if } X_{\mathbf{T}} \text{ is Brownian, page 416}; & (6.26) \\
\mathcal{S}_{\mathbf{T}}^{g_7}(X_{\mathbf{T}}) & = & \dfrac{1}{3}X_t^3 - \displaystyle\int_0^t X_s\,ds & \text{if } X_{\mathbf{T}} \text{ is Brownian, page 417}; & (6.27) \\
\mathbf{S}_{\mathbf{T}}^{g_8}(X_{\mathbf{T}}) & = & X_t^3 & \text{for any process } X_{\mathbf{T}}, \text{ page 418}; & (6.28) \\
\mathbf{S}_{\mathbf{T}}^{g_9}(X_{\mathbf{T}}) & = & X_t^p & \text{for any process } X_{\mathbf{T}}, \text{ page 418}; & (6.29)
\end{array}
$$

The meaning of such evaluations is summarized in (5.33) and (5.34) above.

For integrals (6.21) to (6.29), variants of the letter S are substituted for $\int$, while the "dX" symbol is replaced by a superscript integrand g. The component

---

[7]Note the error in statement of $\int_0^t s\,dX_s$ in theorem 229 ([MTRV] page 396), repeated in line 9 of page 419. This result should be $\int_0^t s\,dX_s = tX_t - \int_0^t X_s\,ds$.

$(X_{\mathbf{T}})$ in notation $S(X_{\mathbf{T}})$ and $\mathcal{S}(X_{\mathbf{T}})$ accords with the notation (X) in $f(X)$ where $f(X)$ denotes an observable which is contingent on basic observable X.

The latter aspect of notation is retained in the formulation $\mathit{f}_{\mathbf{T}}^{g}(X_{\mathbf{T}},\mathcal{N})$ which aims to replace[8] stochastic integrals $\mathbf{S}$ and $\mathcal{S}$ by simpler stochastic sums. In addition to the random variability of process $X_{\mathbf{T}}$, stochastic sums include random variability of the sample set of times $N = \{s_1, \ldots, s_{n-1}, s_n\}$ at which corresponding sample process values $\{x(s_j)\}$ are observed.

This manoeuvre evades integral formulation such as $\int_0^t X(s)\, dX(s)$ which, on the face of it, involve **all** occurrence values $x(s)$ at **all** times s.

Example 26 *For stochastic summand g_3 or dX_s^3 in (5.31) and (5.32), the result is*

$$\mathit{f}_{\mathbf{T}}^{g_3}(X_{\mathbf{T}}, \mathcal{N}) \overset{G}{=} 0, \quad or \quad \int_{\mathbf{R}^{\mathbf{T}}} \left| \sum_{j=1}^{n} (x_j - x_{j-1})^3 \right| G(I[N]) = 0.$$

The method of Example 25 can be adapted to prove this. With $\varepsilon > 0$ given, a gauge γ_1 can be chosen so that, for all γ_1-fine divisions $\mathcal{D}_1$ of $\mathbf{R}^{\mathbf{T}}$,

$$(\mathcal{D}_1) \sum_{j=1}^{n} (x_j - x_{j-1})^2 G(I[N]) < t + \varepsilon.$$

Now choose $\gamma < \gamma_1$ so that, except for the G-null set of $(x_s) \in \mathbf{R}^{\mathbf{T}}$ which are not uniformly continuous in $\mathbf{T}$ with respect to $s \in \mathbf{T}$,

$$|x_j - x_{j-1}| < \varepsilon$$

for all $s_j \in N$. Then, for each γ-fine division $\mathcal{D}$ of $\mathbf{R}^{\mathbf{T}}$,

$$(\mathcal{D}) \sum_{j=1}^{n} |x_j - x_{j-1}|^3 G(I[N]) < \varepsilon\, (\mathcal{D}) \sum_{j=1}^{n} (x_j - x_{j-1})^2 G(I[N]) < \varepsilon\, (t + \varepsilon),$$

giving the result.

The next case is stochastic summand g_4, frequently used in applications. The evaluation of $\mathit{f}_{\mathbf{T}}^{g_4}(X_T, \mathcal{N})$ is similar to the evaluation of the corresponding weak stochastic integral $\int_0^t X_s\, dX_s$.

Example 27 *For Brownian $X_{\mathbf{T}}$ ($\mathbf{T} = \,]0, t]$), (6.24) is the weak stochastic integral*

$$\int_0^t X_s\, dX_s = S_{\mathbf{T}}^{g_4}(X_{\mathbf{T}}) \overset{G}{=} \frac{1}{2}X_t^2 - \frac{1}{2}t.$$

In stochastic sum format, with $g_4(x_s) = x_s(x_{s'} - x_s)$, this becomes

$$\mathit{f}_{\mathbf{T}}^{g_4}(X_{\mathbf{T}}, \mathcal{N}) \simeq \mathit{f}_{\mathbf{T}}^{g_4}(x_{\mathbf{T}}, N)\,[\mathbf{R}^{\mathbf{T}}, G], \qquad \mathit{f}_{\mathbf{T}}^{g_4}(X_{\mathbf{T}}, \mathcal{N}) \overset{G}{=} \frac{1}{2}X_t^2 - \frac{1}{2}t.$$

---

[8]Sometimes there is no practical advantage in replacing strong stochastic integrals $\mathbf{S}$ by stochastic sums f. Neither of these concepts is available in the classical Lebesgue-Itô theory of stochastic integration.

The latter implies, for instance,

$$\mathrm{E}\left[\oint_{\mathbf{T}}^{g_4}(X_{\mathbf{T}},\mathcal{N})\right] \;=\; \int_{\mathbf{R}^{\mathbf{T}}}\left(\oint_{\mathbf{T}}^{g_3}(x_{\mathbf{T}},N)\right)G(I[N]) \;=\; \mathrm{E}\left[\tfrac{1}{2}X_t^2 - \tfrac{1}{2}t\right] \;=\; 0;$$

with similar equalities for other random variation parameters, as explained in (5.33) and (5.34) above. As in [MTRV], page 405,

$$\oint_{\mathbf{T}}^{g_4}(x_{\mathbf{T}},N) \;=\; \sum_{j=1}^{n}g_4(s_j) \;=\; \sum_{j=1}^{n}x(s_{j-1})\left(x(s_j)-x(s_{j-1})\right)$$

$$=\; \sum_{j=1}^{n}\tfrac{1}{2}\left(x_j+x_{j-1}\right)\left(x_j-x_{j-1}\right) \;-\; \sum_{j=1}^{n}\tfrac{1}{2}\left(x_j-x_{j-1}\right)\left(x_j-x_{j-1}\right)$$

$$=\; \tfrac{1}{2}x_n^2 \;-\; \tfrac{1}{2}\sum_{j=1}^{n}\left(x_j-x_{j-1}\right)^2 \;=\; \tfrac{1}{2}x_t^2 \;-\; \tfrac{1}{2}\oint_{\mathcal{T}}^{g_2}(x_{\mathbf{T}},N).$$

The evaluation in Example 25 can then be used to give

$$\oint_{\mathbf{T}}^{g_4}(x_{\mathbf{T}},N) \;\overset{G}{=}\; \tfrac{1}{2}X_t^2 - \tfrac{1}{2}t.$$

It remains to evaluate the remaining stochastic sums arising from (5.32) above. The background ideas are in the literature of this subject, and -complete versions are in chapter 8 of [MTRV] using the strong/weak stochastic integral approach.

Therefore, rather than working through each stochastic sum evaluation in detail, they can be listed giving some key steps in each case, with further details left to the reader.

Example 28 *In (6.25),* $\mathbf{S}_{\mathbf{T}}^{y_5}(X_{\mathbf{T}}) = tX_t - \int_0^t X_s\,ds$ *if* $X_{\mathbf{T}}$ *is Brownian.[9] The integrand* $g_5(s)$ *is* $s\,dX_s$, *or* $s(X_{s'}-X_s)$, *and the stochastic sum format is*

$$\oint_{\mathbf{T}}^{g_5}(X_{\mathbf{T}},\mathcal{N}) \qquad or \qquad \sum_{j=1}^{n}s_{j-1}\left(X(s_j)-X(s_{j-1})\right),$$

with

$$\oint_{\mathbf{T}}^{g_5}(X_{\mathbf{T}},\mathcal{N}) \;\simeq\; \oint_{\mathbf{T}}^{g_5}(x_{\mathbf{T}},N)\left[\mathbf{R}^{\mathbf{T}},G\right].$$

We must prove $\oint_{\mathbf{T}}^{g_5}(x_{\mathbf{T}},N) \overset{G}{=} tX_t - \int_0^t x_s\,ds$, *or*

$$\int_{\mathbf{R}^{\mathbf{T}}}\left|\sum_{j=1}^{n}s_{j-1}\left(x(s_j)-x(s_{j-1})\right) - \left(tX_t - \int_0^t x_s\,ds\right)\right|G(I[N]) \;=\; 0$$

By re-arranging terms,

$$\sum_{j=1}^{n}s_{j-1}\left(x(s_j)-x(s_{j-1})\right) \;=\; s_{n-1}x_t \;-\; \sum_{j=1}^{n-1}x_j\left(s_j-s_{j-1}\right),$$

---

[9]Note the typo in the statement of this in [MTRV] page 419.

and, provided $x_{\mathbf{T}}$ is continuous, the result is obtained by suitable choices of gauge $\gamma = (L, \delta_{\mathcal{N}})$ in $\mathbf{R}^{\mathbf{T}}$. The only use of Brownian motion properties is the assumption of continuity of $x_{\mathbf{T}}$; otherwise the result holds for arbitrary joint distribution function $F(I[N])$.

Example 29 *Summand g_7 is $x_s^2(x_{s'} - x_s)$ and, for $N = \{s_1, \ldots, s_{n-1}, s_n\}$,*

$$\oint_{\mathbf{T}}^{g_7}(X_{\mathbf{T}}, \mathcal{N}) = \sum_{j=1}^{n} X_{j-1}^2 (X_j - X_{j-1});$$

with

$$\int_{\mathbf{R}^{\mathbf{T}}} \left| \sum_{j=1}^{n} x_{j-1}^2 (x_j - X_{j-1}) - \left(\frac{1}{3}x_t^3 - \int_{\mathbf{T}} x_s\, ds\right) \right| G(I[N]) = 0$$

to be proven. Re-arrangement of the stochastic sum gives

$$\oint_{\mathbf{T}}^{g_7}(x_{\mathbf{T}}, N) = \frac{1}{3}\left(x_t^3 - \sum_{j=1}^{n}(x_j - x_{j-1})^3 - 3\sum_{j=1}^{n} x_{j-1}(x_j - x_{j-1})^2 \right).$$

Using appropriate gauges γ in $\bar{\mathbf{R}}^{\mathbf{T}}$, and provided the process $X_{\mathbf{T}}$ is Brownian (so continuity and isometry can be invoked),

$$\oint_{\mathbf{T}}^{g_7}(x_{\mathbf{T}}, N) \overset{G}{=} \frac{1}{3}x_t^3 - \int_{\mathbf{T}} x(s)\, ds.$$

Example 30 *Here are two further stochastic sums,*

$$\oint_{\mathbf{T}}^{g_9}(X_{\mathbf{T}}, \mathcal{N}), \qquad \oint_{\mathbf{T}}^{g_0}(X_{\mathbf{T}}, \mathcal{N}),$$

corresponding to $\int_{\mathbf{T}} d(X_s^p)$ and $\int_{\mathbf{T}} (dX_s)^q$, and representing

$$\sum_{j=1}^{n}\left(x_j^p - x_{j-1}^p\right), \qquad \sum_{j=1}^{n}(x_j - x_{j-1})^q$$

respectively. For the first one, $\oint_{\mathbf{T}}^{g_9}(x_{\mathbf{T}}, \mathcal{N}) = x_t^p$, so

$$\oint_{\mathbf{T}}^{g_9}(x_{\mathbf{T}}, \mathcal{N}) \overset{F}{=} x_t^p, \qquad \int_{\mathbf{T}} d(X_s^p) = X_t^p,$$

the latter being a strong stochastic integral for arbitrary process $X_{\mathbf{T}}$ (not necessarily Brownian), and for arbitrary $p \neq 0$. In the second case, provided the process is Brownian, and provided $q > 2$, the isometry argument gives

$$\oint_{\mathbf{T}}^{g_0}(x_{\mathbf{T}}, \mathcal{N}) \overset{G}{=} 0, \qquad \int_{\mathbf{T}} (dX_s)^q = 0,$$

the latter integral being weak stochastic. (The case $q = 2$ gives isometry, see Example 25 above.)

These examples show that the calculations

$$\oint_{\mathcal{J}_{\mathbf{T}}} \cdots, \qquad \mathbf{S}_{\mathbf{T}} \cdots, \qquad \mathcal{S}_{\mathbf{T}} \cdots$$

may yield different numerical values in each case, but the differences come from a set of processes $x_{\mathbf{T}}$ which constitute a G-null set in $\bar{\mathbf{R}}^{\mathbf{T}}$; and the values yielded by each of the calculations have the same distribution function in $\mathbf{R}$ (theorem 83, [MTRV] page 200).

This has been established on a case-by-case basis in the examples[10] above. Is it true in general? The following result shows that strong stochastic integrability implies stochastic summability (that is, the stochastic sum is a random variable).

Theorem 1 *If* $X_{\mathbf{T}}$ *is Brownian and if* $\mathbf{S}_{\mathbf{T}}^{g}(X_{\mathbf{T}})$ *exists then* $\oint_{\mathcal{J}_{\mathbf{T}}}^{g}(X_{\mathbf{T}}, \mathcal{N})$ *satisfies*

$$\oint_{\mathcal{J}_{\mathbf{T}}}^{g}(x_{\mathbf{T}}, N) \overset{G}{=} \mathbf{S}_{\mathbf{T}}^{g}(x_{\mathbf{T}}).$$

Proof. (The assumption of Brownian motion ensures that the set of discontinuous $x_{\mathbf{T}}$ is G-null. It is not needed otherwise.) For each continuous $x_{\mathbf{T}}$ choose $L(x_{\mathbf{T}})$ so that $N \supseteq L(x_{\mathbf{T}})$ implies

$$\left| \oint_{\mathcal{J}_{\mathbf{T}}}^{g}(x_{\mathbf{T}}, N) - \mathbf{S}_{\mathbf{T}}^{g}(x_{\mathbf{T}}) \right| < \varepsilon.$$

With $\delta_{\mathcal{N}}$ arbitrary and $\gamma = (L, \delta_{\mathcal{N}})$, let $\mathcal{D}$ be a γ-fine division of $\mathbf{R}^{\mathbf{T}}$. Then

$$(\mathcal{D}) \sum \left| \oint_{\mathcal{J}_{\mathbf{T}}}^{g}(x_{\mathbf{T}}, N) - \mathbf{S}_{\mathbf{T}}^{g}(x_{\mathbf{T}}) \right| G(I[N]) < \varepsilon (\mathcal{D}) \sum G(I[N]) < \varepsilon,$$

giving the result. ○

Likewise for weak stochastic integrals:

Theorem 2 *If* $X_{\mathbf{T}}$ *is Brownian and if* $\mathcal{S}_{\mathbf{T}}^{g}(X_{\mathbf{T}})$ *exists then* $\oint_{\mathcal{J}_{\mathbf{T}}}^{g}(X_{\mathbf{T}}, \mathcal{N})$ *satisfies*

$$\oint_{\mathcal{J}_{\mathbf{T}}}^{g}(x_{\mathbf{T}}, N) \overset{G}{=} \mathcal{S}_{\mathbf{T}}^{g}(x_{\mathbf{T}}).$$

Proof. By definition ([MTRV] page 407), $\mathcal{S}_{\mathbf{T}}^{g}(x_{\mathbf{T}})$ is a function $f(x_{\mathbf{T}})$ such that

$$\int_{\mathbf{R}^{\mathbf{T}}} \left| \mathcal{R}_{\mathbf{T}}^{g,r}(x_{\mathbf{T}}) - \mathcal{S}_{\mathbf{T}}^{g}(x_{\mathbf{T}}) \right| G(I[N]) \rightarrow 0$$

as $r \rightarrow \infty$. (This does not define $\mathcal{S}_{\mathbf{T}}^{g}(x_{\mathbf{T}})$ uniquely. But if there is another function $\bar{f}(x_{\mathbf{T}})$ satisfying this condition, it is straightforward to show that $f \overset{G}{=} \bar{f}$; so $\mathcal{S}_{\mathbf{T}}^{g}$ is defined up to G-equivalence.) With $\varepsilon > 0$ given, it is then possible to find

---

[10]The strong stochastic integral calculations in the examples above yielded exactly the same numerical values as the corresponding stochastic sum calculations. But, in principle, this need not hold universally.

a positive integer r_ε and a gauge $\gamma = (L, \delta_N)$ such that, for all γ-fine divisions $\mathcal{D}$ of $\mathbf{R^T}$, and for all $r \geq r_\varepsilon$,

$$(\mathcal{D}) \sum \left| \mathcal{R}_\mathbf{T}^{g,r}(x_\mathbf{T}) - \mathcal{S}_\mathbf{T}^g(x_\mathbf{T}) \right| G(I[N]) \;<\; \varepsilon.$$

Next, use the fact that, for Brownian motion, the set of $x_\mathbf{T} \in \mathbf{R^T}$ which fail to be uniformly continuous is G-null ([MTRV] theorem 184, page 296, with continuous modification in $\mathbf{R^T}$). For uniformly continuous $x_\mathbf{T}(s)$ ($s \in \mathbf{T}$), choose $r_1 \geq r_\varepsilon$ and $L^{(1)}(x_\mathbf{T}) \geq M_{r_\varepsilon}$, $L^{(1)}(x_\mathbf{T}) \geq L(x_\mathbf{T})$, so that, for $N \geq L^{(1)}(x_\mathbf{T})$,

$$\left| \oint_\mathbf{T}^g (x_\mathbf{T}, N) - \mathcal{R}_\mathbf{T}^{g,r_1}(x_\mathbf{T}) \right| \;<\; \varepsilon.$$

Define gauge $\gamma^{(1)}$ as $\gamma^{(1)} = (L^{(1)}, \delta_N)$, and choose any $\gamma^{(1)}$-fine division $\mathcal{D}_1$ of $\mathbf{R^T}$. Then $(\mathcal{D}_1) \sum \left| \oint_\mathbf{T}^g(x_\mathbf{T}, N) - \mathcal{S}_\mathbf{T}^g(x_\mathbf{T}) \right| G(I[N]) \;=$

$$
\begin{aligned}
&= (\mathcal{D}_1) \sum \left| \left(\oint_\mathbf{T}^g(x_\mathbf{T}, N) - \mathcal{R}_\mathbf{T}^{g,r_1}(x_\mathbf{T}) \right) + \left(\mathcal{R}_\mathbf{T}^{g,r_1}(x_\mathbf{T}) - \mathcal{S}_\mathbf{T}^g(x_\mathbf{T}) \right) \right| G(I[N]) \\
&\leq (\mathcal{D}_1) \sum \left| \left(\oint_\mathbf{T}^g(x_\mathbf{T}, N) - \mathcal{R}_\mathbf{T}^{g,r_1}(x_\mathbf{T}) \right) \right| G(I[N]) \;+ \\
&\qquad\qquad\qquad + (\mathcal{D}_1) \sum \left| \left(\mathcal{R}_\mathbf{T}^{g,r_1}(x_\mathbf{T}) - \mathcal{S}_\mathbf{T}^g(x_\mathbf{T}) \right) \right| G(I[N]) \\
&< \varepsilon \, (\mathcal{D}_1) \sum G(I[N]) + \varepsilon \;=\; 2\varepsilon;
\end{aligned}
$$

giving $\oint_\mathbf{T}^g(x_\mathbf{T}, N) \overset{G}{=} \mathcal{S}_\mathbf{T}^g(x_\mathbf{T})$ as required. $\bigcirc$

Theorems 1 and 2 show that stochastic sums $\oint$, based on finite samples of times $s \in \mathbf{T}$ (so there are no convergence issues in $\mathbf{T}$), can be used instead of stochastic integrals $\mathcal{S}$, or $\mathbf{S}$, whose convergence (weak or strong, respectively) in $\mathbf{T}$ may be hard to establish.

6.9   Introduction to Itô's Formula

Some of the stochastic integrals discussed above have the form $\int_\mathbf{T} f(X_s) dX_s$. The outcome of such a calculation is a single, unpredictable real number w. If the range of such possible values w has a probability distribution function F, then the result of this procedure can be regarded as an *observable*. The range Ω of possible values of w can be taken to be $\mathbf{R}$, and then, denoting the distribution function F by F_W, the observable is

$$W \;\simeq\; w\,[\mathbf{R}, F_W]. \tag{6.30}$$

The stochastic integral $\int_\mathbf{T} f(X_s) dX_s$ has the form $h(X_\mathbf{T})$, so it has the form of a *contingent observable*; and, in this alternative form, it can be represented as

$$h(X_\mathbf{T}) \;\simeq\; h(x_\mathbf{T})\,[\mathbf{R^T}, F_{X_\mathbf{T}}], \qquad h(X_\mathbf{T}) \;=\; \int_\mathbf{T} f(X_s) dX_s. \tag{6.31}$$

Each sample outcome $w \in \mathbf{R}$ has $w = \int_{\mathbf{T}} f(x_s) dx_s = h(x_{\mathbf{T}})$, and this relationship between the elementary and contingent forms of the observable is denoted by

$$W = \int_{\mathbf{T}} f(X_s) dX_s.$$

For an observable to qualify as a random variable it must be possible to analyse it in terms of probabilities. In [MTRV] this is reduced to existence of expected value of the observable. In this case, expected value of the random variable is

$$E[W] = \int_{\mathbf{R}} w \, F_W(I), \qquad E[h(X)] = \int_{\mathbf{R}^{\mathbf{T}}} h(X_{\mathbf{T}}) F_{X_{\mathbf{T}}}(I[N])$$

where $h(X_{\mathbf{T}}) = \int_{\mathbf{T}} f(X_s) dX_s$; and since the same experiment underlies the two calculations, we should have equivalence

$$E[W] = E[h(X)], = E\left[\int_{\mathbf{T}} f(X_s) dX_s \right].$$

Sufficient conditions for the mathematical equivalence of the elementary and contingent representations are given in theorem 78 (page 195 of [MTRV]; see also page 198).

Which version of the observable is the most amenable to mathematical analysis? Which of the calculations, $E[W]$ or $E[h(X)]$ is the easiest to calculate?

The preceding discussion does not assume that the process $X_{\mathbf{T}}$ is Brownian. In other words $F_{X_{\mathbf{T}}}$ is not assumed to be the Brownian distribution function G.

But when $X_{\mathbf{T}}$ is assumed to be Brownian, the traditional version of stochastic calculus has a device called *Itô's formula* which provides a way to calculate expected value when F_W—the distribution function of the elementary form of the stochastic integral $\int_{\mathbf{T}} f(X_s) dX_s$—is unknown or difficult to analyse.

So if, for instance, F_W is difficult to work with, Itô's formula provides a way of using the known distribution functions for the underlying normally distributed random variables X_s ($s \in \mathbf{T}$).

Suppose f depends explicitly on s as well as X_s ($0 \le s \le t$), and, writing

$$f(X_s, s) = Y_s, \qquad f(x_s, s) = y_s \in \mathbf{R},$$

assume that all relevant partial derivatives of f exist, with notation

$$\frac{\partial f(w, s)}{\partial w} = f'(w, s), \qquad \frac{\partial f(w, s)}{\partial s} = \dot{f}(w, s), \qquad \frac{\partial^2 f(w, s)}{\partial w^2} = f''(w, s).$$

Here is a version of Itô's formula, with $Y_s = f(x_s, s)$:

$$dY_s = \dot{f}(X_s, s) ds + \frac{1}{2} f''(X_s, s) ds + f'(X_s, s) dX_s, \quad \text{or} \quad (6.32)$$

$$\int_{\mathbf{T}} dY_s = \int_{\mathbf{T}} \dot{f}(X_s, s) ds + \int_{\mathbf{T}} \frac{1}{2} f''(X_s, s) ds + \int_{\mathbf{T}} f'(X_s, s) dX_s. \quad (6.33)$$

Note that $Y_t = \int_{\mathbf{T}} dY_s$ by telescoping (or cancellation of terms) in the Riemann sums. Applications of Itô's formula are demonstrated in [MTRV] section 8.12 (pages 429–433).

6.10   Itô's Formula for Stochastic Sums

Suppose the stochastic integrand $f(X_s, s)dX_s$ of Itô's formula is replaced by stochastic summand $f(X_{j-1}, s_{j-1})(X_j - X_{j-1})$, with corresponding stochastic integral and stochastic sum

$$h(X_T) = \int_T f(X_s, s)\, dX_s, \qquad h(X_T, \mathcal{N}) = \sum_{j=1}^{n} f(X_{j-1}, s_{j-1})(X_j - X_{j-1})$$

respectively. The corresponding sample values are

$$h(x_T) = \int_T f(x_s, s)\, dx_s, \qquad h(x_T, N) = \sum_{j=1}^{n} f(x_{j-1}, s_{j-1})(x_j - x_{j-1}).$$

Note that the stochastic integral $\int_T f(x_s, s)\, dx_s$ (strong or weak version) may not always exist, whereas the corresponding stochastic sum always exists. That is to say,

$$\oint_T^g (x_T, N) = \sum_{j=1}^{n} f(x_{j-1}, s_{j-1})(x_j - x_{j-1})$$

always exists, where g denotes stochastic summand $f(x_s, s)(x_{s'} - x_s)$.

Again writing $Y_s = f(x_s, s)$, (6.32) is valid provided f is sufficiently "well behaved". With stochastic sum replacing stochastic integral, (6.33) becomes

$$
\begin{aligned}
Y_n &= \sum_{j=1}^{n} (Y_j - Y_{j-1}) \\
&= \sum_{j=1}^{n} \dot{f}(X(s_{j-1}), s_{j-1})(s_j - s_{j-1}) \; + \\
&\quad + \frac{1}{2} \sum_{j=1}^{n} f''(X(s_{j-1}), s_{j-1})(s_j - s_{j-1}) \; + \\
&\quad + \sum_{j=1}^{n} f'(X(s_{j-1}), s_{j-1})(X(s_j) - X(s_{j-1})). \qquad (6.34)
\end{aligned}
$$

Writing

$$
\begin{aligned}
g(s) &= f(X_s, s)(s' - s), \\
\dot{g}(s) &= \dot{f}(X_s, s)(s' - s), \\
g''(s) &= f''(X_s, s)(s' - s), \\
g'(s) &= f'(X_s, s)(X_{s'} - X_s),
\end{aligned}
$$

then, in the $\oint$ notation, (6.34) becomes

$$Y_t = \oint_T^{\dot{g}} (X_T, \mathcal{N}) + \frac{1}{2} \oint_T^{g''} (X_T, \mathcal{N}) + \oint_T^{g'} (X_T, \mathcal{N}). \qquad (6.35)$$

6.11 Proof of Itô's Formula

As in [MTRV] (section 8.11, pages 426–429), suppose initially that X is an arbitrary process $X_{\mathbf{T}} \simeq x_{\mathbf{T}} [\mathbf{R}^{\mathbf{T}}, F_{X_{\mathbf{T}}}]$, and suppose $f(X_s, s)$ is a contingent observable, where f is a deterministic function of two variables w and s, with $w \in \mathbf{R}$ and $0 \le s \le t$.

As in [MTRV], apply Taylor's theorem (with remainder) to the function $f(w, s)$. Assume that the $(m+1)$-th order partial derivatives of f exist. Suppose $0 < s < s' \le t$ and $x_{\mathbf{T}} \in \mathbf{R}^{\mathbf{T}}$. Taylor's theorem states that $f(x_{s'}, s') - f(x_s, s)$ is equal to

$$\left((x_{s'} - x_s) \frac{\partial}{\partial x_s} + (s' - s) \frac{\partial}{\partial s} \right) f(x_s, s) \;+\; \cdots \tag{6.36}$$

$$\cdots \;+\; \frac{1}{m!} \left((x_{s'} - x_s) \frac{\partial}{\partial x_s} + (s' - s) \frac{\partial}{\partial s} \right)^m f(x_s, s) \;+\; R_m,$$

with

$$R_m \;=\; \frac{1}{(m+1)!} \left((x_{s'} - x_s) \frac{\partial}{\partial \xi} + (s' - s) \frac{\partial}{\partial u} \right)^{m+1} f(\xi, u), \tag{6.37}$$

and

$$\xi \;=\; x_s + \lambda (x_{s'} - x_s), \qquad u \;=\; s + \lambda(s' - s), \qquad 0 < \lambda < 1, \tag{6.38}$$

where λ depends on f, and on the values x_s and $x_{s'}$. (The parameter λ does not depend on values x_ς for $\varsigma \ne s$, $\varsigma \ne s'$.) With $m = 1$, write

$$\frac{\partial f(x_s, s)}{\partial x_s} \;=\; f'_s, \qquad \frac{\partial f(x_s, s)}{\partial s} \;=\; \dot{f}_s, \tag{6.39}$$

$$\frac{\partial^2 f(\xi, u)}{\partial \xi^2} \;=\; f''_\lambda, \qquad \frac{\partial^2 f(\xi, u)}{\partial u^2} \;=\; \ddot{f}_\lambda, \qquad \frac{\partial^2 f(\xi, u)}{\partial \xi \partial u} \;=\; \dot{f}'_\lambda.$$

With $m = 1$, and writing $y_s = f(x_s, s)$, (6.36) becomes

$$\begin{aligned} y_{s'} - y_s \;=\;\; & (x_{s'} - x_s) f'_s + (s' - s) \dot{f}_s + \tfrac{1}{2}(x_{s'} - x_s)^2 f''_\lambda + \\ & + (x_{s'} - x_s)(s' - s) \dot{f}'_\lambda + \tfrac{1}{2}(s' - s)^2 \ddot{f}_\lambda. \end{aligned} \tag{6.40}$$

First consider the implication for stochastic integrals. Assume $f(0,0) = 0$. The strong stochastic integral of the left-hand side of (6.40) exists, with

$$\int_{\mathbf{T}} (y_{s'} - y_s) \;=\; \int_{\mathbf{T}} dy_s \;=\; y_t - y_0 \;=\; f(x_t, t) - f(x_0, 0) \;=\; f(x_t, t) - f(0, 0) \;=\; f(x_t, t);$$

so the strong stochastic integral of the right hand side of (6.40) exists, and

$$Y_t \;=\; \int_{\mathbf{T}} \left(\dot{f}_s \, ds + \frac{1}{2} f''_\lambda \, dX_s^2 + f'_s \, dX_s + \dot{f}'_\lambda \, dX_s ds + \frac{1}{2} \ddot{f}_\lambda \, ds^2 \right). \tag{6.41}$$

When (X_s) is Brownian, it is shown in [MTRV] that the final two terms can be omitted, and dX_s^2 can be replaced by ds, and Itô's formula finally emerges as

$$ Y_t \;=\; \int_T \dot f_s\, ds \;+\; \frac{1}{2}\int_T f_s''\, ds \;+\; \int_T f_s'\, dX_s $$

provided $Y_0 = f(0,0) = 0$. This change of variable expresses Y_t in terms of (X_s).

This has been dealt with in section 8.11 of [MTRV]. But the task in hand is stochastic summation, not stochastic integration.

Accordingly, with $N = \{s_1,\ldots,s_{n-1},s_n\}$, the Taylor's theorem equation (6.40) can be written as $\quad y_j - y_{j-1} \;=$

$$
\begin{aligned}
=\;& (x_j - x_{j-1})f_{j-1}' + (s_j - s_{j-1})\dot f_{j-1} + &&\text{(6.42)}\\
&+ \frac{1}{2}(x_j - x_{j-1})^2 f_{\lambda_j}'' + (x_j - x_{j-1})(s_j - s_{j-1})\dot f_{\lambda_j}' + \frac{1}{2}(s_j - s_{j-1})^2 \ddot f_{\lambda_j},
\end{aligned}
$$

where λ_j is the λ of (8.25) in page 427 of [MTRV].

Then, if $\sum_{j=1}^{n}$ is placed in front of each term of (6.42), the result is a stochastic sum version of Itô's formula. The first three terms of the right hand side of equation (6.42) correspond to the three terms of the right hand side of (6.35):

$$ Y_t \;=\; f(X_t,t) \;=\; \oint_T^{\dot g}(X_T,N) + \frac{1}{2}\oint_T^{g''}(X_T,N) + \oint_T^{g'}(X_T,N); \quad \text{(6.43)} $$

and this in turn matches the classical version (6.33), with $\oint$ replacing $\int$. Note that $\oint$ involves calculating sums of finite numbers of random variables and does not entail convergence issues. The $\int$ formulation in (6.33) is more problematical than the finite sums $\oint$ of random variables in (6.43).

But the validity of (6.43) remains to be confirmed. Rewrite it as

$$ f(x_t,t) \;=\; \sum_{j=1}^{n} f_{j-1}'(x_j - x_{j-1}) + \sum_{j=1}^{n} \dot f_{j-1}(s_j - s_{j-1}) + \frac{1}{2}\sum_{j=1}^{n} f_{j-1}''(s_j - s_{j-1}). \quad \text{(6.44)} $$

Define

$$ k_1(x_T,N) \;=\; \sum_{j=1}^{n}\left((x_j - x_{j-1})f_{j-1}' + (s_j - s_{j-1})\dot f_{j-1} + \frac{1}{2}(s_j - s_{j-1})f_{j-1}'' \right) $$

$$
\begin{aligned}
k_2(x_T,N) \;=\; \sum_{j=1}^{n}\Big(&(x_j - x_{j-1})f_{j-1}' + (s_j - s_{j-1})\dot f_{j-1} + \frac{1}{2}(x_j - x_{j-1})^2 f_{\lambda_{j-1}}''\\
&+ (x_j - x_{j-1})(s_j - s_{j-1})\dot f_{\lambda_j}' + \frac{1}{2}(s_j - s_{j-1})^2 \ddot f_{\lambda_{j-1}} \Big),
\end{aligned}
$$

If we can show that

$$ k_1(x_T,N) \;\overset{G}{=}\; k_2(x_T,N), $$

or $\int_{\mathbf{R}^{\mathbf{T}}} |k_1(x_{\mathbf{T}}, N) - k_2(x_{\mathbf{T}}, N)| \, G(I[N]) = 0$, then (6.43) is proved. Write

$$
\begin{aligned}
k(x_{\mathbf{T}}, N) \;=\;& k_1(x_{\mathbf{T}}, N) - k_2(x_{\mathbf{T}}, N) \\
\;=\;& \sum_{j=1}^{n} \frac{1}{2} \left((s_j - s_{j-1}) f''_{j-1} - (x_j - x_{j-1})^2 f''_{\lambda_{j-1}} \right) - \\
& - \sum_{j=1}^{n} \left((x_j - x_{j-1})(s_j - s_{j-1}) \dot{f}'_{\lambda_j} + \frac{1}{2}(s_j - s_{j-1})^2 \ddot{f}_{\lambda_{j-1}} \right).
\end{aligned}
$$

A sufficient condition for completion of proof is to assume that the derivatives of f are continuous on $[0, t]$, and thus bounded by some positive number α. Then continuity of Brownian $x_{\mathbf{T}}(s)$, along with the isometry argument of Example 25, means that, with $\varepsilon > 0$ given, $L(\mathbf{x}_T)$ and gauge γ can be chosen so that for any γ-fine division $\mathcal{D}$ of $\mathbf{R}^{\mathbf{T}}$,

$$
(\mathcal{D}) \sum |k(x_{\mathbf{T}}, N)| \, G(I[N]) \;<\; \varepsilon \beta \, (\mathcal{D}) \sum G(I[N]) \;=\; \beta \varepsilon
$$

where β is some fixed positive bound. $\qquad \bigcirc$

6.12 Stochastic Sums or Stochastic Integrals?

To conclude, the following exemplar has been used to motivate the underlying aggregative ideas of stochastic integration and stochastic summation:

$$
\sum f(X_s) \, \Delta X_s, \quad \text{or} \quad \sum_{j=1}^{n-1} f(X(s_{j-1})) \, (X(s_j) - X(s_{j-1})), \qquad (6.45)
$$

with sample path formulation

$$
\sum f(x_s) \, \Delta x_s, \quad \text{or} \quad \sum_{j=1}^{n-1} f(x(s_{j-1})) \, (x(s_j) - x(s_{j-1})). \qquad (6.46)
$$

To visualise (6.45) and (6.46), think of a financial portfolio consisting of a variable amount $f(X_s)$ (at time s) of shares whose value at time s is X_s. If the underlying share price changes from $x(s)$ at time s to $x(s')$ at later time s', the portfolio value changes by the approximate amount $f(x(s_{j-1})) \, (x(s_j) - x(s_{j-1}))$, and we then aggregate or accumulate these individual changes by means of the Riemann sum-type calculations in (6.45) and (6.46), in order to estimate the overall change w in portfolio value over the whole time range $s = 0$ to $s = t$.

For the moment denote the Riemann sums of (6.45) and (6.46) by $h(X_{\mathbf{T}})$ and $h(x_{\mathbf{T}})$, respectively; and write $w = h(x_{\mathbf{T}})$, so $w \in \mathbf{R}$.

The underlying parameters such as $x(s)$ are unpredictable and are regarded as random variables $X(s)$, or process (X_s), $= X_{\mathbf{T}}$. Then, provided a distribution function F_W can be found for it, the unpredictable real numbers w may constitute a random variable W with domain[11] $\mathbf{R}$. This gives an elementary observable formulation for aggregate change in value:

$$W \simeq w[\mathbf{R}, F_W].$$

The aggregate change in value can be looked at from two different points of view:

$$W \simeq w[\mathbf{R}, F_W], \quad \text{or} \quad h(X_{\mathbf{T}}) \simeq h(x_{\mathbf{T}})\left[\mathbf{R}^{\mathbf{T}}, G\right], \quad \text{with} \quad w = h(x_{\mathbf{T}})$$

since each of these represents the same thing—the aggregate change in portfolio value.

The two different representations W and $h(X_{\mathbf{T}})$ are, respectively, elementary observable and joint contingent[12] observable. They point to two alternative ways of calculating the random variability properties of the "aggregate change of value" phenomenon, based around the alternative integration calculations

$$\mathrm{E}[W] = \int_{\mathbf{R}} w\, F_W(J), \qquad \mathrm{E}[h(X_{\mathbf{T}})] = \int_{\mathbf{R}^{\mathbf{T}}} h(x_{\mathbf{T}})G(I[N]).$$

The "$\mathrm{E}[W]$" approach involves one-dimensional analysis in sample domain $\mathbf{R}$ with probability distribution F_W, while the "$\mathrm{E}[h(X_{\mathbf{T}})]$" approach involves infinite-dimensional analysis in sample domain $\mathbf{R}^{\mathbf{T}}$ with probability[13] distribution G.

Which of the two is best? [MTRV] has a comprehensive calculus of integrals $\int_{\mathbf{R}^{\mathbf{T}}} \cdots$ (infinite-dimensional integrals), with primacy given to "$\mathrm{E}[h(X_{\mathbf{T}})]$". The classical approach, on the other hand, focusses on "$\mathrm{E}[W]$". Either way the starting point has to be the probability measure (such as G or $F_{X_{\mathbf{T}}}$) in $\mathbf{R}^{\mathbf{T}}$. Classically, the corresponding probability measure F_W in $\mathbf{R}$ has to be deduced from G (or F_{X_T}). This can be quite difficult, but Itô's formula can resolve the problem in some cases.

The observable W is based not just on the distribution function F_W but also on the sample elements w. This presents another problem with "$\mathrm{E}[W]$".

Empirically it seems natural that a portfolio containing a variable quantity (denoted $f(x_s)$) of shares will undergo changes in total portfolio value at different times s, corresponding to the changes in the share value $x(s)$; and therefore it is empirically or intuitively meaningful to think of the overall change in value

---

[11]If values such as x_s or w are to be interpreted as prices, their domain should be $\mathbf{R}_+$, the non-negative real numbers. Also, while we continue to use the relatively familiar Brownian G as distribution function in $\mathbf{R}^{\mathbf{T}}$, it need have no special standing, and may be inappropriate. These aspects are dealt with in [MTRV], pages 433–489.

[12]In [MTRV] pages 433-444, the valuation of share options introduces more contingency. A share option is an asset whose value depends on the underlying share value. In this case the option value can be represented as a contingent observable $\phi(W)$, depending on the elementary observable W with sample domains $\mathbf{R}$. Or it can be represented as a "contingent-contingent" observable $\phi(f(X_{\mathbf{T}}))$ with contingency $\phi \circ f$ on joint-basic observables (X_s) in domain $\mathbf{R}^{\mathbf{T}}$. (Strictly speaking the sample values are prices so these domains are $\mathbf{R}_+$ and $\mathbf{R}_+^{\mathbf{T}}$.)

[13]If $X_{\mathbf{T}}$ is standard Brownian the joint distribution function is $G(I[N])$. For general process $X_{\mathbf{T}}$, denote general distribution function by $F_{X_{\mathbf{T}}}(I[N])$.

between times $s = 0$ and $s = t$. The question is, how to formulate this change mathematically.

The symbol w is used to denote the overall change (growth or loss) in the portfolio value at terminal time t. As discussed earlier, this is not to be understood as $x(t)f(x(t)) - x(0)f(x(0))$. Instead, an investor intelligence is assumed to operate at each instant s of time, $0 \le s < t$, continuously amending the quantity $f(x(s))$ of shares in accordance with a particular financial logic called "*no-arbitrage risk-neutrality*". (See [MTRV] pages 433–440.)

Accordingly, a calculation on the lines of (6.46) is used to determine w. The trouble with this is that (6.46) may yield a different value w depending on the choice of the times $s_1, \ldots, s_{n-1}, s_n$. So w is not well-defined by (6.46).

Also, the "*no-arbitrage, risk-neutral*" portfolio[14] quantities f depend on share values x_s at each and every time s ($0 \le s < t$), not just on some finite selection of times. Since (6.46) can be viewed as a Riemann sum, this suggests that the value w should be obtained by means of an integral

$$w := \int_0^t f(x(s))\, dx(s) \tag{6.47}$$

to which (6.46) is an approximation. This is the **stochastic integral** solution.

It is true that such a construction involves each and every time s in $[0, t]$. But in general there is no guarantee that the integral (6.47) exists for every possible sample path $(x(s))$, so by itself (6.47) does not resolve the problem[15] of securing a mathematical definition of w. Even when $(x(s))$ is continuous Brownian it will "typically" have unbounded variation, so the integral construction (6.47) can be problematic.

A classical solution to this is the *Itô integral* which uses weak convergence of Riemann sums such as (6.46). The resulting limit (when it exists) is, by the character of the Itô construction, a random variable in the classical Kolmogorov sense. But the limit is not a natural limit of Riemann sums and, though called a stochastic integral, the limit is not actually an integral; so the machinery and theorems of integration are not generally available to it.

The -complete system of integration of [MTRV] guarantees existence of limits of Riemann sums in many cases (see [MTRV] section 8.3, pages 391–398), so the integrals $\int_0^t f(x_s)\, dx_s$ exist, and in those cases the integral is called the **strong stochastic integral**. Writing integrand $f(x_s)\, dx_s$ as $g(s)$, the strong stochastic integral is

$$\mathbf{S}_{\mathbf{T}}^g(x_{\mathbf{T}}), \ = \ \int_0^t f(x_s)\, dx_s, \tag{6.48}$$

obtained by the usual gauge-Riemann sum definition of the Burkill-complete integral; and the full power of integration theory is available, including limit theorems and Fubini's theorem.

---

[14] See [MTRV], section 8.14

[15] Example 45 (page 181 of [MTRV]) illustrates an extreme "pathology" in which Stieltjes integrability fails.

The $\mathbf{S}_{\mathbf{T}}^g(\cdots)$ notation for strong stochastic integral emphasises the integral as a functional of sample paths $x_{\mathbf{T}} \in \mathbf{R}^{\mathbf{T}}$. With distribution function such as $G(I[N])$ in $\mathbf{R}^{\mathbf{T}}$, this form of the stochastic integral is a contingent observable

$$\mathbf{S}_{\mathbf{T}}^g(X_{\mathbf{T}}) \simeq \mathbf{S}_{\mathbf{T}}^g(x_{\mathbf{T}})[\mathbf{R}^{\mathbf{T}}, G].$$

To qualify as a random variable in the -complete system, the functional $\mathbf{S}_{\mathbf{T}}^g(x_{\mathbf{T}})$ must be G-integrable on $\mathbf{R}^{\mathbf{T}}$, with

$$\mathrm{E}\left[\mathbf{S}_{\mathbf{T}}^g(X_{\mathbf{T}})\right] = \int_{\mathbf{R}^{\mathbf{T}}} \mathbf{S}_{\mathbf{T}}^g(x_{\mathbf{T}})G(I[N]).$$

This construction contrasts with the classical Itô integral whose definition[16] ensures that it is a random variable in the classical Kolmogorov sense.

As in the Itô theory, in some significant instances the Riemann sums of (6.46) fail to converge in the gauge or -complete system. The most important case is

$$\int_0^t (dx_s)^2, \quad \text{or} \quad \mathcal{S}_{\mathbf{T}}^{g_2}(x_{\mathbf{T}});$$

where the latter notation refers to the -complete version of weak stochastic integral given in section 8.5, [MTRV] pages 407–409. The weak limit in this case is t (see theorem 234, [MTRV] page 403).

Existence of the weak limit $\mathcal{S}_{\mathbf{T}}^g(x_{\mathbf{T}})$ in this sense does not guarantee that the weak limit is a random variable in the -complete sense. For that to hold,

$$\mathrm{E}\left[\mathcal{S}_{\mathbf{T}}^g(X_{\mathbf{T}})\right], = \int_{\mathbf{R}^{\mathbf{T}}} \left(\mathcal{S}_{\mathbf{T}}^g(x_{\mathbf{T}})\right)G(I[N]),$$

must exist (Brownian case). For $g = g_2$, with $\mathcal{S}_{\mathbf{T}}^{g_2}(x_{\mathbf{T}}) = t$,

$$\mathrm{E}\left[\mathcal{S}_{\mathbf{T}}^{g_2}(X_{\mathbf{T}})\right] = \int_{\mathbf{R}^{\mathbf{T}}} \left(\mathcal{S}_{\mathbf{T}}^{g_2}(x_{\mathbf{T}})\right)G(I[N]) = t\int_{\mathbf{R}^{\mathbf{T}}} G(I[N]) = t.$$

Measurability—in the -complete sense of existence of $\mathrm{E}\left[\mathcal{S}_{\mathbf{T}}^g(X_{\mathbf{T}})\right]$—is generally fairly easy to establish for those strong and weak stochastic integrals discussed in [MTRV]. So, by this standard, these are random variables in the -complete sense; just as they are in the Lebesgue/Kolmogorov/Itô sense in which measurability is built into the definition.

To sum up, two fundamental issues have to be considered in relation to stochastic integrals, whether in the -complete or the Itô sense:

1. Convergence of expressions (6.46), and

2. Measurability of these limits whenever they exist in the strong, weak, or Itô senses.

---

[16]The contrast is a superficial one. In most cases the measure theoretical method of Kolmogorov yields the same result as the -complete construction using Riemann sums (6.46).

Stochastic sums $\oint_{\mathbf{T}}^{g}(X_{\mathbf{T}}, \mathcal{N})$ have been introduced as an alternative to the strong, weak, and Itô stochastic integrals. To what extent do the preceding issues apply to stochastic sums?

With $g(s) = f(x_s)(x_{s'} - x_s)$, the expression $\oint_{\mathbf{T}}^{g}(x_{\mathbf{T}}, N)$ has the same form as the Riemann sum (6.46). And $\oint_{\mathbf{T}}^{g}(x_{\mathbf{T}}, N)$ has no *a priori* convergence requirement of the stochastic integral kind, whether strong, weak or Itô. Therefore the first of the two issues mentioned above does not apply to stochastic sums.

The second issue is whether $\oint_{\mathbf{T}}^{g}(X_{\mathbf{T}}, \mathcal{N})$ is a random variable—in other words, whether $\mathrm{E}\left[\oint_{\mathbf{T}}^{g}(X_{\mathbf{T}}, \mathcal{N})\right]$ exists. In the -complete system this means existence of the infinite-dimensional integral

$$\int_{\mathbf{R}^{\mathbf{T}}} \left(\oint_{\mathbf{T}}^{g}(x_{\mathcal{T}}, N)\right) G(I[N]) \text{ in the Brownian case; or}$$

$$\int_{\mathbf{R}^{\mathbf{T}}} \left(\oint_{\mathbf{T}}^{g}(x_{\mathcal{T}}, N)\right) F_{X_{\mathcal{T}}}(I[N]) \text{ in the general case.}$$

As demonstrated in the preceding sections, this can be done by examining the summand-integrand g and its stochastic sum or Riemann sum $\oint_{\mathbf{T}}^{g}(x_{\mathcal{T}}, N)$.

Since only the second of the above two fundamental issues applies to stochastic sums, these are simpler to use than stochastic integrals.

A point made at the beginning of this section is that (6.46) only refers to a finite selection $N = \{s_1, \ldots, s_{n-1}, s_n\}$ of the uncountably infinite number of times $s \in \mathbf{T}$, even though the value w depends on every $s \in \mathbf{T}$ and not on a finite number of them.

However the selection N is variable in the infinite-dimensional integral $\int_{\mathbf{R}^{\mathbf{T}}} \cdots$. The possible outcomes w, estimated as stochastic sums $\oint_{\mathbf{T}}^{g}(x_{\mathbf{T}}, N)$, depend on infinitely many random outcomes x_s at infinitely many possible selections of times $N = \{s_1, \ldots, s_{n-1}, s_n\}$.

When the random variability of w is analysed by means of $\int_{\mathbf{R}^{\mathbf{T}}} \cdots$, the infinite variability of samples of values of x_s and s is intrinsic to that mode of analysis.

Part II

Field Theory

Chapter 7

Gauges for Product Spaces

7.1   Introduction

Product domains are the underlying theme of [MTRV], along with their gauges and integrals. The products involved are, essentially, $\mathbf{R} \times \mathbf{R} \times \mathbf{R} \times \cdots$, finitely or infinitely. The product factor is $\mathbf{R}$, but other domains can easily be substituted as factors in such products.

In fact, any domain which lends itself to -complete integral construction can be used as a component of a product space, for formation of -complete integration in the latter. For instance, product spaces can themselves be component spaces of product spaces. So instead of $\mathbf{R} \times \mathbf{R} \times \mathbf{R}$, we could have $\mathbf{R} \times \mathbf{R}^3 \times \mathbf{R}^2$.

The latter is essentially the same as $\mathbf{R} \times \mathbf{R} \times \cdots \times \mathbf{R}, = \mathbf{R}^6$. What is the point of restructuring this product as $\mathbf{R} \times \mathbf{R}^3 \times \mathbf{R}^2$? This chapter addresses this and similar issues.

The main point of the chapter is how to construct gauges for structured product domains. Along the way, it is helpful to take a close look at the proof of theorem 4 (divisibility theorem) in [MTRV].

7.2   Three-dimensional Brownian Motion

In [MTRV] a sample space $\Omega = \mathbf{R}^{\mathbf{T}}$, usually with $\mathbf{T}$ denoting an interval of time such as $]0,t]$, is used to represent stochastic processes $X_{\mathbf{T}}$, as in

$$X_{\mathbf{T}} \simeq x_{\mathbf{T}} \left[\mathbf{R}^{\mathbf{T}}, F_{X_{\mathbf{T}}} \right].$$

A sample path $x_{\mathbf{T}}$ then consists of real numbers $x(s)$ (or x_s) for $0 < s \le t$,

$$x_{\mathbf{T}} = (x(s) : 0 < s \le t) = (x_s : 0 < s \le t) = (x_s)_{0 < s \le t}.$$

For example, a Brownian motion process is represented as

$$X_{\mathbf{T}} \simeq x_{\mathbf{T}} \left[\mathbf{R}^{\mathbf{T}}, G \right]$$

Gauge Integral Structures for Stochastic Calculus and Quantum Electrodynamics, First Edition. Patrick Muldowney.
© 2021 John Wiley & Sons, Inc. Published 2021 by John Wiley & Sons, Inc.

where the distribution function $G(I[N])$ is defined on cells $I[N] \subseteq \mathbf{R}^\mathbf{T}$. In this case the variable, or outcome, at time s is $x(s)$, where the latter denotes the displacement in one dimension of the Brownian particle from the point of origin of the particle at time $s = 0$.

So $x(0) = 0$ is the origin or starting point of the particle, and $x(s)$ is the distance (at later time s) of the particle from the origin. This distance is unpredictable and indefinite, but is assumed to follow some rule of normal distribution; so it is "observable" in the sense used in [MTRV]. In fact $x(s)$ is a possible outcome of a random variable X_s.

In other words, $x(s)$ is a "sample value" of the observable (or random variable) $X(s)$, and $x = (x(s))_{s \in \mathbf{T}}$ is a "sample path" of the process (joint observable, joint random variable $X = (X(s))_{s \in \mathbf{T}}$. The collection $X_\mathbf{T}$ ($= X$) of such X_s (or $X(s)$), subject to a joint probability distribution

$$G(I[N]) \;=\; P\left(x(s) \in I(s),\; s \in N\right),$$

gives the Brownian process $X_\mathbf{T} \simeq x_\mathbf{T}\left[\mathbf{R}^\mathbf{T}, F_{X_\mathbf{T}}\right]$.

As it stands, however, this mathematical formulation describes the random motion in one direction only; whereas, in reality, such a particle is moving randomly, not in one dimension but in three.

Example 31 *In order to have a mathematical representation of Brownian particle motion in three dimensions, the following scheme can be used. Suppose the motion is in mutually perpendicular spatial directions $k = 1$, $k = 2$, $k = 3$ (that is, horizontal and vertical dimensions of a plane, and a third dimension of height/depth); and suppose the origin of the motion at time $s = 0$ is $(0,0,0)$,*

$$(x(1,0),\, x(2,0),\, x(3,0)) \;=\; (0,0,0), \quad or \quad (x_{1,0}, x_{2,0}, x_{3,0}) \;=\; (0,0,0).$$

Likewise, at time $s > 0$, the particle displacement $x(k,s)$ in each of the three dimensions $k = 1, 2, 3$ is

$$
\begin{aligned}
(x(1,s),\, x(2,s),\, x(3,s)) \;&=\; (x(k,s) : k = 1,2,3;\ 0 < s \le t)\\
&=\; (x_{ks} : k = 1,2,3;\ 0 < s \le t)\\
\text{with}\quad X_{ks} = X(k,s) \;&\simeq\; x_{ks}\,[\Omega,\, G]
\end{aligned}
$$

for each $k = 1, 2, 3$ and each s, $0 < s \le t$, where sample space $\Omega = \mathbf{R}$, and where distribution function G satisfies (among other properties)

$$G(I_{ks}) \;=\; P\left(x_{ks} \in I_{ks}\right)$$

for each cell I_{ks} in $\mathbf{R}$ for each k and each s; the probability being given by the normal distribution Φ with mean zero and variance s, for $k = 1, 2, 3$ and $0 < s \le t$.

Three-dimensional Brownian motion is not adequately described in Example 31 above. There are some further features which should be taken into account.

(For instance, the actual probabilities governing the process have been left unspecified.) At time s a Brownian particle is located at position

$$\mathbf{x}(s) \;=\; \mathbf{x}_s \;=\; (x(1,s), x(2,s), x(3,s)) \;=\; (x_{1s}, x_{2s}, x_{3s}) \in \mathbf{R}^3.$$

For times s_j, $0 = s_0 < s_1 < s_2 < \cdots < s_{n-1} < s_n = t$, denote position by

$$\mathbf{x}(s_j) \;=\; \mathbf{x}_j \;=\; (x(1,s_j), x(2,s_j), x(3,s_j)) \;=\; (x_{1j}, x_{2j}, x_{3j}) \;=\; (x_{kj})_{k=1,2,3} \in \mathbf{R}^3.$$

Let $\mathbf{I} = I_1 \times I_2 \times I_3$ denote cells in R^3 where each I_k $(k = 1, 2, 3)$ is a cell in $\mathbf{R}$. For $1 \le j \le n$ let

$$\mathbf{I}_j \;=\; I_{1j} \times I_{2j} \times I_{3j}.$$

For Brownian particle moving in three dimensions the joint event $\mathbf{I}_j$, with $1 \le j \le n$, denotes the possibility that the particle position $\mathbf{x}_j$ at times s_j is in $\mathbf{I}_j$; so

$$x_{kj} \in I_{kj}, \quad k = 1, 2, 3, \quad j = 1, 2, \ldots n.$$

The next step is to formulate a probability distribution, or probability values for the joint event $\mathbf{I}_j$, $1 \le j \le n$; in other words, the probabilities

$$P(\mathbf{x}_j \in \mathbf{I}_j)_{1 \le j \le n}, \;=\; P(x_{kj} \in I_{kj})_{k=1,2,3,\ 1 \le j \le n}$$

The conditions BM1 to BM7 ([MTRV], pages 305–306) are applicable to the increments $x_{kj} - x_{k,j-1}$ for $1 \le j \le n$ and for $k = 1, 2, 3$, even though the increments are now expressed in all three dimensions of physical space rather than the one-dimensional simplification of [MTRV].

In particular, for any fixed j (or s_j), the increments

$$x_{1j} - x_{1,j-1}, \qquad x_{2j} - x_{2,j-1}, \qquad x_{3j} - x_{3,j-1},$$

in each of the three physical dimensions, are each normally distributed random variables, statistically independent of each other, and of the other increments, at other times $s_{j'} \ne s_j$.

Thus, using the same reasoning as in the one-dimensional case in [MTRV], and writing $\mathbf{I} = \mathbf{I}_1 \times \cdots \times \mathbf{I}_n$, the probability distribution function G for the three-dimensional Brownian motion is $G(\mathbf{I}) =$

$$= P(\mathbf{x}_j \in \mathbf{I}_j : 1 \le j \le n) = \prod_{k=1}^{3} \left(\prod_{j=1}^{n} \left(\frac{\int_{I_{kj}} \exp\left(\frac{-(x_{kj} - x_{k,j-1})^2}{2(t_j - t_{j-1})} \right) dx_{kj}}{\sqrt{2\pi (t_j - t_{j-1})}} \right) \right). \tag{7.1}$$

This expression is a three-dimensional analogue of the one-dimensional construction in section 6.8 (pages 284–288) of [MTRV].

As discussed in pages 87–88 of [MTRV], a sample path $x_{\mathbf{T}}$ can be thought of as a displacement-time graph **or** as a point of an infinite-dimensional Cartesian product space $\mathbf{R}^{\mathbf{T}}$. But "dimension", in "three-dimensional analogue" above, refers, not to this issue, but to the difference between $\mathbf{R}$ (in $\mathbf{R}^{\mathbf{T}}$) and $\mathbf{R}^3$ (in

$\left(\mathbf{R}^3\right)^{\mathbf{T}}$). In other words, it relates to the difference between $x_j - x_{j-1}$ and $x_{kj} - x_{k,j-1}$ $(k = 1, 2, 3)$.

The analysis of one-dimensional Brownian motion $X_{\mathbf{T}}$ in [MTRV] used the stochastic process (defined in the -complete sense) $X_{\mathbf{T}} \simeq x_{\mathbf{T}}[\mathbf{R}^{\mathbf{T}}, G]$ with distribution function G defined in section 6.8 of [MTRV]. In the above discussion, some of the elements of a mathematical representation of three-dimensional Brownian motion have been introduced; in particular, a function $\mathbf{x}_{\mathbf{T}}$ which specifies the position in three-dimensional space, at each time t, of a Brownian particle, and a new version of a probability function $G(\mathbf{I})$ in (7.1).

The question then arises, whether it is possible to build on these elements (that is, $\mathbf{x}_{\mathbf{T}}$ and $G(\mathbf{I})$) to form a -complete stochastic process

$$\mathbf{X}_{\mathbf{T}} \simeq \mathbf{x}_{\mathbf{T}}[\Omega, G],$$

where the sample space Ω is the set $\left(\mathbf{R}^3\right)^{\mathbf{T}}$ of all $\mathbf{x}_{\mathbf{T}}$.

To achieve this objective it is necessary to establish that a -complete integration system of gauge-constrained point-cell elements $(\mathbf{x}, \mathbf{I})$ can be constructed in the domain $\Omega = (\mathbf{R}^3)^{\mathbf{T}}$, in accordance with axioms DS1 to DS8 of section 4.1 in [MTRV]; and in particular to establish existence of gauge-constrained divisions of domain $(\mathbf{R}^3)^{\mathbf{T}}$, as in theorem 4, pages 121–124 of [MTRV].

In preparation for this, some illustrative examples follow.

7.3   A Structured Cartesian Product Space

Example 32 *Here is an example of multi-dimensional random variation which involves measurements which depend on parameters other than time s and location in three-dimensional space. Suppose a single experiment on a moving object involves multiple measurements, with unpredictable outcomes, as follows:*

1. *Spatial orientation o of the object, given by two angles o_1 and o_2 measured in radians; so there are two unpredictable outcomes o_1 and o_2 where $0 \leq o_i < 2\pi$ $(i = 1, 2)$, or $o_i \in S = [0, 2\pi[$, $i = 1, 2$.*

2. *Location l of the object in 3-dimensional space, given by three distance values which measure (in centimetres, say) the co-ordinates of the location of the object relative to some point of origin $(0, 0, 0)$; giving $l_j \in \mathbf{R}$, $j = 1, 2, 3$.*

3. *Energy value of the object (measured in joules, say); giving unpredictable outcome e where $e \geq 0$, or $e \in \mathbf{R}_+$*

To summarize, write

$$x = (x_o, x_l, x_e) \in S^2 \times \mathbf{R}^3 \times \mathbf{R}_+.$$

If appropriate distribution functions exist, this scenario can be formulated in -complete terms, or Riemann-observable terms (i.e. as in [MTRV]) as follows.

If outcomes e (or x_e, measurement of energy) have a probability distribution F_{X_e} then $X_e \simeq x_e[\mathbf{R}_+, F_{X_e}]$ is a basic observable. If joint outcomes $x_l = \left(x_{l_j}\right)$ ($j = 1, 2, 3$) have a joint probability distribution F_{X_l} then $X_l \simeq x_l[\mathbf{R}^3, F_{X_l}]$ is a joint basic observable. If joint outcomes $x_o = (x_{o_i})$ ($i = 1, 2$) have a joint probability distribution F_{X_o} then $X_o \simeq x_o[S^2, F_{X_o}]$ is a joint basic observable.

However, the experiment consists of a single joint measurement of each of the variables. This suggests a single joint observable X involving all of the parameters of the experiment. Let

$$M_o = \{1, 2\}, \quad M_l = \{1, 2, 3\}, \quad M_e = \{1\}, \quad M = (M_o, M_l, M_e) ,$$
$$x_M = ((x_{o_1}, x_{o_2}), (x_{l_1}, x_{l_2}, x_{l_3}), (x_e))) \in (S \times S) \times (\mathbf{R} \times \mathbf{R} \times \mathbf{R}) \times \mathbf{R}_+ ;$$
$$\text{and let} \quad X_M \simeq x_M [\Omega, F_{X_M}] , \tag{7.2}$$

assuming a joint distribution function F_{X_M} exists, defined on cells

$$I = I(M) = I(M_o) \times I(M_l) \times I(M_e)$$

of sample space

$$\Omega = (S \times S) \times (\mathbf{R} \times \mathbf{R} \times \mathbf{R}) \times \mathbf{R}_+.$$

The symbols variously denoted by M are labels corresponding to sets of dimensions N in the theory formulated in [MTRV] for domain $\mathbf{R}^T$.

In terms of the Riemann-observable (or -complete) theory in [MTRV], the *joint basic* observable x_M cannot be understood as a random variable; but a *contingent* observable $f(X_M)$ may, under certain circumstances, qualify as a random variable.

To illustrate, suppose the value of some imagined physical property, call it the *bentropia* of the system (in order to "*simulate realism*"), is measured as a real number $\mathcal{B}$ obtained by means of a deterministic calculation f on the orientation angles, the displacement distances, and the energy of the object,

$$\mathcal{B} = f(x_M) = f((x_{o_1}, x_{o_2}), (x_{l_1}, x_{l_2}, x_{l_3}), (x_e)). \tag{7.3}$$

This distinguishes between joint-basic observables and contingent observable:

$$X_M = (X_o, X_l, X_e), = ((X_{o_1}, X_{o_2}), (X_{l_1}, X_{l_2}, X_{l_3}), X_e) \quad \text{and}$$
$$f(X_M) = f((X_{o_1}, X_{o_2}), (X_{l_1}, X_{l_2}, X_{l_3}), X_e), \quad \text{respectively.}$$

If $\int_\Omega f(x_M) F_{X_M}(I(M))$ exists then $\mathcal{B}$ is a contingent random variable, with expected value

$$E[\mathcal{B}] = \int_\Omega f(x_M) F_{X_M}(I(M)) \tag{7.4}$$
$$= \int_{R_M} f((x_{o_1}, x_{o_2}), (x_{l_1}, x_{l_2}, x_{l_3}), (x_e)) F_{X_M}(I(M)),$$

where $R_M, = \Omega$, denotes the structured product domain

$$(S \times S) \times (\mathbf{R} \times \mathbf{R} \times \mathbf{R}) \times \mathbf{R}_+$$

and $M = (M_o, M_l, M_e)$. For (7.4) to be meaningful in the Riemann-observable sense, $\int_{R_M}$ must qualify as a -complete integral, as defined in chapter 4 of [MTRV]. Essentially, this means there must be gauges, divisions and Riemann sums which satisfy the conditions of chapter 4.

The single most important step in this, from which most of the others follow directly, is the divisibility property which is established in theorem 4, page 121 of [MTRV].

In this case, (7.4) can simply be treated as an integral on $\mathbf{R}^6$—in fact a sub-domain of $\mathbf{R}^6$, with $x = (x_1, x_2, x_3, x_4, x_5, x_6)$ replacing $\mathbf{x} = (x_o, x_l, x_e)$. But it is instructive to treat it instead as an integral on a Cartesian product of the three domains

$$S \times S, \quad \mathbf{R} \times \mathbf{R} \times \mathbf{R}, \quad \text{and} \quad \mathbf{R}_+ .$$

The first two of these are themselves Cartesian products, and the associated elements in the three domains are, respectively,

$$(x_o, I(M_o)), \quad (x_l, I(M_l)), \quad (x_e, I(M_e)),$$

with gauges $\delta(x_o)$, $\delta(x_l)$, and $\delta(x_e)$.

The structured product domain $\Omega = (S \times S) \times (\mathbf{R} \times \mathbf{R} \times \mathbf{R}) \times \mathbf{R}_+$ is to be denoted by R_M. A point $\mathbf{x} = x_M$ of R_M is (x_o, x_l, x_e). This is not quite the same as $(x_{1,o}, x_{2,o}, x_{1,l}, x_{2,l}, x_{3,l}, x_e) \in \mathbf{R}^6$, though the latter is a bit simpler and works equally well.

Cells $I = I_M$ in R_M are $I_o \times I_l \times I_e$. (The more straightforward Cartesian product of six one-dimensional component cells can also be used in this case.)

The point-cell elements of R_M are (x_M, I_M), and these elements are associated if each of x_o, x_l, x_e is associated with, respectively, I_o, I_l, I_e in $S \times S$, $\mathbf{R} \times \mathbf{R} \times \mathbf{R}$, and $\mathbf{R}_+$. The point x_M is an associated point (or tag point) of cell I_M if x_M is a vertex of I_M.

A gauge in R_M is a function δ_M,

$$\delta_M = ((\delta_o(o_1), \delta_o(o_2)), (\delta_l(l_1), \delta_l(l_2), \delta_l(l_3)), \delta_e(e)),$$

defined for each tag point x_M of R_M. An associated point-cell pair (x_M, I_M) is δ_M-fine if each of the components (x_o, I_o), (x_l, I_l) (x_e, I_e) is δ-fine. That is, each

$$(x_{o_i}, I_{o_i}), \quad (x_{l_j}, I_{l_j}) \quad (x_e, I_e)$$

is δ_{o_i}-fine, δ_{l_j}-fine, δ_e-fine, respectively.

For this to be effective, the conditions and properties of the integration theory of [MTRV] chapter 4 must be satisfied by the structured elements (x_M, I_M) of R_M. The single biggest issue here is theorem 4 (pages 121–124 of [MTRV]). We must prove δ_M-fine divisibility of R_M and its sub-domains. That is, given any gauge δ_M, the integration domain R_M has a δ_M-fine division $\mathcal{D} = \{(x_M, I_M)\}$.

But, unlike theorem 4, R_M involves only finite Cartesian products. Therefore the proof of divisibility in R_M is not the same as the proof of theorem 4 of [MTRV]. Instead, it is similar to the standard or classical proof which is outlined in section 2.3 of [MTRV] (page 45). The well-known proof[1] uses iterated bisection, as follows.

Theorem 3 *Given a gauge* δ_M *on* R_M, *there exists a* δ_M-*fine division* $\mathcal{D}$ *of* R_M, $\mathcal{D} = \{(x_M, I_M)\}$. *Likewise if* R_M *is replaced as domain by any cell or figure in* R_M.

Proof. Assume, for contradiction, that the statement is false. Each component in the Cartesian product

$$R_M = (S \times S) \times (\mathbf{R} \times \mathbf{R} \times \mathbf{R}) \times \mathbf{R}_+,$$

can be bisected successively. At each stage a sub-domain is found to be non-divisible. As in page 45 of [MTRV], this gives a contradiction. ○

With this result to hand, it is not difficult to reproduce the theory of Riemann-observables for structured domains such as R_M. Therefore equation (7.4) is valid, and it is possible to incorporate this kind of random variable into the theory of [MTRV].

The preceding examples present some contrasting problems. In Example 32 there are three "dimensions", labelled o, l, and e; representing, respectively, the two-, three-, and one-dimensional domains $S \times S$, $\mathbf{R} \times \mathbf{R} \times \mathbf{R}$, and $\mathbf{R}_+$.

In contrast, Example 31 has infinitely many dimensions, each labelled s for $s \in {]}0, t]$, and each dimension s representing a three-dimensional Cartesian product space $\mathbf{R}^3 = \mathbf{R} \times \mathbf{R} \times \mathbf{R}$, where, for $k = 1, 2, 3$, the unpredictable measurement $x(k, t)$ (or $x_{k,t}$) is the kth co-ordinate of a Brownian particle at time s $(0 < s \leq t)$; so $x(s) = (x_{k,s})_{k=1,2,3} = (x_1(s), x_2(s), x_3(s)) \in \mathbf{R}^3$.

Divisibility of $(\mathbf{R}^3)^{\mathbf{T}}$ is proved in Theorem 6 below. But first, here are some background issues involving gauges for product spaces.

7.4   Gauges for Product Spaces

Example 33 *A gauge in a one-dimensional domain (such as* $\mathbf{R}$*) is a function* $\delta(y) > 0$ *defined for* $y \in \bar{\mathbf{R}}$, $= [-\infty, \infty]$. *Section 2.17 (pages 79–81 of [MTRV]) introduces gauges* δ *for the finite-dimensional Cartesian product domain* $\mathbf{R}^n$ *where* n *is any positive integer. Such a gauge* δ *is simply* $\delta(x) > 0$ *for* $x \in \bar{\mathbf{R}}^n$. *Use the symbol* δ^a *to distinguish such a gauge from* δ^b *in (7.5) below. To keep track of dimensions, denote the domain* $\mathbf{R}^n$ *by*

$$\mathbf{R}^n = \prod_{j=1}^{n} \mathbf{R}_j$$

where $\mathbf{R}_j = \mathbf{R}$ *for each* j, *and where a typical element of* $\bar{\mathbf{R}}_j$ *is denoted by* x_j, *and* $x = (x_1, \ldots, x_n)$ *is a typical element of* $\bar{\mathbf{R}}^n$. *Assume that a gauge* $\delta_j(x_j) > 0$

---

[1] [69] contains an unusual version of the proof; which is replicated in Example 13.

is defined for each $\mathbf{R}_j$. The question posed here is whether a gauge

$$\delta^b(x) \;=\; \delta^b(x_1,\ldots,x_n)$$

can be constructed from component gauges δ_j for the product space $\mathbf{R}^n = \prod_1^n \mathbf{R}_j$. In [MTRV] pages 79–81, there is no presumption of elements $\delta_j(x_j)$. Instead, a function $\delta(x) = \delta(x_1,\ldots,x_n)$ is defined for $x \in \bar{\mathbf{R}}^n$, without considering any kind of one-dimensional elements $\delta_j(x_j)$. However, it is feasible to construct a gauge δ in $\mathbf{R}^n$ as the following composite of the functions δ_j:

$$\delta^b(x) \;=\; (\delta_1(x_1),\ldots,\delta_n(x_n)), \qquad (7.5)$$

where component gauges $\delta_j(x_j)$ are assumed to be given for each $\mathbf{R}_j = \mathbf{R}$.

With this definition of gauge δ^b, the next question is to attribute meaning to "δ^b-fine in $\mathbf{R}^n$". Therefore suppose

$$(x,I), \;=\; ((x_1,\ldots,x_n),\, I_1 \times \cdots \times I_n),$$

is an associated pair in $\mathbf{R}^n$ (so (x_j,I_j) are associated in $\mathbf{R}_j, = \mathbf{R}$, for $j = 1,\ldots n$). Then, with gauge δ^b given by (7.5), we declare that (x,I) is δ^b-fine in $\mathbf{R}^n$ if (x_j,I_j) is δ_j-fine in $\mathbf{R}_j$ for $1 \le j \le n$.

Here are some properties of δ^a and δ^b gauges:

1. Given δ^b, if (x,I) is δ^b-fine then there exists a gauge δ^a such that (x,I) is δ^a-fine. For each $x = (x_1,\ldots,x_n)$, simply take

$$\delta^a(x) \;=\; \min\{\delta_1(x_1),\ldots,\delta_n(x_n)\}.$$

2. Thus if $h(x,I)$ is defined on $\mathbf{R}^n$, and if h is integrable on $\mathbf{R}^n$ with respect to δ^b gauges (with integral value α), then h is integrable on $\mathbf{R}^n$ with respect to δ^a gauges. For, by hypothesis, with $\varepsilon > 0$ given, there exists a gauge δ^b so that, for every δ^b-fine division $\mathcal{D}^b$ of $\mathbf{R}^n$,

$$\left|\alpha - (\mathcal{D}^b)\sum h(x,I)\right| \;<\; \varepsilon.$$

Now choose δ^a as in 1, to complete the proof.

3. Conversely, with δ^a given, it is **not** possible to define a gauge δ^b which ensures that every δ^a-fine pair (x,I) is also δ^b-fine. (This is analogous[2] to the extra "discrimination" that Riemann-complete gauge functions $\delta(y)$ ($y \in \bar{\mathbf{R}}$) provide, in comparison with the constant δ of standard Riemann integration.) To see this, consider any δ^b-gauge in a bounded two-dimensional domain $S = [0,1] \times [0,1]$,

$$\delta^b(x) \;=\; \delta^b(x_1,x_2) \;=\; (\delta_1(x_1),\delta_2(x_2))$$

where δ_j is a gauge in $[0,1]$ with $1 > \delta_j(x_j) > 0$ ($j = 1,2$). Let $\delta^a(x)$ be

$$\delta^a(x) \;=\; \delta^a(x_1,x_2) \;=\; \delta_1(x_1)\delta_2(x_2).$$

---

[2]The analogy: $\delta^a(x)$ corresponds to variable $\delta(y)$, and $\delta^b(x)$ corresponds to constant δ.

Then, for each $x = (x_1, x_2) \in S$, there are cells I_j $(j = 1, 2)$ with

$$\delta^a(x) = \delta_1(x_1)\delta_2(x_2) < |I_j| < \delta_j(x_j)$$

for $j = 1, 2$, so (x, I) is δ^b-fine, but not δ^a-fine.

4. *This means that—just like Riemann integration (with constant $\delta > 0$) and Riemann-complete integration (with variable function $\delta(y) > 0$)—there are functions $h(x, I), = h((x_1, x_2), I_1 \times I_2)$ which are integrable if the Riemann sums are formed with gauges δ^a, but are not integrable with gauges δ^b.*

   We can say that δ^a-integration is **stronger** *than δ^b-integration, in the way that Riemann-complete integration is stronger than Riemann integration. (That is, a function f which fails to be Riemann integrable may be Riemann-complete integrable.)*

   The role of a gauge (such as δ^a or δ^b) is to restrict the associated pairs (x, I) which can be admitted as members of division $\mathcal{D}$ in the formation of Riemann sums $(\mathcal{D}) \sum h(x, I)$ which are used to estimate $\alpha = \int h(x, I)$, with

$$|\alpha - (\mathcal{D}) \sum h(x, I)| < \varepsilon$$

   as integrability criterion.

   The more restrictive the gauge, the stronger the integral.

5. *Taking this to extremes, if the gauge is such that* **every** *(x, I) is admissible then (in a bounded domain) only the constant function $h = \alpha$ is integrable.[3] At the other extreme, if* **no** *(x, I) is admissible, then* **every** *function h is integrable. The mathematically useful integration scenarios lie between these two extremes.*

6. *Despite the preceding issues, Fubini's theorem ([MTRV], pages 160–165) sometimes enables us to get a connection between δ^a-integrals and δ^b-integrals in $\mathbf{R}^n$. For instance, suppose $(x, I) = ((x_1, x_2), I_1 \times I_2)$ in $\mathbf{R}^2$, and suppose*

$$h(x, I) = h_1(x_1, I_1)h_2(x_2, I_2).$$

   Now suppose h is δ^a-integrable in $\mathbf{R}^2$. Then Fubini's theorem implies

$$\int_{\mathbf{R}^2} h(x, I) = \int_{\mathbf{R}_2} \left(\int_{\mathbf{R}} h_1(x_1, I_1) \right) h_2(x_2, I_2)$$

$$= \int_{\mathbf{R}} h_1(x_1, I_1) \int_{\mathbf{R}} h_2(x_2, I_2).$$

   Since the latter two integrals are one-dimensional, the δ^a gauges reduce to δ^b gauges in these two integrals.

---

[3] Theorem 67 (page 180 of [MTRV]) gives an example of a non-constant integrand $h(x, I)$, $= f(x)D(I)$, which is integrable if and only if factor $f(x)$ is constant. See Example 13.

7. *The preceding argument based on Fubini's theorem (for δ^a-integrals) can be extended to integrals in $\mathbf{R}^n$; and it can be extended to some other kinds of integrands provided the iterated integrals of Fubini's theorem are expressible as successive integrals with respect to variables x_j $(j = 1, \ldots, n)$, each of which is reducible to a δ^b-integral.*

8. **The δ^a-integral is stronger than the δ^b-integral. But a survey of the results and properties of the -complete integral $\int_{\mathbf{R}^n} h(x, I)$ (δ^a version) in [MTRV] shows that all results are equally valid for corresponding statements expressed in terms of gauges δ^b.**

Recapitulating, the definition of the gauge integral is as follows. A function $h(x, I)$ is integrable on $\mathbf{R}^n$, with integral value α if, given $\varepsilon > 0$, there exists a gauge δ such that, if $\mathcal{D}$ is a δ-fine division of $\mathbf{R}^n$, then

$$\left| \alpha - (\mathcal{D}) \sum h(x, I) \right| < \varepsilon.$$

If there are no δ-fine divisions $\mathcal{D}$ of $\mathbf{R}^n$, then this definition is vacuous, implying (in accordance with point 5. above) that **every** function h is integrable. Existence of δ-fine divisions is therefore the foundation of every form of -complete integration.

For gauges of form δ^a, this issue is addressed in various places in [MTRV] (such as section 2.3, page 45) using the method of successive bisection. Existence of divisions relative to gauges of form δ^b can be easily established by the same method.

Theorem 4 *Given a gauge δ^b in $\mathbf{R}^n$, there exists a δ^b-fine division of $\mathbf{R}^n$.*

Proof. Assuming non-divisibility (for contradiction), successive bisection delivers a point $x = (x_1, \ldots, x_n) \in \bar{\mathbf{R}}^n$, and, using the gauge values $\delta_j(x_j)$ for this point, a δ^b-fine cell I can be found, and this proves divisibility of $\mathbf{R}^n$. $\bigcirc$

7.5   Gauges for Infinite-dimensional Spaces

The preceding discussion demonstrates that there are various alternative ways of setting up a -complete system of integration for product spaces. These ideas are also helpful in formulating such a system when the product is infinite; where, for instance, the domain is "$\mathbf{R}^\infty$" rather than $\mathbf{R}^n$. This is illustrated as follows.

Example 34 *Suppose $\mathbf{T} = \,]0, t]$, and the domain of integration is the infinite-dimensional Cartesian product space*

$$\mathbf{R}^{\mathbf{T}} = \prod_{s \in \mathbf{T}} \mathbf{R} = \prod_{s \in \mathbf{T}} \mathbf{R}_s$$

where $\mathbf{R}_s = \mathbf{R}$ for each $s \in \mathbf{T}$. (The following account works equally well for $\mathbf{R}^T$ where T is any infinite set.) In [MTRV], several approaches to this are *discussed; but the method which is selected for use throughout the rest of that book is described in section 4.2, pages 116–119. This can be summarized as follows.*

$$L: \qquad \bar{\mathbf{R}}^{\mathbf{T}} \rightarrow \mathcal{N}(\mathbf{T}), \qquad x_{\mathbf{T}} \mapsto L(x_{\mathbf{T}}) \quad \in \ \mathcal{N}(\mathbf{T}),$$
$$\delta: \ \bar{\mathbf{R}}^{\mathbf{T}} \times \mathcal{N}(\mathbf{T}) \ \rightarrow \ ]0,\infty[, \qquad (x_{\mathbf{T}}, N) \ \mapsto \ \delta(x_{\mathbf{T}}, N) \ > \ 0.$$

$$(7.6)$$

With finite dimension-set $N \supset L(x_{\mathbf{T}})$ selected from the preceding line, the purpose of δ is to regulate the lengths $|I(s)|$ of the restricted edges $I(s)$ ($s \in N$) of associated pairs $(x_{\mathbf{T}}, I[N])$ in $\bar{\mathbf{R}}^{\mathbf{T}}$.

Only a finite number of edges of $I[N]$ is restricted. As discussed in [MTRV], this is suggestive of gauge-restriction of edges of finite-dimensional cells $I(N)$ in the finite-dimensional Cartesian product domain $\mathbf{R}^N$. Example 33 above describes two ways of providing gauge-restriction for $I(N)$:

(a) A gauge $\delta^a(x(N)) > 0$, with $|I(s)| < \delta^a(x(N))$ for each $s \in N$;

(b) A gauge $\delta^b(x(N)) = (\delta_s(x(s)) : s \in N)$, with $|I(s)| < \delta^b(x(s))$ for each $s \in N$, where $\delta^b(y) > 0$ is a gauge in $\mathbf{R} = \mathbf{R}_s$ for each $s \in N$.

But while [MTRV] mentions option (a) above, that book chooses instead the following form of gauge-restriction for $I[N]$:

(c) A gauge $\delta^c(x_{\mathbf{T}}, N) > 0$, with $|I(s)| < \delta^c(x_{\mathbf{T}}, N)$ for each $s \in N$.

*In other words, for each $s \in N$, the restriction on the edge $I(s)$ of $I[N]$ depends on **every** component $x(s)$ ($s \in \mathbf{T}$, infinite) of $x(\mathbf{T}) \in \bar{\mathbf{R}}^{\mathbf{T}}$; and not just on the finite number of components $x(s)$ for $s \in N$, as described in (a) above.*

It is easily seen that δ^c gives a "stronger" integral than δ^a (which is, in turn, stronger than δ^b, as discussed in Example 33.) But the "extra strength" of δ^c is superfluous[4] for present purposes. Examination of the results in [MTRV] (which use δ^c) show that they can also be obtained with gauges δ^a or δ^b.

For present purposes, it is more convenient here to use gauges δ^b in $\mathbf{R}^{\mathbf{T}}$, $=]0, t]$, and similarly in other relevant infinite-dimensional Cartesian product spaces such as $\mathbf{R}^T$ where T is an arbitrary infinite set.

Write $\mathbf{R}_+ =]0, \infty[$. Here is a description of the chosen gauge for $\mathbf{R}^{\mathbf{T}}$, or $\mathbf{R}^T$:

$$L: \ \bar{\mathbf{R}}^{\mathbf{T}} \ \rightarrow \ \mathcal{N}(\mathbf{T}), \qquad x_{\mathbf{T}} \mapsto \ L(x_{\mathbf{T}}) \qquad \in \ \mathcal{N}(T),$$
$$\delta_N: \ \bar{\mathbf{R}}^N \ \rightarrow \ \mathbf{R}_+^N, \qquad x(N) \mapsto \ (\delta_s(x(s)))_{s \in N} \quad \in \ \mathbf{R}_+^N,$$
$$\delta_{\mathcal{N}} \ := \ \{\delta_N : N \in \mathcal{N}\},$$
$$\gamma \ := \ (L, \delta_{\mathcal{N}}).$$

$$(7.7)$$

---

[4]The reason for this is that the integrands of interest in this book are mostly of the "sampling" kind, depending on a finite number of coordinates x_s with dimension $s \in N$ for variable N.

The gauge for $\mathbf{R^T}$*, or* $\mathbf{R}^T$*, is the pair* (L, δ_N)*, which can be denoted by* γ^b *since the* δ_N *component has the form* δ^b *of Example 33. But whenever it is not likely to be confused with the gauges of [MTRV], notation* $\gamma = (L, \delta_N)$ *can be used without superscript b.*

Thus essential components of a gauge in $\mathbf{R}^T$*, or* $\mathbf{R^T}$*, are as follows.*

- *For each* $s \in N \in \mathcal{N}(T)$*, there is a gauge* δ_s *in the corresponding domain* $\mathbf{R}$.

- *If* T *contains only a* **finite** *number of elements* s*, then a gauge* δ *in* $\mathbf{R}^T$ *is*

$$(\delta_s(x(s)))_{s \in T}$$

  where (x_T, I_T) *is* δ*-fine if* (x_s, I_s) *is* δ_s*-fine for each* $s \in T$.

- *If* T *(or* $\mathbf{T}$*) contains* **infinitely many** *elements* s *(as is the case for* $\mathbf{T} =]0, t]$*), then, as in (7.7) above, a function* L *is used to select, for each* $x_T \in \bar{\mathbf{R}}^T$,

$$L: \bar{\mathbf{R}}^T \rightarrow \mathcal{N}(T), \qquad x_T \mapsto L(x_T) \in \mathcal{N}(T).$$

- *For infinite* T *a gauge* γ *in* $\mathbf{R}^T$ *can then be written*

$$\gamma = \left(L, ((\delta_s)_{s \in N})_{N \in \mathcal{N}(T)} \right) \quad \text{or alternatively} \quad \gamma = (L, \delta_N), \qquad (7.8)$$

or simply $\gamma = (L, \delta)$*. An associated pair* $(x, I[N])$ *in* $\mathbf{R}^T$ *is* γ*-fine if*

$$N \supseteq L(x) \quad \text{and} \quad (x_s, I_s) \text{ is } \delta_s\text{-fine for each } s \in N.$$

The above definitions of gauges work for any infinite set T**, not just** $\mathbf{T} =]0, t]$**.**

As divisibility is the basis of -complete integration, γ-fine divisibility must be established as in Theorem 3. But since $\mathbf{T}$ (or T) is infinite, the simple bisection argument of section 2.3, page 45 of [MTRV], is not sufficient and a proof on the lines of theorem 4 (pages 121–124 of [MTRV]) is needed.

The formulation of the latter proof is, of necessity, burdened with quite a lot of technical detail and notation. The following example seeks to give the underlying idea in a simpler, less technical way.

Example 35 *Suppose* $S = [0, 1]$ *and the domain of integration is*

$$S^3 =]0, 1] \times]0, 1] \times]0, 1],$$

with associated elements

$$(x, I) = ((x_1, x_2, x_3), I_1 \times I_2 \times I_3)$$

where, for $i = 1, 2, 3$*,* x_i *is an element of* $[0, 1]$ *and* I_i *is a cell* $]u, v] \subseteq S =]0, 1]$*. In terms of the preceding discussion, a gauge can be defined in various ways:*

[a] A function $\delta(x_1, x_2, x_3) > 0$ defined for $0 \le x_i \le 1$, $(i = 1, 2, 3)$, or

[b] A triple $(\delta_1(x_1), \delta_2(x_2), \delta_3(x_3))$, with $\delta_j(x_j) > 0$, $j = 1, 2, 3$.

The following explanation works equally well for both, but the first one is closer to the kind of gauge used in theorem 4 of [MTRV]. The usual way to prove δ-divisibility is by a process of successive bisection:

$$\left] \frac{q_{i_1}}{2^r}, \frac{q_{i_1}+1}{2^r} \right] \times \left] \frac{q_{i_2}}{2^r}, \frac{q_{i_2}+1}{2^r} \right] \times \left] \frac{q_{i_3}}{2^r}, \frac{q_{i_3}+1}{2^r} \right],$$

$q_{i_k} = 0, 1, \ldots, 2^r - 1$, $k = 1, 2, 3$, and $r = 1, 2, 3, \ldots$. Then an initial assumption of non-divisibility eventually produces a contradiction. But, in line with the proof of theorem 4 (pages 121–124 of [MTRV]), and dropping much of the technical notation, denote the three dimensions by t_1, t_2, t_3 instead of $1, 2, 3$. Think of the domain S^3 as a cubical block of cheese, each of whose edges has length 1.

1. *Instead of bisecting the unit cube into successively smaller cubes, select one dimension only, t_1, for successive bisection. The initially cubical block of cheese is converted, by successively slicing along one dimension, into successively thinner slices of cheese, of which, at each stage, at least one slice, $J^r = I_{t_1}^r \times S \times S$, must be non-$\delta$-divisible from the initial assumption of non-divisibility (—system [a]. For system [b], J^r must be non-δ_1-divisible). As the bisection value r tends to infinity a value y_1 (or y_{t_1}) is arrived at by the usual bisection argument, and, at each stage of bisection, y_{t_1} is the fixed t_1 co-ordinate of a point $(y_{t_1}, x_{t_2}, x_{t_3})$ which is the associated point or tag point of slice J^r.*

2. *The next step is to consider a domain $S \times S$ with dimension t_1 removed. We take a paper-thin "slice" of cheese (actually so thin that it has thickness zero) consisting, in space, of the points $(y_{t_1}, x_{t_2}, x_{t_3})$ with y_{t_1} fixed and $0 \le x_{t_i} \le 1$ for $i = 1, 2$. This is no longer really a slice of cheese; it is mathematically a two-dimensional Cartesian product domain $[0, 1] \times [0, 1]$. Define a gauge (system [a]) on this domain by*

$$\delta(x_2, x_3) = \delta(x_{t_2}, x_{t_3}) := \delta(y_{t_1}, x_{t_2}, x_{t_3}).$$

(For the alternative kind of gauge (system [b]), use $(\delta_2(x_2), \delta_3(x_3))$ from the original gauge $(\delta_j(x_j))_{j=1,2,3}$.) By the original assumption, this two-dimensional domain must be non-δ-divisible (with the new, two-dimensional meaning of δ). Because if there is a δ-fine division $\mathcal{D}$ of $S \times S$ (or, if preferred, of $\{y_{t_1}\} \times S \times S$), with

$$\mathcal{D} = \{(x_{t_2}, x_{t_3}, I_{t_1} \times I_{t_2})\};$$

or, alternatively, $\{((y_{t_1}, x_{t_2}, x_{t_3}), \{y_{t_1}\} \times I_{t_1} \times I_{t_2})\}$,

then r can be chosen so that 2^{-r} is less than $\delta(y_{t_1}, x_{t_2}, x_{t_3})$ for each of the finite number of tag points $(y_{t_1}, x_{t_2}, x_{t_3})$ of $\mathcal{D}$ with fixed y_{t_1}. But then a δ-fine division of $I^r \times S \times S$ can be produced from the terms of $\mathcal{D}$:

$$\{((y_{t_1}, x_{t_2}, x_{t_3}), I_{t_1}^r \times I_{t_1} \times I_{t_2})\}.$$

So if the two-dimensional "paper-thin cheese sheet" is divisible then for sufficiently large r the "somewhat thin" (three-dimensional) cheese slices of 1 above are divisible, which contradicts the original hypothesis of non-divisibility.

3. *Now select a second dimension t_2, and again commence successive bisection of the (vanishingly thin) "cheese-sheet"; this time in the direction t_2, leaving direction t_3 unbisected. This produces a succession of non-divisible two-dimensional strips of width 2^{-r} and length 1, leading as before to a fixed value y_{t_2} from the intersection of the strips.*

4. *Finally, we get a non-divisible (one-dimensional) line in direction t_3, and further bisection gives fixed y_{t_3}, giving a fixed point $y = (y_{t_1}, y_{t_2}, y_{t_3})$ in $S \times S \times S$. Taking r large enough, a cell $I_{t_1}^r \times I_{t_2}^r \times I_{t_3}^r$ is found to be divisible, with y as tag point, and this is found to contradict the original assumption of non-divisibility of $S \times S \times S$.*

As pointed out at the start of Example 35 the exact same result can be obtained directly by simultaneously bisecting in all three dimensions. So why take the rather more complicated route of bisecting in one dimension at a time? The next example demonstrates that this method can be useful.

Example 36 *Instead of $S \times S \times S$ (with $S = [0,1]$), suppose T represents a countably infinite number of dimensions $\{t_1, t_2, t_3, \ldots\}$, and suppose $R = S^T$ is the domain to be considered. In a finite-dimensional domain such as S^3 or S^n, a gauge is a function $\delta(x) > 0$. But in infinite dimensions $\prod_t \{S : t \in T\}$, there is an additional condition. For each finite subset $N \subset T$, the following kinds of δ can be used:*

$$\delta(x(N)), \qquad \delta(x_T, N) \qquad or \qquad \{(\delta_s(x_s))_{s \in N}\}.$$

The third δ is the one chosen for this book. But the divisibility argument that follows works equally well for all three.

This kind of gauge δ makes a supposedly infinite-dimensional $\int_{S^T}$ look like a straightforward finite-dimensional $\int_{S^N}$. But there is a catch. It is true that something like $\int_{S^N}$ occurs—in the sense of something resembling a Riemann sum estimate of $\int_{S^N}$. However there is a further condition on the gauges: for each x_T, a minimal finite set $L(x_T) \subset T$ is specified, and, in forming Riemann sum estimates of $\int_{S^T}$, we allow only those finite-dimensional cells $I(N)$ for which $N \supseteq L(x_T)$. The definition of a gauge $\gamma, = (L, \delta_N)$, is such that, in choosing gauges, the cardinality of the finite sets $N \supseteq L(x)$ can be arbitrarily large, while the corresponding numbers $\delta_s(x_s)$ can be arbitrarily small. These ideas are illustrated graphically in [MTRV] pages 81, 87, and (especially) 102.

The issue here is, with any gauge $\gamma = (L, \delta_N)$ given, that the domain S^T must be γ-divisible for the theory to work. There is a proof of this kind in [MTRV] theorem 4, pages 121–124. In Example 35 above, aspects of this proof are illustrated, but only for T of finite cardinality. The underlying intent of the *proof in Example 35 (in which T is a finite set) is demonstrated in the present example in which T is infinite.*

1. *Assume S^T is non-γ-divisible.*

2. *As in Example 35 above, choose any dimension label. That is, choose any element of T. We have already enumerated T as $\{t_1, t_2, t_3, \ldots\}$, so let the choice be t_1.*

3. *There is no harm in continuing to visualise S^T as a cube or "block of cheese". (Anyway, our powers of geometric visualisation do not easily extend to more than three dimensions.) In reality S^T is, so to speak, a hyper-cube—a "hyper-block of cheese".*

4. *As in Example 35, bisect successively S^T in dimension t_1 only. (That is, successively re-slice the hyper-block of cheese into ever-thinner slivers.) As before this produces cells (slices) $J^1 \supset J^2 \supset J^3 \supset \cdots$, each of which must be non-divisible by assumption 1 above. As before, a common, fixed number y_{t_1} is arrived at.*

5. *Now get rid of dimension t_1, so the hyper-block of cheese becomes a "flat hyper-block" or hyper-plane $S^{T \setminus \{t_1\}}$. Choose a second dimension t_2, bisect again, and find another fixed co-ordinate y_{t_2}.*

6. *Thus far, the argument is similar to Example 35. In that case we arrived at dimension t_3, bisected in that dimension, found y_{t_3}, and then stopped there because there were no more dimensions to slice. The proof was then completed by finding a contradiction to assumption 1. But in the case of S^T there are further dimensions $\{t_4, t_5, \ldots\}$, and we can **never stop!** In other words, for $j = 1, 2, 3, \ldots$, we find (inductively) fixed co-ordinates y_{t_j}, giving (by induction) a single fixed point y_T in S^T.*

7. *Thus the final (or t_3) step in Example 35 is not available to us when T is infinite. To see how to salvage the argument, remember that the gauge γ is not just $\delta_\mathcal{N}$, but also includes a factor $L(x_T)$, a finite subset of T; a possibly different finite set for each x_T. Essentially, it is the finiteness of the set $L(y_T)$ (see step 6) that gives a contradiction to 1, forcing an eventual stop to the iteration. Return to steps 2, 3, and 4 above, where the fixed co-ordinate y_{t_1} has been found. Now consider the hyperplane*

$$Z_T^1 \;=\; \{z_T\} \;=\; \left\{ (y_{t_1}, z_{t_2}, z_{t_3}, \ldots) : z_{t_j} \in S, \; j = 2, 3, 4, \ldots \right\}.$$

8. *Suppose (for contradiction) that there is, in Z_T^1, a point*

$$z^1 \;=\; (y_{t_1}, z_{t_2}, z_{t_3}, \ldots)$$

for which

$$L(z^1) \cap \{t_2, t_3, \ldots\} \;=\; \varnothing, \qquad so \;\; L(z^1) \;=\; \{t_1\}.$$

Choose r so that

$$2^{-r} \quad < \quad \delta_{t_1}(y_{t_1}).\tag{7.9}$$

From this it is easy to find a cell or cells I^r which form a γ-fine division of the cell J^r in step 4 above, giving a contradiction.

9. *Therefore, if 1 is valid, then for each point $z_T = (y_{t_1}, z_{t_2}, z_{t_3}, \dots)$ of the hyperplane Z_T^1 the set $L(z_T)$ of the gauge γ contains a co-ordinate label t_k with $k > 1$.*

10. *Likewise, with step 5 completed, a "hyper-hyper-plane" Z_T^2 of the hyperplane Z_T^1 is considered, and the assumption of non-divisibility in 1 above implies that the set*

$$L(z_T^2) \;=\; L\left(y_{t_1}, y_{t_2}, z_{t_3}, z_{t_4}, \dots\right)$$

contains a co-ordinate label t_k with $k > 2$.

11. *But in step 6, a fixed point $y_T = (y_{t_1}, y_{t_2}, y_{t_3}, \dots)$ is arrived at. The set $L(y_T)$ is defined in the gauge γ, and consists of a finite set*

$$\{s_1, s_2, \dots, s_n\} \;\subset\; T,$$

with $s_1 < s_2 < \cdots < s_n = t_m$ for some m. So $k > m$ implies $t_k \notin L(y_T)$.

12. *Assumption 1 implies that iteration m of the single-coordinate bisection process delivers a coordinate label t_k, with $k > m$, for which*

$$t_k \in L(y_T) \;=\; L(y_{t_1}, \dots, y_{t_m}, y_{t_{m+1}}, y_{t_{m+2}}, \dots),$$

with $(y_{t_1}, \dots, y_{t_m}, y_{t_{m+1}}, y_{t_{m+2}}, \dots) \in Z_T^m$; contradicting step 11.

Thus assumption 1 gives a contradiction, and must therefore be false. The argument can be adapted for uncountable T, as demonstrated in the proof of theorem 4 in [MTRV].   ◯

The steps above provide a method for establishing divisibility for all three versions of gauge γ in $\mathbf{R}^T$, including the version γ^b defined in (7.7). The condition labelled (7.9) above is the critical step. Thus the following corresponds to theorem 4 (page 120–124 of [MTRV]) and, as shown above, has almost identical proof.

Theorem 5 *Given a gauge γ (or γ^b) defined as (7.7), there exists a γ-fine division of $\mathbf{R}^T$.*

Proof. With minor adaptations, the proof follows that of Example 36. ○

In (7.8) a gauge γ on infinite-dimensional domain $\mathbf{R}^T$ is given as

$$\gamma = \left(L, ((\delta_s)_{s \in N})_{N \in \mathcal{N}(T)} \right) = (L, \delta_N).$$

This means that one-dimension gauges $\delta_s(x(s))$ depend, not just on $s \in T$, but also on the sample N of elements to which s belongs in the Riemann sum selection for the integral. This may seem unnecessarily complicated. Why not just allow δ_s to depend on s and $x(s)$? In that case, γ would be

$$(L, (\delta_s)_{s \in T}).$$

The following example demonstrates why the more delicate and discriminating construction in (7.8) may sometimes be needed.

Example 37 *For simplicity, replace domain $\mathbf{R}^T$ by $\Omega =]0,1]^T$, with $T = [0,1]$. Let $d(y)$ denote the Dirichlet function $d(y) = 1$ if y is rational, $d(y) = 0$ otherwise. Let s be a fixed a fixed element of T, say $s = \frac{1}{2}$. With volume function $|I[N]|$ as in Example 11, define an integrand $f(x(N))|I[N]|$ by*

$$f(x(N)) = \begin{cases} d\left(x\left(\frac{1}{2}\right)\right) & \text{if } N = \left\{\frac{1}{2}\right\}, \\ 0 & \text{otherwise;} \end{cases} \qquad |I[N]| = \prod_{s \in N} |I(s)|.$$

It seems that $\int_\Omega f(x(N))|I[N]| = 0$. But some attention to the gauge γ may be required. If N has more than one element, then $\delta_s(x(s)) > 0$ is arbitrary; but if N contains just one element s then $\delta_s(x(s))$ should be defined as in section 2.7 of [MTRV], page 51. (In this case the difficulty can be avoided simply by ensuring that each N contains more than one element.)

7.6 Higher-dimensional Brownian Motion

One reason for the preceding review of product gauges is to seek to formulate three-dimensional Brownian motion in terms of the system of -complete integration. Returning to this issue, and following the scheme established in Example 32, let $\mathbf{T}$ denote $]0,t]$, and let sample space Ω be

$$R_{3,\mathbf{T}} := \left(\mathbf{R}^3\right)^{\mathbf{T}} = \prod\left\{\mathbf{R}^3 : s \in \mathbf{T}\right\}.$$

If N is any finite subset of $\mathbf{T}$, let $R_{3,N}$ denote $\left(\mathbf{R}^3\right)^N$, $= \prod\left\{\mathbf{R}^3 : s \in N\right\}$, with finite-dimensional points $\mathbf{x}_N$, cells $\mathbf{I}_N$, association $(\mathbf{x}_N, \mathbf{I}_N)$ and gauges δ defined as in Example 32 and (7.5). For any finite set $N \in \mathcal{N}(\mathbf{T})$, a cell $\mathbf{I} = \mathbf{I}_\mathbf{T} = \mathbf{I}_\mathbf{T}[N]$ in $R_{3,\mathbf{T}}$ is

$$\mathbf{I}_\mathbf{T} = \mathbf{I}_N \times R_{3,\mathbf{T} \setminus N} = \mathbf{I}_N \times \left(\mathbf{R}^3\right)^{\mathbf{T} \setminus N}.$$

So if $N = \{t_1, \ldots, t_n\} \subset \,]0,t]$, with $t_0 = 0$ and $t_n = t$,

$$\mathbf{I_T} \;=\; \mathbf{I_T}[N] \;=\; \left(\prod_{j=1}^{n} \big(I_1(t_j) \times I_2(t_j) \times I_3(t_j)\big)\right) \times R_{3,\mathbf{T}\setminus N},$$

where each component term of the Cartesian product $R_{3,\mathbf{T}\setminus N}$ is $\mathbf{R} \times \mathbf{R} \times \mathbf{R}$. In physical terms, to say that $\mathbf{x}_T \in \mathbf{I}$ means that, for dimensions $i = 1,2,3$ at time t_j $(1 \leq j \leq n)$, the ith co-ordinate of the spatial position of the particle lies in the real interval $I_i(t_j)$; and for any time $s \notin N$, the ith co-ordinate of the particle in space is unspecified, or arbitrary, for each of $i = 1,2,3$.

A point-cell pair $(\mathbf{x}_T, \mathbf{I_T}[N])$ are *associated* in $R_{3,\mathbf{T}}$ if $(\mathbf{x}_N, \mathbf{I}_N)$ are associated in R_N. That means that, as one-dimensional objects, $(x_i(t_j), I_i(t_j))$, $=$ $(x_{i,j}, I_{i,j})$, are associated in $\mathbf{R}$ for $i = 1,2,3$ and for $t_j \in N$.

Next, define a gauge in $R_{3,\mathbf{T}}$, as follows. For each $\mathbf{x_T},= (x_{1s}, x_{2s}, x_{3s})_{s\in\mathbf{T}}$, in $\bar{R}_{3,\mathbf{T}}$ (that is, $R_{3,\mathbf{T}}$ with points at infinity added), and for each finite set $N \in \mathcal{N}(\mathbf{T})$, let

$$L: \; \big(\bar{\mathbf{R}} \times \bar{\mathbf{R}} \times \bar{\mathbf{R}}\big)^{\mathbf{T}} \;\to\; \mathcal{N}(\mathbf{T}) \quad \text{where } \mathbf{x_T} \mapsto L(\mathbf{x_T}) \in \mathcal{N}(\mathbf{T});$$

$$\text{and let } \; \delta_N : \; \big(\bar{\mathbf{R}} \times \bar{\mathbf{R}} \times \bar{\mathbf{R}}\big)^{N} \;\to\; \big(\mathbf{R}_+ \times \mathbf{R}_+ \times \mathbf{R}_+\big)^{N} \qquad (7.10)$$

$$\text{where } \; \mathbf{x}_N \mapsto \big(\delta_s(x_{1s}),\, \delta_s(x_{2s}),\, \delta_s(x_{3s})\big)_{s\in N}.$$

A *gauge* γ in $R_{3,\mathbf{T}}$ consists of

$$\big(L, \; \{\delta_N : N \supseteq L(\mathbf{x_T}),\ N \in \mathcal{N}(\mathbf{T})\}\big); \qquad (7.11)$$

which, for brevity, can be written

$$\gamma \;=\; (L, \delta_{\mathcal{N}}), \quad \text{or simply} \quad \gamma \;=\; (L, \delta)$$

Given a gauge γ, an associated point-cell pair $(\mathbf{x_T}, \mathbf{I_T}[N])$ in $R_{3,\mathbf{T}}$ is γ-*fine* if $N \supseteq L(\mathbf{x}_T)$ and if $(\mathbf{x}_N, \mathbf{I}_N)$ is δ_N-fine in $R_{3,N}$. (The latter means that, for each $s \in N$, and for $i = 1,2,3$, the one-dimensional pair $(x_i(s), I_i(s))$, $= (x_{i,s}, I_{i,s})$, is δ_s-fine in $\mathbf{R}$.)

Armed with understanding from Example 36, a proof of γ-divisibility of the structured domain $R_{3,\mathbf{T}},= (\mathbf{R} \times \mathbf{R} \times \mathbf{R})^{\mathbf{T}}$, can be undertaken as follows. The proof works for any infinite set T.

Theorem 6 *Given a gauge* $\gamma = (L, \delta_{\mathcal{N}})$ *of* $R_{3,\mathbf{T}}$, *there exists a* γ-*fine division* $\mathcal{D}, = \{(\mathbf{x}_T, \mathbf{I}_T[N])\}$, *of* $R_{3,\mathbf{T}}$.

Proof. Assume (for contradiction) that there is no γ-division of $R_{3,\mathbf{T}}$. The first lines (page 121 of [MTRV]) of the proof of theorem 4 carry forward unchanged. But the bisection in dimension t_1, indicated by cells I^r at the bottom of page 121, needs to be amended as follows. In dimension t_1 of $R_{3,\mathbf{T}}$, for $i = 1,2,3$ let each of $I_i^r(t_1)$ denote one of the one-dimensional cells in the last line of page

121; and, with additional subscript labels $1, 2, 3$ relating to the three Cartesian components of the single dimension t_1, let

$$I^r(t_1) \ = \ I_1^r(t_1) \times I_2^r(t_1) \times I_3^r(t_1), \quad (\text{or } \ I_{t_1}^r \ = \ I_{1,t_1}^r \times I_{2,t_1}^r \times I_{3,t_1}^r),$$

so $I^r(t_1)$ is a cell of $\mathbf{R}^3$. (For emphasis the label t_1 can be in-line rather than subscripted.) Line 2 of page 122 in [MTRV] therefore gives a fixed

$$y(t_1) \ = \ (y_1(t_1), y_2(t_1), y_3(t_1)) \ \in \ \bar{\mathbf{R}}_{t_1}^3 \ = \ \bar{\mathbf{R}} \times \bar{\mathbf{R}} \times \bar{\mathbf{R}}.$$

The crux of the γ-divisibility proof occurs in lines 3 to 13 of page 122. Fortunately, for the new and more structured domain $R_{3,\mathbf{T}}$, this part of the proof carries forward practically unchanged. Note that, for domain $R_{3,\mathbf{T}}$, line 6 of page 122 becomes

$$2^{-r} \ < \ \min\{\delta_{t_1}(y_1(t_1)),\, \delta_{t_2}(y_2(t_1)),\, \delta_{t_3}(y_3(t_1))\}.$$

The re-definition of the gauge in lines 14 to 19 of page 122 of [MTRV] needs similar small modifications. The rest of the proof consists of iterations of these steps, leading to contradiction of the initial assumption. $\quad\bigcirc$

This result is the first step in providing a "Riemann-observable" domain or sample space $(\mathbf{R} \times \mathbf{R} \times \mathbf{R})^{\mathbf{T}}$ for three-dimensional Brownian motion. The proof for m-dimensional Brownian motion, $(\mathbf{R}^m)^{\mathbf{T}}$ is essentially the same. It also works for domain

$$\Omega \ = \ \prod(\mathbf{R}^{m_t} : t \in T)$$

where T is infinite and each m_t is a positive integer depending on t. (In Theorem 6, which is aimed at three-dimensional Brownian motion, $m_t = 3$ for each $t \in T$.) Likewise for other Cartesian product domains such as those proposed for quantum electrodynamics in Chapter 9 below.

Theorem 6 deals with divisibility of domain $\Omega = (\mathbf{R} \times \mathbf{R} \times \mathbf{R})^{\mathbf{T}}$ with T infinite. The following Example describes another structure, similar to $(\mathbf{R}^3)^{\mathbf{T}}$ with infinite $\mathbf{T}$.

Example 38 *Let T be an infinite set and*

$$\Omega \ = \ \mathbf{R}^T \times \mathbf{R}^T \times \mathbf{R}^T \ = \ (\mathbf{R}^T)^3,$$

denoted $R_{T,3}$ for short. Elements of $R_{T,3}$ have the form $\mathbf{x} \ =$

$$= \ (x_T^1, x_T^2, x_T^3) \ = \ ((x_{t,j})_{t\in T})_{j=1,2,3} \ = \ ((x_{t,1})_{t\in T}, (x_{t,2})_{t\in T}, (x_{t,3})_{t\in T}), \quad (7.12)$$

with $x_{t,j}, = x_{tj}, \in \mathbf{R}$ for $j = 1, 2, 3$; such that a component

$$(x_{t_1,1},\, x_{t_2,2},\, x_{t_3,3}) \ \not\Rightarrow \ t_1 = t_2 = t_3; \quad so \ (\mathbf{R}^T)^3 \ \neq \ (\mathbf{R}^3)^T. \quad (7.13)$$

It is sometimes easier to perceive the meaning if the label j ($j = 1, 2, 3$) is written as superscript instead of subscript; so the following representation:

$$\mathbf{x} \ = \ ((x_t^1)_{t\in T},\, (x_t^2)_{t\in T},\, (x_t^3)_{t\in T})$$

can be used. For $j = 1, 2, 3$ let

$$N^j \;=\; \{t_1^j, \ldots, t_{n^j}^j\} \;\in\; \mathcal{N}(T),$$

and[5] write $\mathbf{N} = (N^1, N^2, N^3)$. *A cell* $\mathbf{I}$, $= \mathbf{I}[\mathbf{N}]$, *in the domain* $\left(\mathbf{R}^T\right)^3$, $= R_{t,3}$, *is*

$$\mathbf{I}[\mathbf{N}] \;=\; I^1 \times I^2 \times I^3 \;=\; I^1[N^1] \times I^2[N^2] \times I^3[N^3];$$

with $I^j(t) \in \mathcal{I}(\mathbf{R})$ *for* $t \in N^j$; *so each such* $I^j(t)$ *is a one-dimensional real interval* $]u, v]$, $]u, \infty[$, *or* $]-\infty, v]$. *A pair* $(\mathbf{x}, \mathbf{I})$ *are associated in* $\left(\mathbf{R}^T\right)^3$, $= R_{T,3}$, *if, for* $j = 1, 2, 3$ *and for each* $t^j \in N^j$, *the one-dimensional pair* $(x_{t^j}^j, I_{t^j}^j)$ *are associated in* $\mathbf{R}$.

The domain $\left(\mathbf{R}^T\right)^3$, $= R_{T,3}$, is a (finite) product of (infinite) products. The next step is to construct a gauge for the domain $R_{T,3}$. **The underlying method of construction of gauges for product spaces is to use already-defined gauges for the component domains of the product domain, and form a product of such gauges.** This works even when the new gauge is a "product of product-gauges". So it is assumed that, for each $t \in T$ and each $y \in \bar{\mathbf{R}}$, one-dimensional gauges $\delta_t(y) > 0$ are given. The domain $R_{T,3}$ is

$$R_{T,3} \;=\; \mathbf{R}^T \times \mathbf{R}^T \times \mathbf{R}^T \;=\; \mathbf{R}_1^T \times \mathbf{R}_2^T \times \mathbf{R}_3^T$$

where $\mathbf{R}_j = \mathbf{R}$ for $j = 1, 2, 3$, the subscripts being added in order to label the gauges on the component spaces $\mathbf{R}_j^T$. Suppose a gauge $\gamma^j = (L^j, \delta_{\mathcal{N}}^j)$ is a gauge on $\mathbf{R}_j^T$ for $j = 1, 2, 3$. Then a gauge γ for $R_{T,3}$ is

$$\gamma \;=\; (\gamma_1, \gamma_2, \gamma_3) \;=\; \left(\left(L^j, \delta_{\mathcal{N}}^j\right)\right)_{j=1,2,3};$$

and an associated pair $(\mathbf{x}, \mathbf{I}(\mathbf{N}))$ is γ-fine if, for $j = 1, 2, 3$, $\left(x^j, I^j[N^j]\right)$ is γ_j-fine. In other words, for $j = 1, 2, 3$,

$$N^j \supseteq L^j(x^j) \quad \text{and} \quad (x_t^j, I_t^j) \text{ is } \delta_{\{t\}}^j\text{-fine for } t \in N^j. \tag{7.14}$$

Example 39 *For (7.14) to be effective in constructing an integration in* $R_{T,3}$, $= \left(\mathbf{R}^T\right)^3$, *the basic theorem of -complete integration must hold, corresponding to theorem 4 on page 121 of [MTRV]; or to divisibility Axiom DS3 on page 112. So, for any given gauge*

$$\gamma \;=\; (\gamma_1, \gamma_2, \gamma_3) \;=\; \left(\left(L^j, \delta_{\mathcal{N}}^j\right)\right)_{j=1,2,3}$$

and for any cell $\mathbf{J}$ *of* $R_{T,3}$, *there must exist a* γ-*fine division of* $\mathbf{J}$. *As usual, it is sufficient to prove this for* $\mathbf{J} = R_{T,3}$. *Once over this hurdle[6] the general*

---

[5]In [MTRV] the symbol $\mathbf{N}$ is used to denote the normal distribution function.

[6]Strictly speaking the other Axioms (1, 2, 4, 5, 6, 7, 8) also need to be confirmed. But in the "rectangular" Cartesian product domains under consideration here, this generally presents no difficulty. See Example 55.

theory of -complete integration for division systems takes effect, and the full theory of chapter 4 ([MTRV] pages 111–181) is valid for this system. To prove γ-divisibility of $R_{T,3}$, $= \mathbf{R}_1^T \times \mathbf{R}_2^T \times \mathbf{R}_3^T$, a version of the proof of theorem 4 ([MTRV] pages 120–124) can be used. Essentially, the steps of this proof are carried out on $\mathbf{R}_j^T$, simultaneously (or "in parallel") for $j = 1, 2, 3$, in Theorem 7 below.

Before proceeding to Theorem 7, here is a reminder of how the -complete integral of a function h is defined by means of gauges. If $h(\mathbf{x}_T, N, \mathbf{I}[N])$ is a real- or complex-valued function, then h is integrable on $\left(\mathbf{R}^T\right)^3$, with integral α, if the following holds. Given $\varepsilon > 0$, there exists a gauge $\gamma = (L^j, \delta^j)$ $(j = 1, 2, 3)$ such that, for every γ-fine division $\mathcal{D}$ of $\left(\mathbf{R}^T\right)^3$,

$$\left| \alpha - (\mathcal{D}) \sum h(\mathbf{x}_T, N, \mathbf{I}[N]) \right| < \varepsilon.$$

This definition is empty and meaningless unless the required divisions $\mathcal{D}$ of $\left(\mathbf{R}^T\right)^3$ exists. When γ-fine divisions are shown to exist, elaboration of the properties of the -complete integral $\int h$ follows a common pattern, and the theory described in chapter 4 of [MTRV] is generally applicable.

Thus the basis for integration in $\left(\mathbf{R}^T\right)^3$ is provided by the following theorem.

Theorem 7 *Given a gauge γ for $\left(\mathbf{R}^T\right)^3$, as defined in (7.14), there exists a γ-fine division of $\left(\mathbf{R}^T\right)^3$.*

Proof. This theorem is also valid if $\left(\mathbf{R}^T\right)^3$ is replaced by any cell $\mathbf{I}$ of $\left(\mathbf{R}^T\right)^3$. Assume (for contradiction) that no γ-fine division of $\left(\mathbf{R}^T\right)^3$ exists. For simplicity, assume also that T is a countable set, $T = \{\tau_1, \tau_2, \ldots\}$, so

$$\begin{aligned}
\left(\mathbf{R}^T\right)^3 &= \mathbf{R}^T \times \mathbf{R}^T \times \mathbf{R}^T \\
&= (\mathbf{R} \times \mathbf{R} \times \mathbf{R} \times \cdots) \times (\mathbf{R} \times \mathbf{R} \times \mathbf{R} \times \cdots) \times (\mathbf{R} \times \mathbf{R} \times \mathbf{R} \times \cdots) \\
&= \left(\mathbf{R} \times \mathbf{R}^{T \smallsetminus \{\tau_1\}}\right) \times \left(\mathbf{R} \times \mathbf{R}^{T \smallsetminus \{\tau_1\}}\right) \times \left(\mathbf{R} \times \mathbf{R}^{T \smallsetminus \{\tau_1\}}\right).
\end{aligned}$$

(The proof can be adapted for uncountable T, as in theorem 4 of [MTRV].) The product domain has three components, corresponding to $j = 1, 2, 3$; and each of the three components is a composite of factors

$$\left(\mathbf{R} \times \mathbf{R}^{T \smallsetminus \{\tau_1\}}\right), \quad = \quad \left(\mathbf{R}_{\tau_1}^j \times \mathbf{R}^{T \smallsetminus \{\tau_1\}}\right)$$

for $j = 1, 2, 3$. Now bisect, successively, the term $\mathbf{R}_{\tau_1}^j$ (jointly for $j = 1, 2, 3$) of the three factors; so that, at each bisection, a non-γ-divisible cell is obtained. As in Example 36, the successive bisections yield

$$\left(\{y_{\tau_1}^1\} \times \mathbf{R}^{T \smallsetminus \{\tau_1\}}\right) \times \left(\{y_{\tau_1}^2\} \times \mathbf{R}^{T \smallsetminus \{\tau_1\}}\right) \times \left(\{y_{\tau_1}^3\} \times \mathbf{R}^{T \smallsetminus \{\tau_1\}}\right).$$

This procedure can be repeated successively for $j = 2, 3, \ldots$, leading to a succession of non-divisible cells in domains

$$\left(\mathbf{R}^{T \smallsetminus \{\tau_1, \tau_2, \ldots, \tau_k\}}\right)$$

for $k = 2, 3, \ldots$. Thus a point

$$y = \left(\left(y_{\tau_1}^1, y_{\tau_1}^2, y_{\tau_1}^3\right), \left(y_{\tau_2}^1, y_{\tau_2}^2, y_{\tau_2}^3\right), \left(y_{\tau_3}^1, y_{\tau_3}^2, y_{\tau_3}^3\right) \ldots\right) \in \left(\mathbf{R}^T\right)^3$$

is arrived at by iteration. And, as in Example 36, if $L(y) = M$ and if $m = \max\{i : \tau_i \in M\}$, the original assumption of non-γ-divisibility fails at the m-th stage in the preceding iteration.   ○

With corresponding division structures the preceding results can be established for domains $\left(\mathbf{R}^T\right)^n$ for any positive integer n. Also, variants of (7.14) can be substituted, without causing difficulty in the preceding proof of γ-divisibility.

The domain $\left(\mathbf{R}^3\right)^T$ (or $R_{3,\mathbf{T}}$, with $\mathbf{T}$ of infinite cardinality) was introduced in Example 31 as a sample space domain for three-dimensional Brownian motion. This has sample values (x_{1s}, x_{2s}, x_{3s}) for $s \in \mathbf{T}$, representing the coordinates in $\mathbf{R} \times \mathbf{R} \times \mathbf{R}$ of a Brownian particle at any time s, $s \in T =]0, t]$. The difference between elements of $\left(\mathbf{R}^3\right)^T$ and $\left(\mathbf{R}^T\right)^3$ was shown in (7.13). The following Example compares the cells of these two domains.

Example 40 *The cells of $R_{3,T}$ are different from the cells of $R_{T,3}$. First consider $R_{3,T}, = \left(\mathbf{R}^3\right)^T$. For any domain Ω with cells I, if $N = \{t_1, \ldots, t_n\}$ is a finite subset of T, then a cell $\mathbf{I}$ of Ω^T is*

$$\mathbf{I} = I[N] = I(N) \times \Omega^{T \smallsetminus N} = (I(t_1) \times \cdots \times I(t_n)) \times \Omega^{T \smallsetminus N},$$

where each $I(t_j)$ is a cell of Ω. When Ω is $\mathbf{R}^3$, it has cells $I = I_1 \times I_2 \times I_3$, where I_1, I_2, I_3 are cells of $\mathbf{R}$. Therefore a cell of $R_{3,T}$ is

$$\mathbf{I} = \left(\prod_{j=1}^n (I_1(t_j) \times I_2(t_j) \times I_3(t_j))\right) \times \left(\mathbf{R}^3\right)^{T \smallsetminus N},$$

where each I_{it_j} is a cell of $\mathbf{R}$.

For comparison, consider cells $\mathbf{I}$ of $R_{T,3}, = \left(\mathbf{R}^T\right)^3$. For any domain Ω a cell of Ω^3 is $I_1 \times I_2 \times I_3$ where each I_i is a cell of Ω. With $\Omega = \mathbf{R}^T$ and $N^i = \{t_1^i, \ldots, t_{n^i}^i\}$, a cell I_i is

$$I_i = I_i[N^i] = I_i(N^i) \times \mathbf{R}^{T \smallsetminus N^i} = \left(I_i(t_1^i) \times I_i(t_1^i) \times \cdots \times I_i(t_{n^i}^i)\right) \times \mathbf{R}^{T \smallsetminus N^i};$$

and a cell $\mathbf{I}$ of $R_{T,3}, = \left(\mathbf{R}^T\right)^3$, is

$$\mathbf{I} = \prod_{i=1}^3 \left(I_i(N^i) \times \mathbf{R}^{T \smallsetminus N^i}\right).$$

7.7 Infinite Products of Infinite Products

It is straightforward to extend the details of Example 38 to include domains $\left(\mathbf{R}^T\right)^n$, $= R_{T,n}$, where n is any positive integer.

In the development of the theory, finite-dimensional domains $\mathbf{R}^n$ have been extended to infinite-dimensional domains $\mathbf{R}^T$. This suggests that in $\left(\mathbf{R}^T\right)^n$ the finite n may be replaced by an infinite set P of dimension labels, giving domain

$$\left(\mathbf{R}^T\right)^P, \ = R_{T,P} \quad \text{for short.}$$

The next step is to provide some meaning to this. For simplicity it is assumed that both T and P are countable.

Example 41 *Since P is assumed countable for the moment, so $P = \{p_1, p_2, \dots\}$, the members $\mathbf{x}$ of the set $\Omega = \mathbf{R}^{T^P}, = R_{T,P}$, consist of sequences*

$$\mathbf{x} \ = \ \left(x_{p_1}, x_{p_2}, \dots, x_{p_j}, \dots\right),$$

where, for $j = 1, 2, 3, \dots$, each term x_{p_j} is itself a sequence

$$
\begin{aligned}
x_{p_j} \ &= \ \left(x_{t_1, p_j}, x_{t_2, p_j}, \dots, x_{t_i, p_j}, x_{t_{i+1}, p_j}, \dots\right) \\
&= \ \left(x_{1j}, x_{2j}, \dots, x_{ij}, x_{i+1,j}, \dots\right) \\
&= \ \left(x_1^j, x_2^j, \dots, x_i^j, x_{i+1}^j, \dots\right).
\end{aligned}
$$

These three equivalent notations for x_{p_j} are used in various contexts. (Putting j in superscript rather than subscript position can sometimes be clearer.) In other words, for $\mathbf{x} \in \Omega = R_{T,P}$,

$$\mathbf{x}, \ = \ \mathbf{x}_{T,P}, \ = \ \left(\left(x_{ij}\right)_{t_i \in T}\right)_{p_j \in P}, \quad \text{with} \quad \Omega \ = \ \mathbf{R}^{T^P} \ = \ \prod_p\left(\prod_t (\mathbf{R} : t \in T) : p \in P\right).$$

Adapting the notation for uncountable T and/or P is straightforward:

$$\mathbf{x}, \ = \ \mathbf{x}_{T,P}, \ = \ \left(\left(x_{tp}\right)_{t \in T}\right)_{p \in P}, \quad \mathbf{x}_{T,p} \ = \ \left(x_{tp}\right)_{t \in T} \quad \text{for any } p \in P.$$

The next step is to specify cells $\mathbf{I}$ in Ω. For the moment, represent the domain Ω as S^P with $S = \mathbf{R}^T$. For $N_P = \{q_1, \dots, q_n\} \in \mathcal{N}(P)$, a cell $\mathbf{I}$ of S^P is

$$\mathbf{I} \ = \ \left(J_{q_1} \times \dots \times J_{q_n}\right) \times S^{P \smallsetminus N_P} \ = \ J(N_P) \times \left(\mathbf{R}^T\right)^{P \smallsetminus N_P},$$

where each J_{q_j} is a cell $\mathbf{I}_{T,j}$ (or $\mathbf{I}_T^j$) in $\mathbf{R}^T$. So if

$$N_{T,j} \ = \ N_T^j \ = \ \{s_{1j}, \dots, s_{m_j,j}\}, \ = \ \{s_1^j, \dots, s_{m_j}^j\}, \ \in \mathcal{N}(P),$$

a cell J_{p_j} of $\mathbf{R}^T$ is

$$
\begin{aligned}
J_{q_j} \ &= \ \mathbf{I}_{T,j} \ = \ I(N_{T,j}) \times \mathbf{R}^{T \smallsetminus N_{T,j}} \\
&= \ I_{s_1, q_j} \times I_{s_2, q_j} \times \dots \times I_{s_{m_j}, q_j} \times \mathbf{R}^{T \smallsetminus N_{T,j}}, \\
&= \ I_{s_1}^{q_j} \times I_{s_2}^{q_j} \times \dots \times I_{s_{m_j}}^{q_j} \times \mathbf{R}^{T \smallsetminus N_{T,j}}.
\end{aligned}
$$

Thus, for $N_P \in \mathcal{N}(P)$ and $N_{T,j} \in \mathcal{N}(T)$, a cell in the domain $\Omega = R_{T,P}$ is

$$
\begin{aligned}
\mathbf{I}_{T,P} &= \left(\prod_{j=1}^{n} \mathbf{I}_T[N_{T,j}] \right) \times \mathbf{R}^{T^{P \smallsetminus N_P}} \qquad (7.15) \\
&= \left(\prod_{j=1}^{n} \left(I_{s_1,q_j} \times I_{s_2,q_j} \times \cdots \times I_{s_{m_j},q_j} \times \mathbf{R}^{T \smallsetminus N_{T,j}} \right) \right) \times \mathbf{R}^{T^{P \smallsetminus N_P}} \\
&= \left(\prod_{q_j \in N_P} \left(\left(\prod_{s_{ij} \in N_{T,j}} I_{s_{ij},q_j} \right) \times \left(\prod_{t \in T \smallsetminus N_{T,j}} \mathbf{R} \right) \right) \right) \times \left(\prod_{p \in P \smallsetminus N_P} \mathbf{R}^T \right).
\end{aligned}
$$

In a loose or suggestive notation, $\mathbf{I}_{T,P}$ can be written $\mathbf{I}[N_{T_P} : N_P]$.

Suppose $N_P \in \mathcal{N}(P)$ is a finite subset of P, and, for each $p \in N_P$, suppose $N_{T,p} \in \mathcal{N}(T)$ is a finite subset of T. Consider a point $\mathbf{x} = \mathbf{x}_{T,P}$ and a cell $\mathbf{I} = \mathbf{I}_{T,P}$ of $\mathbf{R}^{T^P}$, so

$$
\mathbf{x}_{T,P} = \left((x_{t,p})_{t \in T} \right)_{p \in P}, \quad \mathbf{I}_{T,P} = \left(\prod_{p \in N_P} \mathbf{I}_T[N_{T,p}] \right) \times \mathbf{R}^{T^{P \smallsetminus N_P}}, \qquad (7.16)
$$

Then the pair $(\mathbf{x}_{T,P}, \mathbf{I}_{T,P})$ are associated in $R_{T,P}$ if, for each $p \in N_P$, and each corresponding $N_{T,p}$, the cell

$$
I_T(N_{T,p}) = \left(\prod_{t \in N_{T,p}} I_{tp} \right) \times \mathbf{R}^{T \smallsetminus N_{T,p}}
$$

has the finite-dimensional point

$$
x(N_{T,p}) = (x_{tp})_{t \in N_{T,p}}
$$

as a vertex in the finite-dimensional domain $\bar{\mathbf{R}}^{N_{T,p}}$. Expressed as points and intervals of $\mathbf{R}$, this means that $x_{tp}, = x_t^p, \in \bar{\mathbf{R}}$ is a vertex of the corresponding restricted cell $I_{tp} = I_t^p \subset \mathbf{R}$ for $p \in N_P$, $t \in N_{T,p}$.

These are the points and cells of $\left(\mathbf{R}^T \right)^P$. Are the points and cells of $\mathbf{R}^{(T^P)}$ the same, or are they different? Write $\Omega' = \mathbf{R}^{(T^P)}$. An element $\mathbf{x}$ of Ω' is a sequence $\left(x_{\mathbf{t}_j} \right)_{j=1,2,3,\ldots}$ where each $\mathbf{t}_j$ is a sequence

$$
\left(t_{j,p_{j_1}}, t_{j,p_{j_2}}, \ldots, t_{j,p_{j_k}}, \ldots \right),
$$

so $\mathbf{x} = \mathbf{x}_{T,P}$, and $\mathbf{R}^{(T^P)} = \left(\mathbf{R}^T \right)^P$. Similarly for cells.

To define -complete integrals on $\mathbf{R}^{T^P}$, gauges must be formulated to regulate the associated pairs $(\mathbf{x}_{T,P}, \mathbf{I}_{T,P}[N_{T_P} : N_P])$ used in forming Riemann sums.

It is assumed that, for each $p \in P$ and each component $\mathbf{R}^T, = \left(\mathbf{R}^T \right)^{\{p\}}$, of the compound product space $\mathbf{R}^{T^P}$, there is a gauge $\gamma^p = \left(L^p, \delta_{\mathcal{N}(T)}^p \right)$ for that particular factor domain $\mathbf{R}^T$,

$$
\gamma^p = \left(L^p, \delta_{\mathcal{N}(T)}^p \right) = \left(L^p, (\delta_t^p)_{t \in N_T, N_T \in \mathcal{N}(T)} \right). \qquad (7.17)
$$

in $\mathbf{R}^T$, constructed in a standard way (4.2), for product space $\mathbf{R}^{T^{\{p\}}}, = \mathbf{R}^T$.

A gauge γ for the compound product space $\mathbf{R}^{T^P}, = R_{T,P}$, is formed from the component gauges γ^p as follows. For each $\mathbf{x}$ of $\left(\bar{\mathbf{R}}^T\right)^P$, let $\mathbf{L}(\mathbf{x}) \in \mathcal{N}(P)$ be a finite subset of P, and for each $N_P \in \mathcal{N}(P)$ and each $p \in N_P$, let γ^p be a gauge for $\mathbf{R}^{T^{\{p\}}}$. Then a gauge γ for $R_{T,P}$ is

$$
\begin{aligned}
\gamma = \gamma_{T,P} &:= \left(\mathbf{L}, (\gamma^p)_{p \in N_P,\, N_P \in \mathcal{N}(P)}\right) \\
&= \left(\mathbf{L}, \left(L^p, (\delta_t^p)_{t \in N_{T,p},\, N_{T,p} \in \mathcal{N}(T)}\right)_{p \in N_P,\, N_P \in \mathcal{N}(P)}\right) \quad (7.18) \\
&= \left(\mathbf{L}, (L^{T,P}, \delta^{T,P})\right)
\end{aligned}
$$

for short. Then the associated pair $(\mathbf{x}, \mathbf{I}) = (\mathbf{x}_{T,P}, \mathbf{I}_{T,P})$ of $R_{T,P}$ is γ-fine if

$$
\begin{aligned}
N_P &\supseteq \mathbf{L}(\mathbf{x}), \\
(x_{T,p}, I_{T,p}) \text{ is } &\gamma^p\text{-fine in } \mathbf{R}^T \text{ for each } p \in N_P \quad (7.19)
\end{aligned}
$$

in (7.16). In other words, $(\mathbf{x}, \mathbf{I})$ is γ-fine if

$$
\begin{aligned}
N_P &\supseteq \mathbf{L}(\mathbf{x}), \\
N_{T,p} &\supseteq L^p(x_{T,p}) \text{ for each } p \in N_P, \quad \text{and} \quad (7.20) \\
(x_t^p, I_t^p), = (x_{tp}, I_{tp}), \text{ is } &\delta_t^p\text{-fine in } \mathbf{R} \text{ for each } t \in N_{T,p}, \ p \in N_P.
\end{aligned}
$$

The purpose of the gauge construction is to be able to define the -complete integral of a function $h(\mathbf{x}, \mathbf{I})$ on the domain $\mathbf{R}^{T^P}$. For this to work there have to be γ-fine divisions of the domain in order to be able to form Riemann sum estimates of the integral. Thus the following divisibility result is needed. (Note that, by theorem 4 of [MTRV], for any $p \in P$, domain $\mathbf{R}^T$ is divisible with respect to each γ^p of (7.17) regardless of whether T is countable or uncountable. Definition (7.17) is slightly different from the one used in [MTRV], but the workings of the proof are essentially the same.)

Theorem 8 *Given a gauge γ in $\mathbf{R}^{T^P}$, there exists a γ-fine division of $\mathbf{R}^{T^P}$.*

Proof. The general idea of the proof is a development or extension of the method of successive bisection combined with successive slicing described in Example 36. Assume, for contradiction, that there is no γ-fine division of the domain $\mathbf{R}^{T^P}$. To simplify the argument and notation a little, the sets T and P have been assumed[7] countable, with $P = \{p_1, p_2, \ldots, \}$, $T = \{t_1, t_2, \ldots\}$. Write $S = \mathbf{R}^T$ and $P_1 = P \setminus \{p_1\}$, so

$$
\mathbf{R}^{T^P} = S \times \left(\left(\mathbf{R}^T\right)^{P \setminus \{p_1\}}\right), = S \times R_{T,P_1}.
$$

---

[7]If T and P are uncountable, just select countable subsets from each, and proceed as in theorem 4 of [MTRV].

Successively bisect $S = \mathbf{R}^T = \mathbf{R}^{\{t_1, t_2, \dots\}}$ as follows. For $s =$

$$-q2^q, \quad -q2^q + 1, \quad -q2^q + 2, \quad \dots, \quad q2^{-q} - 1,$$

write I_q as one of the q-binary one-dimensional cells

$$] - \infty, -q], \qquad]s2^{-q}, (s+1)2^{-q}], \qquad]q, \infty[, \qquad (7.21)$$

corresponding to each choice of s. These one-dimensional cells can be used to form binary partitions of finite-dimensional domains $\mathbf{R}^N$ and the infinite-dimensional domain $\mathbf{R}^T$, as described in [MTRV], pages 95–97. For $r = 1, 2, 3, \dots$, let $N_r = \{t_1, t_2, \dots, t_r\}$, and for each q let J^{qr} be a cell in $S = \mathbf{R}^T$,

$$J^{qr} = J^{qr}[N_r] = (I_{1,q} \times I_{2,q} \times \cdots \times I_{r,q}) \times \mathbf{R}^{T \setminus N_r}$$

where each $I_{j,q}$ ($j = 1, 2, \dots, r$) is a choice of one of the q-binary cells in (7.21). Combining all such choices, the resulting finite collection of cells $\{J^{qr}\}$ form a partition of $\mathbf{R}^T$ (called an (r, q)-binary partition in [MTRV], page 96). At least one of the cells

$$J^{qr} \times R_{T, P \setminus \{p_1\}} = J^{qr} \times R_{T, P_1} \subset R_{T, P} = \mathbf{R}^{T^P}$$

must be non-γ-divisible because, if each of them has a γ-division, then the union of their γ-divisions is a γ-division of full domain $R_{T, P}$. Choose one such non-divisible cell $J^{qr} \times R_{T, P_1}$, and form a binary partition of J^{qr} with binary sub-cells $J^{q+1, r+1}$. As before, at least one of

$$J^{q+1, r+1} \times R_{T, P_1} \subset R_{T, P} = \mathbf{R}^{T^P}$$

must again be non-γ-divisible. Take the intersection of the closure of the nested sequence of the chosen cells

$$\cdots \supset \bar{J}^{q, r} \supset \bar{J}^{q+1, r+1} \supset \cdots, \quad \text{so there exists} \quad y_{T,1} \in \bigcap \bar{J}^{q, r},$$

with $y_{T,1} \in S = \mathbf{R}^T$. Consider the elements $\mathbf{x} = \mathbf{x}_{T,P} \in R_{T,P}$ whose initial component is $y_{T,1}$, so

$$\mathbf{x} = \mathbf{x}_{T,P} = \left(y_{T,1}, \mathbf{x}_{T, P \setminus \{p_1\}} \right), \quad \mathbf{x}_{T, P_1} = \mathbf{x}_{T, P \setminus \{p_1\}} \in R_{T, P \setminus \{p_1\}} = R_{T, P_1}.$$

Suppose there is no $\mathbf{x}_{T, P_1}, = \mathbf{x}_{T, P \setminus \{p_1\}}, \in R_{T, P \setminus \{p_1\}} = R_{T, P_1}$, such that

$$\mathbf{L} \left((y_{T,1}, \mathbf{x}_{T, P_1}) \right) \bigcap P_1 \neq \varnothing.$$

Then $\mathbf{L} \left((y_{T,1}, \mathbf{x}_{T, P \setminus \{p_1\}}) \right) = \{p_1\}$ for all $\mathbf{x}_{T, P \setminus \{p_1\}} \in R_{T, P \setminus \{p_1\}}$. In that event, a contradiction is obtained because, for some q and r large enough, a cell $J^{qr} \times R_{T, P \setminus \{p_1\}}$ in $S \times R_{T, P \setminus \{p_1\}}$ is a γ^{p_1}-fine division of itself (see (7.19) above), contrary to the way the cells J^{qr} are selected. Following the logic of the proof of theorem 4 in [MTRV] (pages 122–123 in particular), a gauge γ_{T, P_1} is then arrived at for which, as a consequence of the original assumption, R_{T, P_1} is non-divisible; and then a new bisection iteration is performed for dimension p_2. After a sufficient number of such bisection iterations for $p_3, p_4, \ldots$ a contradiction of the original assumption is eventually obtained as in theorem 4 of [MTRV], because the sets $\mathbf{L}(\mathbf{x}_{T,P})$ cannot be infinite. $\bigcirc$

Example 42 *As in theorem 4 of [MTRV] the above proof can be adapted for uncountable sets T and P. The case T uncountable is already covered in [MTRV] theorem 4. For P uncountable, elements $p_1, p_2, \ldots$ have to be successively chosen from P in order to make the preceding argument; and then the method of the last seven lines (pages 123–4) of the proof of theorem 4 can be used. If required the proof can be further extended to domains*

$$\mathbf{R}^{T^{P^{\cdots Q}}},$$

where the labelling sets $T, P, \ldots, Q$ are finite or infinite; provided points, cells, association, and gauges are defined appropriately.

The logic of such extensions is in the preceding account of $\mathbf{R}^{T^P}$. It can be summarized as follows. Suppose $T_1, T_2, \ldots, T_m$ are sets. Our aim is to define a gauge γ on domain

$$\mathbf{R}^{T_1^{T_2 \cdots T_m}}, \;=\; R_{T_1, T_2, \ldots, T_m}.$$

A gauge γ_1 is defined on $\mathbf{R}^{T_1}$. Suppose, for $1 < k < m$, a gauge γ_k is defined on $R_{T_1, \ldots, T_k}$. Then the method used for $\mathbf{R}^{T^P}$ above can be used inductively to produce a gauge

$$\gamma_{k+1} \quad for \quad \mathbf{R}^{T_1^{\cdots \cdots T_k^{T_{k+1}}}}, \;=\; R_{T_1, \ldots, T_k, T_{k+1}}.$$

A gauge enables us to produce a system of integration in the domain. The preceding examples illustrate gauges in various product spaces.

But what about the corresponding integrals? No definition, examples, properties, theorems, proofs have been given above for any actual integral. Fortunately, the required integration theory, including variation theory, Fubini's theorem, and limit theorems, follows directly from the abstract definition and development of the -complete integral given in [77], and in [72, 73, 74, 76].

The basic principles of this abstract *division system*, or Henstock integral, are outlined in the Axioms of [MTRV], pages 111–113. The fundamental point is DS3 (the *Division Axiom*), which has been established for the various product domains above. Because the cells I of these product spaces are essentially rectangular, the other axioms (DS1, DS2, DS4–DS8) apply in a straightforward way. (See also Example 55.)

Therefore, since the abstract definition and proofs are already available, it is not necessary to repeat them here for the product space domains under discussion. Similarly for domains such as $\left(\mathbf{R}^3 \times \left(\mathbf{R}^3 \times \mathbf{R} \right)^{\mathbf{R}^3} \right)^T$ which feature in the Feynman theory of quantum electrodynamics in Chapter 9 below.

Chapter 8

Quantum Field Theory

In [MTRV] (or [121] in the Bibliography) a mathematical theory of Feynman path integrals was presented. This took the form of $\int_{\mathbf{R}^T} \cdots$ where T is an interval of time and $\mathbf{R}^T$ is a Cartesian product of the real line $\mathbf{R}$ infinitely many times; and where the integrand involves some function of a quantity in physics called *action*.

The formulation is based on an analysis by Feynman, as presented in [F1] and [FH], for instance[1], which provides a mathematical framework for describing physical phenomena of the quantum mechanical kind.

The physical scenario in [F1] is a mechanical force acting on a particle possessing mass. Classical physics has a mathematical representation of the resulting particle motion, based essentially on the Newtonian formula: *particle acceleration equals applied force divided by particle mass*, or $F = ma$.

An alternative to the Newtonian view is provided by the Lagrangian method in dynamics [98, 97], including Euler's equations of motion, Hamilton's equations, and Hamilton's *principle of least action* (PLA).

In [F1] and [FH] Feynman's path integral theory, based on classical action, provides an analysis of quantum mechanical phenomena in the single particle system; phenomena such as interference patterns which are not associated with classical particle motion but which are typically observed in wave motion.

Feynman's analysis is a work of physics which uses mathematical language and methods. Feynman's path integration is investigated in [MTRV] by means of -complete integration in $\int_{\mathbf{R}^T} \cdots$; which is mathematical theory, not physics.

In this context any mathematical theory is useful only to the extent that it is effective in describing actual physical phenomena. The mathematical formulation in [MTRV] succeeded in deducing some physical consequences such as Schrödinger's equation ([MTRV] and Feynman's interaction diagrams.

This is the spirit in which the integration structures of quantum electrody-

---

[1][39] and [46], respectively.

Gauge Integral Structures for Stochastic Calculus and Quantum Electrodynamics, First Edition. Patrick Muldowney.
© 2021 John Wiley & Sons, Inc. Published 2021 by John Wiley & Sons, Inc.

namics are presented in Chapter 9 below as

$$\int_{\Omega} f(\chi)\, d\chi, \quad \text{where } \Omega = \left(\mathbf{R}^3 \times \left(\mathbf{R}^3 \times \mathbf{R} \right)^{\mathbf{R}^3} \right)^T$$

is a product of product spaces, and where $\chi \in \Omega$ is a "history" of the interaction of an electric current with an electromagnetic field.

Again, this is mathematical theory, not physics. A reader who wishes to learn physics should use physics sources. What is offered here is a particular kind of mathematics which aspires to be useful in physics.

Nevertheless, in a spirit of analogy and conjecture, this book also presents some interpretations of physics which seem to support and motivate the product-of-product-spaces approach to quantum field theory.

To recapitulate, in [F1] Feynman introduced path integrals to describe quantum mechanical wave aspects of the motion of a single particle subject to interaction with a conservative mechanical force with potential V. In [43] and other works Feynman used his integral method to describe quantum effects in a field subjected to interaction with some external agency.

Classically the single particle system has a fairly familiar mathematical expression in equations such as $F = ma$ which apply at particular spatial locations. In contrast, a field can be described by formulations such as Maxwell's equations for electromagnetic waves in a region of space, with no electric charges or currents present:

$$\nabla \cdot \mathbf{E} = 0, \quad \nabla \cdot \mathbf{B} = 0, \quad \nabla \times \mathbf{E} = -\frac{\partial \mathbf{B}}{\partial t}, \quad \nabla \times \mathbf{B} = \epsilon_0 \mu_0 \frac{\partial \mathbf{E}}{\partial t}.$$

It is not proposed to delve into the physical meaning of field equations like these, other than to select some features for further discussion, such as the kind of variables present, and their role in wave motion, in order to provide some context and motivation for mathematical modelling.

Physical waves (such as water waves) have amplitude and oscillation. The following are plane-wave solutions of the above equations:

$$\mathbf{E}(\mathbf{r}, t) = \mathbf{E}_0 \cos(\mathbf{k}.\mathbf{r} - \omega t), \quad \mathbf{B}(\mathbf{r}, t) = \mathbf{B}_0 \cos(\mathbf{k}.\mathbf{r} - \omega t + \varphi)$$

where φ is the phase difference (retardation or advancement) between electric field $\mathbf{E}$ and magnetic field $\mathbf{B}$, so if $\varphi = -\frac{\pi}{2}$ then $\mathbf{B}(\mathbf{r}, t) = \mathbf{B}_0 \sin(\mathbf{k}.\mathbf{r} - \omega t)$. Absorbing ϕ into $\mathbf{B}_0$, the above solutions can be written

$$\mathbf{E}(\mathbf{r}, t) = \mathbf{E}_0 e^{\iota(\mathbf{k}.\mathbf{r} - \omega t)}, \quad \mathbf{B}(\mathbf{r}, t) = \mathbf{B}_0 e^{\iota(\mathbf{k}.\mathbf{r} - \omega t)},$$

where $\iota = \sqrt{-1}$, and the physical solutions are the real parts of these expressions. There are two factors in each case, corresponding to wave amplitude and wave oscillation. The wave amplitudes given by these solutions are $\mathbf{E}_0$ and $\mathbf{B}_0$. The oscillation factor in each case is $e^{\iota(\cdots)}$ in which other features of the electromagnetic wave motion at location $\mathbf{r}$ and time t, such as wavelength and velocity, are given by the variable parameters of $\mathbf{k}.\mathbf{r} - \omega t$.

Though the physical parameters are real numbers it is convenient mathematically to use complex numbers in the wave oscillation factor. And while the above wave amplitudes $\mathbf{E}_0$ and $\mathbf{B}_0$ are real, it may also be mathematically convenient to allow complex-valued wave "amplitudes".

Just as a particle motion can exhibit wave-like phenomena at micro-level, an electromagnetic wave can exhibit particle-like phenomena. This is the wave-particle duality of quantum mechanics and quantum field theory.

The basic building block of Feynman's method is the **action wave functional**, defined on each of all possible histories of a quantum system. If this functional is aggregated for all histories, the result (in the form of a Feynman integral) is a single, combined, or "average" **wave function** for the system. The wave function encompasses the wave-particle duality of the quantum-level phenomena of the system as a whole.

In the terminology of this book Feynman's action wave functional can be interpreted as a **sampling sum**—a version of the stochastic sums of preceding chapters; which, in turn, are a new and more fundamental version of the stochastic integrals of probability theory.

In these pages we seek to tease out a reliable mathematical meaning for some aspects of the quantum field theory of the 1965 book [FH] (*Quantum Mechanics and Path Integrals* by Feynman and Hibbs [46]), focussing primarily on construction of integrals. [FH] is primarily a book of physics. It is therefore occasionally necessary to venture into those waters, even if only in a metaphorical spirit of analogy, conjecture, and suggestion, in order to try to convey motivation and a sense of physical reality to the mathematical concepts and methods of this book.

The equation $F = ma$ above belongs to the Newtonian theory of mechanical motion. This theory developed in the first place on the basis of a physical sense of velocity, acceleration, and related concepts such as momentum, force, and energy.

Several centuries of investigation eventually delivered an acceptable mathematical understanding of these underlying physical phenomena, in the form of a mathematically rigorous theory of calculus—rates of change and their inverses; differentiation and integration.

Nowadays this branch of mathematics is sometimes called *mathematical analysis*. Its badge or symbol is the ε-δ *method*. After the work of centuries, and having solved the problem of the Newton-Leibniz calculus in physics, it was applied successfully (by Kolmogorov and others) to even older problems of probability and random variation.

The advantages and disadvantages of this development of mathematical analysis in probability are commented on in Part I of this book. From the beginning of the twentieth century, the development associated with H. Lebesgue [100, 101, 102] have been central to mainstream mathematical analysis. But this development uses absolute convergence which is unsuitable for certain areas of probability such as stochastic calculus. (See also Chapter 10 below.)

In physics, the Feynman theory of quantum mechanics is much more recent than Newtonian mechanics, and is still problematical in mathematics terms. As

in Part I, Part II of this book seeks to gain insight into some of these mathematical issues by returning to a pre-Lebesgue, Riemannian, or ε-δ approach.

8.1   Overview of Feynman Integrals

Here is an outline of the Feynman integral analysis of quantum level phenomena
of any given physical scenario or system:

1. At each time s of a time period $T = \,]0, \tau[$ (or $T = \,]\tau', \tau[$), there is a **state
   of the system** at time $s \in T$, to be denoted as $\chi(s)$.

2. For instance, the system described in [MTRV] is a single particle of mass
   m in a domain of space (or space-time), throughout which there is a
   conservative field with potential energy function denoted by $V(y, s)$ for
   each space location y and each time s. In [MTRV] the spatial domain
   (or region of space in which the particle can be located) is, for simplicity,
   taken to be $\mathbf{R}$, as in Feynman's original paper [F1]. More realistically, this
   domain is a subset (proper or otherwise) of $\mathbf{R}^3$, $= \mathbf{R} \times \mathbf{R} \times \mathbf{R}$, so at time s
   the particle is located at position $x(s) = (x_1(s), x_2(s), x_3(s))$, where the
   force field has potential

   $$V(x, s) \;=\; V(x_1(s), x_2(s), x_3(s));$$

   *and **the state of the system at time s is the position $x(s)$ of the
   particle**. So, in this case,*

   $$\chi(s) \;=\; x(s) \;=\; (x_1(s), x_2(s), x_3(s)). \tag{8.1}$$

   The alternate designation $\chi(s)$ is intended to be more generally applicable
   to a class of phenomena broader than particle motion.

3. Construct a **lagrangian function** $\mathcal{L}$ for the given physical scenario. The
   lagrangian is, at any given time s, a particular function depending on
   certain parameters of the system. For instance, in the preceding single
   particle scenario, the parameters include the position $x(s)$ and velocity
   $\dot{x}(s)$ of the particle at time s.

4. Calculate the **action** S for the system. The meaning of "action" is explained in broad terms in [48], and there is a comprehensive account in
   [98], including its origins in the works of Lagrange [97] and Hamilton
   [65]. The general idea is that a physical system operates in such a way as
   to minimize the action of the system. In anthropomorphic terms action
   might be thought of as "effort". The anthropomorphic analogy is that
   a person faced with some physical task generally prefers to perform the
   task with the least expenditure of physical effort. In this case human will
   or intention is involved. A physical system operates in accordance with a
   Principle of Least Action, but without intervention of will, intention,
   or foresight.

5. The origins and role of the theory of action in quantum mechanics are described in [10], which includes Feynman's PhD thesis [35], his 1948 paper [39], and Dirac's seminal paper [26]. The latter begins as follows:

> *Quantum mechanics was built up on a foundation of analogy with the Hamiltonian theory of classical mechanics....Now there is an alternative foundation for classical dynamics, provided by the lagrangian.... [T]he Lagrangian method allows one to collect together all the equations of motion and express them as the stationary property of a certain action function...(the time-integral of the lagrangian).*

Dirac continues: *"These equations involve partial derivatives of the Lagrangian with respect to the coordinates and velocities and no meaning can be given to such derivatives in quantum mechanics."* Some difficulties of this kind can be evaded by the aggregation method used in this book and in [MTRV]. For instance, given a function $\mathcal{L}(x(s), \dot{x}(s))$, partial derivatives

$$\frac{\partial \mathcal{L}}{\partial x(s)}, \qquad \frac{\partial \mathcal{L}}{\partial \dot{x}(s)}$$

can appear in sampling form with differential elements $\partial \mathcal{L}$, $\partial x(s)$, and $\partial \dot{x}(s)$ replaced by finite differences based on $N = \{s_j\}$. The differences and quotients are to be "averaged out" in the integration process $\int_{\mathbf{R}^T} \cdots$, with sampling differences $s_j - s_{j-1}$ tending to zero in aggregate. Aggregation or averaging of such sampling functions is demonstrated in (8.7) and (8.8), page 211 below.

6. The single particle system of 2. above involves the interaction of a particle of mass m subject to a mechanical force with potential $V(y)$ (or $V(y, s)$) which varies with position $y = x(s)$. The operation, process, trajectory, or *history* of this particular system consists of the motion of the particle during the time period $T =]0, \tau[$ (or $T =]\tau', \tau[$), with given initial state or position $x(0) = \xi'$ (or $x(\tau') = \xi'$) and given final state or position $x(\tau) = \xi$. The action of the system depends on the path $\{x(s)\}$ taken by the particle between given initial and final times and states (or positions); so

$$S = S\left(\left(x(s)_{x(0)=\xi'}^{x(\tau)=\xi}\right)_{0 < s < \tau}\right).$$

The action $S = S(\cdot)$ in this case is the integral with respect to time s of the lagrangian function $\mathcal{L}$ where $\mathcal{L}(y(s))$ is the kinetic energy (or K.E.) of the particle at time s minus the potential energy (or P.E.) V of the particle at time s,

$$S = S(\cdot) = S\left(x(s)_{\xi'}^{\xi}\right) = \int_{\xi'}^{\xi} \mathcal{L}(x(s))ds$$

$$= \int_{\xi'}^{\xi} (\text{K.E.} - \text{P.E.}) \, ds$$

$$= \int_{\xi'}^{\xi} \left(\frac{1}{2}m\left(\frac{dx(s)}{ds}\right)^2 - V(x(s))\right) ds,$$

the integral being taken along the path (or trajectory or history) $x(s)$ for $0 \leq s \leq \tau$ followed by the particle between times $s = 0$ and $s = \tau$. The PLA (Principle of Least Action) states that the actual path $\bar{x}$ "chosen" by the system is the one for which

$$S\left(x(s)_{\xi'}^{\xi}\right) = S(\bar{x}) = \bar{S}$$

is a local minimum, turning point, or point of inflexion. In classical physics the **calculus of variations** [168] may be used to determine $\bar{x}$.

7. In [F1], Feynman showed how to model quantum mechanical phenomena of a physical system using the action S of the system:

   (a) Construct a wave formula $\alpha(\cdot) \exp \frac{\iota}{\hbar} S(\cdot)$ where $\iota = \sqrt{-1}$ and action S depends on the history $(\chi(s))_{0 \leq s \leq \tau}$ of the system. This is the basic step in the Feynman approach to quantum physics. It is the integrand for construction of a **path integral** (or **integral over histories**).

   (b) Construct a weighting or integrator function (— a "measure", so to speak) from the joint state function $(\chi((s)_{0 \leq s \leq \tau})$, so a "path integral" (or "integral over histories") emerges from

$$\int_{\Omega} \left(\alpha(\cdot) \exp \frac{\iota}{\hbar} S(\cdot)\right) d\chi \left((s)_{0 < s < \tau}\right), \tag{8.2}$$

   where domain Ω corresponds to "degrees of freedom" of possible system histories $(\chi((s)_{0 < s < \tau}))$.

   (c) In [F1] Feynman starts with oscillatory factors $\exp \frac{\iota}{\hbar} S(\cdot)$ and then adds amplitudes $\alpha(\cdot)$ as "normalization factors", giving integrand $\alpha(\cdot) \exp \frac{\iota}{\hbar} S(\cdot)$ which ensures:

   i. convergence of the integral $\int_{\Omega} \cdots$ in (8.2) to a

   ii. value $\psi(\chi(\tau))$ called[2] the **wave function** of the system.

The core of the Feynman theory of quantum physics is the expression

$$\exp \frac{\iota}{\hbar} S(\chi(s)_{s=0}^{s=\tau}) \tag{8.3}$$

_____

[2]In [F1] Feynman designated the complex number ψ as *probability amplitude* (see Section 9.10 below). In [MTRV] the designation *state function* is also used, but is to be avoided because of possible ambiguity.

which has an oscillatory character, suggestive of wave-like behaviour, with consequent wave interference due to the system simultaneously manifesting all possible histories χ, each of which has "its own history action wave" suggested by (8.3)—with amplitude factor $\alpha(\cdot)$ added. (In this book $\alpha(\cdot)$ is used for the mechanical system of [F1], while $\eta(\cdot)$ is used for the quantum field theory of [FH].)

The action functional S is central to the Feynman path integral theory. For the single particle interaction case of 2. above (described in [F1] and in [MTRV]) the calculation of the action is familiar from the study of elementary physics. The theory of this system in [F1] and [MTRV] is about the effects on the particle system of a conservative force field whose potential energy field is $V(y,s)$ at each point (y,s) of the space-time domain $\mathbf{R} \times T$. While V can have an effect on the particle motion, it is assumed in [F1] and in [MTRV] that the particle does not have any reciprocal effect on the potential energy values $V(y,s)$. This is realistic if, for instance, the particle mass is so small that it has no effect on, say, an ambient gravitational field with potential V.

But a charged particle moving in an electromagnetic field can produce changes in the electric and/or magnetic potentials in the field. To apply the Feynman integral approach to the resulting quantum phenomena of an interactive physical system such as this, suppose a state function $\chi(s)$ exists for the system for each time s, and suppose the system has lagrangian and action functions $\mathcal{L}$ and S, respectively. Then a "probability amplitude" or system wave function $\psi(\chi(\tau))$ for the state of the system at time τ, and all histories χ, is defined by the integral calculation

$$\psi(\chi(\tau)) = \int_{\Omega} \left(\eta(\cdot) \exp \frac{\iota}{\hbar} S(\cdot) \right) d\chi \left((s)_{0 < s < \tau} \right). \tag{8.4}$$

This calculation is formally the same as (8.2) which describes a particle, not a field. But while the form of the path integral is the same, different physical systems may have different state functions (or histories) χ, different domains Ω, different representations of action S, and different integrators $d\chi \left((s)_{0 < s < \tau} \right)$. In general other parameters such as history action wave amplitudes $\alpha(\cdot)$ will also differ. (In Chapter 9, the history action wave amplitude is denoted by $\eta(\cdot)$ to distinguish it from the single particle theory which uses notation $\alpha(\cdot)$. As a term for $\psi(\xi,\tau)$, "probability amplitude" is avoided as far as possible.[3])

The above is a broad brush outline of the Feynman integral method, in which neither the physical nor the mathematical aspects are specified in any detail. The general idea here is to extend the mathematical system of [MTRV] in order to provide a possible mathematical framework for quantum interaction in fields using the model of path integrals introduced in [MTRV]. The emphasis and focus are on mathematical structures; leaving the physical meaning (if any) to "look out for itself".

---

[3] [167] includes a discussion of the meaning and history of the "probability amplitude" interpretation of the wave function. See Section 9.10 below.

8.2   Path Integral for Particle Motion

In [MTRV] a mathematical theory of path integrals is given for a physical system consisting of a single particle moving in a single dimension in a field of potential V, showing how the field affects the "state" (or motion) of the particle.

The following is a review of some salient features of path integrals for particle systems as presented in [F1] and [MTRV], in order to obtain some orientation on how to develop this theory into a theory of Feynman integrals for fields.

1. [F1] describes a physical experiment in which particle motion exhibits wave-like interference when passing through narrow slits; as if the single particle is associated with a wave or waves which can pass through several slits at the same time, simultaneously generating multiple waves which can "cancel each other out" when they are out of phase, and which can "reinforce each other" when they are in phase. In other words at micro-level a single particle exhibits destructive and constructive wave interference phenomena which are not detectable at macro-level.

2. [F1] then introduces a harmonic functional $\exp\frac{\iota}{\hbar}S(x)$ which varies for different particle paths (or "histories") $x = (x(s)_{s=0}^{s=\tau})$, where S is the action of the system for the path x.

3. In Brownian motion, infinitely many possible paths are available to the diffusion of Brownian particles, but each particular particle traverses a single definite (if unpredictable) path of its own. The diffusion properties of the Brownian system as a whole are established by aggregating contributions from each of the possible paths or histories, weighted proportionally in accordance with a Gaussian joint probability distribution function. (See [MTRV], chapter 7.)

4. In quantum mechanics the experimental observation is that the motion of a single particle has wave-like characteristics which are not present in Brownian motion. Feynman's theory implies that every possible particle path or history—every possible combination of possible successive states of the particle system—contributes proportionately to an overall aggregate. Accordingly, the terms $\exp\frac{\iota}{\hbar}S(x)$ are summed over all paths x, with proportionate contributions of wave-form action $\exp\frac{\iota}{\hbar}S(x)$ from all possible paths (or histories) x.

5. At this stage of his exposition in [F1] Feynman suggested that some kind of mathematical integration of the functional $\exp\frac{\iota}{\hbar}S(x)$ is required. This is indicated above by the integrator $d\chi\left((s)_{0<s<\tau}\right)$, which, for motion of a particle in a field of potential $V(x(s))$ (or $V(x(s),s)$) becomes $dx\left((s)_{0<s<\tau}\right)$. If $x(s)$ is a point of three-dimensional space, $x(s) = (x_1(s), x_2(s), x_3(s)) = (x_{1s}, x_{2s}, x_{3s})$, then

$$d\chi\left((s)_{0<s<\tau}\right) = \left(dx_1(s)\,dx_2(s)\,dx_3(s)\right)_{0<s<\tau}.$$

(But in [F1], as in [MTRV], the description of particle motion is restricted to one dimension only, so $d\chi\left((s)_{0<s<\tau}\right) = dx\left((s)_{0<s<\tau}\right)$, where dx is dx_1.)

6. Feynman stated in [F1] that he did not have at his disposal any way of performing an integration of the kind

$$\int \exp\left(\frac{\iota}{\hbar}S(x)\right) dx\left((s)_{0<s<\tau}\right). \tag{8.5}$$

This looks like integration with respect to a joint variable $(x(s))_{0<s<\tau}$ in a product domain $\prod\{\mathbf{R}: 0 < s < \tau\}$. Fubini's theorem (if applicable) states that integrals with respect to a joint variable (in a joint or compound domain) can sometimes be evaluated as iterated one-dimensional integrals with respect to successive components of the joint variable. So, as an *ad hoc* solution, Feynman proposed an iteration, for $j = 1, 2, \ldots, n, \ldots$, of one-dimensional integrals in the form

$$\cdots\left(\int \cdots \exp\left(\frac{\iota}{\hbar}S(x)\right) \cdots dx_{s_j}\right)\cdots \tag{8.6}$$

where $S(x)$ is to be regarded as $S(x_{s_1}, x_{s_2}, x_{s_3}, \ldots)$ for the purpose of the calculation. The details of this construction can be seen in [F1]; section 2-4 of [FH] says "...*divide the independent variable time into steps of width ε*". In this book and in [MTRV], this corresponds to using *cylinder function approximations to the integrand* which will be examined later (in Section 8.7, for instance).

7. Recall that, for a particle path x, the action $S(x)$ is the integral, with respect to time s, along the particle path x from $x(0) = \xi'$ to $x(\tau) = \xi$, of the lagrangian function $\mathcal{L}(x(s)) = $ Kinetic Energy at $x(s)$ – Potential Energy at $x(s)$;

$$S(x) = \int_{s=0,\, x(0)=\xi'}^{s=\tau,\, x(\tau)=\xi} \left(\frac{1}{2}m\left(\frac{dx(s)}{ds}\right)^2 - V(x(s))\right) ds. \tag{8.7}$$

This functional has to be inserted in (8.6). The form given to this in [F1] is

$$S_n(x) = \sum_{j=1}^{n} \left(\frac{1}{2}m\left(\frac{x_{s_j} - x_{s_{j-1}}}{s_j - s_{j-1}}\right)^2 - V(x_{s_{j-1}})\right)\left(x_{s_j} - x_{s_{j-1}}\right), \tag{8.8}$$

as if each path x consists of straight line segments joining $x(s_{j-1})$ with $x(s_j)$ for $j = 1, 2, \ldots n$, with equal time intervals

$$s_j - s_{j-1} = \frac{\tau}{n} \quad \left(\text{or} \quad \frac{\tau - \tau'}{n}\right) \text{ for each } j.$$

8. The idea is that, as the number n of iterations increases, the sequence of n-dimensional integrals converges to (8.5):

$$\cdots \left(\int \cdots \exp\left(\frac{\iota}{\hbar} S_n(x)\right) \cdots dx_{s_j} \right) \cdots \quad \longrightarrow \quad \int \exp\left(\frac{\iota}{\hbar} S(x)\right) dx \left((s)_{0<s<\tau}\right).$$

$$(8.9)$$

9. If (8.9) and (8.4) are compared, it will be seen that the latter does not contain factors $\alpha(\cdot)$. These appear in Feynman's presentation in the form of normalization factors A^{-1}. Section 2-4 of [FH] states: "*Unfortunately, to define such a normalization factor seems to be a very difficult problem and we do not know how to do it in general.*" In section 4-1 of [FH], equation (4.8) establishes that

$$A = \left(\frac{2\pi\iota\hbar\varepsilon}{m}\right)^{\frac{1}{2}} \qquad\qquad (8.10)$$

With $\alpha(\cdot) = \left(A^{-1}\right)^n$, this agrees with the action wave amplitude expression for paths in [MTRV], and in this book, for the system consisting of a single particle interacting with a conservative mechanical force (see Section 8.6 below). Feynman does not specify factors A in other systems such as quantum electrodynamics.

This is an outline of some of the mathematical aspects of the theory presented in [F1] and [FH].

8.3   Action Waves

Broadly speaking, at any given time a physical system exists in a **state** described by various physical parameters of the system. The state of a system is changeable over time and space. There is a characteristic of the system which is known as its **action**. At any given time, the action of the system can be calculated from the **history** of certain physical parameters over some prior period of time. The values of the action of a system change in correspondence with the changes of state of the system.

To illustrate these ideas, consider the physical system of [MTRV] chapter 7. This system consists of a single particle acted on by a conservative force with potential V, in accordance with the Newtonian mechanical laws of motion (so force equals mass times acceleration), in unrestricted space, over a fixed time period $[\tau', \tau]$.

At time t ($\tau' < t \le \tau$) the state of the single particle system is the position $x(t)$ of the particle, and the history of the system at time t is the prior positions $x(s)$ ($\tau' \le s < t$) of the particle, with given initial position $x(\tau') = \xi'$. The mechanical action S of the single particle system at time t is the integral of the lagrangian (—the lagrangian of the particle is kinetic energy minus potential energy) over the history $x(\cdot)$, $= (x(s))_{\tau' \leq s < t}$, of the system,

$$S(x(\cdot)) \;=\; \int_{x(s)_{\tau' \leq s < t}} \left(\frac{1}{2}m \left(\frac{dx(s)}{ds} \right)^2 - V(x(s)) \right) ds \tag{8.11}$$

$$=\; \int_{x(s)_{\tau' \leq s < t}} \frac{1}{2}m \left(\frac{dx(s)}{ds} \right)^2 ds + \int_{x(s)_{\tau' \leq s < t}} (-V(x(s)))\, ds$$

$$=\; S_0(x(\cdot)) + S'(x(\cdot));$$

where S_0 is the "ground state" action of the system in the absence of the external "disturbance" V, and S' is the contribution made by the disturbance (excitation or perturbation) V to the action S. Ground action S_0 has alternative representations

$$S_0(x(\cdot)) \;=\; \int_{\tau'}^{t} \frac{1}{2}m \left(\frac{dx(s)}{ds} \right)^2 ds \;=\; \int_{\tau'}^{t} \frac{1}{2}m \left(\frac{dx(s)}{ds} \right) dx(s). \tag{8.12}$$

In classical mechanics the successive positions $\{x(t)\}_{\tau' < t \leq \tau}$, and the related particle velocities $\{\dot{x}(t)\}_{\tau' < t \leq \tau}$, can both be measured; and the path $x(\cdot), = (x(t))_{\tau' < t \leq \tau}$, of the particle is the unique path $x(\cdot)$ for which the action $S(x(\cdot))$ is minimal or stationary.

But, in quantum mechanics, particle position and particle velocity cannot both be determined simultaneously, so the particle path (or history) is indefinite. For instance, if at time s the position $x(s)$ of the particle is known then its velocity $\dot{x}(s)$ is unknown. Therefore, at any given time s, the particle position cannot be determined in advance of some "infinitesimally later" time.

Such "indefiniteness" suggests unpredictability, or randomness.[4] From this perspective a path $x(\cdot)$ may be thought of as an unpredictable occurrence of a random variable $X(\cdot)$ with domain, or sample space, consisting of all possible paths $\{x(t)\}_{\tau' < t \leq \tau}$ where $x(t) \in \mathbf{R}$ for each t; so $X(\cdot)$ is a stochastic process[5] $(X(t))_{\tau' < t \leq \tau}$ with sample space $\mathbf{R}^{]\tau', \tau]}$. Write

$$\mathbf{T} =]\tau', \tau], \qquad\qquad T =]\tau', \tau[, = \mathbf{T} \setminus \{\tau\}.$$

(In [MTRV] the domain $\mathbf{R}^T$ is described as *marginal domain* relative to $\mathbf{R}^\mathbf{T}$, and superscript $^-$ is used to distinguish elements x^-, I^-, etc., of $\mathbf{R}^T$ from the corresponding elements x, I of $\mathbf{R}^\mathbf{T}$. In the present work the domain $\mathbf{R}^T$ is more prominent so, unless otherwise indicated, elements x, I, etc., are taken to belong to $\mathbf{R}^T$.)

For a single particle system it is shown in chapter 7 of [MTRV] that there is a function G (written G_c in [MTRV]) defined on subsets of $\mathbf{R}^\mathbf{T}$ which has some

---

[4]But in quantum mechanics, more than randomness is involved. At some later time there are "many different particle positions"; not just "many different **possible** positions".

[5]From that perspective, the S_0 component of (8.11) is a stochastic integral, $S_0(X) = \int_{\tau'}^{t} \left(\frac{dX(s)}{ds} \right)^2 ds = \int_{\tau'}^{t} \frac{dX(s)}{ds}\, dX(s)$.

of the properties of a probability function; so, in those terms, $X, = X(\cdot)$, may in some sense be an observable $X \simeq x\left[\mathbf{R^T}, G\right]$, so that a function $f(X)$ may be a random variable with mean or expected value

$$\mathbf{E}[f(X)] = \int_{\mathbf{R^T}} f(x)\, dG. \tag{8.13}$$

This interpretation is not intended to be physically meaningful. Rather, it is intended to give some perspective to the mathematical calculations which are used to describe the evolution of the single particle system in quantum mechanical terms.

The marginal domain $\mathbf{R}^T, = \mathbf{R}^{]0,\tau[}$ is used to calculate the wave function[6]

$$\psi = \int_{\mathbf{R}^{]0,\tau[}} \cdots. \tag{8.14}$$

The following is a summary of the calculation of wave function ψ, as presented in chapter 7 of [MTRV].

Physical waves can combine together to form a wave by constructive and destructive interference of the separate waves. In the Feynman integral analysis of quantum mechanics, action waves $\alpha(\cdot)e^{\frac{i}{\hbar}S(x(\cdot))}$ for distinct histories or paths $x(\cdot)$ are combined, and the combined path waves yield a system wave:

$$\psi(x(t),t) = \int_{\{x(\cdot)\}} e^{\frac{i}{\hbar}S(x(\cdot))}\left(\alpha(\cdot)dx(\cdot)\right) = \int_\Omega \left(\alpha(\cdot)e^{\frac{i}{\hbar}S(x(\cdot))}\right)dx(\cdot), \tag{8.15}$$

and, for $0 < t \le \tau$ and $x(t) \in \mathbf{R}$, $\psi(x(t),t)$ is the **wave function** of the system as a whole. In (8.15) the domain of integration is $\Omega = \mathbf{R}^T = \mathbf{R}^{]0,t[}$,

$$\Omega = \{x(\cdot)\} = \{x : x(s) \in \mathbf{R},\ 0 < s < t \le \tau\}, \quad x(0) = \xi', \quad x(t) = \xi, \tag{8.16}$$

where t is given, and where ξ', ξ are given real numbers.

Expressions $\alpha e^{\iota f(y,t)}$ are used in mathematical physics to describe physical wave motion, the second factor (real or imaginary part) imparting information about frequency of wave oscillation at location y and time t, and the first factor α giving the amplitude of the wave motion. So α is generally a positive real number.

In contrast, expressions such as $\alpha(\cdot)$ and $\bar\alpha$ in (8.2), (8.25), and (8.26) are complex numbers (just as the oscillation factor $\exp\left(\frac{i}{\hbar}S\right)$ is complex-valued), and therefore do not conform to standard mathematical descriptions of wave motion. Nevertheless, the term "action wave" may be helpful, not necessarily in terms of physical phenomena (which take precedence over any mathematical description), but as motivation for mathematical analysis of path integrals and integrals over histories. Thus the description of action waves in Section 8.4 below is offered as a possible way of interpreting the mathematical description.

---

[6]The relationship of (8.14) to (8.13) is described as *marginal density of expectation* in [MTRV] (page 325). This is contradictory to Feynman's term *probability amplitude*. See Section 9.10.

To connect with the presentation in chapter 7 of [MTRV], here is a summary of the notation used. A time sample is $N = \{s_1, \ldots, s_{n-1}\}$ with $s_0 = 0$ and $s_n = t$. For a path $x = x_T \in \mathbf{R}^T$, a time-sampled path is

$$x(N) = (x(s_1), \ldots, x(s_{n-1})) = (x_1, \ldots, x_{n-1})$$

with $x(s_0) = x_0 = \xi'$ and $x(s_n) = x_n = x(t) = \xi$. Corresponding to (8.10), an action wave amplitude for a time-sampled path $x(N)$ is

$$\alpha(N) = \prod_{j=1}^{n} \left(\sqrt{\frac{1}{2\pi \iota \hbar m^{-1}(s_j - s_{j-1})}} \right), \tag{8.17}$$

$$\bar{\alpha} = \sqrt{\frac{1}{2\pi \iota \hbar m^{-1}(s_n - s_0)}}, \qquad s_0 = 0, \ s_n = t. \tag{8.18}$$

If $s_j - s_{j-1} = \frac{s_n - s_0}{n} = \varepsilon$ for all j, then $\alpha(N) = A^{-n}$ in (8.10). From (8.8) write

$$S_n(x) = S(x(N)) = S_0(x(N)) + S'(x(N)), \tag{8.19}$$

$$S_0(x(N)) = \frac{m}{2} \sum_{j=1}^{n} \frac{(x_j - x_{j-1})^2}{s_j - s_{j-1}}, \tag{8.20}$$

$$S'(x(N)) = - \sum_{j=1}^{n} V(x_{j-1})(s_j - s_{j-1}), \tag{8.21}$$

$$g(x(N)) = \frac{\exp\left(\frac{\iota m}{\hbar 2} \sum_{j=1}^{n} \frac{(x_j - x_{j-1})^2}{s_j - s_{j-1}} \right)}{\prod_{j=1}^{n} \sqrt{2\pi \iota m^{-1} \hbar (s_j - s_{j-1})}}$$

$$= \alpha(N) e^{\frac{\iota}{\hbar} S_0(x(N))}, \tag{8.22}$$

$$G(I[N]) = \int_{I(N)} \frac{\exp\left(\frac{\iota m}{\hbar 2} \sum_{j=1}^{n} \frac{(x_j - x_{j-1})^2}{s_j - s_{j-1}} \right)}{\prod_{j=1}^{n} \sqrt{2\pi \iota m^{-1} \hbar (s_j - s_{j-1})}} dx_1 \cdots dx_{n-1}$$

$$= \int_{I(N)} \alpha(N) g(x(N)) \, dx(N) \tag{8.23}$$

for cells $I(N)$, $I[N]$ of $\mathbf{R}^N$ and $\mathbf{R}^T$, respectively. Note that the sampling function $g(x(N))$ of (8.22) is different from the point function $g(s)$ of Chapter 6; also, that the symbol G in (8.23) is complex-valued, while G in (5.10) of Part I (page 115) is real-valued. (In [MTRV] the symbol G_c is used with c allowed to be complex or real. But in this book the distinction between the two usages is to be ascertained from the context. So in Part I the Brownian motion G is real-valued, while in Part II the Feynman integral G is complex-valued.)

8.4 Interpretation of Action Waves

For a single particle system, $x(t)$ is a possible state of the system (or location of the particle) at time t. The aggregate action wave of the system is located throughout the domain of all possible paths. As quantum mechanical system, a single particle has the same dispersed character (or multi-location character) as the aggregate action wave. In classical physics the particle at time t occupies some particular location $x(t)$. In quantum mechanics the wave-particle at time t manifests itself in **every** location (or state) $x(t) \in \mathbf{R}$.

The system wave function ψ makes it possible to establish a quasi-probabilistic[7] estimate $x(t)$ for the state of the system (position of the particle) at time t. The wave function ψ also makes it possible to estimate other characteristics of the system such as particle momentum. The two interpretations of the system— as a wave described by wave function $\psi(x(t), t)$, and as a particle with particle properties such as position $x(t)$ or velocity $\dot{x}(t)$—are inseparable.

The integral (8.15) over the domain Ω (as in (8.16)) of all possible paths or system histories $x(\cdot)$ is the Feynman path integral described in [MTRV]. Broadly speaking, it represents an aggregation of the values of $e^{\frac{i}{\hbar}S(x(\cdot))}$ for all possible paths or histories $x(\cdot)$, with a weighting represented above by $\alpha(\cdot)\,dx(\cdot)$. The latter is an "integrator function" on subsets S of paths in sampling domain Ω,

$$S = \{x(\cdot)\} \subset \Omega = \mathbf{R}^T$$

for this physical system. The weighting $\alpha(\cdot)dx(\cdot)$ for the integration $\int_\Omega \cdots$ (the aggregation or summation of the action waves for individual paths), is often represented in the literature of the subject by means of some indicative notation, such as

$$d\mu(x(\cdot)) \quad \text{or} \quad \mathcal{D}x(\cdot).$$

(The precise meaning of these notations, and of the aggregation operation "$\int_\Omega \cdots$", have to be specified.)

For any physical system or process which is **not** subject to any disturbance, interaction, or potential V, denote by S_0 the action of the system, corresponding to its history (succession of "undisturbed" states) over the prior time period. If some external disturbance or constraint or perturbation is applied, the system under perturbation (or constraint) then exhibits a history with a modified value S of action. This may involve addition of extra terms S' in the calculation of the new value of action for the constrained history, so

$$S = S_0 + S'$$

in that case. In the theory, quantum mechanical phenomena in physical systems are described by postulating, for each path or history x, an *action wave* whose oscillatory factor is $e^{\frac{i}{\hbar}S}$. So if the action of the unconstrained system is $S_0(x)$ for each available history

$$x = x(\cdot) = (x(s))_{\tau' \le s < t},$$

then the oscillation factor of the corresponding action wave is $e^{\frac{i}{\hbar}S_0(x)}$. When a constraint is applied to the system, resulting in addition of a term $S'(x)$ to the value of action, the result is a history action wave with oscillation factor

$$e^{\frac{i}{\hbar}S(x)} = e^{\frac{i}{\hbar}(S_0(x)+S'(x))} = e^{\frac{i}{\hbar}S_0(x)}e^{\frac{i}{\hbar}S'(x)};$$

---

[7]In [F1] ψ is called a **probability amplitude**. See Section 9.10 below.

so $e^{\frac{i}{\hbar}S(x)}$ is a product of two exponential factors, the first being the oscillation factor $e^{\frac{i}{\hbar}S_0(x)}$ for the unconstrained system, and the second factor being a contribution $e^{\frac{i}{\hbar}S'(x)}$ arising from the application of a constraint in the system.

When an amplitude[8] factor $\alpha(\cdot)$ is included for each history x, the formula for the action wave becomes $\alpha(\cdot)e^{\frac{i}{\hbar}S(x)}$. The quantum mechanical features of the system emerge from superposition of waves for all possible histories with their corresponding action values

$$\sum \alpha(\cdot)e^{\frac{i}{\hbar}S(x)}, \quad = \quad \sum \left\{ \alpha(\cdot)e^{\frac{i}{\hbar}S(x)} : \text{all histories } x \right\};$$

(with a further weighting factor corresponding to $dx(\cdot)$) so that wave interference (constructive and destructive) may occur, yielding a single or aggregate wave for the system as a whole. This calculation incorporates the counter-intuitive notion that the system simultaneously manifests all possible histories or trajectories, along with their corresponding values $S(x)$ for the action.

There are generally infinitely many admissible "histories" available to the system. If the aim is to superimpose, by addition, all the corresponding action waves then some weighting is required in order to yield some average (aggregate or resultant) value for the infinitely many wave functions to be summed; and if the aggregate value is itself to be a resultant wave function with its own amplitude and oscillation properties, then the weighting factor should be appropriately determined. Accordingly, the aggregation idea in $\sum \alpha(\cdot)e^{\frac{i}{\hbar}S(\cdot)}d(\cdot)$ is rendered as an integral

$$\psi \quad = \quad \int \alpha(\cdot)e^{\frac{i}{\hbar}S(\cdot)}d(\cdot)$$

where $\sum$ over all histories is replaced by $\int$ over all histories (the Feynman integral), with $d(\cdot)$ indicating an appropriate weighting in the aggregation.

The value ψ—if and when it is successfully formulated—is the wave function for the system, from which the characteristics of the system in space-time can be deduced. Using $S = S_0 + S'$, the wave function can be expressed as

$$\psi \quad = \quad \int_\Omega \alpha(\cdot)e^{\frac{i}{\hbar}S(\cdot)}d(\cdot) \quad = \quad \int_{\mathbf{R}^T} e^{\frac{i}{\hbar}S'(\cdot)}\left(\alpha(\cdot)e^{\frac{i}{\hbar}S_0(\cdot)}d(\cdot)\right) \quad = \quad \int_{\mathbf{R}^T} e^{\frac{i}{\hbar}S'(\cdot)}dG$$

where G is a marginal version of a Brownian-like cell function in $\mathbf{R}^T$, and

$$\text{``} dG \quad = \quad \alpha(\cdot)e^{\frac{i}{\hbar}S_0(\cdot)}d(\cdot) \text{''}$$

in some sense. Details of this -complete approach are summarized in (8.29) and (8.30) in Section 8.6 below. But first we consider how $\psi = \int_\Omega \cdots$ is handled in [F1] and [FH] using the calculus of variations.

8.5 Calculus of Variations

The theory of lagrangian and action (Hamilton's principle) originated in the works of J.-L. Lagrange [97] and W.R. Hamilton [65]. There is a good account

---

[8]This is called *normalizing factor* in [F1] and [FH].

by C. Lanczos in [98] . The Euler-Lagrange equation of the calculus of variations
[60, 168] is used to establish the classical **path of least action**.

Referring to [50, 162] as sources in his PhD thesis [35] (with normalization
factors A^{-1} in place of $\alpha(\cdot)$), Feynman gives wave function ψ as

$$\psi = \int_\Omega \alpha(\cdot)e^{\frac{\iota}{\hbar}S(\cdot)}d(\cdot) = \bar{\alpha}e^{\frac{\iota}{\hbar}\bar{S}}. \tag{8.24}$$

In [FH] it is argued that, provided action S is "large" relative to $\hbar$, ψ has
the form $\bar{\alpha}e^{\frac{\iota}{\hbar}\bar{S}}$ where $\bar{S} = S(\bar{x})$ and where $\bar{x}(\cdot)$ is the classical path of least
action, the unique path for which, in classical physics, the action is least (or
stationary) relative to all other possible paths or histories of the system. Some
intuitive support for this is provided by figure 6.3, page 302 of [MTRV].

In [FH], the calculus of variations leads to the value $\bar{S}$ in (8.24).[9]   But
calculus of variations is less helpful in determining $\bar{\alpha}$.

In [MTRV] values $\bar{\alpha}$ and $\bar{S}$ are established by evaluation of the integral
$\int_\Omega \cdots$ for simple cases; and this procedure is confirmed in general (whenever the
integral exists) by deducing Schrödinger's equation from $\int_\Omega \cdots$. (See [MTRV]
theorem 220, page 350.)

Suppose a free particle (potential $V = 0$) with mass m is observed at position
ξ' at time τ', and at position ξ at later time τ; so $x(\tau') = \xi'$ and $x(\tau) = \xi$. Then
the classical path of least action is $\bar{x}$ where, for $\tau' \le t \le \tau$,

$$\bar{x}(t) = \xi' + \frac{\xi - \xi'}{\tau - \tau'}(t - \tau').$$

For this path the particle velocity is constant, with $\dot{\bar{x}}(t) = \frac{\xi - \xi'}{\tau - \tau'}$, and the action
$\bar{S}, = S(\bar{x})$, is

$$\bar{S} = \left(\frac{m(\xi - \xi')}{2(\tau - \tau')}\right)^2 (\tau - \tau') = \frac{m(\xi - \xi')^2}{2(\tau - \tau')}.$$

This value for $\bar{S}$ provides confirmation of the calculus of variations argument for
(8.24). It also agrees with the evaluation of the path integral $\int \alpha(\cdot)e^{\frac{\iota}{\hbar}S(\cdot)}d(\cdot)$
in theorem 202 (page 324 of [MTRV]) which gives system wave function

$$\psi = \psi(\xi, \tau; \xi', \tau') = \int_\Omega \alpha(\cdot)e^{\frac{\iota}{\hbar}S(\cdot)}d(\cdot) = \bar{\alpha}e^{\frac{\iota}{\hbar}\bar{S}},$$

$$\text{with } \bar{\alpha} = \sqrt{\frac{m}{2\pi\hbar\iota(\tau - \tau')}} \quad \text{and} \quad \bar{S} = \frac{m(\xi - \xi')^2}{2(\tau - \tau')}. \tag{8.25}$$

Thus the aggregation of action waves $\alpha(\cdot)e^{\frac{\iota}{\hbar}S(x(\cdot))}$ for the individual paths $x(\cdot)$
gives system action wave

$$\sqrt{\frac{m}{2\pi\iota\hbar(\tau - \tau')}}e^{\frac{\iota}{\hbar}\bar{S}}$$

---

[9]A mathematical proof of existence of the integral is absent; Feynman's account deals with
physics.

with amplitude factor $\sqrt{\frac{m}{2\pi\iota\hbar(\tau-\tau')}}$ and oscillation factor $e^{\frac{\iota}{\hbar}\bar{S}}$. In [MTRV] the evaluation of this path integral is based on the evaluation of Fresnel's integral using the Riemann-complete system of Kurzweil and Henstock. (See section 6.5 of [MTRV].)

What if V is non-zero? Like kinetic energy $\frac{m}{2}\left(\frac{dx(s)}{ds}\right)^2$, potential energy V is non-negative. Suppose, for instance, that V (= $V(y,s)$) has non-zero constant value $\beta > 0$ for all $y \in \mathbf{R}$ and all s ($\tau' \leq s \leq \tau$. For each path x with given $x(\tau') = \xi'$, $x(\tau) = \xi$, the contribution of potential energy to the action is

$$S'(x) \;=\; -\int_{x(\cdot)} V(x(s))\,ds \;=\; -\beta(\tau-\tau'),$$

the same negative value for all possible paths x. The kinetic energy $\frac{m}{2}\left(\frac{dx(s)}{ds}\right)^2$ can only increase (or stay the same) for any path x other than the straight line joining the points

$$(\tau',\ x(\tau')),\qquad (\tau,\ x(\tau)).$$

Therefore, for constant potential $V = \beta$, and for all $\beta \geq 0$, the straight line is the path $\bar{x}$ of least action; and, according to the physics reasoning of Feynman, the oscillation factor of the wave function ψ for this system is $\exp\left(\frac{\iota}{\hbar}S(\bar{x})\right)$ =

$$=\ \exp\left(\frac{\iota}{\hbar}\left(\frac{m(\xi-\xi')^2}{2(\tau-\tau')}\right) - \beta(\tau-\tau')\right),\ =\ \exp\left(\frac{\iota}{\hbar}(S_0(\bar{x}) + S'(\bar{x}))\right).$$

Feynman's reasoning about the normalizing factor A^{-1} (or amplitude factor $\bar{\alpha}$) is less clear. But, noting that

$$S'(x),\ =\ -\beta(\tau-\tau'),$$

is constant for all $x \in \mathbf{R}^T$, the path integral approach gives the following in this case: $\psi\ =\ \psi(\xi,\tau;\xi',\tau')\ =\ \int_{\mathbf{R}^T}\alpha(\cdot)e^{\frac{\iota}{\hbar}S(\cdot)}d(\cdot)\ =$

$$=\ \int_{\mathbf{R}^T}\alpha(\cdot)e^{\frac{\iota}{\hbar}S_0(x(\cdot)) + S'(x(\cdot))}dx(\cdot)$$

$$=\ e^{\frac{\iota}{\hbar}S'(x(\cdot))}\int_{\mathbf{R}^T}\alpha(\cdot)e^{\frac{\iota}{\hbar}S_0(x(\cdot))}dx(\cdot)$$

$$=\ e^{-\frac{\iota}{\hbar}\beta(\tau-\tau')}\int_{\mathbf{R}^T}\alpha(\cdot)e^{\frac{\iota}{\hbar}S_0(x(\cdot))}dx(\cdot)$$

$$=\ e^{-\frac{\iota}{\hbar}\beta(\tau-\tau')}\sqrt{\frac{m}{2\pi\iota\hbar(\tau-\tau')}}e^{\frac{\iota}{\hbar}\left(\frac{m}{2}\left(\frac{(\xi-\xi')^2}{\tau-\tau'}\right)\right)} \tag{8.26}$$

$$=\ \sqrt{\frac{m}{2\pi\iota\hbar(\tau-\tau')}}e^{\frac{\iota}{\hbar}\left(\frac{m}{2}\left(\frac{\xi-\xi'}{\tau-\tau'}\right)^2 - \beta\right)(\tau-\tau')},\ =\ \bar{\alpha}e^{\frac{\iota}{\hbar}S(\bar{x})},$$

confirming that the action wave amplitude $\bar{\alpha}$ (corresponding to Feynman's normalizing factor) is the same as in the free particle case.

This is consistent with the result obtained in [FH] by means of the calculus of variations. (The meaning of (8.24), including integrand terms $\alpha(\cdot)$, $S(\cdot)$, $d(\cdot)$, and $\int_\Omega$, is provided in [MTRV], and is discussed further in Section 8.6 below.)

It is possible to continue in this vein, taking V to be a step function in the sense of [MTRV], section 7.19, pages 360–366. In that case the calculations are still of the Fresnel type encountered in the cases $V = 0$ and $V = \beta > 0$, but there are many separate terms to be added up at the end.

As in traditional analysis the next stage (after step function V) might be to proceed to continuous $V(y,t)$. This can be done if the variability of the function V is severely restricted. As indicated in [MTRV] section 7.19, pages 360–366, this further advance from step function V to continuous V is more problematic. The kind of reasoning involved is demonstrated in Section 8.10 below (dealing with simple harmonic oscillation), in which the potential energy function $V(x(t))$ is $\frac{m}{2}\varpi^2(x(t))^2$.

Nevertheless the path integral evaluations of [MTRV] concur with the calculus of variations argument of [FH], at least when $\hbar$ is "small" relative to system action S. That is to say, in the particular cases considered in [MTRV], the path integral—when it is evaluated as an integral—works out as $e^{\frac{i}{\hbar}\bar{S}}$ (wave oscillation factor) multiplied by the factor $\bar{\alpha}$ (wave amplitude factor), a result which corresponds to what is predicted by the calculus of variations argument of [FH].

So why bother trying to define and evaluate path integrals if the more traditional and more highly developed calculus of variations gives the same results (up to $e^{\frac{i}{\hbar}\bar{S}}$)?

The calculus of variations technique in [FH] does not shed any light on how to determine aggregate wave amplitude $\bar{\alpha}$; which, in cases (8.25) and (8.26), is a complex number. This issue was raised by Feynman in his PhD thesis, [35] page 73 with A^{-1} corresponding to $\alpha(\cdot)$ above:

> *A point of vagueness is the normalization factor, A. No rule has been given to determine it for a given action expression. This question is related to the difficult mathematical question as to the conditions under which the limiting process of subdividing the time scale ... actually converges.*

The [MTRV] version of path integral theory addresses both of these issues; see theorem 202 (page 324) and 223 (page 365).

Also, while $\bar{x}$ is important, neighbouring paths x are also important. Much of [FH] is concerned with deducing insights into the physical phenomena from the path integral model itself, rather than discovering particular integral values or evaluations involving $\bar{x}$. For example, a notable consequence of the path integral model is the Feynman diagram technique. Feynman diagrams are a major physics resource and these are based, not on calculus of variations, but on the representation of the system wave function ψ as a path integral or integral over histories, from which the diagrams are extracted or deduced. The following is an outline.

If the interaction exponential factor $e^{\frac{i}{\hbar}S'}$ is expanded as a series, the resulting re-formulation of "the average/aggregate of all possible action waves" leads to the Feynman diagrams interpretation of the path integral, by which physicists are able to model the effects of the constraints or perturbations acting on the original unconstrained system. These effects may take the form of emission or absorption of "interaction particles", often illustrated by wavy lines in the Feynman diagrams.

Broadly, the re-formulation goes on these lines:

$$\int \alpha e^{\frac{\iota}{\hbar}S} d(\cdot) \;=\; \int \alpha e^{\frac{\iota}{\hbar}S_0 + S'} d(\cdot) \;=\; \int \left(\alpha e^{\frac{\iota}{\hbar}S_0} e^{\frac{\iota}{\hbar}S'} \right) d(\cdot)$$

$$=\; \int \alpha e^{\frac{\iota}{\hbar}S_0} \left(1 + \frac{\iota}{\hbar}S' + \frac{1}{2!}\left(\frac{\iota}{\hbar}S'\right)^2 + \cdots \right) d(\cdot)$$

$$=\; \sum_{n=0}^{\infty} \left(\int \alpha e^{\frac{\iota}{\hbar}S_0} \left(\frac{1}{n!}\left(\frac{\iota}{\hbar}S'\right)^n \right) d(\cdot) \right);$$

where the order of $\int$ and $\sum_{n=0}^{\infty}$ is reversed in the final step.

When some further re-calculation is performed it turns out that, for each n, the term $\int \alpha e^{\frac{\iota}{\hbar}S_0} \left(\frac{1}{n!}\left(\frac{\iota}{\hbar}S'\right)^n \right) d(\cdot)$ of the series expansion (or perturbation expansion) can be interpreted as one of the Feynman diagrams portraying the physical interaction effects such as emission or absorption of a particle. Even if factor α is unknown or omitted, it seems that this procedure provides valuable physical insight.

A more detailed account is provided in [MTRV] pages 366–382. The term $n = 0$ represents the "ground state" with no interaction effect, so there are no emission/absorption lines in the Feynman diagram for this case. The physically important effects are given by the first few terms such as $n = 1$ and $n = 2$.

Approximation of this kind is central to the physics of path integrals. In integration theory a distinction is made between integrals defined constructively (as, for instance, by using Riemann sum estimates or approximations) or implicitly (such as definition of integral as antiderivative). Unlike the Riemann sum method, the implicit or indirect approach to path integrals provided by the calculus of variations is not so amenable to approximation.

Various path integrals are evaluated in [FH] using a variety of methods such as arguments from physics and from the calculus of variations. See, for instance, *General Results for Quadratic Action* in section 7-4 of [FH].

8.6 Integration Issues

In [F1] Feynman noted mathematical gaps in his theory. Is (8.6) valid? Can some definite mathematical meaning be given to (8.5)? Notation such as

$$\int \cdots dx \, ((s)_{0 < s < \tau})$$

is used in (8.15)—on page 214 above and various other places. But some definite meaning needs to be provided for such notation, as shown in [MTRV].

It is also worthwhile, for reasons of physics as well as mathematics, to try to give mathematical meaning to Feynman's integrals, so they can be treated as actual integrals rather than simply suggestion or intuition. As shown in [MTRV], rigorous mathematical construction of the integrals enables us to provide rigorous interpretations of important physical consequences such as Schrödinger's equation and the Feynman diagram technique. And it is possible to do this without departing very far from Feynman's original logic and presentation.

The coverage in [MTRV] deals with the quantum mechanical effects of a field on a particle, using the method of Riemann sums (Henstock-Kurzweil or -complete integrals) on the subject-matter of Feynman's paper [F1]. In subsequent papers Feynman developed the action wave method in order to analyse quantum electrodynamics—the quantum mechanical effects of a charged particle on an electromagnetic field. And in [FH] the material of these papers is brought together in review, in such a way as to explain the train of thought leading from [F1] to quantum electrodynamics.

The purpose of this book is to develop the Riemann structure of -complete integration, in order to encompass the integral-over-histories of quantum electrodynamics, as formulated in [FH]. In doing so, it is useful also to review the intermediate steps as described in [FH].

Section 4.1 above provides an overview of the relevant integration theory; see also Chapter 10 below. A reader who is new to the subject may benefit from the introduction to the Riemann sum method presented in chapter 3 of [MTRV], entitled *Infinite-Dimensional Integration*. (A full mathematical exposition is provided in chapters 4, 6, and 7 of that book, which are the parts of most relevance to Feynman integration.)

In the traditional and widely used integral notation $\int_\Omega f(\cdot)\, d(\cdot)$, the symbol $d(\cdot)$ refers to some function of subsets of Ω. In Lebesgue integration it refers to $\mu(A)$, a measure function μ on measurable subsets A of Ω. In Riemann-type integration it refers to a length-, area- or volume-type function $|I|$ on interval-type subsets I of Ω.

Feynman was unable to find a measure function corresponding to $d(\cdot)$, $dx(\cdot)$, or $dx\left((s)_{0<s<\tau}\right)$, and instead used the indirect approach of (8.6), in which the sequence of iterated one-dimensional integrals

$$\int \cdots \int \cdots dx_{s_j} \cdots$$

is supposed to converge to some limiting value; and the latter is taken as the meaning of $\int \cdots dx\left((s)_{0<s<\tau}\right)$.

In [MTRV] the order is reversed. The expression $\int \cdots dx\left((s)_{0<s<\tau}\right)$ is given a definite meaning as an integral, and conditions are established for which (8.6) is valid. Here is an outline of integration issues:

1. In the more traditional systems of integration such as Lebesgue and Riemann integration, the definition of the integral $\int_\Omega f(\cdot)\, d(\cdot)$ requires certain conditions to be satisfied by the integrand elements $f(\cdot)$ and/or $d(\cdot)$. Henstock's definition of the -complete system of integration places no condition

or restriction on elements $f(\cdot)$ and/or $d(\cdot)$. Instead his definition requires a structure to be satisfied by those elements $x \in \Omega$ and $I \subset \Omega$ which are to be admissible in the Riemann sums $\sum f(x)g(I)$ which define the -complete integral

$$\int_\Omega f\, dg \quad \text{or} \quad \int_\Omega f(x)g(I).$$

(In Chapter 10 of Part III below, Lebesgue's integral is expressed as a Riemann-Stieltjes integral, for which a -complete extension is easily formulated.)

2. With the appropriate structure of (x, I) in Ω, it is possible to obtain directly a **definition** of the integral $\int_\Omega f(x\,((s)_{0<s<\tau}))\,dx\,((s)_{0<s<\tau})$ posited by Feynman in [F1], with no concern for any measure function on subsets of elements $x\,((s)_{0<s<\tau})$ of Ω, nor for any special properties (such as measurability) of integrand functions $f(\cdot)$. This is what is presented in [MTRV].

3. The preceding remark relates to **definition** only. Actual **existence** of any particular Feynman integral, and the properties it may satisfy, will often depend on properties satisfied by the integrand functions, such as continuity and the like. But while such properties enter into subsequent parts of the theory, in the -complete system of [MTRV] they do not have to be laid down as requirements in advance of the definition of the integral itself.

4. The -complete integral of [MTRV] provides flexibility in structuring the elements of the integrand. A detailed exposition of that approach is provided there. For present purposes, some salient points can be highlighted as follows.

5. In (8.2) and (8.4) the integrand consists of factors

$$\exp\left(\frac{\iota}{\hbar}S(\chi)\right) \quad \text{and} \quad d\chi\,((s)_{0<s<\tau})$$

along with additional elements designated by $\alpha(\cdot)$ or $\eta(\cdot)$ which, in [F1], Feynman refers to as normalization factors. In [F1] a history $\chi\,((s)_{0<s<\tau})$ of the elementary states of the system is a possible particle path $x = x(\cdot) = x\,((s)_{0<s<\tau})$; and the domain Ω of integration consists of all possible particle paths x. So, in [F1] and [MTRV],

$$\Omega \;=\; \{x\,((s)_{0<s<\tau})\} \;=\; \mathbf{R}^T,$$

which is a Cartesian product $\prod\{\mathbf{R}: 0 < s < \tau\}$ with $T = \{s: 0 < s < \tau\}$.

6. The subsets of Ω to which a weighting or integrator $dx\,((s)_{0<s<\tau})$ is applied are intervals or cells I in $\Omega = \mathbf{R}^T$, where I_s is a one-dimensional cell in $\mathbf{R}$ for any s $(0 < s < \tau)$;

$$I \;=\; I_{s_1} \times I_{s_2} \times \cdots \times I_{s_{n-1}} \times \mathbf{R}^{T \setminus \{s_1,\dots,s_{n-1}\}};$$

I is denoted as

$$I = I[N] \quad \text{where} \quad N = \{s_1, \ldots, s_{n-1}\}, \quad \text{with} \quad 0 < s_1 < s_2 < \cdots < s_{n-1} < \tau;$$

and N is a variable in the integration process, just as $x\left((s)_{0<s<\tau}\right)$ is variable.

7. If $|I_s|$ denotes the length of the one-dimensional interval I_s in $\mathbf{R}$, the weighting $dx\left((s)_{0<s<\tau}\right)$ in $\int_\Omega \cdots dx\left((s)_{0<s<\tau}\right)$ is the "cell volume" weighting

$$|I| \;=\; |I[N]| \;=\; \prod_{j=1}^{n-1} |I_{s_j}|$$

which is the volume of the cell $I(N) := I_{s_1} \times I_{s_2} \times \cdots \times I_{s_{n-1}}$ in the $n-1$-dimensional Cartesian product space $\mathbf{R} \times \cdots \times \mathbf{R}$. This corresponds to the way the familiar Riemann integral is weighted in finite-dimensional space. Since the set of dimensions N is variable and arbitrarily large, this is **not** a reversion from infinite-dimensional $\mathbf{R}^T$ to finite-dimensional $\mathbf{R}^{n-1}$.

8. The reason for using subscripts $s_1, \ldots, s_{n-1}$ rather than $s_1, \ldots, s_{n-1}, s_n$ is because it is notationally convenient to reserve the nth subscript for $s_n = \tau$. Likewise, $s_0 = 0$ or τ'. In [MTRV], and in this book, symbols s, s_j, t, t_j are used to denote instants of time in T.

9. The "point-function integrand" f in

$$\int_\Omega f\left(\cdot\right) d(\cdot) \;=\; \int_\Omega f\left(x\left((s)_{0<s<\tau}\right)\right) dx\left((s)_{0<s<\tau}\right)$$

includes a factor

$$\exp\left(\frac{\iota}{\hbar}S\right), \;=\; \exp\left(\frac{\iota}{\hbar}S\left(x\left((s)_{0<s<\tau}\right)\right)\right).$$

In [MTRV] the form given to this point-function integrand is similar to the form (8.8) used by Feynman in [F1]. Accordingly, the form of the path integral in [MTRV] is $\int_\Omega f\left(\cdot\right) d(\cdot)$ where $f(\cdot)$ includes a factor

$$\exp\left(\frac{\iota}{\hbar}\sum_{j=1}^n \left(\frac{1}{2}m\left(\frac{x_{s_j} - x_{s_{j-1}}}{s_j - s_{j-1}}\right)^2 - V(x_{s_{j-1}})\right)(s_j - s_{j-1})\right);$$

and $d(\cdot)$ represents the weighting function $|I|$ on cell subsets of the domain $\Omega = \mathbf{R}^T$, with

$$|I| \;=\; |I[N]| \;=\; |I_{s_1}||I_{s_2}|\cdots|I_{s_{n-1}}|; \quad \text{giving}$$
$$d(\cdot) \;=\; dx\left((s)_{0<s<\tau}\right) \;=\; dx_{s_1}\, dx_{s_2}\cdots dx_{s_{n-1}} \quad \text{for } N = \{s_1, s_2, \ldots, s_{n-1}\}.$$

Then the path integral $\int_\Omega f\left(\cdot\right) d(\cdot)$, without normalization factor (or "history action wave amplitude") $\alpha(\cdot)$, is

$$\int_{\mathbf{R}^T} \exp\left(\frac{\iota}{\hbar}\sum_{j=1}^n \left(\frac{1}{2}m\left(\frac{x_{s_j} - x_{s_{j-1}}}{s_j - s_{j-1}}\right)^2 - V(x_{s_{j-1}})\right)(s_j - s_{j-1})\right) dx_{s_1}\cdots dx_{s_{n-1}}.$$

$$(8.27)$$

10. The latter integral looks at first sight like a finite-dimensional integral. If the elements $s_1, \ldots, s_{n-1}$ were fixed, then it would indeed be an extended Riemann integral in $\mathbf{R}^{n-1}$; and $\int_{\mathbf{R}^T}$ should be $\int_{\mathbf{R}^{n-1}}$.

11. In fact n is variable and can be arbitrarily large in this calculation; and the elements $N = \{s_1, \ldots, s_{n-1}\}$ are variable, just as the elements x_{s_j} are. It is customary to think of elements x_j being "integrated out". These elements or variables appear in Riemann sums, but not in the final integral. Elements s_j are additional variables in infinite-dimensional integrals, and can also be thought of as being "integrated out".

12. Reverting to the random variation perspective, the choices x and N in $\int_{\mathbf{R}^T} \cdots$ can be thought of as occurrences of a single joint random occurrence (x, N) in sampling domain $\mathbf{R}^T$. Integrals such as

$$\int_{\tau'}^{\tau} \frac{1}{2} m \left(\frac{dx(s)}{ds} \right)^2 ds, \qquad \int_{\tau'}^{\tau} \frac{1}{2} m \left(\frac{dx(s)}{ds} \right) dx(s)$$

of (8.12) appear as integrand functional in $\mathbf{R}^T$. But, as discussed in earlier chapters on stochastic integration, these integrals do not exist except for null sets of x in $\mathbf{R}^T$. Instead, integrals such as $\int f(x(s)) \, dx(s)$ are replaced by stochastic sums; with similar sampling construct for functionals $\int f(x(s)) \, ds$. Thus action $S(x(\cdot))$ appears in (8.27) as *sampling sum* $S(x, N)$ (or $S(x(N), N)$, or simply $S(x(N))$).

13. Thus, from the perspective of random variation, and using random variable notation and terminology, the integrand in (8.27) can be regarded as a function of stochastic sum (or sampling sum) $\exp \left(\frac{\iota}{\hbar} S(X, \mathcal{N}) \right)$, where the stochastic/sampling sum is

$$\oint (X, \mathcal{N}) = \sum_{j=1}^{n} \left(\frac{1}{2} m \left(\frac{X_{s_j} - X_{s_{j-1}}}{s_j - s_{j-1}} \right)^2 - V(X_{s_{j-1}}) \right) (s_j - s_{j-1}).$$

The random variation perspective is incomplete since no probability function has been specified for sampling domain $\mathbf{R}^T$. (These issues are described more fully in chapter 7 of [MTRV].) Therefore sampling sum notation $S(x, N)$ is preferable to stochastic sum notation $\oint (X, \mathcal{N})$.

There is a problem with (8.27). As it stands, Feynman's normalization factor (or action wave amplitude) $\alpha(\cdot)$ of (8.4) is missing from it; and the calculation given in (8.27) does not actually converge to a system wave function. To give the correct calculation we use some standard arguments of -complete integration, as follows. (Full details are in [MTRV], chapters 6 and 7.) The integrand

$$\exp \left(\frac{\iota}{\hbar} \sum_{j=1}^{n} \left(\frac{1}{2} m \left(\frac{x_{s_j} - x_{s_{j-1}}}{s_j - s_{j-1}} \right)^2 - V(x_{s_{j-1}}) \right) (s_j - s_{j-1}) \right) dx_{s_1} \cdots dx_{s_{n-1}}$$

can be written as

$$\exp\left(\frac{\iota}{\hbar}\sum_{j=1}^{n}\left(-V(x_{s_{j-1}})\right)(s_j-s_{j-1})\right)$$

multiplied by

$$\exp\left(\frac{\iota}{\hbar}\sum_{j=1}^{n}\left(\frac{1}{2}m\left(\frac{(x_{s_j}-x_{s_{j-1}})^2}{s_j-s_{j-1}}\right)\right)\right)dx_{s_1}\cdots dx_{s_{n-1}}.$$

Given random choice of $N = \{s_1, \ldots, s_{n-1}\}$, let $N^+ = \{s_1, \ldots, s_{n-1}, s_n\}$ where $s_n = \tau$. For any cell $I(N)$, $= I_{s_1} \times I_{s_2} \times \cdots \times I_{s_{n-1}}$, in the $n-1$-dimensional space $\mathbf{R}^N = \mathbf{R} \times \cdots \times \mathbf{R}$, let I_{s_n} be a cell of $\mathbf{R}$ and let

$$I(N^+) = I(N) \times I_n = I_{s_1} \times I_{s_2} \times \cdots \times I_{s_{n-1}} \times I_{s_n}, \quad I[N^+] = I(N^+) \times \mathbf{R}^{\mathbf{T}\setminus N^+}$$

where $\mathbf{T} =]\tau', \tau]$, and cells $I[N^+]$ form the structure of Riemann sums for integrals $\int_{\mathbf{R}^{\mathbf{T}}} \cdots$. For some given N^+ write

$$H(I(N^+)) = \int_{I(N^+)} \exp\left(\frac{\iota}{\hbar}\sum_{j=1}^{n}\left(\frac{1}{2}m\left(\frac{(x_{s_j}-x_{s_{j-1}})^2}{s_j-s_{j-1}}\right)\right)\right)dx_{s_1}\cdots dx_{s_{n-1}}\,dx_{s_n}$$

if this finite-dimensional integral exists. In fact the Fresnel integral

$$\int_{\mathbf{R}^{N^+}} \exp\left(\frac{\iota}{\hbar}\sum_{j=1}^{n}\left(\frac{1}{2}m\left(\frac{(x_{s_j}-x_{s_{j-1}})^2}{s_j-s_{j-1}}\right)\right)\right)dx_{s_1}\cdots dx_{s_{n-1}}\,dx_{s_n}$$

exists and equals

$$\prod_{j=1}^{n}\sqrt{\frac{2\pi\iota\hbar(s_j-s_{j-1})}{m}},$$

giving

$$\int_{\mathbf{R}^{N^+}} \exp\left(\frac{\iota}{\hbar}\sum_{j=1}^{n}\left(\frac{1}{2}m\left(\frac{(x_{s_j}-x_{s_{j-1}})^2}{s_j-s_{j-1}}\right)\right)\right)\prod_{j=1}^{n}\frac{dx_{s_j}}{\sqrt{m^{-1}2\pi\iota\hbar(s_j-s_{j-1})}} = 1;$$

$$(8.28)$$

so $H(I(N^+))$ exists for each cell $I[N^+]$ of $\mathbf{R}^{\mathbf{T}}$. (See [MTRV] theorem 16, page 130.) This is where the "normalization factors" $\alpha(\cdot)$ come from. These factors are needed[10] in order to get convergence (or existence) of the Fresnel-based integral in (8.27).

Note that (8.27) has $\int_{\mathbf{R}^{\mathbf{T}}} \cdots, = \int_{\mathbf{R}^{]\tau',\tau[}} \cdots$, so there is no integration on dimension $s_n = \tau$, and, with $N = \{s_1, \ldots, s_{n-1}\}$, the cells used are $I[N]$, not $I[N^+]$. Accordingly, in (8.27), the factor

$$\alpha(\cdot) = \alpha(N) = \prod_{j=1}^{n}\sqrt{\frac{m}{2\pi\iota\hbar(s_j-s_{j-1})}}$$

---

[10]In his 1965 Nobel Prize lecture [47] Feynman explained how he got the idea of the normalization factors which are interpreted in Section 8.4 above as "amplitudes for action waves".

should be included. Thus a corrected version of (8.27) is

$$\int_{\mathbf{R}^T} \frac{\exp\left(\frac{\iota}{\hbar}\sum_{j=1}^n \left(\frac{1}{2}m\left(\frac{x_{s_j}-x_{s_{j-1}}}{s_j-s_{j-1}}\right)^2 - V(x_{s_{j-1}})\right)(s_j-s_{j-1})\right)}{\prod_{j=1}^n \sqrt{2\pi\iota\hbar m^{-1}(s_j-s_{j-1})}} \prod_{j=1}^{n-1} dx_{s_j}. \quad (8.29)$$

Also, for every cell $I(N)$ in $\mathbf{R}^N$, (8.28) implies that the finite-dimensional cell function

$$G(I(N)) = \int_{I(N)} \frac{\exp\left(\frac{\iota m}{2\hbar}\sum_{j=1}^n \frac{\left(x_{s_j}-x_{s_{j-1}}\right)^2}{(s_j-s_{j-1})}\right)}{\prod_{j=1}^n \sqrt{2\pi\iota\hbar m^{-1}(s_j-s_{j-1})}} \prod_{j=1}^{n-1} dx_{s_j} \quad (8.30)$$

exists; and, for every cell $I[N]$ in the infinite-dimensional space $\mathbf{R}^T, = \mathbf{R}^{]\tau',\tau[}$, we can then define the infinite-dimensional cell function $G(I[N])$ as

$$G(I[N]) := G(I(N)). \quad (8.31)$$

The following results from [MTRV] are also relevant.

1. Theorem 167, page 285: $\int_{\mathbf{R}^{]\tau',\tau]}} G(I[N]) = 1$.

2. Theorem 202, page 324:

$$\int_{\mathbf{R}^{]\tau',\tau[}} G(I[N]) = \frac{\exp\frac{\iota m(\xi-\xi')^2}{2\hbar(\tau-\tau')}}{\sqrt{2m^{-1}\pi\iota\hbar(\tau-\tau')}},$$

$$\int_{\mathbf{R}}\left(\int_{\mathbf{R}^{]\tau',\tau[}} G(I[N])\right)d\xi = \int_{\mathbf{R}^{]\tau',\tau]}} G(I[N]) = 1.$$

3. Theorem 169 (page 285) provides a link between the -complete integral and the versions such as (8.2), (8.4), and (8.9) above.

4. Theorem 223 (page 365 of [MTRV]) addresses Feynman's definition (corresponding to (8.9) above).

5. Pages 353–366 of [MTRV] examine various issues connecting the analyses in [F1] and [MTRV].

To sum up, in [F1] Feynman noted that there was a difficulty in formulating an appropriate measure in $\mathbf{R}^T$ which, for the purpose of defining an integral of $\exp\frac{\iota S(x(\cdot))}{\hbar}$, would somehow provide a valid meaning for an integrator $dx(s)_{0<s<\tau}$ (taking $T =]0,\tau]$, $\tau' = 0$). The solution prescribed in [MTRV] is to separate a multiplicative factor

$$\exp\left(\frac{\iota m}{2\hbar}\sum\frac{(x_{s'}-x_s)^2}{s'-s}\right)$$

(corresponding to ground state action) from the integrand $\exp\frac{\iota S(x(\cdot))}{\hbar}$ above, giving an expression of the form

$$\int_{\mathbf{R}^T}\left(\exp\left(\frac{-\iota}{\hbar}\sum V(x(s))(s'-s)\right)\right)\left(\exp\left(\frac{\iota m}{2\hbar}\sum\frac{(x_{s'}-x_s)^2}{s'-s}\right)dx(s)_{0<s<\tau}\right)$$

or

$$\int_{\mathbf{R}^T}\left(\exp\left(\frac{-\iota}{\hbar}\sum_{j=1}^n V(x_{j-1})(s_j-s_{j-1})\right)\right)\left(\exp\left(\frac{\iota m}{2\hbar}\sum_{j=1}^n\frac{(x_j-x_{j-1})^2}{s_j-s_{j-1}}\right)dx_1\cdots dx_{n-1}\right)$$

$$(8.32)$$

for the path integral. In corrected (or "normalized") form (8.29), this in turn is represented as

$$\int_{\mathbf{R}^T}\left(\exp\left(\frac{-\iota}{\hbar}\sum V(x(s))(s'-s)\right)\right)dG,\qquad\qquad(8.33)$$

since, using some -complete integration theory ([MTRV] theorem 170, page 285), the integral (8.29) exists whenever the integral (8.33) exists, and then the two are equal.

If $\iota=\sqrt{-1}$ is replaced by -1, the corresponding "path integral" problem is solved in the classical Lebesgue integral theory of Brownian motion. In this case the integrator dG corresponds to a Lebesgue-Stieltjes measure in domain $\mathbf{R}^T$.

This justifies the comparison which is sometimes made between the theories of Brownian motion and Feynman path integration. In chapters 6 and 7 of [MTRV], the symbol c appears in place of $\frac{\iota m}{2\hbar}$ in (8.32) and (8.33) above. Taking c to be $-\frac{1}{2}$ gives a description of Brownian motion. Taking c to be $\frac{\iota m}{2\hbar}$ gives the Feynman path integral construction.

This shows that there is more than just a parallelism or analogy between Brownian motion and the Feynman theory of particle motion in a field of conservative mechanical force; since both can be formulated as special cases of a single mathematical theory using -complete integrals, without invoking a measure function in either case.

8.7   Numerical Estimate of Path Integral

A common practice in integration is to deduce integrability of a function h from integrability of functions of a simpler kind, such as step functions which approximate to h. The following is a summary of progression or gradation of integrands $h(x)|I|$, or $h(x,I)$, in domains such as $\mathbf{R}^T$:

1. *Step functions* h which can take only a finite number of different values;

2. *Cylinder functions* $h_M(x_T)$ which depend only on the values taken by f at $x(M)$ where M is a fixed, finite number of elements of T, and is independent of $x(s)$ where $s\in T\smallsetminus M$;

3. *Sampling functions* $h(x(N))$, or $h(x(N), N)$, which, for any given x_T, depend only on $x(s)$ for $s \in N$, where N is variable, not fixed, and which can be any subset of T provided N is finite.[11]

This book and its companion [MTRV] are concerned largely with such integrands, along with particular integrators such as $|I[N]|$ or $G(I[N])$.

Example 43 *The -complete system of integration can deal with various unusual integrands which may be useful in testing out solutions and theories. One such is an integrand which is its own integrator, depending on points x, but not on cells I. (See Example 7, page 65 above.) Another is the incremental Dirichlet function $D(I)$ of [MTRV] section 4.14 (pages 178–181), which is itself integrable, but for which $f(x)D(I)$ is non-integrable unless f is constant—see also Examples 13 and 14 above.*

Some further examples, including some which illustrate relevant aspects of the background integration theory, are presented here in order to deal with some counter-intuitive pitfalls which can arise when integrating step functions.

Example 44 *To define a step function on domain $R =]0, 1]$, let $\mathcal{P} = \{J_i\}$ ($i = 1, \ldots, m$) be a partition of $]0, 1]$, and suppose a function $f(s)$ has constant value a_i for $s \in J_i$, $i = 1, \ldots, m$. A step function can be similarly defined for a product domain $R_1 \times R_2 =]0, 1] \times]0, 1]$. Furthermore, if the component domains R_1 and R_2 are partitioned by $\mathcal{P}_1 = \{J_i^1 : j = 1, \ldots, m_2\}$ and $\mathcal{P}_2 = \{J_j^2 : j = 1, \ldots, m_2\}$, respectively, then a product of the two partitions gives a partition*

$$\mathcal{P} = \left\{ J_{ij} = J_i^1 \times J_j^2 : 1 \leq i \leq m_1, \ 1 \leq j \leq m_2 \right\}$$

of $R_1 \times R_2$. This is a regular partition, like the parallel rows and columns of sub-intervals in the right hand diagram of figure 3.3, page 91 of [MTRV]. It is particularly easy to calculate Riemann sums of step functions defined on the cells of regular partitions such as $\mathcal{P}$. If $f(s) = a_{ij}$ for $s \in J_{ij}$, then iterated sums give

$$(\mathcal{P}) \sum f(s)|I_{ij}| = \sum_{j=1}^{m_2} \left(\sum_{i=1}^{m_1} c_{ij}|J_{ij}| \right).$$

If a partition $\mathcal{P}$ is not regular it is sometimes helpful to produce regular or partially regular sub-partitions as in figure 3.4, page 91 of [MTRV].

Example 45 *Suppose $f(x)$ has constant value 1 for $x \in]0, 1]$. It is then to be expected that $\int_0^1 f(x) \, dx$ exists and equals 1. But there is a possibility that this can fail, with Riemann sums not adding up to value 1.*

For example, a division $\mathcal{D}$ of $]0, 1]$ consists of pairs $(x,]u, v])$ with $x = u$ or $x = v$. One of these is $(x,]0, v])$ with $x = 0$ or $x = v$. Therefore the value $f(0)$ must be defined. If f is discontinuous at $x = 0$, with $f(0) = 0$, say, then a

---

[11]The argument of a sampling function h is generally $x(N)$, or $(x(N), N)$, with elements of N appearing explicitly. It is usually sufficient to write it as $h(x(N))$ instead of $h(x(N), N)$.

Riemann sum can include a term $f(0)(v-0)$, $= 0$, and in that case the Riemann sum as a whole has value less than 1.

To ensure convergence of the Riemann sums to integral value 1, define a gauge δ as follows. For $\varepsilon > 0$ let $\delta(x)$ satisfy

$$0 < \delta(x) < x \text{ for } x > 0; \quad \text{and} \quad 0 < \delta(x) < \varepsilon \text{ for } x = 0. \qquad (8.34)$$

Then every δ-fine division $\mathcal{D}$ of $]0,1]$ includes a pair $(0,]0,v])$ with $v < \varepsilon$, and

$$1 - \varepsilon < (\mathcal{D})\sum f(x)|I| < 1$$

for every δ-fine $\mathcal{D}$. The conditions in (8.34) provide a basic illustration of conformance of gauges.

The next Example uses double integrals and iterated integrals, and the theorems of Fubini and Tonelli which relate such integrals to each other. An integral on a product domain $K =]0,1] \times]0,1]$ is $\int_K \cdots$; also written

$$\int_{]0,1] \times]0,1]} \cdots, \quad \text{or} \quad \int_0^1 \int_0^1 \cdots.$$

This is a double integral; whereas an iterated integral is the integral of an integral, $\int_0^1 \left(\int_0^1 \cdots \right)$. Fubini's theorem (section 4.11 of [MTRV]) says that a double integral equals the corresponding iterated integral. Tonelli's theorem says that if an iterated integral exists, then, under sufficient conditions, the corresponding double integral also exists and the double integral equals the iterated integral.

Example 46 *In this Example variables x_1 and x_2 are used in the manner of elementary algebra or calculus, to label a pair of axes in two dimensions. Suppose a function $h_1(x_1, x_2)$ is defined on domain $K =]0,1] \times]0,1]$. Suppose h_1 is integrable, so $\int_K h_1(x_1, x_2)\, dx_1 dx_2$ exists. Now suppose h_1 is "cylindrical", in the sense that, for each x_1, the function $h_1(x_1, x_2)$ depends on x_1 only, and is independent of x_2; so $h_1(x_1, x_2) = h(x_1)$ for all x_2. Then*

$$\int_K h_1(x_1, x_2)\, dx_1 dx_2 = \int_K h(x_1)\, dx_1 dx_2$$
$$= \int_0^1 \left(\int_0^1 h(x_1)\, dx_1 \right) dx_2 = \int_0^1 h(x_1)\, dx_1$$

using Fubini's theorem, so h is integrable on $]0,1]$; and

$$\int_0^1 \int_0^1 h_1(x_1, x_2)\, dx_1 dx_2 = \int_0^1 h(x_1)\, dx_1,$$

Conversely, suppose $h(x_1)$ is integrable on $]0,1]$ with $\int_0^1 h(x_1)\, dx_1 = \beta$. For $(x_1, x_2) \in K$ define $h_1(x_1, x_2) := h(x_1)$. (Alternatively, define $h_2(x_2) = 1$ for $0 \le x \le 1$, and define $h_1(x_1, x_2) = h(x_1)h_2(x_2)$.) Then

$$\int_0^1 \left(\int_0^1 h_1(x_1, x_2)\, dx_1 \right) dx_2 = \int_0^1 \beta\, dx_1 = \beta,$$

so the iterated integral exists. Then, under sufficient conditions, Tonelli's theorem implies that the double integral of h_1 on K exists. (See [MTRV] page 155.) The double integral certainly exists if h is constant, or a step function, in $]0,1]$.

Fubini-type arguments require careful choice of gauges, divisions, and Riemann sums, as discussed in [MTRV], section 4.10 pages 154–159. In this case the integrands h_1 and h are less demanding.

Here are some Riemann sum constructions for the iterated and double integrals above. Suppose gauges $\delta_1(x_1)$ and $\delta_2(x_2)$ are defined, with joint gauge $\delta = (\delta_1, \delta_2)$ for the two-dimensional domain K. Let $\mathcal{D}$ be a δ-fine division of K,

$$\mathcal{D} = \{(x_1, x_2), I_1 \times I_2)\},$$

with Riemann sum $(\mathcal{D}) \sum h_1(x_1,x_2)|I_1||I_2|, = (\mathcal{D}) \sum h(x_1)|I_1||I_2|$. Then the domain $]0,1]\times]0,1]$ is partitioned by cells $I_1 \times I_2$, the first cell being horizontal and the second vertical. For each $I_1 \times I_2$ of $\mathcal{D}$, extend I_2 vertically from $x_2 = 0$ to $x_2 = 1$. These extensions subdivide each $I_1 \times I_2$ of $\mathcal{D}$, further partitioning K into stacks of cells $J_1 \times J_2$ arranged in vertical columns. (There may be no horizontal columns if the horizontal cells I_1 are not also extended.)

Each vertical column $J\times]0,1]$ has a factor $h(x_1)$ where x_1 comes from $\mathcal{D}$ and is not necessarily a vertex of J. But the individual cells J combine to give a horizontal cell J_1 which

- *has an x_1 from $\mathcal{D}$ as a vertex, and*

- *is $\delta_1(x_1)$-fine because D is (δ_1,δ_2)-fine.*

The pairs (x_1, J_1) therefore form a δ_1-fine division of $]0,1]$. A fuller account of the gauge technique for Fubini's theorem is in [MTRV] section 4.10, pages 154–159. Figures 3.3 and 3.4 (page 91 of [MTRV]) show how to extend edges and faces of a partition in order to give a regular partition containing only columns and rows of cells, aligned vertically and horizontally.

Here is another example of Riemann summation using regularization.

Example 47 *Write $R =]0,1]$. Let Υ denote the Cartesian product space R^R. As in (4.1), Chapter 4.1 above, a cell in R^R is*

$$I = I[N] = \prod_{t\in N} I_t \times \prod_{t\in R\setminus N} R$$

where N is any finite subset of R, and, for $t \in N$, I_t is a one-dimensional cell $]u_t, v_t]$ in R. With $|I_t| = v_t - u_t$, the volume function $|I|$ is

$$|I| = |I[N]| = |I(N)| = \prod_{t\in N}|I_t| = \prod_{t\in N}(v_t - u_t);$$

and if $\mathcal{P} = \{I\}$ is a partition of $\Upsilon = R^R$ then

$$(\mathcal{P})\sum|I| = (\mathcal{P})\sum|I[N]| = 1;$$

	$t = \frac{1}{3}$		$t = \frac{2}{3}$		
$I^{(1)}$	$]0, \frac{1}{2}]$	$\times$	$]0, \frac{1}{2}]$	$\times$	$\Pi_{t \in T \setminus N} R$
$I^{(2)}$	$]0, \frac{1}{2}]$	$\times$	$]\frac{1}{2}, 1]$	$\times$	$\Pi_{t \in T \setminus N} R$
$I^{(3)}$	$]\frac{1}{2}, 1]$	$\times$	$]0, \frac{1}{2}]$	$\times$	$\Pi_{t \in T \setminus N} R$
$I^{(4)}$	$]\frac{1}{2}, 1]$	$\times$	$]\frac{1}{2}, 1]$	$\times$	$\Pi_{t \in T \setminus N} R$

Table 8.1: List of four cells forming a regular partition $\mathcal{P}$ of Υ.

so $|I[N]|$ is a distribution function on R^R. To prove this, construct a regular sub-partition $\mathcal{P}' = \{J\}$ of $\mathcal{P}$ by extending the edges and faces of the cells I in $\mathcal{P}$. (See figures 3.3 and 3.4, page 91 of [MTRV].) It is then easy to see that

$$(\mathcal{P}) \sum |I| = (\mathcal{P}') \sum |J| = 1.$$

Theorems 153 and 168 (pages 276 and 285 of [MTRV]) involve the more "technical" volume functions $\mathbf{G}(I[N])$ and $G(I[N])$. But the same regularization technique is used as proof.

As illustration of the use of step functions and regular partitions, here is a numerical calculation for a simplified single particle system. Referring to Section 8.2, this is an evaluation of a Feynman-type path integral for a free particle, with ground state action functional S_0 in the form of a step function, and with simplified domain.

Example 48 *Consider the system of Section 8.2 consisting of a single particle of mass m subject to a conservative mechanical force, with particle motion in one dimension only. Suppose initial time $\tau' = 0$ and final time $\tau = 1$ (so $T =]0,1[$) with initial and final particle positions $x(0) = \xi' = 0$ and $x(1) = \xi = 0$. For simplicity suppose, at time s, $0 < s < 1$, the range of possible particle positions is $0 < x(s) < 1$, or $x(s) \in R =]0,1[$. The domain for this system is therefore $\Upsilon = R^T =]0, 1[^{]0, 1[}$. Let*

$$N = N_T = \{t_1, t_2\} = \left\{ \frac{1}{3}, \frac{2}{3} \right\}$$

be fixed sample times, with $t_0 = \tau' = 0$ and $t_3 = \tau = 1$; and in dimensions t $(t = \frac{1}{3}$ and $t = \frac{2}{3})$ partition the domain of $x(t)$ with a single partition point $\frac{1}{2}$. Then Υ is partitioned by the four cells $I^{(j)}$ $(j = 1, 2, 3, 4)$ listed in Table 8.1. The four two-dimensional cells $I(N)$ form a regular partition of R^2, and $\mathcal{P} = \{I[N]\}$ is a regular partition of the infinite-dimensional domain Υ, with

$$|I(N)| = \frac{1}{2} \times \frac{1}{2} = \frac{1}{4} = |I[N]|$$

for each of the four cells. In the ground state (with potential energy $V = 0$) the action for a path $x(\cdot)$ is $\int_0^1 \frac{m}{2} \left(\frac{dx}{dt} \right)^2 dt$, and for sampling path

$$\left(0, \, x\left(\frac{1}{3}\right), \, x\left(\frac{2}{3}\right), \, 0 \right),$$

	$t = \frac{1}{3}$	$t = \frac{2}{3}$	$x\left(\{0,\frac{1}{3},\frac{2}{3},1\}\right)$
$x^{(1)}$	0	0	$(0,0,0,1)$
$x^{(2)}$	0	$\frac{1}{2}$	$(0,0,\frac{1}{2},1)$
$x^{(3)}$	$\frac{1}{2}$	0	$(0,\frac{1}{2},0,1)$
$x^{(4)}$	$\frac{1}{2}$	$\frac{1}{2}$	$(0,\frac{1}{2},\frac{1}{2},1)$

Table 8.2: List of representative co-ordinates for step function calculation.

the sampling value $S_0(x(N))$ of the ground state action is

$$\frac{m}{2}\left(\left(\frac{x\left(\frac{1}{3}\right)-0}{\frac{1}{3}-0}\right)^2\left(\frac{1}{3}-0\right) + \left(\frac{x\left(\frac{2}{3}\right)-x\left(\frac{1}{3}\right)}{\frac{2}{3}-\frac{1}{3}}\right)^2\left(\frac{2}{3}-\frac{1}{3}\right) + \left(\frac{1-x\left(\frac{2}{3}\right)}{1-\frac{2}{3}}\right)^2\left(1-\frac{2}{3}\right)\right).$$

$$(8.35)$$

Choose a sample path $x^{(j)}$ ($j = 1,2,3,4$) from each of the four cells of $\mathcal{P}$. For the two "random" components $x\left(\frac{1}{3}\right)$ and $x\left(\frac{2}{3}\right)$ of each sample path $x^{(j)}$, take (for instance) the lower left hand vertex of each of the four cells $I(N)$ ($N = \{\frac{1}{3},\frac{2}{3}\}$) in Table 8.2. Taking particle mass $m = 1$ for convenience, four sampling values for the ground state action $S_0(x(N))$ are then obtained; one for each $x^{(j)}$:

$$S_0(x(N)) \quad = \quad 3, \quad \frac{3}{2}, \quad \frac{15}{4}, \quad \frac{3}{2}, \quad respectively. \qquad (8.36)$$

Omitting the normalization factor (or path action wave amplitude) $\alpha(N)$, the Riemann sum estimate of the path integral on Υ (using only the oscillation factor as integrand) is $(\mathcal{P})\sum \exp\left(\frac{\iota}{\hbar}S_0(x(N))\right)|I[N]| \quad =$

$$= \quad \frac{1}{4}\left(\exp\left(\frac{3\iota}{\hbar}\right) + \exp\left(\frac{3\iota}{2\hbar}\right) + \exp\left(\frac{15\iota}{4\hbar}\right) + \exp\left(\frac{3\iota}{2\hbar}\right)\right)$$

which reduces to a complex value when the numerical value of $\hbar$ is inserted. For simplicity, take $\hbar$ to be 1. Then the value of the "path integral" (without the $\alpha(N)$ factor) is

$$\int_{R^T} \exp\left(\frac{\iota}{\hbar}S_0(x(N))\right)|I[N]| \quad = \quad -0.4 + 0.4\iota,$$

approximately.

Example 48 is a Riemann sum for a step function estimate of an integrand. No gauges γ appear in it. This is not unusual in integration—gauge restrictions are needed in order to create a robust theory, but need not necessarily be invoked in every routine procedure. But, even when integrating step functions, Example 45 shows that gauge restriction may be necessary in order to ensure convergence.

The calculation in Example 48 deals only with the oscillation factor

$$\omega_0(x(N))) \quad = \quad \exp\left(\frac{\iota}{\hbar}S_0(x(N))\right)$$

of the integrand, with amplitude factor $\alpha(N)$ omitted. For the single particle system of [F1] and [FH], (8.27) and (8.29) resolve this issue; with $N = \left\{0, \frac{1}{3}, \frac{2}{3}, 1\right\}$, and with $\alpha(N)$ and $\bar{\alpha}$ equal to, respectively,

$$\prod_{j=1}^{n} \left(2\pi\iota\hbar m^{-1}(t_j - t_{j-1})\right)^{-\frac{1}{2}}, \qquad \left(2\pi\iota\hbar m^{-1}(\tau - \tau')\right)^{-\frac{1}{2}},$$

where $\alpha(N)$ is the amplitude of the action wave for a sample path, and $\bar{\alpha}$ is the amplitude for the aggregate action wave for the system.

Feynman's comments in [35] and [FH] on this issue have already been mentioned. But the following Example looks again at (8.27) and (8.29) to see whether they may provide any further insight into $\alpha(N)$ and $\bar{\alpha}$.

Example 49 *Sample path action wave amplitude $\alpha(N)$ may be deduced from*

$$\int_{\mathbf{R}^T} \omega_0(x(N))\, dx(N)$$

where ω_0 is the oscillation factor for ground state action $S_0(x(N))$,

$$\omega_0(x(N)) \;=\; e^{\frac{\iota}{\hbar}\sum_{j=1}^{n}\left(\frac{1}{2}m\left(\frac{x_j - x_{j-1}}{s_j - s_{j-1}}\right)^2\right)(s_j - s_{j-1})}.$$

Consider the finite-dimensional integral version of (8.27), with $V = 0$, with $N = \{s_1, \ldots, s_{n-1}\}$ fixed, and with $\mathbf{R}^N$ replacing $\mathbf{R}^T$:

$$\int_{\mathbf{R}^N} \exp\left(\frac{\iota}{\hbar}\sum_{j=1}^{n}\left(\frac{1}{2}m\left(\frac{x_j - x_{j-1}}{s_j - s_{j-1}}\right)^2\right)(s_j - s_{j-1})\right) dx_1 \cdots dx_{n-1}.$$

This integral exists (see, for instance, [MTRV] theorems 140 and 202, pages 268 and 324), so Fubini's theorem is applicable, and it can be evaluated by iterated integration on variables $x_1, x_2, \ldots, x_{n-1}$. Integration on x_1 yields

$$\frac{\left(2\pi\iota\hbar m^{-1}(s_1 - s_0)\right)^{\frac{1}{2}}\left(2\pi\iota\hbar m^{-1}(s_2 - s_1)\right)^{\frac{1}{2}}}{\left(2\pi\iota\hbar m^{-1}(s_2 - s_0)\right)^{\frac{1}{2}}} \int_{\mathbf{R}^{N\setminus\{s_1\}}} (\cdots)\ dx_2\, dx_3 \cdots dx_{n-1}.$$

(See lemma 13, [MTRV] page 262.) Integrating this in variable x_2 gives

$$\frac{\prod_{j=1}^{3}\left(2\pi\iota\hbar m^{-1}(s_j - s_{j-1})\right)^{\frac{1}{2}}}{\left(2\pi\iota\hbar m^{-1}(s_3 - s_0)\right)^{\frac{1}{2}}} \int_{\mathbf{R}^{N\setminus\{s_1\}}} (\cdots)\ dx_3 \cdots dx_{n-1}.$$

Continuing with integration on variables x_3, $\ldots$, x_{n-1}, the factors $(s_j - s_{j-1})^{\frac{1}{2}}$ of $\alpha(N)$ appear successively; and after $n-1$ iterations the integral value

$$\frac{\prod_{j=1}^{n}\left(2\pi\iota\hbar m^{-1}(s_j - s_{j-1})\right)^{\frac{1}{2}}}{\left(2\pi\iota\hbar m^{-1}(\tau - \tau')\right)^{\frac{1}{2}}} \exp\left(\frac{\iota m(\xi - \xi')^2}{2\hbar(\tau - \tau')}\right) \qquad (8.37)$$

appears, where $s_0 = \tau'$, $s_n = \tau$, $x_0 = \xi'$, and $x_n = \xi$. Since N is fixed, the integral

$$\int_{\mathbf{R}^N} \frac{\exp\left(\frac{\iota}{\hbar}\sum_{j=1}^n \left(\frac{1}{2}m\left(\frac{x_{s_j}-x_{s_{j-1}}}{s_j-s_{j-1}}\right)^2\right)(s_j - s_{j-1})\right)}{\prod_{j=1}^n \sqrt{2\pi\iota\hbar m^{-1}(s_j - s_{j-1})}} \prod_{j=1}^{n-1} dx_{s_j}$$

exists, with value

$$\frac{1}{(2\pi\iota\hbar m^{-1}(\tau - \tau'))^{\frac{1}{2}}} \exp\left(\frac{\iota m(\xi - \xi')^2}{2\hbar(\tau - \tau')}\right).$$

Writing

$$\alpha(N) = \prod_{j=1}^n (2\pi\iota\hbar m^{-1}(s_j - s_{j-1}))^{\frac{1}{2}}, \tag{8.38}$$

the expression (8.37) and the successive one-dimensional integration steps leading up to it have the appearance of an algorithm for finding $\alpha(N)$, the amplitude for the action wave of sample path $x(N)$.

It has already been established that, for ground state $\omega = \omega_0$, the path integral $\int_{\mathbf{R}^T} \alpha(N)\omega_0(x(N))dx(N)$ exists (see [MTRV] theorem 202, page 324), with

$$\int_{\mathbf{R}^T} \alpha(N)\omega_0(x(N))dx(N) = \frac{1}{(2\pi\iota\hbar m^{-1}(\tau - \tau'))^{\frac{1}{2}}} \exp\left(\frac{\iota m(\xi - \xi')^2}{2\hbar(\tau - \tau')}\right)$$

$$= \bar{\alpha}\,\omega_0(\bar{x}(\cdot)), \quad \text{so}$$

$$\bar{\alpha} = (2\pi\iota\hbar m^{-1}(\tau - \tau'))^{-\frac{1}{2}} \quad \text{and}$$

$$S_0(\bar{x}(\cdot)) = \frac{m(\xi - \xi')^2}{2(\tau - \tau')}.$$

The latter is the action for the ground state path of least action—the straight line joining the points (τ', ξ') and (τ, ξ). This provides support for the calculus of variations method outlined in Section 8.5.

Note also that the wave function $\psi(\xi, \tau)$, a complex number, can be represented in many different ways as a product of two complex numbers β_1 and β_2, with $|\beta_2| = |\omega(x(\cdot))| = 1$. The correct factorization $\bar{\alpha}\,\omega_0(\bar{x}(\cdot))$ can be found if any two of

$$\psi(\xi, \tau), \quad \bar{\alpha}, \quad \omega(\bar{x}(\cdot)),$$

are known.

Example 50 *If the calculus of variations method is valid for Example 48 above, then (with simplification $m = \hbar = 1$) the oscillation factor value is*

$$\omega_0(S(\bar{x}(\cdot))) = \exp\left(\frac{\iota m(\xi - \xi')}{2\hbar(\tau - \tau')}\right) = \exp\left(\frac{\iota(1-0)}{2(1-0)}\right) = \exp\frac{\iota}{2} = 0.9 + 0.5\iota$$

approximately. Then, in Example 48 with fixed $N = \{\frac{1}{3}, \frac{2}{3}\}$, the amplitudes satisfy (approximately)

$$\frac{-0.4 + 0.4\iota}{\alpha(N)} = \frac{0.9 + 0.5\iota}{\bar{\alpha}}.$$

Unlike the calculations in Example 48, expressions (8.27), (8.28), and (8.29) have a good foundation in the physical theory of quantum mechanics. The illustrative calculations in Examples 48 and 48 are based on assumptions which diverge substantially from that theory. Therefore Example 50 should (in terms of physics) be regarded as conjecture.

More detailed and more realistic numerical evaluations of path integrals can be performed by methods such as those in the pricing calculations of Chapter 12 below.

8.8   Free Particle in Three Dimensions

In order to avoid complication of notation the systems discussed in preceding sections are restricted to motion in one dimension only. Notation for higher dimensional systems is presented here.

To start, motion of a single particle in one dimension only is indicated by subscript 1. In one dimension, the position of the particle at time s $(0 < s < \tau)$ is denoted by $x_1(s)$, or x_{1s}. Suppose the initial position (or initial state of the single free particle system) at time $s = 0$ is $x_1(0), = x_{1,0}, = 0$; and suppose the final state of the system at time $s = \tau$ is $x_1(\tau), = x_{1\tau}, = \xi_1$. A path for this system is

$$x_1(\cdot) \;=\; x_{1,(\cdot)} \;=\; \{x_1(s) : 0 < s < \tau\} \;=\; \{x_{1s} : 0 < s < \tau\}\} = (x_{1s})_{0<s<\tau}.$$

The oscillation factor for the action wave for the one-dimensional path $x_{1,(\cdot)}$ is

$$\exp\left(\frac{\iota}{\hbar}S(x_{1,(\cdot)})\right).$$

In the -complete system, the action wave for sampling elements $(x_{1,(\cdot)}, N)$ has amplitude and oscillation factors

$$\prod_{j=1}^{n}\sqrt{\frac{m}{2\pi\iota\hbar(s_j - s_{j-1})}}, \qquad \exp\left(\frac{\iota}{\hbar}\frac{m}{2}\sum_{j=1}^{n}\frac{(x_{1j} - x_{1,j-1})^2}{s_j - s_{j-1}}\right),$$

respectively, where the variable sampling elements are $N = \{s_1, \ldots, s_{n-1}\}$ with $x_{1j} = x_{1s_j}$ for $j = 1, \ldots, n-1$; and with fixed $s_0 = 0$, $s_n = \tau$, $x_{1,0} = 0$, $x_{1n} = \xi_1$. (In essentials this is the interpretation of the path action wave applied in [FH] and [F1], with the proviso that the time increments $s_j - s_{j-1}$ are the same for each j (—a version of the cylinder function method). The latter condition is removed in [MTRV]. Permitting arbitrary choice of sample times $N = \{s_j\}$ gives direct access to the -complete integral on an infinite-dimensional domain of all paths.)

Denote the product of the two factors as $h(x, N)$. Then, by the evaluation in theorem 202 (page 324 of [MTRV]), the path integral for the system is

$$\int_{\mathbf{R}^T} h(x, N)|I[N]| \;=\; \frac{m}{\sqrt{2\pi\iota\hbar\tau}}\exp\left(\frac{\iota}{\hbar}\frac{m}{2}\frac{\xi_1^2}{\tau}\right).$$

In classical (non-quantum mechanical) physics, the one-dimensional path of least action for this free particle is

$$\left\{ x_{1s} = \frac{s}{\tau}\xi_1 \ : \ 0 < s < \tau \right\},$$

which, in a displacement time graph $\{(s, y)\}$, is the straight line joining the point $s = 0$, $y = 0$ to the point $s = \tau$, $y = \xi$.

The corresponding classical action $\bar{S}$ is obtained by integrating the lagrangian $\frac{m}{2}\left(\frac{dx_1(s)}{ds}\right)^2$ with respect to s along the path of least action, giving

$$\bar{S} \ = \ \frac{m\xi_1^2}{2\tau};$$

so the wave function (or value of the path integral for the system) equals

$$\bar{a}\exp\left(\frac{\iota}{\hbar}\bar{S}\right), \ = \ \frac{1}{\sqrt{2\pi\iota\hbar\tau}} \exp\left(\frac{\iota}{\hbar}\frac{m}{2}\frac{\xi_1^2}{\tau}\right).$$

Thus the path integral aggregate of the path action waves is the particle action wave with amplitude $\bar{a}$ given by the first factor $\frac{1}{\sqrt{2\pi\iota\hbar\tau}}$ and with oscillation factor $\exp\left(\frac{\iota}{\hbar}\frac{m}{2}\frac{\xi_1^2}{\tau}\right)$ given by the classical path of least action.

Now visualize particle motion in three dimensions with initial position (at time $s = 0$)

$$(x_1(0), x_2(0), x_3(0)) \ = \ (x_{1,0}, x_{2,0}, x_{3,0}) \ = \ (0, 0, 0),$$

and with final position (time $s = \tau$)

$$(x_1(\tau), x_2(\tau), x_3(\tau)) \ = \ (x_{1,\tau}, x_{2,\tau}, x_{3,\tau}) \ \blacksquare \ (\xi_1, \xi_2, \xi_3).$$

Suppose sample path $x \in \left(\mathbf{R}^3\right)^T$ and time sample $N \in \mathcal{N}$ are given:

$$x \ = \ (x(s))_{s \in T} \ = \ (x_{1s}, x_{2s}, x_{3s})_{0 < s < \tau}, \qquad N \ = \ \{s_1, \ldots, s_{n-1}\}.$$

For $p = 1, 2, 3$, write co-ordinates $x_p(s_j)$ as x_{pj}, $j = 0, 1, 2, \ldots, n$, where

$$x_{p0} \ = \ x_p(s_0) \ = \ x_p(0) \ = \ 0, \qquad x_{pn} \ = \ x_p(s_n) \ = \ x_p(\tau) = \xi.$$

Then, as in [MTRV] and [F1], sample velocities are estimated as

$$\sqrt{\sum_{p=1}^3 \left(\frac{x_{pj} - x_{p,j-1}}{s_j - s_{j-1}}\right)^2},$$

and the sample estimate for particle kinetic energy for path x at time s_j is

$$\frac{m}{2}\sum_{p=1}^3 \left(\frac{x_{pj} - x_{p,j-1}}{s_j - s_{j-1}}\right)^2,$$

If the particle motion is subject to external force with potential

$$V(x(s), s) \;=\; V\left((x_{1s}, x_{2s}, x_{3s}), s\right),$$

then the lagrangian $\mathcal{L}(\dot{x}(s), x(s), s)$ and sample action $S(x)$ for the system are

$$\frac{m}{2}\left(\frac{dx(s)}{ds}\right)^2 - V(x(s), s),$$

$$\sum_{j=1}^{n}\left(\frac{m}{2}\sum_{p=1}^{3}\left(\frac{x_{pj} - x_{p,j-1}}{s_j - s_{j-1}}\right)^2 - V\left((x_{1,j-1}, x_{2,j-1}, x_{3,j-1}), s_{j-1}\right)\right)(s_j - s_{j-1}),$$

respectively, where the latter is the estimate for sample path x and sample times $N = \{s_1, \ldots, s_{n-1}\}$.

Denote the sample action by $S(x, N)$, and denote the corresponding particle path action wave by

$$\alpha(N)e^{\frac{i}{\hbar}S(x,N)}.$$

Then, provided this is integrable, the system quantum mechanical action wave for the particle system is the wave function path integral

$$\psi(\xi, \tau) \;=\; \int_{\Omega} \alpha(N)e^{\frac{i}{\hbar}S(x,N)}$$

where domain $\Omega = \left(\mathbf{R}^3\right)^T$.

Now suppose the potential V is zero everywhere, so there is no excitation or perturbation of the particle motion, and the particle system is "free". (In other words, it is in the ground state or vacuum state, to use some of the terminology of [FH].) In this case the path integral is

$$\int_{\Omega} \alpha(N)e^{\frac{i}{\hbar}S_0(x,N)}|I[N]| \tag{8.39}$$

where $S_0(x, N)$ is the ground state action for arbitrary sample path/sample times (x, N),

$$= \;\; \sum_{j=1}^{n}\left(\frac{m}{2}\sum_{p=1}^{3}\left(\frac{x_{pj} - x_{p,j-1}}{s_j - s_{j-1}}\right)^2\right)(s_j - s_{j-1}), \;\; = \;\; \sum_{j=1}^{n}\left(\frac{m}{2}\sum_{p=1}^{3}\frac{(x_{pj} - x_{p,j-1})^2}{s_j - s_{j-1}}\right),$$

and $|I[N]| \;=\; \prod_{j=1}^{n-1}|I(s_j)|$. The integral (8.39) exists if the sample path action wave amplitude $\alpha(x, N)$ is taken to be

$$\left(\prod_{j=1}^{n}\frac{m}{\sqrt{2\pi\imath\hbar(s_j - s_{j-1})}}\right)^3,$$

with

$$\int_{\Omega} \alpha(N)e^{\frac{i}{\hbar}S_0(x,N)}|I[N]| \;=\; \int_{\Omega} \alpha(N)e^{\frac{i}{\hbar}S_0(x,N)}dx[N]$$

$$= \;\; \left(\sqrt{\frac{m}{2\pi\imath\hbar\tau}}\right)^3 e^{\frac{i}{\hbar}\frac{m}{2\tau}(\xi_1^2 + \xi_2^2 + \xi_3^2)}. \tag{8.40}$$

The proof is similar to the proof of theorem 202 (page 324 of [MTRV]) for one-dimensional paths x.

Note that $\frac{m}{2\tau}\left(\xi_1^2 + \xi_2^2 + \xi_3^2\right)$ is the action $\bar{S}$ for the classical path of least action for a free particle with initial position $x(0) = (0,0,0)$ at time $s = 0$ and final position $x(\tau) = (\xi_1, \xi_2, \xi_3)$ at time $s = \tau$.

As in [MTRV] the complex-valued function $G(J[N])$ is defined on cells $J[N]$ of $\Omega = \left(\mathbf{R}^3\right)^T$, and is finitely additive on disjoint unions of cells, where

$$G(J[N]) = \int_{J[N]} \left(\prod_{j=1}^{n}\sqrt{\frac{m}{2\pi\iota\hbar(s_j - s_{j-1})}}\right)^3 |I[N]|.$$

The integral $\int \alpha(N)e^{\frac{\iota}{\hbar}S_0(x,N)}|I[N]|$ can be thought of as[12]

$$\int \alpha(N)e^{\frac{\iota}{\hbar}S_0(x,N)}(dx_{1s_1}dx_{2s_1}dx_{3s_1})(dx_{1s_2}dx_{2s_2}dx_{3s_2})\cdots(dx_{1s_{n-1}}dx_{2s_{n-1}}dx_{3s_{n-1}})$$

with sample elements $N = \{s_1, s_2, \ldots, s_{n-1}\} \subset T$ regarded as variables in the integration process, just as the sample elements $x \in \Omega$ (with components x_{ps}) are integration variables. These sample elements are aggregated in $\int_{\Omega}\cdots$, and they do not appear in the evaluated form of (8.40). They are "integrated out", so to speak.

As in theorem 171 (page 286 of [MTRV]), given integrand $f(x, N)$, if either of the integrals

$$\int_{\Omega}f(x,N)G(I[N]), \qquad \int_{\Omega}f(x,N)\alpha(N)e^{\frac{\iota}{\hbar}S_0(x,N)}|I[N]|$$

exists then the other one exists and the two are equal. This is a significant result in the calculus of -complete integrals.

- If T is taken to be $]0,\tau[$, with τ excluded, the time $\tau = s_n$ is omitted from the sample time elements $N = \{s_1, \ldots, s_{n-1}\}$. Then

$$\Omega = \left(\mathbf{R}^3\right)^{]0,\tau[} = \left(\mathbf{R}^3\right)^T$$

and the state function $\psi(\xi,\tau), = \psi((\xi_1,\xi_2,\xi_3),\tau)$ for the single particle system subject to interaction (or excitation) potential V is

$$\begin{aligned}
\psi(\xi,\tau) &= \int_{\Omega}\alpha(N)e^{\frac{\iota}{\hbar}S(x,N)}|I[N]| \\
&= \int_{\Omega}\alpha(N)e^{\frac{\iota}{\hbar}S_0(x,N)+S'(x,N)}|I[N]| \\
&= \int_{\Omega}e^{\frac{\iota}{\hbar}S'(x,N)}\left(\alpha(N)e^{\frac{\iota}{\hbar}S_0(x,N)}|I[N]|\right) \\
&= \int_{\Omega}e^{\frac{\iota}{\hbar}S'(x,N)}G(I[N]),
\end{aligned}$$

---

[12]The notation $\int\cdots|I|$ is used in [MTRV] for -complete integrals, and it corresponds to $\int\cdots dx$ in traditional notation. Likewise, if $\mu(\cdot)$ is defined on subsets of domain Ω, including cells $I \subset \Omega$, the traditional $\int\cdots d\mu$ becomes $\int\cdots\mu(I)$ in the -complete notation.

with existence if any of the above integrals exists. The result is the wave function $\psi(\xi, \tau), = \psi((\xi_1, \xi_2, \xi_3), \tau)$, for the single particle system subject to interaction (or excitation) potential V. (In [MTRV] the notations N^- and N are used for $\{s_1, \ldots, s_{n-1}\}$ and $\{s_1, \ldots, s_{n-1}, s_n = \tau\}$, respectively, with superscript $^-$ added to other symbols as appropriate.)

- If $\mathbf{T}$ is taken to be $]0, \tau]$, with $\Omega = \left(\mathbf{R}^3\right)^{]0,\tau]} = \left(\mathbf{R}^3\right)^{\mathbf{T}}$, then $\int_\Omega G(I[N]) = 1$, using arguments similar to those of theorems 168–170, page 285 of [MTRV].

8.9   From Particle to Field

Following the physical concept of action functional in [F1] and [FH], and using the mathematical method of [MTRV], preceding sections deal with quantum effects of a field on a particle.

Chapter 9 of [FH] deals with quantum electrodynamics, in the sense of interaction of charged particle and field. Leading up to this, [FH] has descriptions of quantum effects in various intermediate systems, using path integration ideas. Mathematical aspects of the path integral method in these intermediate systems can be expressed in terms of the -complete integration of [MTRV]. Some of these systems are now outlined, starting with *simple harmonic oscillation*:

Example 51 *Section 8-1 of [FH] describes a simple harmonic oscillator,*[13] *a special case of (8.6) with $V(x(s)) = \frac{m}{2}\varpi^2 (x(s))^2$. In classical or non-quantum physics, suppose a particle of mass m oscillates in one dimension with angular frequency ϖ. Suppose particle displacement (or position) at initial time $t = 0$ is $x(0) = 0$, with displacement $x(t)$ at time t, without forcing or damping, and with maximum displacement β. One solution is $x(t) = \beta \sin \varpi t$. A conservative external force causing this particle oscillation has potential energy*

$$V(x(s)) \;=\; \frac{m\varpi^2}{2} x(s)^2$$

at location $x(s)$. The kinetic energy of the particle at time s is

$$\frac{m}{2}\left(\frac{dx(s)}{ds}\right)^2,$$

and the lagrangian for arbitrary displacement $x(s)$ is

$$\mathcal{L}(x(s)) \;=\; \frac{m}{2}\left(\left(\frac{dx(s)}{ds}\right)^2 - \varpi^2 x(s)^2\right).$$

---

[13] In describing simple harmonic oscillation, *angular frequency* is denoted here by the symbol ϖ, because the more traditional notation ω has already been used above to denote oscillation factor as a functional of action.

The action $S(x)$ for arbitrary displacement history (or path) $x = (x(s))_{0 < s < \tau}$ is

$$S(x) = \int_{x(0)}^{x(\tau)} \mathcal{L}(x(s))ds.$$

In the version of path integral theory given in Section 8.3 above, the corresponding action wave is $\alpha(\cdot)\exp\left(\frac{\iota}{\hbar}S(x)\right)$. For $T = \,]0,\tau[$, and using indirect methods[14] which do not specify amplitudes $\alpha(\cdot)$ for individual action waves, Problem 3-8 in [FH] gives the value of the path integral for the harmonic oscillator system as $\psi(\xi,\tau;\xi',\tau') = \psi(\xi,\tau;0,0) = \psi(x(\tau),\tau) =$

$$= \int_{\mathbf{R}^T} \alpha(\cdot)\exp\left(\frac{\iota}{\hbar}S(x)\right) dx(s)_{0<s<\tau} \tag{8.41}$$

$$= \left(\frac{m\varpi}{2\pi\iota\hbar\sin\varpi\tau}\right)^{\frac{1}{2}} \exp\left(\varpi\frac{\iota}{\hbar}\frac{1}{2}m\left(x(\tau)\right)^2\cot\varpi\tau\right), \tag{8.42}$$

where $x(\tau) = \xi$, $\xi' = 0$, $\tau' = 0$, and where $\int_{\mathbf{R}^T} \cdots$ is understood in terms of the meaning ascribed to it in [FH]. How does this translate into -complete integral terms? The action functional is

$$S(x) = \int_{s=0,\,x(0)=\xi'}^{s=\tau,\,x(\tau)=\xi} \left(\frac{1}{2}m\left(\frac{dx(s)}{ds}\right)^2 - \varpi^2(x(s))^2\right)ds. \tag{8.43}$$

This functional is meaningful only for everywhere differentiable paths x. Just as stochastic integrals can be replaced by stochastic sums, (8.43) can be replaced by a sample expression. With sampling dimensions $N = \{s_1,\ldots,s_{n-1}\}$, $(s_0 = \tau', s_n = \tau)$, the sample action for arbitrary $x \in \mathbf{R}^T$ is $S(x,N) =$

$$= \sum_{j=1}^{n}\left(\frac{1}{2}m\left(\frac{x(s_j)-x(s_{j-1})}{s_j-s_{j-1}}\right)^2 - \varpi^2\left(x(s_{j-1})\right)^2\right)(s_j-s_{j-1}), \tag{8.44}$$

and the integrand $\alpha(\cdot)\exp(\cdots)dx(\cdot)$ for the path integral version of wave function (8.41) is $h(x,I[N],N) = \alpha(N)\,e^{\frac{\iota}{\hbar}S(x,N)}dx(N) =$

$$= \left(\prod_{j=1}^{n}\sqrt{\frac{m}{2\pi\iota\hbar(s_j-s_{j-1})}}\right)e^{\frac{\iota}{\hbar}\sum_{j=1}^{n}\frac{m}{2}\left(\left(\frac{x_j-x_{j-1}}{s_j-s_{j-1}}\right)^2-\varpi^2x_{j-1}^2\right)(s_j-s_{j-1})}|I[N]|,$$

where the sampling sum $S(x,N)$ replaces integral $S(x)$. With variable sample times N (giving variable sample displacements x_j for variable times s_j in variable N), this is the -complete version of the integrand

$$\alpha(\cdot)\,e^{\frac{\iota}{\hbar}\int_0^\tau \frac{m}{2}\left(\left(\frac{dx(s)}{ds}\right)^2-\varpi^2x(s)^2\right)ds}\,dx(s)_{0<s<\tau},$$

---

[14]Various path integrals are evaluated in [FH] using a variety of methods such as arguments from physics and from the calculus of variations. See, for instance, *General Results for Quadratic Action* in section 7-4 of [FH].

of (8.41), expressed in terms of sampling sum instead of integral $\int \cdots ds$ along the path x. It makes explicit the amplitude $\alpha(N)$ for the action wave for each sample path (or sample history):

$$\alpha(N) = \prod_{j=1}^{n} \sqrt{\frac{m}{2\pi\iota\hbar(s_j - s_{j-1})}} \quad for \quad (x, N) \in \mathbf{R}^T \times \mathcal{N}(T). \qquad (8.45)$$

The system described in Example 8.10 above consists of a free particle interacting with a field of potential energy $\frac{1}{2}m\varpi^2 y^2$ at locations $y \in \mathbf{R}$. Section 8-2 of [FH] describes a *forced harmonic oscillator*. In this case the ground state (or vacuum state) is the simple harmonic oscillator. An external force $f(s)$ acts on the oscillator; its potential energy is $f(s)y(s)$. This system is a forced harmonic oscillator. Its ground state action (in sample form) is (8.44) above, and the forcing contributes a further amount $f(s_{j-1})x(s_{j-1})$, so the sample action for sample path (x, N) is $S(x, N) =$

$$= \sum_{j=1}^{n} \left(\left(\frac{1}{2}m\left(\frac{x_j - x_{j-1}}{s_j - s_{j-1}}\right)^2 - \varpi^2 x_{j-1}^2 \right) + f(s_{j-1})x_{j-1} \right)(s_j - s_{j-1}); \qquad (8.46)$$

and the wave function path integral for the forced system is

$$\psi(\xi, \tau) = \int_{\mathbf{R}^T} \alpha(N) e^{\frac{i}{\hbar}S(x,N)} |I[N]|.$$

Feynman provides further details of this in pages 60–65 of his PhD thesis [35]. (See also pages 55–60 of [10]).

The single particle harmonic oscillator analysis is used in chapter 8 of [FH] to devise models for multi-particle systems, such as a polyatomic molecule and a three-dimensional crystal. Section 8-4 gives a simpler example: a one-dimensional crystal, which can be thought of as a string of equal atoms, equally placed in a line, and moving (or vibrating) only in the direction of the line.

It is assumed that each atom interacts only with its two neighbouring atoms, as if each atom in the line is connected to the two adjoining atoms by a pair of springs; and this is the means by which simple harmonic oscillation can be brought into the picture. And if the wavelength of the vibration is long in comparison with the spacing between the atoms, then it is possible to treat the string of atoms as continuous matter.

Some mathematical aspects of simple harmonic oscillation are considered in Section 8.10 below. Multi-particle and continuous mass systems are in Sections 8.11 and 8.12.

But first, here is an illustration of the effect of an electromagnetic field on a single particle possessing charge as well as mass.

Example 52 *Section 8.2 above deals with the motion of a particle of mass m subject to an external force with potential V. Suppose, in addition to mass, the particle has electric charge e; and suppose the external disturbance to the particle motion is caused by an ambient electromagnetic field (instead of the conservative mechanical force field considered earlier).*

This is the scenario of section 7-6 of [FH], which examines the effect of the electromagnetic field on the particle motion; in particular the quantum mechanical effects revealed by integrating $\exp\frac{\iota}{\hbar}S$ *over the space of particle paths. (The charged particle can, in turn, produce quantum mechanical effects in the form of reciprocal disturbance to the electromagnetic field, and [FH] goes on to address the latter subject—which belongs to quantum field theory—in chapter 9 under the heading Quantum Electrodynamics.)*

Section 7-6 of [FH] shows how to construct an appropriate potential energy function V for a charged particle subject to force from an electromagnetic field. The lagrangian for three-dimensional motion of a particle of mass m and electric charge e is given as

$$\mathcal{L}(\dot{\mathbf{x}}(t),\mathbf{x}(t),t) \;=\; \frac{m}{2}|\dot{\mathbf{x}}(t)|^2 \,-\, e\phi(\mathbf{x}(t),t) \,+\, \frac{e}{c}\dot{\mathbf{x}}(t)\cdot\mathbf{A}(\mathbf{x}(t),t),$$

where boldface vector notation $\mathbf{x}(t)$ *denotes particle position in three-dimensional space. The variable* ϕ *is a scalar potential for the electric field, and* $\mathbf{A}$ *is a vector potential for the magnetic field. Assume, for simplicity, that the system is one-dimensional[15] and write x for* $\mathbf{x}$ *and A for* $\mathbf{A}$. *For the purpose of the discussion [FH] assumes* $\phi = 0$, *and considers* $\mathbf{A}, = A$, *to be a perturbation of an otherwise undisturbed charged particle motion. For convenience assume speed of light* $c = 1$ *(so velocities are expressed as a fraction of c). For each path (or history) x of the particle this gives lagrangian* $\mathcal{L}$ *and action* $S(x)$, *respectively,*

$$\mathcal{L}(\dot{x}(t),x(t),t) \;=\; \frac{m}{2}\dot{x}(t)^2 \,+\, e\dot{x}(t)A(x(t),t),$$

$$S(x)_{0<t<\tau} \;=\; \int_{x(0)}^{x(\tau)}\left(\frac{m}{2}\dot{x}(t)^2 \,+\, e\dot{x}(t)A(x(t),t)\right)dt. \qquad (8.47)$$

This is a description of the system in terms of classical physics. To deal with quantum mechanical phenomena in the system, the oscillation component

$$\exp\left(\frac{\iota}{\hbar}S(x)_{0<t<\tau}\right)$$

of a wave function for arbitrary particle path x is constructed. This, in turn, resolves into a product of factors

$$\exp\frac{\iota}{\hbar}S(x) \;=\; \exp\frac{\iota}{\hbar}\left(\int_{x(0)}^{x(\tau)}\frac{m}{2}\dot{x}(t)^2 \,+\, e\dot{x}(t)A(x(t),t)dt\right) \qquad (8.48)$$

$$=\; \left(\exp\frac{\iota}{\hbar}\int_{x(0)}^{x(\tau)}\frac{m}{2}\dot{x}(t)^2 dt\right)\left(\exp\frac{\iota}{\hbar}\int_{x(0)}^{x(\tau)} e\dot{x}(t)A(x(t),t)dt\right).$$

In the terminology of [FH] (section 8-8), the ground state of the system corresponds to

$$\exp\frac{\iota}{\hbar}\int_{x(0)}^{x(\tau)}\frac{m}{2}\dot{x}(t)^2 dt,$$

---

[15]This is to simplify the mathematical notation. It is meaningless in terms of physics.

also called the vacuum state. The second factor

$$\exp \frac{\iota}{\hbar} \int_{x(0)}^{x(\tau)} e\,\dot{x}(t) A(x(t),t)\,dt$$

represents perturbation of the system, also called excitation. So, in the notation used earlier,

$$\exp \frac{\iota}{\hbar} S(x) \;=\; \left(\exp \frac{\iota}{\hbar} S_0(x)\right)\left(\exp \frac{\iota}{\hbar} S'(x)\right).$$

In [FH] different versions of the perturbation factor are examined, giving the following two estimates as alternative representations for the oscillation (8.48) in the wave function:

$$\exp \frac{\iota}{\hbar} \left(\sum_{j=1}^{n} \left(\frac{m}{2}\left(\frac{x_j - x_{j-1}}{s_j - s_{j-1}}\right)^2 + e\,\frac{x_j - x_{j-1}}{s_j - s_{j-1}} A(x_{j-1}, s_{j-1}) \right)(s_j - s_{j-1}) \right),$$

$$\exp \frac{\iota}{\hbar} \left(\sum_{j=1}^{n} \left(\frac{m}{2}\left(\frac{x_j - x_{j-1}}{s_j - s_{j-1}}\right)^2 + e\,\frac{x_j - x_{j-1}}{s_j - s_{j-1}} A(x_j, s_j) \right)(s_j - s_{j-1}) \right);$$

where $x_j = x(s_j)$. (In [FH] it is assumed that the time increments $s_j - s_{j-1}$ are all equal. In the -complete integral version of the theory this assumption is not required.)

For a variety of reasons [FH] rejects both of these estimates[16] and advances the following:

$$\exp \frac{\iota}{\hbar} S(x, N) \;=\; \exp \frac{\iota}{\hbar} \left(\sum_{j=1}^{n} \left(\frac{m}{2}\left(\frac{x_j - x_{j-1}}{s_j - s_{j-1}}\right)^2 + \right.\right.$$

$$\left.\left. +\; e\,\frac{x_j - x_{j-1}}{s_j - s_{j-1}} \left(\frac{A(x_{j-1}, s_{j-1}) + A(x_j, s_j)}{2}\right) \right)(s_j - s_{j-1}) \right)$$

$$=\; \exp \frac{\iota}{\hbar} \left(S_0(x, N) + S'(x, N) \right)$$

$$=\; \left(\exp \frac{\iota}{\hbar} S_0(x, N)\right)\left(\exp \frac{\iota}{\hbar} S'(x, N)\right).$$

Then, as before, the path integral for the particle in the ground state is, with $G(I[N])$ defined by (8.31),

$$\int_{\mathbf{R}^T} \alpha(N) \exp\left(\frac{\iota}{\hbar} S_0(x, N)\right) |I[N]| \;=\; \int_{\mathbf{R}^T} \alpha(N) \exp\left(\frac{\iota}{\hbar} S_0(x, N)\right) dx(N)$$

$$=\; \int_{\mathbf{R}^T} G(I[N])$$

$$=\; \frac{e^{\frac{\iota m (\xi - \xi')^2}{2\hbar(\tau - \tau')}}}{\sqrt{2\pi \iota \hbar m^{-1}(\tau - \tau')}} \,;$$

---

[16]Here, each of the estimates uses sampling sum formulation, as in (8.44) and (8.41) above.

*and if either of the following integrals exist, they both exist and they are equal
to the wave function $\psi(\xi, \tau)$ for the particle motion in the electric field:*

$$\int_{\mathbf{R}^T} \alpha(N) e^{\frac{k}{\hbar} S(x,N)} dx(N), \qquad \int_{\mathbf{R}^T} e^{\frac{k}{\hbar} S'(x,N)} dG.$$

This follows from theorem 204 (page 324 of [MTRV]).

8.10 Simple Harmonic Oscillator

The simple harmonic oscillator of Example 51 above is a special case of the single
particle system described in Section 8.2 above (and in chapter 7 of [MTRV]);
with potential energy $V(y) = \frac{m\varpi^2}{2} y^2$.

There are several reasons to consider this particular case in more detail. In
chapter 8 of [FH], Feynman uses a "line of interconnected simple harmonic osc-
illators" as the first step towards a quantum theory of fields, in which minuscule
disturbances of adjoining field elements are the source of quantum phenomena.
And it can be a case study for the calculus of the -complete method of integration.

Example 51 above (page 240) has an introduction to a path integral formulation of a simple harmonic oscillator. In accordance with (8.41), the wave
function in this case is $\psi(\xi, \tau) = \psi_\varpi(\xi, \tau) =$

$$= \int_{\mathbf{R}^T} \alpha(N) e^{\frac{k}{\hbar} S(x(N))} dx_T = \int_{\mathbf{R}^T} \alpha(N) e^{\frac{k}{\hbar} (S_0(x(N)) + S'(x(N)))} dx_T$$

$$= \int_{\mathbf{R}^T} \frac{\exp\left(\frac{\iota}{\hbar} \sum_{j=1}^n \frac{m}{2} \left(\left(\frac{x_j - x_{j-1}}{s_j - s_{j-1}}\right)^2 - \varpi^2 x_j^2\right)(s_j - s_{j-1})\right)}{\prod_{j=1}^n \sqrt{\frac{2\pi \iota \hbar}{m}}(s_j - s_{j-1})} |I[N]|$$

$$= \int_{\mathbf{R}^T} \exp\left(-\frac{\iota m}{2\hbar} \sum_{j=1}^n \varpi^2 x_j^2 (s_j - s_{j-1})\right) \frac{\exp\left(\frac{\iota m}{2\hbar} \sum_{j=1}^n \frac{(x_j - x_{j-1})^2}{s_j - s_{j-1}}\right)}{\prod_{j=1}^n \sqrt{\frac{2\pi \iota \hbar}{m}}(s_j - s_{j-1})} |I[N]|$$

$$= \int_{\mathbf{R}^T} \exp\left(\frac{\iota}{\hbar} S'(x(N))\right) g(x(N)) dx(N)$$

$$= \int_{\mathbf{R}^T} \exp\left(\frac{\iota}{\hbar} S'(x(N))\right) G(I[N]) \tag{8.49}$$

provided the integral exists. (The reason why $g(x(N))|I[N]|$ can be replaced
by $G(I[N])$ in (8.49) is because of variational equivalence—see theorem 204,
[MTRV] page 324; and also Example 54 below.)

Next, consider aspects of this integral. Here are some comments[17] from
section 3-5 of [FH], under the heading of Gaussian Integrals:

---

[17]These comments are under the heading of "*Gaussian Integrals*" in [FH], though "*Fresnel
Integrals*" might be more accurate. Because of their similarity, chapters 6 and 7 of [MTRV]
use a common designation "*c*-Gaussian" or "*c*-Brownian", where parameter $c = a + \iota b$ in the
exponent of $e^{(\cdots)}$ can have value $c = -1$ (Gaussian) or $c = \iota$ (Fresnel).

> So long as the integrand ⋯ contains the path variables [x_j] only
> up to second order, a solution that will be complete except possibly
> for some simple multiplying factors can be obtained. ⋯ [This] factor
> is a function of the times at the endpoints of the paths. For most of
> the path integrals which we shall study, the important information
> is contained in the exponential term rather than in the latter factor.
> In fact in most cases we shall not even find it necessary to evaluate
> this latter factor.

The "simple multiplying factors" are related to the factor $\alpha(N)$ in (8.49) (designated as action wave amplitude in Section 8.3 above). As to the exponential factor, [FH] states:

> It is possible to carry out the integral over all paths ⋯ by dividing
> the region into short time elements, and so on. [This corresponds to
> sampling partitions $\{s_1, \ldots, s_n\}$ in (8.49) above.] Such integrals can
> always be carried out ⋯ so long as the integrand is an exponential
> function which contains the path variables [x_j] only up to second
> order.

In other words, according to [FH], expressions such as that in (8.49), in which the path parameters x_j appear in polynomial form with highest power 2 (so quadratic), can always be integrated.

Continuing in this vein, section 3-5 of [FH] says that if the potential function $V(x_{j-1})$ can be expanded as a Taylor series then terms containing x_{j-1}^n, with exponent $n \geq 3$, are negligible in the integration. [FH] claims that, if such terms can be omitted the path integral exists and can be calculated.

Can the -complete calculus add or subtract anything regarding such claims? In addition to calculus of variations, the calculus used in [F1] and in [FH] consists of estimating the wave function using cylinder function approximation of the integrand (in [MTRV] terminology).[18] In this method the integrand in (8.49) is taken to be constant except at a finite number m of specified times τ_j in T. This converts $\int_{\mathbf{R}^T}$ to $\int_{\mathbf{R}^m}$. As demonstrated in [MTRV] it is sufficient to make this change in just one part of the integrand, namely, the excitation term $\exp\left(\frac{\iota}{\hbar}S'(x(\cdot))\right)$. The integrability of $G(I[N])$ on $\mathbf{R}^{]\tau_{j-1}, \tau_j[}$ for each j then converts $\int_{\mathbf{R}^T}$ to $\int_{\mathbf{R}^m}$. (See theorem 159, page 250 of [MTRV].)

But the preferred approach in [MTRV], and in this book, is the more familiar and more basic method of approximation of the integrand by step functions. In this case integrands are assumed constant, not just in time interval $\tau_{j-1} < s < \tau_j$, but also in space intervals $y_{k-1} < x_j < y_k$. (Again, in this case the step function approximation may be applied only to the excitation factor $\exp\left(\frac{\iota}{\hbar}S'(x(\cdot))\right)$ of the integrand in (8.49), without applying it to the other factors of the integrand.)

One advantage of a step function estimate is that it is always integrable when it is defined on a partition of the domain. In that case the integral of the step function consists of the sum of a finite number of terms; and the step function

---

[18]In section 2-4 of [FH] Feynman writes: "...divide the independent variable time into steps of width ε". Other authors use the term "time-slicing" for this procedure.

is integrable, both absolutely and non-absolutely.[19] In contrast, cylinder functions (or Feynman's *constant time-width* ε estimates) are not always "obviously integrable".

Those x in $\mathbf{R}^T$ which are not uniformly continuous form a G-null subset of $\mathbf{R}^T$. Each continuous x is the limit of a countable sequence of step functions. A step function in some domain (such as $\mathbf{R}$ or $\mathbf{R}^T$) is constant on each cell of a partition of the domain. A partition is a finite number of disjoint cells whose union consists of the whole domain.

Furthermore, a continuous function f of continuous $x \in \mathbf{R}^T$ can be expressed as a limit of step functions, taking constant values on cells of partitions of $\mathbf{R}^T$, as described in chapter 7 of [MTRV].

(It is also possible to resolve a domain into a countably infinite union of disjoint cells—not a partition of the domain—and to describe a function which is constant on each of the countably many cells as a step function. But such a function is not necessarily integrable. It will be sufficient for present purposes to regard a step function as having only a finite number of distinct values, corresponding to the the finite number of distinct cells in a partition.[20])

In infinitely many dimensions, regular partitions are the easiest kind to work with. A regular partition can be thought of as "strips" of cells. To illustrate this in a finite-dimensional Cartesian product space $\mathbf{R} \times \mathbf{R}$, the first factor space $\mathbf{R}$ (written $\mathbf{R}_1$) can be partitioned by disjoint cells $\mathcal{P} = \{J\}$, and the second factor $\mathbf{R}$ ($= \mathbf{R}_2$) left unpartitioned. Then

$$\mathbf{R}^2 = \mathbf{R}_1 \times \mathbf{R}_2 = \bigcup_{J \in \mathcal{P}} J \times \mathbf{R}_2.$$

The general idea is presented in chapter 3 of [MTRV], where *partially regular*, *regular*, and *binary* partitions of $\mathbf{R}^T$ are discussed. Page 81 of [MTRV] has diagrams illustrating partitions (not regular, not binary) of unbounded domains. Page 91 has diagrams of regular and partially regular partitions. A regular (not necessarily binary) partition of $\mathbf{R}^3$, $= \mathbf{R}_1 \times \mathbf{R}_2 \times \mathbf{R}_3$, can be constructed as follows. Let $\mathcal{P}_1 = \{J_1\}$, $\mathcal{P}_2 = \{J_2\}$ be partitions of factor spaces $\mathbf{R} = \mathbf{R}_1$ and $\mathbf{R} = \mathbf{R}_2$, respectively. Then

$$\mathcal{P} = \{J_1 \times J_2 \times \mathbf{R}_3 : J_1 \in \mathcal{P}_1, \ J_2 \in \mathcal{P}_2\},$$

in which factor space $\mathbf{R} = \mathbf{R}_3$ is unpartitioned, is a regular partition of $\mathbf{R}^3$ consisting of "strips". For $\mathcal{P}$ to be a binary partition of $\mathbf{R}^3$ (in the sense of [MTRV]), the partitions $\mathcal{P}_1$ and $\mathcal{P}_2$ should be binary. That means that each $\mathcal{P}_i$ ($i = 1, 2$) should consist of cells K^q of the form

$$]-\infty, -q], \ \cdots, \ ]-q+(k-1)2^{-q}, -q+k2^{-q}], \ \cdots, \ ]q, +\infty[\qquad (8.50)$$

for $j = 1, 2, \ldots, q2^{q+1}$, obtained by successive bisection of $]-q, q]$, as described in section 3.5 of [MTRV] (pages 95–97). This kind of construction can be translated

---

[19]Existence of non-integrable step functions is implied by theorem 67 of [MTRV], page 180; an extreme case which does not apply in domain $\mathbf{R}^T$, except for a G-null set of $x_T \in \mathbf{R}^T$.

[20]But integrability of step functions does not necessarily imply integrability of their limits; see [MTRV] section 7.19, pages 360–366; and also further comments below.

into binary, regular partitions of $\mathbf{R}^T$ (as in section 3.5 of [MTRV]). In this case, cells $J_1 \times J_2$ are replaced by cylindrical cells $J(M)$ where $M = \{\tau_i : i = 1, \ldots m\} \in \mathcal{N}(T)$ is a given finite subset of T, and "strips" are cylindrical cells

$$J[M] \;=\; J(M) \times \mathbf{R}^{T \setminus M} \;=\; \prod_{\tau_i \in M} J_{\tau_i} \times \prod_{t \in T \setminus M} \mathbf{R}_t,$$

with $\mathbf{R}_t = \mathbf{R}$ for each t. An additional step in constructing regular binary partitions $\mathbf{R}^T$ is to successively bisect the interval $T =]\tau', \tau]$; so M is taken as

$$M_r \;=\; \{\tau_{jr}\}_{j=1,2,\ldots,2^r-1} \;=\; \left\{ \tau' + (\tau - \tau')\frac{j}{2^r} : j = 1, 2, \ldots, 2^r - 1 \right\} \qquad (8.51)$$

for $r = 1, 2, 3, \ldots$. An (r,q)-*binary partition* of $\mathbf{R}^T$ consists of cylindrical cells ("strip" cells) $J[M_r]$ where, for each $\tau_j \in M_r$, a cell J_{τ_j} is chosen from the list of one-dimensional cells K^q. **In each cell $J[M_r]$ an (r,q)-*binary step function* $f_{rq}(x_T)$ has the same constant value for all $x_T \in J[M_r]$.**

The constant value α_{rq} in $J[M_r]$ can be chosen in any convenient way. For instance, it could be taken to be the value of f at a vertex of $J[M_r]$. Or, for continuous f and x_T in closed bounded domains, it could be taken to be the infimum value of $f(x_T)$ in $J[M_r]$.

Example 53 *Let $\mathcal{P} = \{J_1, \ldots, J_m\}$ be a partition of $\mathbf{R}^T$, not necessarily binary, regular, nor partially regular. Let f be a step function with constant value α_i on J_i for $i = 1, \ldots, m$. Let $h(x, I)$ be an integrator function, with $H(\mathbf{R}^T) = \int_{\mathbf{R}^T} h(x, I)$. Then $\int_{\mathbf{R}^T} f(x_T) h(x, I)$ exists, with*

$$\int_{\mathbf{R}^T} f(x_T) h(x, I) \;=\; \sum_{i=1}^m \alpha_i H(J_i).$$

To prove this, note that the indefinite integral H of h exists, and H is additive on cells J of $\mathbf{R}^T$, with $H(J) = \int_J h(x, I)$; and H is variationally equivalent to h. (See theorem 18, page 132 of [MTRV].) Therefore, for any function $k(x_T)$,

$$\int_{\mathbf{R}^T} k(x) h(x, I) \;=\; \int_{\mathbf{R}^T} k(x) H(I)$$

if either integral exists. Choose a gauge γ which conforms to the partition $\mathcal{P}$; so if (x, I) is γ-fine then $I \subseteq J_i$ for some i. Let $\mathcal{D}$ be a γ-fine partition of $\mathbf{R}^T$. Then

$$(\mathcal{D}) \sum f(x) H(I) \;=\; \sum_{i=1}^m \{f(x) H(I) : (x, I) \in \mathcal{D}, I \subseteq J_i,\} \;=\; \sum_{i=1}^m \alpha_i H(J_i),$$

from which the result follows. $\bigcirc$

The objective here is to assess the integrability of (8.49),

$$\int_{\mathbf{R}^T} \exp\left(-\frac{\imath m}{2\hbar} \sum_{j=1}^n \varpi^2 x_j^2 (s_j - s_{j-1})\right) \frac{\exp\left(\frac{\imath m}{2\hbar} \sum_{j=1}^n \frac{(x_j - x_{j-1})^2}{s_j - s_{j-1}}\right)}{\prod_{j=1}^n \sqrt{\frac{2\pi \imath \hbar}{m}}(s_j - s_{j-1})} |I[N]|,$$

or $\int_{\mathbf{R}^T} \exp\left(\frac{\iota}{\hbar} S'(x(N))\right) G(I[N])$, which, if it exists, gives the wave function for a simple harmonic oscillator in path integral format. Example 53 implies that the path integral exists when the integrand is replaced by a step function estimate.

Because the set of discontinuous x_T in $\mathbf{R}^T$ can be taken to be G-null, Example 54 shows that

$$\exp\left(\frac{\iota}{\hbar} S'(x(N))\right) \overset{G}{=} \exp\left(-\frac{\iota m \varpi^2}{2\hbar} \int_T x(s)^2 ds\right), \quad \text{or} \quad f(x(N)) \overset{G}{=} f(x)$$

for short; so $f(x)$ can replace $f(x(N))$ in (8.49).

Example 54 *To demonstrate G-variational equivalence of $f(x(N))$ and $f(x)$, note that $G(I[N])$ is $\mathbf{R}^T \times \mathcal{N}(T)$-VBG* in $\mathbf{R}^T \times \mathcal{N}(T)$ ([MTRV] theorem 179, page 288, combined with continuous modification of G; see also theorem 204, page 324 of [MTRV]); so there exist positive numbers α_j, pairs $(A_j, \mathcal{M}_j)$ with*

$$A_j \subset \mathbf{R}^T, \quad \mathcal{M}_j \subset \mathcal{N}(T), \quad j = 1, 2, 3, \ldots,$$

and a gauge $\gamma_0 = (L^{(0)}, \delta_{\mathcal{N}}^{(0)})$ so that, for each γ_0-fine division $\mathcal{D}$ of $\mathbf{R}^T$,

$$(\mathcal{D}) \sum \mathbf{1}_{(A_j, \mathcal{M}_j)}(x, N) |G(I[N])| < \alpha_j$$

for $j = 1, 2, 3, \ldots$. (The function $\mathbf{1}_{(A_j, \mathcal{M}_j)}(x, N)$ is equal to 1 if $x \in A_j$ and $N \in \mathcal{M}_j$, and equal to 0 otherwise.) Let $\varepsilon > 0$ be given. Using continuity of sample paths x_T ([MTRV] theorem 188, page 300), define $L_1(x_T) \supseteq L_0(x_T)$, $\gamma = (L, \delta_N) < \gamma_0$, so that, for continuous $x_T \in A_j$ and $N \supseteq L(x_T)$, $N \in \mathcal{M}_j$,

$$|f(x(N)) - f(x)| < \frac{\varepsilon}{\alpha_j 2^j}, \quad j = 1, 2, 3, \ldots.$$

Then, for any γ-fine division $\mathcal{D}$ of $\mathbf{R}^T$, $(\mathcal{D}) \sum |f(x(N)) - f(x)| |G(I[N])| =$

$$= \sum_{j=1}^{\infty} (\mathcal{D}) \sum \mathbf{1}_{(A_j, \mathcal{M}_j)}(x, N) |f(x(N)) - f(x)| |G(I[N])| < \sum_{j=1}^{\infty} \frac{\varepsilon}{\alpha_j 2^j} \alpha_j = \varepsilon,$$

giving the result. ○

The step function idea can be used to consider existence of the path integral version of the wave function for a simple harmonic oscillator,

$$\int_{\mathbf{R}^T} f(x) G(I[N]) \quad \text{with} \quad f(x) = \exp\left(-\frac{\iota m \varpi^2}{2\hbar} \int_{T'}^{\tau} x(s)^2 ds\right) = e^{\frac{\iota}{\hbar} S'(x)}.$$

A fairly standard practice in the integration of a function $h(x)$ is to start with a step function approximation or estimate of integrand $h(x)$. A step function in $\mathbf{R}^T$ has some of the characteristics of a function in a finite-dimensional space

$\mathbf{R}^M$. This opens up some of the problems, mentioned above, of projection from a higher dimensional space to a lower dimensional one.[21]

Consider integrand $f(x) = e^{\frac{i}{\hbar}S'(x)}$, with integrator $G(I[N])$ in $\mathbf{R}^T$. The set of $x \in \mathbf{R}^T$ which are not uniformly continuous form a G-null subset of $\mathbf{R}^T$, provided the continuous modification is applied. Alternatively (and equivalently), the uncountable set $T =]\tau', \tau[$ can be replaced by a countable, dense subset of T, such as the set of rational numbers t for which $\tau' < t < \tau$. In the latter case, an appropriate modification (the *continuous modification*) of the meaning of $\int_{\tau'}^{\tau} x(t)^2 \, dt$ must be applied to the function $f(x)$. (These issues are discussed in [MTRV] sections 7.5 and 7.6, pages 315–320.)

Thus all x_T which fail to be uniformly continuous can be ignored in the integration with respect to G; and the integral $\int_T x(s)^2 ds$—which appears in the action functional $S'(x)$—exists for all except a G-null set of $x(\cdot)$.

Suppose (r, q)-binary step function approximations $\alpha_{rq} = f_{rq}(x_T)$ are chosen for $x_T \in J[M_r] \in \mathcal{P}_{rq}$, with

$$f_{rq}(x_T) \;\rightarrow\; f(x_T) \quad \text{as} \quad r, q \;\rightarrow\; \infty$$

for continuous x_t and continuous f. In assessing integrability of (8.49), the following holds, as a first step:

$$f_{rq}(x_T) g(x(N)) \;\rightarrow\; f(x_T) g(x(N)) \quad \text{as} \quad r, q \;\rightarrow\; \infty.$$

As demonstrated in [MTRV], sections 4.12 and 4.13, what is then needed is some relation

$$|f_{rq}(x_T) g(x(N)) - f(x_T) g(x(N))| \;<\; \varepsilon w(x(N))$$

where $w(x(N)) > 0$ is integrable on $\mathbf{R}^T$, with

$$\int_{\mathbf{R}^T} w(x(N)) |I[N]|, \;=\; \int_{\mathbf{R}^T} w(x(N)) \, dx(N), \quad \text{finite}.$$

However, from (8.49), $|f_{rq}(x_T) g(x(N)) - f(x_T) g(x(N))| \;=\;$

$$= \;|f_{rq}(x_T) - f(x_T)| \, |g(x(N))|$$

$$= \;|f_{rq}(x_T) - f(x_T)| \left(\frac{1}{\sqrt{2\pi m^{-1}\hbar}} \right)^n \left(\prod_{j=1}^{n} \frac{1}{\sqrt{s_j - s_{j-1}}} \right)$$

$$= \;|f_{rq}(x_T) - f(x_T)| \left(\prod_{j=1}^{n} \frac{1}{\sqrt{2\pi m^{-1}\hbar(s_j - s_{j-1})}} \right). \qquad (8.52)$$

The first factor becomes small as $r, q \rightarrow \infty$. The second factor becomes very large as the sample of times N increases in the process $\int_{\mathbf{R}^T} \cdots$ of integration on $\mathbf{R}^T$. But these two processes are independent of each other, and one process

---

[21] Points are depicted in figure 3.1 (page 87 of [MTRV]) depicts points, while figure 3.2 depicts cells. The "*Flatland* challenge" [1] applies.

need not necessarily compensate for the other to produce a finite result. (The desired result is zero.) So, while the aim is to deduce

$$\int_{\mathbf{R}^T} f_{rq}(x_T) g(x(N)) \, dx(N) \; \to \; \int_{\mathbf{R}^T} f(x_T) g(x(N)) \, dx(N) \quad \text{as} \quad r, q \; \to \; \infty,$$
$$(8.53)$$

it is hard to see how the integral convergence theorems of [MTRV] sections 4.12 and 4.13 (pages 165–178) can be applied without making further assumptions about the limit function $f(x_T)$.

This is discussed in more detail in [MTRV] section 7.19, pages 360–366, where it is argued that, to deduce (8.53), the limit function f must itself be somewhat "step-function-like".

Such a conclusion is a consequence of the "continuity" in (8.53). But in Section 2.5 above, a case was put forward that price processes, though assumed continuous for purposes of mathematical analysis, are step-wise constant in the "real world", with discrete up-ticks and down-ticks occurring at discrete times.

Could it also be the case that, at "quantum-level", physical processes and phenomena are similarly discrete rather than continuous, and could it be that this has some bearing on any mathematical description of them? If that is so, then the conundrums of (8.52) and (8.53) may evaporate and disappear.

Another feature of the path integral theory is the ubiquitousness of $\int_{\mathbf{R}} \cdots$, or $\int_{-\infty}^{\infty}$. The action wave interference phenomena are a consequence of indeterminacy, or degrees of freedom, in values of physical variables. And there is justification for allowing such mathematical variables to have theoretical range $]-\infty, +\infty[$.

But indefiniteness amounting to $]-\infty, +\infty[$ appears somewhat "unphysical". The reality seems to be that physical variability of actual realities such as particle position, or field element values, amounts to the most minute of tremors. Should there be finite cut-off points, not $-\infty$ or $+\infty$? For instance, Examples 48 and 49 above use $]-1, 1[$ instead of $]-\infty, \infty[$.

In principle, such modifications can be included in the theory presented here. The -complete integral construction on $]-\infty, +\infty[$ uses finite cut-off points,[22] as indicated in (3.4) above. (See also section 2.15, pages 73–78 of [MTRV].)

Nonetheless, $\int_{-\infty}^{\infty}$ seems to be a necessary starting point in order to get the mathematical representations up and running.

8.11 A Finite Number of Particles

Unlike a particle located at a single point in space, a field (in physics) is an entity or phenomenon which manifests itself throughout a region of space. Just like a particle acted on by a force causing it to change its state or position, a field can be subject to external disturbance which may alter the state of the field at the infinitely many locations in space where the field is present.

---

[22]Corresponding physical cut-off points are not mentioned in [F1] or [FH], possibly because it is expected that the reader should be aware of an obvious physical necessity for these.

Over time a succession of such states constitutes a *history* of the field, corresponding to a *path* of a particle. So instead of a succession of single points and the resulting path, quantum field theory deals with a whole region of such points and the resulting history of the region.

In advance of engaging with quantum field theory, chapter 8 of [FH] describes a number of preliminary scenarios which help to set the scene and build up physical theory. These scenarios include simple harmonic oscillator (single particle), the polyatomic molecule, the one-dimensional crystal (involving a finite number of oscillating particles), a "line of atoms", and a three-dimensional crystal.

There is a wealth of physical insight and information in [FH]. But our focus here is on mathematical construction rather than physical principle. So the following case of a finite number of particles is more of a "thought experiment" (in the limited sense of speculation or hypothesis), without much concern as to whether it possesses physical reality or meaning.

For example, physics textbooks describe multi-particle systems in which the particles cannot be distinguished from each other. The following examples disregard important, if subtle, points of physics such as these; focussing instead on mathematical issues.

Likewise the continuous mass field in Section 8.12 below, which builds on the mathematical construction put forward in this section.

Consider again the single free particle with no external mechanical force; so potential V is zero. Suppose the particle of mass μ consists of two parts with mass $\mu^{(1)}$ and $\mu^{(2)}$, respectively; joined together to form a single particle with $\mu = \mu^{(1)} + \mu^{(2)}$; whose joint motion combines as single particle motion.

The motion of each sub-particle generates an action value. These two values combine to give the value of the action for the motion of the single combined particle. For $k = 1, 2$, each part contributes $\frac{\mu^{(k)}}{2} \frac{(x(t_j)-x(t_{j-1}))^2}{t_j-t_{j-1}}$ to the action of the system. The sum of these two contributions is

$$\frac{\mu^{(1)}}{2} \frac{(x(t_j) - x(t_{j-1}))^2}{t_j - t_{j-1}} + \frac{\mu^{(2)}}{2} \frac{(x(t_j) - x(t_{j-1}))^2}{t_j - t_{j-1}} = \frac{\mu}{2} \frac{(x(t_j) - x(t_{j-1}))^2}{t_j - t_{j-1}}$$

since each of the two parts follows the same particle path x; and the particle path action wave remains $\alpha(N)e^{\frac{i}{\hbar}S(x(N))}$, with system aggregate action wave given by the integral over paths $\int_{\mathbf{R}^T} \alpha(N)e^{\frac{i}{\hbar}S(x(N))}dx(N)$ as before.

Now suppose the system consists of a finite number ν of particles with mass $\mu^{(1)}, \mu^{(2)}, \dots, \mu^{(\nu)}$, each subject to conservative force (such as gravity) with potential V. Suppose V depends on mass μ and position y, $V = V(\mu, y)$. (For instance potential energy due to gravity acting on mass μ is $\mu g y$ where y is height and g is acceleration due to gravity.)

Suppose each particle has the same initial position $x(0) = (0,0,0)$, as if they were a single particle of mass

$$\mu = \mu^{(1)} + \mu^{(2)} + \dots + \mu^{(\nu)},$$

but they are **not** combined as one particle; and, for $s > 0$, each particle follows its own separate path in three dimensions. For the kth particle $(1 \le k \le \nu)$ the path is

$$x^{(k)} = \left(x_1^{(k)}(s), x_2^{(k)}(s), x_3^{(k)}(s)\right)_{0<s\le\tau} = \left(x_{1s}^{(k)}, x_{2s}^{(k)}, x_{3s}^{(k)}\right)_{0<s\le\tau};$$

with final positions

$$\xi^{(k)}(\tau) = \left(\xi_1^{(k)}(\tau), \xi_2^{(k)}(\tau), \xi_3^{(k)}(\tau)\right) = \left(\xi_{j\tau}^{(k)}\right)_{j=1,2,3}$$

for $k = 1, 2, \ldots, \nu$. In $x_{js}^{(k)}$ the superscript k identifies individual particle number, the second subscript s denotes time, and the first subscript j denotes dimension.

Section 8.8 shows that three-dimensional analysis is easily deducible from one dimension. Therefore, for simplicity, we proceed in one dimension only; so the subscript j is eliminated, and the position at time s $(0 < s \le \tau)$ of particle k is given by the number $x^{(k)}(s), = x_s^{(k)}$, with final destination $\xi^{(k)}(\tau), = \xi_\tau^{(k)}$.

The initial state (time $s = 0$) and final state (time $s = \tau$) of this system consist of the initial and final positions of the ν particles:

$$(0, 0, \ldots, 0), \qquad \left(\xi_\tau^{(1)}, \xi_\tau^{(2)}, \ldots, \xi_\tau^{(\nu)}\right).$$

Writing $\xi = \left(\xi_\tau^{(1)}, \xi_\tau^{(2)}, \ldots, \xi_\tau^{(\nu)}\right)$, the objective is to formulate a wave function

$$\psi(\xi, \tau) := \psi\left(\left(\xi_\tau^{(1)}, \xi_\tau^{(2)}, \ldots, \xi_\tau^{(\nu)}\right), \tau\right)$$

for the system from which the physical characteristics of the system can be deduced. In previous examples this was accomplished as follows:

- Formulate an action wave $\alpha(N)e^{\frac{i}{\hbar}S(x,N)}$ for each possible history (x, N) of the system;

- Formulate a system action wave $\psi(\xi, \tau)$ by aggregating the history action waves $\alpha(N)e^{\frac{i}{\hbar}S(x,N)}$ as an integral over histories

$$\int_\Omega \alpha(N)e^{\frac{i}{\hbar}S(x,N)}|I[N]|, = \int_\Omega \alpha(N)e^{\frac{i}{\hbar}S(x,N)}dx(N).$$

For a system of ν particles moving independently, denote by $x^{(k)}$ a sample path for the kth particle $(1 \le k \le \nu)$; so $x^{(k)}(s) \in \mathbf{R}$ for $s \in \,]0, \tau[\,= T$; and $x^{(k)} \in \mathbf{R}^T$. (If three-dimensional $x^{(k)}(s)$ is $\left(x_i^{(k)}(s)\right)_{i=1,2,3}$, then $x^{(k)} \in \left(\mathbf{R}^3\right)^T$.)

At time s, a state of the ν-particle system is

$$\mathbf{x}(s) = \left(x^{(1)}(s), x^{(2)}(s), \ldots, x^{(\nu)}(s)\right) = \left(x^{(k)}(s)\right)_{k=1,2,\ldots,\nu}; \quad \text{with} \quad \mathbf{x} \in (\mathbf{R}^\nu)^T.$$

(If $x^{(k)}(s) \in \mathbf{R}^3$ is three-dimensional then $\mathbf{x} \in \left((\mathbf{R}^3)^\nu\right)^T$; but for greater simplicity we are taking $x^{(k)}(s)$ to be one-dimensional.) For domains

$$(\mathbf{R}^\nu)^{]0,\tau[} = (\mathbf{R}^\nu)^T, \qquad (\mathbf{R}^\nu)^{]0,\tau]} = (\mathbf{R}^\nu)^T,$$

sample times N are denoted by, respectively,

$$N = \{s_1, \ldots, s_{n-1}\} \quad \text{and} \quad N = \{s_1, \ldots, s_n\}$$

with $0 = s_0 < s_1 < \cdots < s_{n-1} < s_n = \tau$. It is easy to replace time interval $T =]0, \tau[$ by $T =]\tau', \tau[$ whenever necessary. With

$$\mathbf{x} = \left(x^k(s)\right)_{0 < s < \tau, \ k=1,2,\ldots,\nu}, \quad \mathbf{x}_j = \mathbf{x}(s_j) = \left(x_j^{(k)}\right)_{k=1,\ldots,\nu},$$

denote a sample history for the ν-particle system by

$$(\mathbf{x}, N) = \left(\left(x^{(k)}\right)_{k=1,2,\ldots,\nu}, \ N\right) \quad \text{or, whenever appropriate,}$$

$$(\mathbf{x}(N), N) = \left(\left(x_j^{(k)}\right)_{j=1,2,\ldots,n-1, \ k=1,2,\ldots,\nu}, \ N\right)$$

The time interval T is the same for all particles, suggesting that sample times N should also be the same. But if it is necessary to allocate different sample times N_k to each particle $\mu^{(k)}$, then

$$N_k = \left\{s_1^{(k)}, s_2^{(k)}, \ldots, s_{n(k)-1}^{(k)}\right\},$$

with $\mathbf{N} = (N_k)_{k=1,2,\ldots,\nu}$; so the sample history for the system as a whole is $(\mathbf{x}, \mathbf{N})$. But here we proceed with the same sample times N for each $\mu^{(k)}$. For a single particle $\mu^{(k)}$ the action for sample path $(x^{(k)}, \{s_1, \ldots, s_{n-1}\})$ is $S(x^{(k)}(N)) =$

$$= \sum_{j=1}^{n} \left(\frac{\mu^{(k)}}{2} \left(\frac{x^{(k)}(s_j) - x^{(k)}(s_{j-1})}{s_j - s_{j-1}}\right)^2 - V\left(\mu^{(k)}, x^{(k)}(t_j)\right)\right)(s_j - s_{j-1}),$$

where $s_{j-1} \le t_j \le s_j$. (For the moment simply take $t_j = s_j$ for each j.) An estimate of the action $S(\mathbf{x}(N))$ for a single sample history $(\mathbf{x}, N)$ is the sum of the sample actions for the ν particles, $S(\mathbf{x}(N)) = \sum_{k=1}^{\nu} S(x^{(k)}(N)) =$

$$= \sum_{k=1}^{\nu} \left(\sum_{j=1}^{n} \left(\frac{\mu^{(k)}}{2}\left(\frac{x^{(k)}(s_j) - x^{(k)}(s_{j-1})}{s_j - s_{j-1}}\right)^2 - V\left(\mu^{(k)}, x^{(k)}(s_j)\right)\right)(s_j - s_{j-1})\right)$$

$$= \sum_{k=1}^{\nu} \left(\sum_{j=1}^{n} \frac{\mu^{(k)}}{2} \frac{\left(x^{(k)}(s_j) - x^{(k)}(s_{j-1})\right)^2}{s_j - s_{j-1}}\right) +$$

$$+ \sum_{k=1}^{\nu} \left(\sum_{j=1}^{n} -V\left(\mu^{(k)}, x_{s_j}^{(k)}\right)(s_j - s_{j-1})\right) \qquad (8.54)$$

$$= S_0(\mathbf{x}(N)) + S'(\mathbf{x}(N))$$

where S_0 is the ground state ($V = 0$) action and S' is the interaction or excitation contribution from potential V in domain $\Omega = (\mathbf{R}^\nu)^T$.

Like previous systems, the form

$$\alpha(N) e^{\frac{i}{\hbar} S(\mathbf{x}(N))}$$

is proposed for the sample history action wave; with system action wave for the ν-particle system (the wave function ψ for the system as a whole) obtained by a weighted superposition of the sample history action waves,

$$\int_\Omega \alpha(N)e^{\frac{i}{\hbar}S(\mathbf{x}(N))}|\mathbf{I}[N]|, \quad \text{or} \quad \int_\Omega \alpha(N)e^{\frac{i}{\hbar}S(\mathbf{x}(N))}d\mathbf{x}(N)$$

where $\Omega = (\mathbf{R}^\nu)^T$ and

$$\mathbf{I}(N) = \prod_{j=1}^{n-1}\mathbf{I}(s_j) = \prod_{j=1}^{n-1}\prod_{k=1}^{\nu}I^k(s_j) \in \mathcal{I}(\mathbf{R}^\nu),$$

$$d\mathbf{x}(N) = \prod_{j=1}^{n-1}\prod_{k=1}^{\nu}dx_j^{(k)}. \tag{8.55}$$

If this integral exists it is the "integral over histories", corresponding to the path integral described in [MTRV], and Section 8.2 above, for the single particle system with $\nu = 1$. Similarly, for integrals on the augmented domain $(\mathbf{R}^\nu)^{\mathbf{T}}$, instead of fixed value $x(\tau) = x^k(\tau) = \xi$, $x(\tau)$ is variable in $\mathbf{R}$, and an additional integration variable $x^{(k)}$ appears in the integral, with symbol $dx^{(k)}$ for $k = 1,\ldots,\nu$.

Next consider the ground state wave function ψ_0 (with $V = 0$) for the ν-particle system. Write $\Omega = (\mathbf{R}^\nu)^T$ and

$$\psi_0(\xi,\tau) = \int_\Omega \alpha(N)e^{\frac{i}{\hbar}S_0(\mathbf{x}(N))}|\mathbf{I}[N]| \tag{8.56}$$

if the latter integral exists. In this case $\alpha(N)$ is the ν-particle history action wave amplitude factor for oscillation factor

$$e^{\frac{i}{\hbar}S_0(\mathbf{x}(N))}, \quad = e^{\frac{i}{\hbar}\sum_{k=1}^{\nu}S_0(x^{(k)}(N))} = \exp\left(\sum_{k=1}^{\nu}\sum_{j=1}^{n}\left(\frac{\iota\mu^{(k)}\left(x_j^{(k)} - x_{j-1}^{(k)}\right)^2}{2\hbar\left(s_j - s_{j-1}\right)}\right)\right),$$

aggregated over the domain Ω by integrating:

$$\int_\Omega \cdots |\mathbf{I}[N]| = \int_\Omega \cdots \prod_{k=1}^{\nu}\prod_{j=1}^{n-1}dx_j^{(k)}, \quad \text{with}$$

$$\alpha(N) = \prod_{k=1}^{\nu}\prod_{j=1}^{n}\sqrt{\frac{\mu^{(k)}}{2\pi\iota\hbar(s_j - s_{j-1})}},$$

$$e^{\frac{i}{\hbar}S_0(\mathbf{x}(N))} = \prod_{k=1}^{\nu}\prod_{j=1}^{n}\exp\left(\frac{\iota\mu^{(k)}\left(x_j^{(k)} - x_{j-1}^{(k)}\right)^2}{2\hbar\left(s_j - s_{j-1}\right)}\right).$$

As in theorem 202 (page 324 of [MTRV]), the integral in (8.56) exists, with value

$$\psi_0(\xi,\tau) = \prod_{k=1}^{\nu}\sqrt{\frac{\mu^{(k)}}{2\pi\iota\hbar(\tau - \tau')}}\exp\left(\frac{\iota\mu^{(k)}\left(\xi^{(k)} - \xi'^{(k)}\right)^2}{2\hbar\left(\tau - \tau'\right)}\right). \tag{8.57}$$

Initial time τ' and initial positions $\xi'^{(k)}$ have been assumed to be zero, but the result is valid for non-zero τ' and $\xi'^{(k)}$.

The time interval T is $]0,\tau[$ (or $]\tau',\tau[$ if required), with domain $(\mathbf{R}^\nu)^T$. For the augmented domain, $\mathbf{T}$ is $T \cup \{\tau\} =]0,\tau]$, and the augmented domain $\Omega = (\mathbf{R}^\nu)^{\mathbf{T}}$ has integrators

$$\mathbf{I}(N) = \prod_{j=1}^{n} \mathbf{I}(s_j) = \prod_{j=1}^{n}\prod_{k=1}^{\nu} I^k(s_j) \in \mathcal{I}(\mathbf{R}^\nu),$$

$$d\mathbf{x}(N) = \prod_{j=1}^{n}\prod_{k=1}^{\nu} dx_j^{(k)}, \tag{8.58}$$

so an extra $dx_n^{(k)}, = d\xi^{(k)}$, is added to (8.55) for each k. As in theorem 170 (page 285 of [MTRV]), $\alpha(N)e^{\frac{i}{\hbar}S_0(\mathbf{x}(N))}$ is integrable on the augmented domain $\Omega = (\mathbf{R}^\nu)^{\mathbf{T}}$ relative to augmented $|\mathbf{I}[N]|$, with

$$\int_\Omega \alpha(N)e^{\frac{i}{\hbar}S_0(\mathbf{x},N)}|\mathbf{I}[N]| = \int_{(\mathbf{R}^\nu)^{\mathbf{T}}} \psi_0(\xi,\tau)\,d\xi = 1;$$

and the corresponding indefinite integral $G(\mathbf{I}[N])$ is integrable, with

$$\int_\Omega G(\mathbf{I}[N]) = 1$$

(corresponding to theorem 168, page 285 of [MTRV]). Thus $G(\mathbf{I}[N])$ is a (complex-valued) distribution function in sample space Ω, and if $\alpha(N)e^{\frac{i}{\hbar}S(\mathbf{x}(N))}$ is integrable with respect to $|\mathbf{I}[N]|$ then $e^{\frac{i}{\hbar}S'(\mathbf{x}(N))}$ is integrable with respect to $G(\mathbf{I}[N])$ and the integral values are equal. (A sufficient condition for this is continuity of $e^{\frac{i}{\hbar}S'(\mathbf{x}(N))}$), as in theorem 171 on page 286 of [MTRV]. In that case $e^{\frac{i}{\hbar}S'(\mathbf{X}(N))}$ is a G-random variable in the augmented sample space domain Ω, with G-expectation value

$$\mathrm{E}\left[e^{\frac{i}{\hbar}S'(\mathbf{X}(N))}\right] = \int_\Omega e^{\frac{i}{\hbar}S'(\mathbf{x}(N))}G(\mathbf{I}[N]) = \int_{(\mathbf{R}^\nu)^{\mathbf{T}}} \psi(\xi,\tau)\,d\xi.$$

The latter step requires use of Fubini's theorem (theorem 54, page 160 of [MTRV]).

Thus, in the excited state of the ν-particle system (with $V \neq 0$), the wave function $\psi(\xi,\tau)$ is the marginal density of expectation of the random variable $e^{\frac{i}{\hbar}S'(\mathbf{X},N)}$. (See *Marginal Densities*, pages 320–325 of [MTRV].)

One-dimensional system states $\left(x^{(k)}(s)\right)$ are used above for simplicity. If the more realistic three-dimensional states

$$x^{(k)}(s) = \left(x_i^{(k)}(s)\right)_{i=1,2,3}, \qquad \xi^{(k)} = \left(\xi_i^{(k)}\right)_{i=1,2,3}, \qquad \xi'^{(k)} = \left(\xi_i'^{(k)}\right)_{i=1,2,3}$$

are introduced, with $\xi = \left(\left(\xi_i^{(k)}\right)_{i=1,2,3}\right)_{k=1,\dots\nu}$ and $\xi' = \left(\left(\xi_i'^{(k)}\right)_{i=1,2,3}\right)_{k=1,\dots\nu}$, then

$$\psi(\xi,\xi';\tau,\tau') = \prod_{k=1}^{\nu}\prod_{i=1}^{3}\left(\sqrt{\frac{\mu^{(k)}}{2\pi i\hbar\tau}}\exp\frac{i\mu^{(k)}\left(\xi_i^{(k)} - \xi_i'^{(k)}\right)^2}{2\hbar\left(\tau-\tau'\right)}\right).$$

8.12   Continuous Mass Field

Section 8.11 above deals with a system consisting of a finite number of discrete masses. To proceed, suppose mass is spread continuously through a domain of three-dimensional space $\mathbf{R}^3$ with density $\rho(y)$ for elements of mass located at positions $y = (y_1, y_2, y_3)$ of the domain. With $\rho(y) \geq 0$ for each y, this represents a *field of mass*; with total mass

$$\int_{\mathbf{R}^3} \rho(y_1, y_2, y_3) \, dy_1 dy_2 dy_3$$

assumed to be finite.

For simplicity switch to one dimension, and assume the mass is distributed in a one-dimensional domain, in $\mathbf{R}$ rather than $\mathbf{R}^3$. This can be envisaged as a finite or infinite line or string consisting of a continuum of point-particles of mass, with each mass-point located at a particular space-point of the one-dimensional domain. Thus each mass-point has its own unique individual location, different from each of the other point-particles.

The aim is to focus on some integration structures involved in setting up a version of Feynman's integral method for defining a wave function for such a scenario, paying only little attention to any issues of physics, and without delving too deeply into more complex mathematical issues.

To offset this neglect, here are some comments from [FH] which suggest further issues. Section 8-4 of [FH] analyses a one-dimensional crystal, which can be thought of as " *... mass particles evenly spaced along a line [with] springs connecting neighboring particles.*" The effect of the two adjoining "springs" on each particle is to produce a version of simple harmonic oscillation. Section 8-5 of [FH] describes a continuous mass approximation to a line of atoms, each oscillating with a particular frequency:

" *... each atom oscillates with a phase difference behind the one next in line. There is a wave of oscillation passing down the line of atoms. If the phase difference between adjacent atoms is small, then the wavelength is long. Of special interest is the behavior of the atoms in the long-wavelength modes. If the wavelength greatly exceeds the spacing between atoms, this spacing is unimportant. In this case the motion can be very well described by the fictitious 'continuous medium' concept. A line of atoms can be replaced by a continuous rod with certain average properties, such as the mass per unit length $\rho = m/d$. More physically, a real rod is actually a discrete set of atoms.*"

So, according to Feynman, the "continuous medium" is really a discrete medium. Not only that, the total number of atoms present must be finite, bringing us back to the Section 8.11 model. (This is reminiscent of the situation described in Chapter 12 below, in which prices vary, not continuously, but in discrete steps, or *basis points*; and in which time—quintessentially continuous— is measured in discrete time ticks.)

If this is the physical reality of quantum mechanics then step function calculations, such as those in Example 48 and Section 9.7, may be worthwhile. But, setting aside such physical issues, consider again the continuous-mass system outlined at the start of this section.

Suppose a single point-particle of the field of mass is located initially (time $s = \tau'$) at distance y from the origin of co-ordinates. For one-dimensional (or "string") domain, y is a real number. (For higher dimensions y consists of multiple co-ordinates.) Total mass is $\int_{\mathbf{R}} \rho(y)\,dy$.

It is assumed that each separate, individual point-particle of the field can, over time, change its location. Denote the position of a particular particle at time s by $x(s)$, or x_s, for $\tau' \leq s < \tau$. For any particular point-particle, use its initial position y as label or identifier as it changes its position over time.

Suppose a particular point-particle is initially located at distance y from the origin of coordinates. Then, at time s ($\tau' \leq s < \tau$), the position of that particle is

$$x^{(y)}(s), \quad \text{or} \quad x_s^{(y)}, \quad \text{with} \quad x^{(y)}(\tau') = y.$$

Thus the symbol y is

- initial location of a point-particle, and

- identifier of that point-particle.

In the ν-particle system of Section 8.11, a system history is

$$\mathbf{x} = \left(x^{(k)}\right)_{k=1,2,\ldots,\nu} = \left(\left(x^{(k)}(s)\right)_{\tau' < s < \tau}\right)_{k=1,2,\ldots,\nu}$$

with initial and final states $\xi' = \mathbf{x}(\tau')$ and $\xi = \mathbf{x}(\tau)$, respectively, where

$$\mathbf{x}(\tau') = \xi' = \left(\xi'^{(k)}\right)_{k=1,2,\ldots,\nu}, \qquad \mathbf{x}(\tau) = \xi = \left(\xi^{(k)}\right)_{k=1,2,\ldots,\nu}$$

are ν-tuples with $\xi'^{(k)}, \xi^{(k)} \in \mathbf{R}$ for each k.

Correspondingly, for the field or continuous mass system in the form of a one-dimensional "string", each "point-element" of the mass field has initial location $\xi'^{(y)} = y$ measured as distance of the point-element from the origin of the one-dimensional co-ordinate system. The final position (time $s = \tau$) of the particle originating in position y is $\xi^{(y)} = x^{(y)}(\tau)$.

Let Y be the set of initial positions y of all particles of the (one-dimensional) field, with $Y \subseteq \mathbf{R}$. Assume Y is $\mathbf{R}$ or an interval in $\mathbf{R}$. At any time s ($\tau' \leq s \leq \tau$) denote the position of the particle which originated at position $y \in Y$ by

$$x^{(y)}(s), = x_s^{(y)}, \quad \text{with} \quad x^{(y)}(\tau') = x_{\tau'}^{(y)} = y.$$

In terms of the field as a whole, consisting of all point-mass elements, a possible state of the system at time s is given by the locations $x_s^{(y)}$ for $y \in Y$, denoted

$$\mathbf{x}(s) = \mathbf{x}_s = \left(x^{(y)}(s)\right)_{y \in Y}; \tag{8.59}$$

and, for $\tau' < s < \tau$ a system history or trajectory for the mass field can be denoted

$$\mathbf{x} = \mathbf{x}_{Y,T} = \left(\mathbf{x}_s\right)_{s \in T} = \left(\left(x^{(y)}(s)\right)_{y \in Y}\right)_{s \in T} \in \left(\mathbf{R}^Y\right)^T, \tag{8.60}$$

with initial and final states

$$\mathbf{x}(\tau') = \xi' = (y)_{y \in Y}, \qquad \mathbf{x}(\tau) = \xi = \left(\xi^{(y)}\right)_{y \in Y}, \qquad (8.61)$$

respectively, where, for each $y \in Y$, $\xi^{(y)}$ is a given real number representing final position of point-particle originating at position y. Both ξ' and ξ can be regarded as infinite 'tuples, in analogy with the ν-particle system of Section 8.11. If the mass field is three-dimensional, then states and histories are

$$\mathbf{x}_s \in \left(\mathbf{R}^3\right)^Y, \qquad \mathbf{x} \in \left(\left(\mathbf{R}^3\right)^Y\right)^T$$

where $Y \subseteq \mathbf{R}^3$, and, for each dimension i $(i = 1, 2, 3)$, $x_i^{(y)}(s) \in \mathbf{R}$ for each $y \in Y$ and each $s \in T$; so

$$-\infty < x_i^{(y)}(s) < \infty.$$

Realistically, in three-dimensional space, it may seem physically counter-intuitive that each particle of a block of material can move any distance in any direction. That resembles explosion and disintegration. But material can be flexible. Each particle of a block of rubber may be able to move or vibrate a little. Physically each motion may only involve minuscule distances of travel. But for the sake of the mathematical argument it is convenient, initially at least, to place no definite mathematical bounds or cut-off values on the distances of movement, so $x^{(y)}(s)$ is allowed, in theory, to take any value between $-\infty$ and $+\infty$.

To proceed, assume the mass field at initial time τ' has density $\rho(y)$, with mass $\rho(y)dy$ in any segment $\jmath$ of length dy. (At time $s > \tau'$ individual particle positions may alter to $x_s^{(y)}$, giving altered mass density $\rho_s\left(x_s^{(y)}\right)$. Denote the latter by $\rho_s^{(y)}$.) Again, total mass (assumed finite) for the one-dimensional field is

$$\int_{\mathbf{R}} \rho(y)dy, \qquad \text{or} \qquad \int_{\mathbf{R}} \rho_s^{(y)}\left(x_s^{(y)}\right) dx_s^{(y)},$$

and this can be estimated by Riemann sums such as

$$(\mathcal{D}) \sum \rho(y)|\jmath^{(y)}|, \quad = \sum_{k=1}^{\nu} \rho(y^{(k)})|\jmath^{(k)}|.$$

As in the discrete or ν-particle system estimates, each such Riemann sum contains only a finite number $\nu = \nu(\mathcal{D})$ of terms, depending on the choice of division $\mathcal{D}$ of Y,

$$\mathcal{D} = \mathcal{D}_Y = \left\{\left(y^{(k)}, \jmath^{(k)}\right) : k = 1, 2, \ldots, \nu(\mathcal{D})\right\}.$$

Section 7.7 shows how to construct the elements of -complete integration on domains $\Omega = \left(\mathbf{R}^Y\right)^T$ which are infinite products of infinite products. Those elements are points $\mathbf{x}, = \mathbf{x}_\Omega$, cells $\mathbf{I}, = \mathbf{I}_\Omega$, divisions $\mathcal{D}, = \mathcal{D}_\Omega$, and gauges $\gamma, = \gamma_\Omega$.

As described in (7.15) a cell $\mathbf{I}$ of Ω is $\mathbf{I}_{Y,T}$, designated loosely as $\mathbf{I}[N_{Y_T} : N_T]$, and formulated as follows. With $T =]\tau',\tau[$, $t_0 = \tau'$, and $t_n = \tau$, let

$$N_T = \{t_1,\ldots,t_{n-1}\} \in \mathcal{N}(T) \quad \text{and, for each } t = t_j \in N_T, \text{ let}$$

$$N_Y^t = N_{Y,t} = \{y_{1,t}, y_{2,t}, \ldots, y_{\nu_t,t}\} = \{y_1^t, y_2^t, \ldots, y_{\nu_t}^t\} \in \mathcal{N}(Y), \text{ or}$$

$$N_Y^j = N_{Y,t_j} = \{y_{1,t_j}, y_{2,t_j}, \ldots, y_{\nu_j,t_j}\} = \{y_1^j, y_2^j, \ldots, y_{\nu_j}^j\} \text{ for short;}$$

also allowing notation y_k^j instead of y_{kj} whenever convenient. Then a cell $\mathbf{I}$ $= \mathbf{I}_{Y,T}$ of the domain $\Omega = \mathbf{R}^{Y^T}$ is

$$\mathbf{I}_{Y,T} = \left(\prod_{j=1}^{n-1} I_{T,j}[N_{Y,j}]\right) \times \left(\mathbf{R}^Y\right)^{T \setminus N_T} \tag{8.62}$$

$$= \left(\prod_{j=1}^{n-1}\left(I_{y_1,t_j} \times I_{y_2,t_j} \times \cdots \times I_{y_{\nu_j},t_j} \times \mathbf{R}^{T \setminus N_{T,j}}\right)\right) \times \left(\mathbf{R}^Y\right)^{T \setminus N_T}$$

$$= \left(\prod_{t_j \in N_T}\left(\left(\prod_{y_{kj} \in N_{Y,j}} I_{kj}\right) \times \left(\prod_{y \in Y \setminus N_{Y,j}} \mathbf{R}\right)\right)\right) \times \left(\prod_{t \in T \setminus N_T} \mathbf{R}^Y\right);$$

where $I_{y_k,t_j} = I_{kj}$. Compositions of sample sets N_T and $N_{Y,t}$ are used in (8.62) to define cells $\mathbf{I}, = \mathbf{I}_{Y,T}$, of $\Omega = \left(\mathbf{R}^Y\right)^T$. As in Section 7.7, an indication of the construction can be provided by notation such as $\mathbf{I}[N_{Y_T}][N_T]$, so cells (8.62) can be denoted as

$$\mathbf{I} = \mathbf{I}_{Y,T} = \mathbf{I}[\mathbf{N}] = \mathbf{I}[N_{Y_T} : N_T] = \mathbf{I}[N_{Y_T}][N_T].$$

As described earlier in Section 7.7, these elements are used to construct divisions and gauges for Ω.

Elements $t_1, t_2, \ldots, t_{n-1}$ of N_T can be regarded as partition points of T, and the partition can be denoted as

$$\mathcal{P}_T = \{\jmath\} = \{\jmath_1, \ldots, \jmath_n\} \tag{8.63}$$

where $\jmath_j =]t_{j-1}, t_j]$ for $j = 1, \ldots, n$, $(t_0 = \tau', \ t_n = \tau)$. A corresponding division $\mathcal{D}_T$ of T can be formed with

$$\mathcal{D}_T = \{(s, \jmath)\} = \{(s_1, \jmath_1), \ldots, (s_n, \jmath_n)\} \tag{8.64}$$

where $s_j = t_{j-1}$ or t_j for each j. Likewise, for each $t = t_j \in N_T$ the elements $y_{kj}, = y_k^j$, of $N_{Y,t} = N_{Y,t_j}$ form a partition $\mathcal{P}_{Y,t}$ and division $\mathcal{D}_{Y,t}$ of Y,

$$N_{Y,t} = \{y_{1t}, y_{2t}, \ldots, y_{\nu^t,t}\} = \{y_1^t, y_2^t, \ldots, y_{\nu^t}^t\},$$

$$\mathcal{P}_{Y,t} = \{\imath\} = \{\imath^t\} = \{]y_{k-1}^t, y_k^t] : k = 1, 2, \ldots, \nu^t\} \quad \text{or}$$

$$\mathcal{P}_{Y,t_j} = \{]y_{k-1}^j, y_k^j] : k = 1, 2, \ldots, \nu^j\}, \quad \text{and} \tag{8.65}$$

$$\mathcal{D}_{Y,t} = \{(z, \imath)\} = \{(z^t, \imath^t)\} \quad \text{or} \tag{8.66}$$

$$\mathcal{D}_{Y,t_j} = \{(z_k^j,]y_{k-1}^j, y_k^j]) : k = 1, 2, \ldots, \nu^j)\}$$

where associated points are $z_k^j = y_{k-1}^j$ or y_k^j in each case.

Recall that a point $\mathbf{x} = \mathbf{x}_{Y,T} = ((x_{yt})_{y \in Y})_{t \in T}$ of Ω is associated with a cell $\mathbf{I} = \mathbf{I}_{Y,T}$ if $x_{yt} \in \bar{\mathbf{R}}$ is a vertex of the one-dimensional cell I_{yt} for each $y \in N_{Y,t}$ and for each $t \in N_T$. Thus,

$\mathbf{x}$ is associated with $\mathbf{I}$ if $x_{y_k, t_j}, = x_{kj}$, is a vertex of I_{kj} for each k, j.

For any given $t \in T$, a gauge γ^t in $\mathbf{R}^Y$ is γ^t where

$$\gamma^t = \left(L^t, \{\delta_y^t\}_{y \in N_Y^t \in \mathcal{N}(Y)} \right),$$
$$L^t(x_{Y,t}) \in \mathcal{N}(Y), \ \forall \ x_{Y,t} \in \bar{\mathbf{R}}^Y,$$
$$\delta_y^t(x_{yt}) > 0, \ \forall \ x_{yt} \in \bar{\mathbf{R}}.$$

Following (7.17) and (7.18) in Section 7.7, a gauge $\gamma, = \gamma_{Y,T}$, in $\Omega = \mathbf{R}^{Y^T}$ is

$$\gamma = \left(\mathbf{L}, (\gamma^t)_{t \in T} \right) = \left(\mathbf{L}, \left(L^t, \{\delta_y^t\}_{y \in N_Y^t \in \mathcal{N}(Y)} \right)_{t \in N_T \in \mathcal{N}(T)} \right) \tag{8.67}$$

where $\mathbf{L}(\mathbf{x}) \in \mathcal{N}(T)$ for all $\mathbf{x} \in (\bar{\mathbf{R}}^Y)^T$. An associated pair $(\mathbf{x}, \mathbf{I})$ of Ω is γ-fine if, in (8.62),

$$N_T \supseteq \mathbf{L}(\mathbf{x}),$$
$$N_Y^t \supseteq L^t(x_{Y,t}) \ \forall t \in N_T, \tag{8.68}$$
$$(x_{yt}, I_{yt}) \ \text{is} \ \delta_y^t\text{-fine in } \mathbf{R} \ \forall y \in N_Y^t \ \text{and} \ \forall t \in N_T.$$

Reverting to the system of ν discrete masses $\mu^{(k)}$ in Section 8.11 for purpose of comparison, the ground state action (as in (8.54)) for sample occurrences $(x^{(k)}, N)$, with $N = \{t_1, \ldots, t_{n-1}\}$ and $t_0 = \tau'$, $t_n = \tau$, is

$$S_0^{(k)}(x^{(k)}(N)) = \frac{\mu^{(k)}}{2} \sum_{j=1}^n \left(\frac{x^{(k)}(t_j) - x^{(k)}(t_{j-1})}{t_j - t_{j-1}} \right)^2 (t_j - t_{j-1}) \tag{8.69}$$

where $x^{(k)}(t)$ denotes the particle distance from the origin of co-ordinates[23] at time $t \in T$. The initial and final positions, $x^{(k)}(\tau')$ and $x^{(k)}(\tau)$, of the kth particle are given values $\xi'^{(k)}$ and $\xi^{(k)}$, respectively. The elements $N = N_T = \{t_1, \ldots, t_{n-1}\}$ form a partition $\mathcal{P} = \{\imath\}$ of $T =]\tau', \tau]$, with $\imath_j =]t_{j-1}, t_j]$ for $j = 1, \ldots, n$. Then (8.69) can be written

$$S_0^{(k)}(x^{(k)}(N)) = \frac{\mu^{(k)}}{2} \sum_{j=1}^n \frac{(x^{(k)}(\imath_j))^2}{|\imath_j|} = \frac{\mu^{(k)}}{2} \sum_{\imath \in \mathcal{P}} \frac{(x^{(k)}(\imath))^2}{|\imath|}.$$

A ν-particle system sample history is

$$(\mathbf{x}, N) = ((x^{(k)}, N)_{k=1,2,\ldots,\nu})$$

---

[23] In the simplified one-dimensional version there is only one co-ordinate.

and the ground state sample action for a system sample history is

$$
\begin{aligned}
S_0^{\text{discrete}}(\mathbf{x}(N)) &= \sum_{k=1}^{\nu} \frac{\mu^{(k)}}{2} \left(\sum_{j=1}^{n} \left(\frac{x^{(k)}(t_j) - x^{(k)}(t_{j-1})}{t_j - t_{j-1}} \right)^2 (t_j - t_{j-1}) \right) \\
&= \sum_{i \in \mathcal{P}} \sum_{k=1}^{\nu} \frac{\mu^{(k)}}{2} \frac{\left(x^{(k)}(i) \right)^2}{|i|}.
\end{aligned}
\tag{8.70}
$$

In the continuous mass field the superscript labels k (with $k = 1, \ldots \nu$) are replaced by superscript y, where $y \in Y \subseteq \mathbf{R}$. Instead of discrete particle $\mu^{(k)}$, a particle initially located at point y is part of an element of continuous mass $\rho(y)|i|$ (or $\rho(y)dy$). At later times s the location of the particle is $x^{(y)}(s)$ (or $x_s^{(y)}$) where mass density is $\rho_s^y = \rho\left(x_s^{(y)} \right)$; and the element of mass[24] is $\rho_s^y |i|$.

Thus the discrete, fixed masses $\mu^{(k)}$ of Section 8.11 are replaced by continuous quantities of mass $\rho_s^y |i|$ which depend on:

- the current time s ($\tau' \leq s < \tau$), and

- representative position y at original time $s = \tau'$ where $x_{\tau'}^{(y)} = x^{(y)}(\tau') = y$.

Thus, for the one-dimensional continuous mass system, and in analogy with (8.69), ground state sample action for a sample mass $\rho(y)|i|$ (at time $s = \tau'$), with sample times N_T, can be estimated as

$$
S_0^{(y)}(x^{(y)}(N_T)) = \sum_{j=1}^{n} \frac{\rho\left(x_{t_j}^{(y)} \right) |i|}{2} \left(\frac{x^{(y)}(t_j) - x^{(y)}(t_{j-1})}{t_j - t_{j-1}} \right)^2 (t_j - t_{j-1})
\tag{8.71}
$$

with sample path $x^{(y)}$ satisfying $x^{(y)}(\tau') = y$ and $x^{(y)}(\tau) = \xi^{(y)}$ as in (8.61); and where $n-1$ is the number of elements in N_T. Next, proceed in analogy with (8.70). For sample times $y \in N_T$ and sample elements of continuous mass centred at locations N_Y^t ($t \in N_T$), the corresponding sample history is $(\mathbf{x}, N_Y^t : N_T)$ and the ground state sample action $S_0, = S_0^{\text{continuous}}, = S_0(\mathbf{x}(N))$ is

$$
\begin{aligned}
S_0(\mathbf{x}, N_{Y,t} : N_T) &= \sum_{j=1}^{n} \left(\sum_{y \in N_{Y,t}} \frac{\rho^{(y)}\left(x_{t_j}^{(y)} \right) |i^{(y)}|}{2} \left(\frac{(x^{(y)}(t_j) - x^{(y)}(t_{j-1}))^2}{t_j - t_{j-1}} \right) \right) \\
&= \frac{1}{2} \sum_{i \in \mathcal{P}_T} \left(\sum_{i^{(y)} \in \mathcal{P}_{Y,t} : \mathcal{P}_T} \rho_t^{(y)} \frac{\left(x^{(y)}(i) \right)^2}{|i|} |i^{(y)}| \right)
\end{aligned}
\tag{8.72}
$$

Thus, for given

$$
\mathbf{x} = \mathbf{x}_{Y,T} = (\mathbf{x}_t)_{t \in T} = \left(\left(x^{(y)}(t) \right)_{y \in Y} \right)_{t \in T} \in \Omega = \left(\mathbf{R}^Y \right)^T
$$

---

[24] Various continuity assumptions are implied here.

(as in (8.60)), the sample value S_0 of ground state action is given by a composition of Riemann sums of a function defined on a composition of domains Y and T; suggesting an integral

$$S_0(\mathbf{x}) \;=\; \int_T \left(\int_Y h_0\left(x_t^{(y)}, \jmath^{(y)}, \imath \right) \right), \qquad \text{where}$$

$$h_0\left(y_t, t, \jmath^{(y)}, \imath \right) \;=\; \frac{1}{2} \rho_t^{(y)} \left(x^{(y)}(t) \right) \left(x^{(y)}(\imath) \right)^2 \frac{\left| \jmath^{(y)} \right|}{|\imath|}.$$

Whenever it exists, this integral is a Burkill-complete integral, as described in chapter 4 of [MTRV]. But, like the stochastic integrals described in Chapter 1 above, the integral $S_0(\mathbf{x})$ does not generally exist.[25] Instead, like the stochastic sums of Chapter 6, the sampling sum version $S_0(\mathbf{x}, \mathbf{N})$ is used to find the integral over histories of sample history action wave

$$\int_\Omega \alpha(\mathbf{N}) e^{\frac{\imath}{\hbar} S_0(\mathbf{x}(\mathbf{N}))} |\mathbf{I}[\mathbf{N}]|. \tag{8.73}$$

This is a kind of average or aggregate value of $\exp\left(\frac{\imath}{\hbar} S_0(\mathbf{x}(\mathbf{N})) \right)$ for all system sample histories

$$(\mathbf{x}, \mathbf{N}) \;\in\; \Omega \times \mathcal{N}(\Omega).$$

By analogy with the primary system described in Section 8.2, the integral (8.73) gives the wave function ψ for the "unexcited" state of the system, when there is no interaction with an external force or system.

In order to analyse the effect on the continuous mass field with a conservative, mechanical, external force acting on the continuous mass field, let V be the potential energy of the interaction. Then the simpler system of Section 8.11 can be used as guide or analogue. In (8.54) the sample action for a finite number of discrete particles is

$$\sum_{k=1}^{\nu} \left(\sum_{j=1}^{n} \frac{\mu^{(k)}}{2} \frac{\left(x^{(k)}(t_j) - x^{(k)}(t_{j-1}) \right)^2}{t_j - t_{j-1}} \;-\; V\left(\mu^{(k)}, x_{t_j}^{(k)} \right) (t_j - t_{j-1}) \right).$$

---

[25] Also, unlike the polyatomic molecule and one-dimensional crystal of [FH] (sections 8-2 and 8-4), the system presented here has little physical meaning except as "thought experiment" or conjecture, but has been chosen in order to suggest a mathematical or gauge integral structure for analysing analogous physical systems.

The analogous sample action for excited state of the continuous mass field is

$$S(\mathbf{x}(\mathbf{N})) \;=\; S_0(\mathbf{x}(\mathbf{N})) \;+\; S'(\mathbf{x}(\mathbf{N}))$$

$$= \sum_{j=1}^{n} \left(\sum_{y \in N_{Y,t}} \frac{\rho^{(y)}\left(x_{t_j}^{(y)}\right)\left|j^{(y)}\right|}{2} \left(\frac{\left(x^{(y)}(t_j) - x^{(y)}(t_{j-1})\right)^2}{t_j - t_{j-1}} \right) \right)$$

$$- \sum_{j=1}^{n} \left(\sum_{y \in N_{Y,t}} V\left(\rho^{(y)}\left(x_{t_j}^{(y)}\right)\left|j^{(y)}\right|, \; x_{t_j}^{(y)}\right)(t_j - t_{j-1}) \right)$$

$$= \frac{1}{2} \sum_{\imath \in \mathcal{P}_T} \left(\sum_{j^{(y)} \in \mathcal{P}_{Y,t} : \mathcal{P}_T} \rho_t^{(y)} \frac{\left(x_t^{(y)}(\imath)\right)^2}{|\imath|} \left|j^{(y)}\right| \right) \qquad (8.74)$$

$$- \sum_{\imath \in \mathcal{P}_T} \left(\sum_{j^{(y)} \in \mathcal{P}_{Y,t} : \mathcal{P}_T} V\left(\rho_t^{(y)}\left(x_{t_j}^{(y)}\right)\left|j^{(y)}\right|, \; x_{t_j}^{(y)}\right)|\imath| \right).$$

Again by analogy, the wave function ψ for the interaction is the integral over all histories of the sample action wave function; so $\psi(\xi, \xi'; \tau, \tau') =$

$$= \int_{\Omega} \alpha(\mathbf{N}) e^{\frac{\imath}{\hbar} S(\mathbf{x}(\mathbf{N}))} |\mathbf{I}[\mathbf{N}]|, \; = \int_{\Omega} \left(e^{\frac{\imath}{\hbar} S'(\mathbf{x}(\mathbf{N}))} \right) \left(\alpha(\mathbf{N}) e^{\frac{\imath}{\hbar} S_0(\mathbf{x}(\mathbf{N}))} \right) |\mathbf{I}[\mathbf{N}]|$$

$$(8.75)$$

whenever this integral exists. The integral over histories, $\int_{\Omega} \cdots$, is given meaning as a -complete integral by means of (8.68) above.

Further, if the analogy (with section 7.20 of [MTRV], pages 366–380) holds, a power series expansion of $e^{\frac{\imath}{\hbar} S'(\mathbf{x}(\mathbf{N}))}$ in (8.75) above gives the Feynman diagrams for the interaction of the continuous mass field with external potential V.

Chapter 9

Quantum Electrodynamics

9.1   Electromagnetic Field Interaction

In Section 8.3, action waves were used to describe quantum phenomena in a system consisting of a single particle possessing mass, and subject to external force. In Section 8.12 the single particle was replaced by a continuous field of mass. The idea in both cases was to describe the quantum mechanical effect of the force on the state of motion of the mass. Any reciprocal effect (of the mass on the force) was omitted from consideration.

A particle is located at a point of space. A field is located at every point of a region of space. Electromagnetism exists in a region of space in which electric force and magnetic force are detectable, by means such as their effect on an electric current passing through the field. Variation in space and time of the electric and magnetic potential of the field constitutes electromagnetic wave motion or radiation.

An electrically charged particle in motion is an electric current. The interaction of an electric current with an electromagnetic field is described by Maxwell's equations,

$$\nabla \cdot \mathbf{E} = 4\pi\rho, \quad \nabla \cdot \mathbf{B} = 0, \quad \nabla \times \mathbf{E} = -\frac{\partial \mathbf{B}}{\partial t}, \quad \nabla \times \mathbf{B} = \frac{1}{c}\left(\frac{\partial \mathbf{E}}{\partial t} + 4\pi\mathbf{j}\right),$$

where $\mathbf{E}$ is the electric field vector, $\mathbf{B}$ is the magnetic field vector, c is the speed of light, ρ is the charge density, $\mathbf{j}$ is the current density, and $\nabla \cdot \mathbf{j} = \frac{\partial \rho}{\partial t}$ (so charge is conserved). Other field descriptors $\mathbf{A}$ and ϕ are given by

$$\nabla \times \mathbf{A} = \mathbf{B}, \qquad \nabla \phi = -\mathbf{E} - \frac{1}{c}\frac{\partial \mathbf{A}}{\partial t}.$$

The field equations can be expressed as equations of a wave motion, or electromagnetic radiation.

Assume that the electric current in the field consists of the motion in the field of a single particle of mass μ and charge e. The radiation of the field changes

Gauge Integral Structures for Stochastic Calculus and Quantum Electrodynamics, First Edition. Patrick Muldowney.
© 2021 John Wiley & Sons, Inc. Published 2021 by John Wiley & Sons, Inc.

the state of motion of the particle, and the motion of the particle changes the state of radiation of the field.

The quantum phenomena[1] of this interaction of field and particle (such as emission/absorption of photons) are collectively known as quantum electrodynamics, a part of quantum field theory; and in chapter 9 of [FH] they are analysed by means of aggregated action waves, or integrals over histories.

Chapter 9 of [FH] contains a wealth of physical reasoning. (In contrast this book seeks to contribute, not to physics, but to specific integration issues which arise in [FH].) In [FH] the Summary (section 9-8, pages 262–265 of Dover 2010 edition) gives

$$S \;=\; S_\mu + S_{\mu\mathcal{F}} + S_\mathcal{F} \;=\; S_\mu(\mathbf{x}) + S_{\mu\mathcal{F}}(\mathbf{x}, \mathbf{A}, \phi) + S_\mathcal{F}(\mathbf{A}, \phi) \tag{9.1}$$

as the action for the system, composed of S_μ (action due to matter—charged mass—alone), $S_{\mu\mathcal{F}}$ (action resulting from interaction of matter μ and electromagnetic field $\mathcal{F}$), and $S_\mathcal{F}$ (action of the field alone). The wave function for the system as a whole involves aggregating or integrating the action wave oscillation $\exp \frac{\iota S}{\hbar}$. In the notation of [FH] this is the integral

$$\int_\Omega \exp\left(\frac{\iota}{\hbar} \left(S_\mu(\mathbf{x}) \;+\; S_{\mu\mathcal{F}}(\mathbf{x}, \mathbf{A}, \phi) \;+\; S_\mathcal{F}(\mathbf{A}, \phi) \right) \right) \boldsymbol{\mathcal{D}}\mathbf{x}\, \boldsymbol{\mathcal{D}}\mathbf{A}\, \boldsymbol{\mathcal{D}}\phi \tag{9.2}$$

where Ω is the domain consisting of all possible histories of the matter-electromagnetic system. At any given point y of the field and at any given time s, the variables $\mathbf{E}$, $\mathbf{B}$, $\mathbf{A}$, and ϕ, describe the electromagnetic field $\mathcal{F}$.

Chapter 9 of [FH] deals primarily with the physics issues in (9.1) and (9.2). The meaning of the latter (involving mathematical integration in some domain) is left at the level of intuition. The general idea of $\int_\Omega \cdots \boldsymbol{\mathcal{D}}\mathbf{x}\, \boldsymbol{\mathcal{D}}\mathbf{A}\, \boldsymbol{\mathcal{D}}\phi$ in (9.2) is that each of the "variables" $\mathbf{x}$, $\mathbf{A}$, and ϕ has many degrees of freedom, or possible histories, and the integration process $\int_\Omega$ involves aggregation over all degrees of freedom of all three "variables" in the matter-electromagnetic interaction.

But, as foregoing examples and scenarios illustrate, -complete integration can make this intuition more precise in respect of some parts of the mathematical construction.

For simplicity restrict the matter-electromagnetic system to one dimension only (even if this is incompatible with physical reality). Next, consider the three components of action S_μ, $S_{\mu\mathcal{F}}$ and $S_\mathcal{F}$ in (9.1). The component S_μ is the action of the charged mass. If this is a single charged particle then, since interaction effects are accounted for in $S_{\mu\mathcal{F}}$, the term S_μ is the ground state action for the mass. In other words S_μ involves only the kinetic energy of the mass, corresponding to ground state action of the path integral theory of [MTRV]. (If the mass consists of several charged particles then the energy of Coulomb interaction of the charges must be combined in S_μ with the total kinetic energy of the individual masses.) For simplicity assume there is a single particle of mass μ and charge e.

---

[1] See [FH] section 9-7, "*The Emission of Light*".

Then the lagrangian of the particle motion at time t is the kinetic energy, and the action is the integral along the particle path $\mathbf{x}^{(\mu)} = (\mathbf{x}(s))_{\tau' < s < \tau}$,

$$S_\mu = \int_{\tau'}^{\tau} \frac{1}{2}\mu \, |\dot{\mathbf{x}}(s)|^2 \, ds = \int_{\tau'}^{\tau} \frac{1}{2}\mu \left(\left(\frac{dx_1(s)}{ds} \right)^2 + \left(\frac{dx_2(s)}{ds} \right)^2 + \left(\frac{dx_3(s)}{ds} \right)^2 \right) ds.$$

For simplicity the system is assumed to be one-dimensional, with $x_1(s)$ written as $x(s)$, and with path or history $\mathbf{x}$ written as $x = (x(s))_{\tau' < s < \tau}$. Then

$$S_\mu = \frac{\mu}{2} \int_{x(\tau')}^{x(\tau)} \left(\frac{dx(s)}{ds} \right)^2 ds = \int_{\tau'}^{\tau} \mathcal{L}_\mu \left(\dot{x}^{(\mu)}(s) \right) ds. \qquad (9.3)$$

With $\mathbf{x}(\tau')$ and $\mathbf{x}(\tau)$ fixed, the second component of action is the contribution $S_{\mu\mathcal{F}}$ from interaction of matter (assumed to be a single particle with mass μ and charge e) and electromagnetic field; and (according to [FH]) is the time integral of the lagrangian function for matter-field interaction,

$$\mathcal{L}_{\mu\mathcal{F}}(s) = \mathcal{L}_{\mu\mathcal{F}}\left(\dot{\mathbf{x}}(s), \mathbf{A}(\mathbf{x}(s)), \phi(\mathbf{x}(s)) \right) = -e\left(\phi(\mathbf{x}(s)) - \frac{1}{c}\dot{\mathbf{x}}(s)\mathbf{A}(\mathbf{x}(s)) \right).$$

The action contribution from matter-field interaction is

$$\begin{aligned} S_{\mu\mathcal{F}} &= -e \int_{\tau'}^{\tau} \left(\phi(\mathbf{x}(s)) - \frac{1}{c}\dot{\mathbf{x}}(s) \cdot \mathbf{A}(\mathbf{x}(s)) \right) ds, \qquad (9.4) \\ &= -e \int_{T} \left(\phi(x(s)) - \frac{1}{c}\dot{x}(s)A(x(s)) \right) ds \end{aligned}$$

in the one-dimensional simplification.

Suppose the electromagnetic field $\mathcal{F}$ is located in a region $\mathbf{M}_\mathcal{F}$ of three-dimensional space $\mathbf{R}^3$. (For simplicity this will become one-dimensional $M_\mathcal{F}$ in $\mathbf{R}$. Symbols $\mathbf{M}$ and M are intended as a prompt for the word "electroMagnetic".) At a given instant of time s, the state of the field is given by values

$$(\mathbf{A}(\mathbf{y}, s), \phi(\mathbf{y}, s))_{\mathbf{y} \in \mathbf{M}_\mathcal{F}} \quad \text{or} \quad (\mathbf{E}(\mathbf{y}, s), \mathbf{B}(\mathbf{y}, s))_{\mathbf{y} \in \mathbf{M}_\mathcal{F}}.$$

In the one-dimensional simplification denote the former by

$$(A(M_\mathcal{F}, s), \phi(M_\mathcal{F}, s)) \quad \text{or} \quad (A_{M_\mathcal{F},s}, \phi_{M_\mathcal{F},s});$$

denoting, at a particular time s, and at every location $y \in M_\mathcal{F}$, a particular value A and a particular value ϕ. This gives the state of the field at all locations y at time s.

The expression $E^2 - B^2$ can be written in terms of A and ϕ, with function $\lambda_\mathcal{F}$ such that (in supposedly one-dimensional format)

$$E^2 - B^2 = 8\pi\lambda_\mathcal{F}(A, \phi);$$

and then, at time s, and for a given state of the field $(A(y, s), \phi(y, s))$ $(y \in M_\mathcal{F})$, the lagrangian $\mathcal{L}_\mathcal{F}$ of the field is (in accordance with [FH])

$$\mathcal{L}_\mathcal{F}(s) = \int_{y \in M_\mathcal{F}} \lambda_\mathcal{F}(A(y, s), \phi(y, s)) \, dy. \qquad (9.5)$$

Then the third component of action in the matter-electromagnetic field system
is the action of the field,

$$
\begin{aligned}
S_{\mathcal{F}} &= S_{\mathcal{F}}(A_{M_{\mathcal{F}},T},\ \phi_{M_{\mathcal{F}},T}) \\[2mm]
&= \frac{1}{8\pi}\int_{\tau'}^{\tau}\left(\int_{y\in M_{\mathcal{F}}}\left(\left|-\nabla\phi-\frac{1}{c}\frac{\partial \mathbf{A}}{\partial t}\right|^{2}-|\nabla\times\mathbf{A}|^{2}\right)dy\right)ds \\[2mm]
&= \int_{\tau'}^{\tau}\left(\int_{y\in M_{\mathcal{F}}}\frac{1}{8\pi}\left(|\mathbf{E}|^{2}-|\mathbf{B}|^{2}\right)dy\right)ds, \\[2mm]
&= \int_{\tau'}^{\tau}\left(\int_{y\in M_{\mathcal{F}}}\frac{1}{8\pi}\left(E(y,s)^{2}-B(y,s)^{2}\right)dy\right)ds \quad \text{for one dimension,} \\[2mm]
&= \int_{T}\left(\int_{y\in M_{\mathcal{F}}}\lambda_{\mathcal{F}}(A(y,s),\phi(y,s))\,dy\right)ds. \tag{9.6}
\end{aligned}
$$

Note that the field action $S_{\mathcal{F}}$ is independent of the history (states or positions)
$\mathbf{x}^{(\mu)}$ (or $x^{(\mu)}$) of the particle. Detailed discussion of these physical aspects can
be found in [FH], along with alternative ways of expressing them.

The action S for the interactive particle-electromagnetic field system is obtained by combining the component parts,

$$
S \;=\; S_{\mu}+S_{\mu\mathcal{F}}+S_{\mathcal{F}} \;=\; \int_{T}(\mathcal{L}_{\mu}+\mathcal{L}_{\mu\mathcal{F}}+\mathcal{L}_{\mathcal{F}})\,ds. \tag{9.7}
$$

This involves composition or iteration of integrals $\int_{M_{\mathcal{F}}}\cdots$ and $\int_{T}\cdots$ from (9.3),
(9.4), and (9.6).

Domain $\mathbf{M}_{\mathcal{F}}\subseteq\mathbf{R}^{3}$, or $M_{\mathcal{F}}\subseteq\mathbf{R}$, is the region of space (three- or one-
dimensional, respectively) in which the electromagnetic field extends, with points
$\mathbf{y}=(y_{1},y_{2},y_{3})\in\mathbf{R}^{3}$ (or $y=y_{1}\in\mathbf{R}$); and with initial and final states appropriately specified. (Whenever it is more convenient simply take $M_{\mathcal{F}}=\mathbf{R}$ instead
of $\mathbf{M}_{\mathcal{F}}=\mathbf{R}^{3}$. Also, the notations $\mathbf{x}$, x, $x(s)$ are reserved for particle position,
or succession of particle positions. Therefore the superscript (μ) can, without
creating ambiguity, be omitted from the symbol x.)

Use the one-dimensional version of the field for simplicity. (This is not
really meaningful in physics terms, but our concern here is to try to get the gist
of relevant aspects of the physics theory in order to provide some orientation
for the mathematical constructions required for defining and calculating the
relevant integrals.)

For interaction between the charged particle and the field to take place,
assume the particle motion is contained in the domain $M_{\mathcal{F}}$ of the field. So
$x^{(\mu)}(s)=x_{s}\in M_{\mathcal{F}}$ at any time $s\in T$, with $T=\,]\tau',\tau[$ as before. It may also be
convenient to take the domains of both the field and particle motion to be $\mathbf{R}$,
so $M_{\mathcal{F}}=\mathbf{R}$. (Just take any field integrand to be zero in $\mathbf{R}\setminus M_{\mathcal{F}}$; and similarly
for the particle motion.)

Recapitulating, the state of the electromagnetic field at time $s\in T$ is given
by the values of the functions

$$
\mathbf{A}(\mathbf{y},s),\quad \phi(\mathbf{y},s)\quad \text{for}\ \mathbf{y}=\mathbf{y}^{(\mathcal{F})}\in M_{\mathcal{F}}\subseteq\mathbf{R}^{3}.
$$

If the one-dimensional version is used, then the vector $\mathbf{A}$ becomes scalar A. (In this context "vector" means three-dimensional, while "scalar" is one-dimensional. The variable ϕ is a physical scalar.) With this simplification the field state at time s is

$$(A(y,s),\ \phi(y,s))_{y\in\mathbf{R}},\ =\ (A_{ys},\ \phi_{ys})_{y\in\mathbf{R}}\ =\ (A_{\mathbf{R},s},\ \phi_{\mathbf{R},s})\ =\ (A_{\mathbf{R}_{\mathcal{F}},s},\ \phi_{\mathbf{R}_{\mathcal{F}},s}).$$

Each of the values A_{ys} and ϕ_{ys} is a real number, with ranges $M_A \subseteq \mathbf{R}$ and $M_\phi \subseteq \mathbf{R}$, respectively; so

$$A_{ys}\ \in\ M_A\ \subseteq\ \mathbf{R}, \qquad\qquad \phi_{ys}\ \in\ M_\phi\ \subseteq\ \mathbf{R}.$$

As before, it may sometimes be simpler to take

$$M_A\ =\ \mathbf{R}\ =\ \mathbf{R}_A \qquad \text{and} \qquad M_\phi\ =\ \mathbf{R}\ =\ \mathbf{R}_\phi.$$

Thus, in one dimension, a state of the electromagnetic field at time s is

$$\varsigma(s)\ :=\ (A_{ys},\ \phi_{ys})_{y\in\mathbf{R}_{\mathcal{F}}}\quad \text{with}\quad (A_{ys},\ \phi_{ys})\ \in\ \mathbf{R}_A\times\mathbf{R}_\phi \tag{9.8}$$

for each $y \in \mathbf{R}_{\mathcal{F}}$ and each $s \in T$; so

$$\varsigma(s)\ \in\ (M_A\times M_\phi)^{M_{\mathcal{F}}}\quad \text{or}\quad (\mathbf{R}_A\times\mathbf{R}_\phi)^{\mathbf{R}_{\mathcal{F}}} \tag{9.9}$$

for each $s \in T$. A history ς of the electromagnetic field is a succession of states $\varsigma(s)$ for $s \in T$, so

$$\varsigma\ =\ (\varsigma(s))_{s\in T}\ =\ \left((A_{ys},\ \phi_{ys})_{y\in\mathbf{R}_{\mathcal{F}}}\right)_{s\in T}\ \in\ \left((\mathbf{R}_A\times\mathbf{R}_\phi)^{\mathbf{R}_{\mathcal{F}}}\right)^T, \tag{9.10}$$

so the domain (or set) of histories of the electromagnetic field can be written

$$\{\varsigma\}\ =\ \left((\mathbf{R}\times\mathbf{R})^{\mathbf{R}}\right)^T;\quad \text{or}\quad \{\varsigma\}\ =\ \left((\mathbf{R}^3\times\mathbf{R})^{\mathbf{R}^3}\right)^T \tag{9.11}$$

if a three-dimensional version is required; and using indicator functions such as $1_{M_{\mathcal{F}}}$ and $1_{\mathbf{R}_{\mathcal{F}}\setminus M_{\mathcal{F}}}$ for restriction purposes whenever necessary. (The range of values of ϕ_{ys} is one-dimensional since ϕ is scalar in physics.)

Now consider a system consisting of a single charged particle interacting with the electromagnetic field. At time $s \in T$ the state (or position) of the particle is $x(s)$ where $x(s)$ has values inside the domain of the electromagnetic field (so they interact),

$$x(s)\ \in\ M_\mu\ \subseteq\ \mathbf{R}\ =\ \mathbf{R}_\mu, \qquad \text{or}\qquad \mathbf{x}(s)\ \in\ M_\mu\ \subseteq\ \mathbf{R}^3\ =\ \mathbf{R}^3_\mu$$

for three dimensions. In one dimension the joint state of the particle-field system at time s is $\chi(s)\ =\ (x(s), \varsigma(s))$ with

$$\chi(s)\ \in\ M_\mu\times(M_A\times M_\phi)^{M_{\mathcal{F}}}\ \subseteq\ \mathbf{R}_\mu\times(\mathbf{R}_A\times\mathbf{R}_\phi)^{\mathbf{R}_{\mathcal{F}}}\ =\ \mathbf{R}\times(\mathbf{R}\times\mathbf{R})^{\mathbf{R}};$$

and the history of the system is

$$\chi\ =\ (\chi(s))_{s\in T}\ \in\ \left(\mathbf{R}\times(\mathbf{R}\times\mathbf{R})^{\mathbf{R}}\right)^T, \qquad \text{or}\quad \chi\ \in\ \left(\mathbf{R}^3\times(\mathbf{R}^3\times\mathbf{R})^{\mathbf{R}^3}\right)^T \tag{9.12}$$

in three dimensions.

9.2   Constructing the Field Interaction Integral

The Feynman integral over histories (9.2) has integrand which includes a factor $e^{\frac{i}{\hbar}S}$, which has been outlined in (9.7) above. But as yet a meaning for the aggregation or integration procedure

$$\int_\Omega \cdots \mathcal{D}x\,\mathcal{D}A\,\mathcal{D}\phi, \quad \text{or} \quad \int_\Omega \cdots d\chi,$$

has not been fully specified. From (9.12),

$$\Omega = \left(\mathbf{R} \times (\mathbf{R} \times \mathbf{R})^{\mathbf{R}}\right)^T = \left(\mathbf{R}_\mu \times (\mathbf{R}_A \times \mathbf{R}_\phi)^{\mathbf{R}_\mathcal{F}}\right)^T$$

where (for instance) $\mathbf{R}_\mu$ corresponds to the range M_μ of possible values of charged particle $x(t)$ (= x_t) for any $t \in T$; and with subscript μ (or x), A, ϕ serving as a reminder of which variable of the system is involved. From (9.7), a history χ is $\chi = (\chi(t))_{t\in T} =$

$$= \left(x(t), (A(y,t), \phi(y,t))_{y\in\mathbf{R}_\mathcal{F}}\right)_{t\in T}, \quad = \left(x_t, (A_{yt}, \phi_{yt})_{y\in\mathbf{R}_\mathcal{F}}\right)_{t\in T},$$

$$= (x_t, (A_{\mathbf{R}_\mathcal{F},t}, \phi_{\mathbf{R}_\mathcal{F},t}))_{t\in T}, \quad = (x_T, (A_{\mathbf{R}_\mathcal{F},T}, \phi_{\mathbf{R}_\mathcal{F},T})).$$

Let λ signify lagrangian function. For any given history χ the oscillation factor of the integrand in (9.2) is $\omega(\chi) = \exp\left(\frac{i}{\hbar}S(\chi)\right)$ with

$$\begin{aligned}
S(\chi) &= S_\mu + S_{\mu\mathcal{F}} + S_\mathcal{F} \\
&= \int_T \left(\int_{\mathbf{R}_\mathcal{F}} \lambda(\chi(t))\,dy\right) dt; \quad\quad (9.13) \\
&= \int_T \left(\int_{\mathbf{R}_\mathcal{F}} \lambda(x_t, (A_{yt}, \phi_{yt}))\,|\jmath|\right) |\imath|
\end{aligned}$$

where $\imath$ and $\jmath$ denote cells of T and $\mathbf{R}_\mathcal{F}$, respectively, and where the precise form of function λ is determined by the functionals (9.3), (9.4), and (9.6) above. The Feynman integral over histories (9.2) is then

$$\begin{aligned}
\int_\Omega \omega(\chi)\,d\chi, &= \int_\Omega e^{\frac{i}{\hbar}\left(\int_T \left(\int_{\mathbf{R}_\mathcal{F}} \lambda(x_t,(A_{yt},\phi_{yt}))\,dy\right) dt\right)}\,d\chi, \quad\quad (9.14) \\
&= \int_\Omega e^{\frac{i}{\hbar}\left(\int_T \left(\int_{\mathbf{R}_\mathcal{F}} \lambda(x_t,(A_{yt},\phi_{yt}))\,dy\right) dt\right)}\mathcal{D}x\,\mathcal{D}A\,\mathcal{D}\phi
\end{aligned}$$

using only the one-dimensional simplifications of field variables x and A, and with history wave amplitude factor η (corresponding to Feynman's "normalization factor") omitted for the moment.

For the purpose of the physical theory, system histories $\chi \in \Omega$ have initial and final states $\chi(\tau')$ and $\chi(\tau)$, respectively.

To obtain suitable notation for the initial and final states of the interacting matter-field system, using the one-dimensional simplification, denote initial and final states of the electromagnetic field by

$$s'^{(\mathcal{F})} = (\mathcal{A}', \bar{\varphi}'), \quad s^{(\mathcal{F})} = (\mathcal{A}, \bar{\varphi}),$$

respectively, where

$$\mathcal{A}' = (A_{y\tau'})_{y \in R_{\mathcal{F}}}, \qquad \mathcal{A} = (A_{y\tau})_{y \in R_{\mathcal{F}}},$$
$$\varphi' = (\phi_{y\tau'})_{y \in R_{\mathcal{F}}}, \qquad \varphi = (\phi_{y\tau})_{y \in R_{\mathcal{F}}};$$

and where, for $y \in R_{\mathcal{F}}$, each of

$$A_{y\tau'}, \quad A_{y\tau}, \quad \phi_{y\tau'}, \quad \phi_{y\tau}, \tag{9.15}$$

is an element of R_A or R_ϕ. Recalling that the state of the interacting matter-electromagnetic field system at intermediate time t is $\chi(t)$, $(\tau' < t < \tau)$, denote the initial and final states of the interacting $(\mu, \mathcal{F})$ system by, respectively,

$$\boldsymbol{\chi}' := \chi(\tau') = \left(\xi'^{(\mu)}, \boldsymbol{s}'^{(\mathcal{F})}\right), \qquad \boldsymbol{\chi} := \chi(\tau) = \left(\xi^{(\mu)}, \boldsymbol{s}^{(\mathcal{F})}\right). \tag{9.16}$$

Then, provided the -complete integral construction can be carried out, with appropriate cells $\mathbf{I[N]}$ constructed in the manner of Sections 8.11 and 8.12, the wave function for the interaction is

$$\psi(\boldsymbol{\chi}, \tau)) = \int_\Omega \eta(\mathbf{N})\omega(\chi)\,d\chi = \int_\Omega \eta(\mathbf{N})\omega(\chi)\,|\mathbf{I[N]}|, \tag{9.17}$$

where $\omega(\chi) = e^{\frac{i}{\hbar}\left(\int_T \left(\int_{R_{\mathcal{F}}} \lambda(x_t, (A_{yt}, \phi_{yt}))\,dy\right) dt\right)}$. Consider partitions

$$\mathcal{P}_T, \quad \mathcal{P}_{\mathcal{F}}, \quad \mathcal{P}_\mu, \quad \mathcal{P}_A, \quad \mathcal{P}_\phi, \quad \text{of } T, \ \ R_{\mathcal{F}}, \ \ R_\mu, \ \ R_A, \ \ R_\phi, \tag{9.18}$$

respectively; consisting of cells $\{\imath\}$, $\{\jmath\}$, $\{I_x\}$, $\{I_A\}$, $\{I_\phi\}$. Write

$$\mathcal{P}_{A\phi} = \{I_{A\phi}\} = \{I_A \times I_\phi : I_A \in \mathcal{P}_A, \ I_\phi \in \mathcal{P}_\phi\}, \quad \text{with } |I_{A\phi}| = |I_A||I_\phi|,$$

so $\mathcal{P}_{A\phi}$ is a partition of $R_A \times R_\phi$. Again omitting history wave amplitudes $\eta(\mathbf{N})$, a Riemann sum for (9.14) is

$$\sum_{\mathcal{P}_\mu}\left(\sum_{\mathcal{P}_{A\phi}}\left(e^{\frac{i}{\hbar}\left(\Sigma_{\mathcal{P}_T}\left(\Sigma_{\mathcal{P}_{\mathcal{F}}} \lambda(x_t, (A_{yt}, \phi_{yt}))|\jmath|\right)|\imath|\right)}\right)|I_{A\phi}|\right)|I_x|, \tag{9.19}$$

$$\text{or} \ \ \sum_{\mathcal{P}_\mu}\left(\sum_{\mathcal{P}_{A,t}}\left(\sum_{\mathcal{P}_{\phi,t}}\left(e^{\frac{i}{\hbar}\left(\Sigma_{\mathcal{P}_T}\left(\Sigma_{\mathcal{P}_{\mathcal{F}}} \lambda(x_t, (A_{yt}, \phi_{yt}))|\jmath|\right)|\imath|\right)}\right)|I_{\phi yt}|\right)|I_{A yt}|\right)|I_{x_t}|.$$

With this basic interpretation of the integral over histories, we can proceed to define the gauge integral on Ω, and perform simple Riemann sum calculations using regular partitions of Ω. (Regular partitions are the easiest kind.)

In (9.14) and (9.19) two different versions of the system action S, $(= S_\mu + S_{\mu\mathcal{F}} + S_{\mathcal{F}})$, are used in, respectively, integral-over-histories and Riemann-sum-over-histories:

$$\int_T \left(\int_{R_{\mathcal{F}}} \lambda(x_t, (A_{yt}, \phi_{yt}))\,dy\right) dt, \tag{9.20}$$

$$\sum_{\mathcal{P}_T}\left(\sum_{\mathcal{P}_{\mathcal{F}}} \lambda(x_t, (A_{yt}, \phi_{yt}))|\jmath|\right)|\imath|, \tag{9.21}$$

respectively. The first version (with $\int_T \cdots$) is not well defined for the integrands in Ω; while the second version (with Riemann sums $\sum_T \cdots$) corresponds to the sampling sum method used throughout [MTRV], and this book, for integrands in Cartesian product spaces such as Ω. As a reminder of the issues, consider action

$$S_\mu(x_T) \;=\; \int_T \frac{\mu}{2}\left(\frac{dx(t)}{dt}\right)^2 dt, \tag{9.22}$$

$$S_\mu(x_T(N_T)) \;=\; (N_T)\sum \frac{\mu}{2}\left(\frac{x(t_i)-x(t_{i-1})}{t_i-t_{i-1}}\right)^2 (t_i - t_{i-1}), \tag{9.23}$$

where $N_T = \{t_1,\ldots,t_{n-1}\}$ consists of the partition points t_i of a partition $\mathcal{P}_T$ of T. Then $S_\mu(x_T)$ of (9.22) is not always well defined, since x_T is not generally differentiable for $x_T \in \mathbf{R}^T$; whereas the sampling sum $S_\mu(x_T(N_T))$ which appears in (9.23) is always well defined, and always exists.

This corresponds to the issue of stochastic integrals and stochastic sums of Part I of this book. The sampling approach, using finite sets N of partition points, is part of what the theory of -complete integration contributes to these subjects.

Charge-field interaction (9.4) has a factor $\dot{x}$ whose sampling version is rendered as in (9.23). Action $S_{\mathcal{F}}$ in (9.6) has terms $\nabla\phi$ and $\frac{\partial \mathbf{A}}{\partial t}$. In a (non-physical) one-dimensional simplification, these reduce to $\frac{\partial\phi(y,t)}{\partial y}$, $\frac{\partial A(y,t)}{\partial t}$; so, with partitioning points N_T and $N_{\mathcal{F}}$ of T and $\mathbf{R}_{\mathcal{F}}$, respectively, the sampling sum version $S_{\mathcal{F}}(\chi(N_T, N_{\mathcal{F}}))$ will contain expressions

$$\frac{\phi(y_j,t_i)-\phi(y_{j-1},t_i)}{y_j - y_{j-1}}, \qquad \frac{A(y_j,t_i)-A(y_j,t_{i-1})}{t_i - t_{i-1}}, \tag{9.24}$$

with $y_j \in N_{\mathcal{F},t_i}$ for each $t_i \in N_T$. According to [FH], establishing the precise form of action in a physics context is a difficult and subtle task. It is a task of physics, not mathematics. Therefore, other than pointing out the preceding mathematical technique for sampling sum construction, no further effort will be expended here on the details of the lagrangian and action functions (λ and S, respectively) for the -complete version of integration-over-histories.

The action (or S functional) component of (9.2) is

$$\int_T \left(\int_{\mathbf{R}_{\mathcal{F}}} \lambda\left(x_t, (A_{yt}, \phi_{yt})\right) dy\right) dt.$$

This is reformulated in Riemann sum sampling form (9.19) as

$$\sum_{\mathcal{P}_T}\left(\sum_{\mathcal{P}_{\mathcal{F}}} \lambda\left(x_t, (A_{yt}, \phi_{yt})\right)|y|\right)|t|.$$

Thus the integral-over-histories (9.2) becomes

$$\int_\Omega e^{\frac{i}{\hbar}\left(\sum_{\mathcal{P}_T}\left(\sum_{\mathcal{P}_{\mathcal{F}}} \lambda(x_t,(A_{yt},\phi_{yt}))|y|\right)|t|\right)} d\chi \tag{9.25}$$

or $\displaystyle\int\left(\int\left(\int e^{\frac{i}{\hbar}\left(\sum_{\mathcal{P}_T}\left(\sum_{\mathcal{P}_{\mathcal{F}}} \lambda(x_t,(A_{yt},\phi_{yt}))|y|\right)|t|\right)} \boldsymbol{D}\phi\right) \boldsymbol{D}A\right) \boldsymbol{D}x.$

9.3 -Complete Integral Over Histories

The formulation $\int_\Omega \cdots d\chi$ of (9.25) must now be expressed in -complete integral form. Riemann sums for the domain Ω have the form $(\mathcal{D}) \sum h(\chi)|\mathbf{I}|$ where cells $\mathbf{I}$ in Ω are sets of points (or histories) χ, so $\mathbf{I} = \{\chi\} \subset \Omega$. The symbol I will be used for component factor cells of $\mathbf{I}$, with subscripts x, A, and ϕ distinguishing between different factor domains. For any $t \in T$ let

$$I_{x_t} = I_{x_{t_i}} = I_i \in \mathcal{I}(\mathbf{R}_\mu),$$

so I_{x_t} is a one-dimensional cell of $\mathbf{R} = \mathbf{R}_\mu$. Similarly, for any $t \in T$ and any $y \in \mathbf{R}_\mathcal{F}$, let

$$I_{A_{yt}} \in \mathcal{I}(\mathbf{R}_A), \quad I_{\phi_{yt}} \in \mathcal{I}(\mathbf{R}_\phi),$$

denote cells in $\mathbf{R}_A$ and $\mathbf{R}_\phi$, respectively. Let

$$N_T = \{t_1, t_2, \ldots, t_{n-1}\} \in \mathcal{N}(T), \quad t_0 = \tau', \ t_n = \tau,$$

For each $t \in N_T$ let $N_{\mathcal{F},t}$ be a finite subset of $M_\mathcal{F} \subseteq \mathbf{R}_\mathcal{F}$, so $N_{\mathcal{F},t} \in \mathcal{N}(\mathbf{R}_\mathcal{F})$,

$$N_{\mathcal{F},t} = \{y_{1t}, y_{2t}, \ldots, y_{m_t,t}\}, \quad \text{or}$$
$$N_{\mathcal{F},t_i} = \{y_{j,t_i} : i = 1, 2, \ldots, m_i\}, \quad i = 1, 2, \ldots n - 1,$$
$$= \{y_{ji} : i = 1, 2, \ldots, m_i\}.$$

The sets N_T, $N_{\mathcal{F},t}$ will be regarded as partition points of, respectively, T and $M_\mathcal{F}$ (or of $\mathbf{R} = \mathbf{R}_\mathcal{F}$ in the latter case), corresponding[2] to the partitions $\mathcal{P}_T$ and $\mathcal{P}_\mathcal{F}$ of (9.19). For each $t = t_i \in N_T$ and each $N_{\mathcal{F},t} = N_{\mathcal{F},t_i} = N_{\mathcal{F},i}$, let

$$I_A(N_{\mathcal{F},i}) = \prod_{j=1}^{m_i} I_{A_{y_j,t_i}} = \prod_{j=1}^{m_i} I_{A_{ji}},$$

$$I_\phi(N_{\mathcal{F},i}) = \prod_{k=1}^{m_i} I_{\phi_{y_j,t_i}} = \prod_{k=1}^{m_i} I_{\phi_{ji}},$$

$$I_{A\phi}(N_{\mathcal{F},i}) = I_A(N_{\mathcal{F},i}) \times I_\phi(N_{\mathcal{F},i}),$$

$$I_{A\phi}[N_{\mathcal{F},i}] = I_{A\phi}(N_{\mathcal{F},i}) \times \mathbf{R}^{\mathbf{R}_\mathcal{F} \setminus N_{\mathcal{F},i}},$$

$$I_{x_i A\phi} = I_{x_i} \times I_{A\phi}(N_{\mathcal{F},i}),$$

$$I_{xA\phi} = \prod_{i=1}^{n-1} I_{x_i A\phi}$$

$$\mathbf{I}(\mathbf{N}) = I_{xA\phi}, \tag{9.26}$$

$$\mathbf{I}[\mathbf{N}] = \mathbf{I}[N_T : N_\mathcal{F}]$$

$$= \left(I_{x_t} \times \left(I_{A\phi}(N_{\mathcal{F},t}) \times (\mathbf{R}_A \times \mathbf{R}_\phi)^{\mathbf{R}_\mathcal{F} \setminus N_{\mathcal{F},t}}\right)\right)^{N_T} \times$$

$$\times \left(\mathbf{R}_\mu \times (\mathbf{R}_A \times \mathbf{R}_\phi)^{\mathbf{R}_\mathcal{F}}\right)^{T \setminus N_T}.$$

---

[2] $T =]\tau', \tau[$ is bounded interval of time; and it may be convenient (and physically more realistic) to think of $M_\mathcal{F}$ as a closed, bounded subset of the space $\mathbf{R} = \mathbf{R}_\mathcal{F}$.

In other words, with N_T and $(N_{\mathcal{F},t})_{t\in N_T}$ given, a cell $\mathbf{I}[\mathbf{N}] \in \mathcal{I}(\Omega)$ is

$$\left(\prod_{t\in N_T} \left(I_{x_t} \times \left(\prod_{y\in N_{\mathcal{F},t}} \left(I_{A_{yt}} \times I_{\phi_{yt}} \right) \right) \right. \right. \times \tag{9.27}$$

$$\left. \left. \times \left(\prod_{y\in \mathbf{R}_{\mathcal{F}}\setminus N_{\mathcal{F},t}} \mathbf{R}_A \times \mathbf{R}_\phi \right) \right) \right) \times$$

$$\times \left(\prod_{t\in T\setminus N_T} \left(\mathbf{R}_\mu \times \prod_{y\in \mathbf{R}_{\mathcal{F}}} (\mathbf{R}_A \times \mathbf{R}_\phi) \right) \right).$$

The symbols $\mathbf{N}$ and $N_T : N_{\mathcal{F}}$ above are shorthand for the composition of N_T, $N_{\mathcal{F},t}$ (with $t \in N_T$) in the preceding construction.

The next step is to specify association between points χ and cells $\mathbf{I}[\mathbf{N}]$ in the domain Ω. What conditions must χ and $\mathbf{I}[\mathbf{N}]$ satisfy for χ to be a "tag point" of $\mathbf{I}[\mathbf{N}]$?

A point χ of the integration domain Ω is a history of the charge-field system,

$$\chi = \left(x_t, (A_{yt}, \phi_{yt})_{y\in \mathbf{R}_{\mathcal{F}}} \right)_{t\in T} \in \left(\mathbf{R}_\mu \times (\mathbf{R}_A \times \mathbf{R}_\phi)^{\mathbf{R}_{\mathcal{F}}} \right)^T = \Omega;$$

and a cell-point pair $(\chi, \mathbf{I}[\mathbf{N}])$ are *associated* in Ω if the following one-dimensional associations hold:

$$(x_i, I_{x_i}) \text{ are associated in } \mathbf{R}_\mu \text{ for each } t_i \in N_T;$$
$$\left(A_{ji}, I_{A_{ji}} \right) \text{ are associated in } \mathbf{R}_A \text{ for each } t_i \in N_T \text{ and}$$
$$\text{for each } y_j \in N_{\mathcal{F},t_i};$$
$$\left(\phi_{ji}, I_{\phi_{ji}} \right) \text{ are associated in } \mathbf{R}_\phi \text{ for each } t_i \in N_T \text{ and}$$
$$\text{for each } y_j \in N_{\mathcal{F},t_i}. \tag{9.28}$$

Briefly, $(\chi, \mathbf{I}[\mathbf{N}])$ are associated in Ω if the real numbers x_{t_i}, $A_{y_j t_i}$, $\phi_{y_j t_i}$ are vertices of, respectively, the real intervals $I_{x_{t_i}}$, $I_{A_{y_j t_i}}$, $I_{\phi_{y_j t_i}}$.

The next step in constructing the integral-over-histories (9.17), or (9.25), is to define gauges γ on the domain $\Omega = \left(\mathbf{R}_\mu \times (\mathbf{R}_A \times \mathbf{R}_\phi)^{\mathbf{R}_{\mathcal{F}}} \right)^T$. Section 7.7 provides a general principle for producing appropriate gauges, specifically (7.17), (7.18), and (7.19). Theorem 8 shows how existence of γ-fine divisions can be proved in such cases, and Example 42 shows how these methods can be adapted for various products of product spaces. Section 7.4 describes a number of different methods for defining gauges in Cartesian product domains. A gauge γ for $\Omega = \left(\mathbf{R}_\mu \times (\mathbf{R}_A \times \mathbf{R}_\phi)^{\mathbf{R}_{\mathcal{F}}} \right)^T$ can be constructed as follows. For each $\chi \in \bar{\Omega}$, with

$$\chi = \left(x_t, (A_{yt}, \phi_{yt})_{y\in \mathbf{R}_{\mathcal{F}}} \right)_{t\in T} \in \left(\bar{\mathbf{R}}_\mu \times (\bar{\mathbf{R}}_A \times \bar{\mathbf{R}}_\phi)^{\mathbf{R}_{\mathcal{F}}} \right)^T = \bar{\Omega},$$

let $\mathbf{L}_T(\chi)$ be a finite subset of T, so

$$\mathbf{L}_T : \bar{\Omega} \mapsto \mathcal{N}(T).$$

In (9.8), (9.9), (9.10), (9.11), the symbol ς denotes states and histories of the electromagnetic field. At time $t \in T$ a state $\varsigma(t)$ of the electromagnetic field is

$$\varsigma(t) \;=\; (A_{yt}, \, \phi_{yt})_{y \in \mathbf{R}_{\mathcal{F}}} \quad \text{with} \quad (A_{yt}, \, \phi_{yt}) \;\in\; \mathbf{R}_A \times \mathbf{R}_\phi.$$

With t given, for each $\varsigma(t)$ let $L_{\mathcal{F},t}^{A\phi}(\varsigma(t))$ be a finite subset of $\mathbf{R}_{\mathcal{F}}$, so

$$L_{\mathcal{F},t}^{A\phi} \;:\; \left(\bar{\mathbf{R}}_A \times \bar{\mathbf{R}}_\phi \right)^{\mathbf{R}_{\mathcal{F}}} \;\mapsto\; \mathcal{N}(\mathbf{R}_{\mathcal{F}}).$$

For each $t \in N_T \in \mathcal{N}(T)$ and each $y \in N_{\mathcal{F},t} \in \mathcal{N}(\mathbf{R}_{\mathcal{F}})$ (for $t \in N_T$), let

$$\delta_t^{(\mu)}, \qquad \delta_{yt}^{(A)}, \qquad \delta_{yt}^{(\phi)}$$

be gauges in $\mathbf{R}_\mu$, $\mathbf{R}_A$, $\mathbf{R}_\phi$, so that

$$\delta_t^{(\mu)}(x_t) > 0, \qquad \delta_{yt}^{(A)}(A_{yt}) > 0, \qquad \delta_{yt}^{(\phi)}(\phi_{yt}) > 0$$

for $x_t \in \bar{R}_\mu$, $A_{yt} \in \bar{R}_A$, and $\phi_{yt} \in \bar{R}_\phi$, respectively. Then

$$\delta_{yt}^{(A\phi)} \;:=\; \left(\delta_{yt}^{(A)}, \, \delta_{yt}^{(\phi)} \right)$$

is a gauge in $\mathbf{R}_A \times \mathbf{R}_\phi$, with

$$\delta_{yt}^{(A\phi)} \left(A_{yt}, \, \phi_{yt} \right) \;:=\; \left(\delta_{yt}^{(A)}(A_{yt}), \, \delta_{yt}^{(\phi)}(\phi_{yt}) \right) \;\in\; \mathbf{R}_+ \times \mathbf{R}_+;$$

and the associated pair

$$\left((A_{yt}, \phi_{yt}), \, I_{yt}^{(A)} \times I_{yt}^{(\phi)} \right)$$

is $\delta_{yt}^{(A\phi)}$-fine (or $\left(\delta_{yt}^{(A)}, \delta_{yt}^{(\phi)} \right)$-fine) in $\mathbf{R}_{A\phi}, \, = \mathbf{R}_A \times \mathbf{R}_\phi$, if

$$\left(A_{yt}, \, I_{A_{yt}} \right) \qquad \text{and} \qquad \left(\phi_{yt}, \, I_{\phi_{yt}} \right)$$

are, respectively, $\delta_{yt}^{(A)}$-fine and $\delta_{yt}^{(\phi)}$-fine in $\mathbf{R}_A$ and $\mathbf{R}_\phi$.

A gauge γ in Ω is $\gamma = (\mathbf{L}, L, \delta)$; specifically,

$$\gamma \;:=\; \left(\left(\mathbf{L}_T, \left\{ \delta_t^{(\mu)} \right\}_{t \in N_T} \right), \, \left(L_{\mathcal{F},t}^{(A\phi)}, \left\{ \delta_{yt}^{(A\phi)} \right\}_{y \in N_{\mathcal{F},t}} \right)_{t \in N_T} \right) \tag{9.29}$$

for $N_T \in \mathcal{N}(T)$, $N_{\mathcal{F},t} \in \mathcal{N}(\mathcal{F})$ ($t \in N_T$). Suppose $(\chi, \mathbf{I})$ are associated in Ω,

$$\chi \;=\; \left(x_t, (A_{yt}, \phi_{yt})_{y \in \mathbf{R}_{\mathcal{F}}} \right)_{t \in T},$$

$$\mathbf{I} \;=\; \left(\prod_{t \in N_T} \left(I_{x_t} \times \left(\prod_{y \in N_{\mathcal{F}}^t} (I_{A_{yt}} \times I_{\phi_{yt}}) \right) \times \left(\prod_{y \in \mathbf{R}_{\mathcal{F}} \setminus N_{\mathcal{F}}^t} \mathbf{R}_A \times \mathbf{R}_\phi \right) \right) \right) \times$$

$$\times \left(\prod_{t \in T \setminus N_T} \left(\mathbf{R}_\mu \times \prod_{y \in \mathbf{R}_{\mathcal{F}}} (\mathbf{R}_A \times \mathbf{R}_\phi) \right) \right) \tag{9.30}$$

with $N_{\mathcal{F},t}$ written $N_{\mathcal{F}}^t$. Then $(\chi, \mathbf{I})$ are γ-fine if

- the set inclusions

$$N_T \supseteq \mathbf{L}_T(\chi), \qquad N_{\mathcal{F}}^t \supseteq L_{\mathcal{F},t}(\varsigma_t), \qquad (9.31)$$

hold for each χ and ς_t $(t \in N_T)$; and

- the cells I_{x_t}, $I_{A_{yt}}$, and $I_{\phi_{yt}}$ are, respectively,

$$\delta_t^{(\mu)}\text{-fine}, \qquad \delta_{yt}^{(A)}\text{-fine}, \qquad \text{and} \quad \delta_{yt}^{(\phi)}\text{-fine} \qquad (9.32)$$

in $\mathbf{R}_\mu$, $\mathbf{R}_A$, and $\mathbf{R}_\phi$, for each $y \in N_{\mathcal{F},t}$, and all $t \in N_T$.

Now suppose there is a real- or complex-valued integrand function $h(\chi, \mathbf{I})$ defined for associated $(\chi, \mathbf{I}) \in \bar{\Omega} \times \mathcal{I}(\Omega)$. More generally, the integrand can be

$$h(\chi, \mathbf{N}, \mathbf{I}[\mathbf{N}]), \qquad \text{where} \quad (\chi, \mathbf{N}, \mathbf{I}[\mathbf{N}]) \in \bar{\Omega} \times \mathcal{N}(\Omega) \times \mathcal{I}(\Omega),$$

and where each $[\mathbf{N}]$ is a composition $[N_T : N_{\mathbf{R}_{\mathcal{F}}}]$ of finite sets N_T and $N_{\mathbf{R}_{\mathcal{F},t}}$ $(t \in N_T)$, as they appear in (9.26) and (9.30) above.

The -complete integral of h on Ω is a Burkill-type integral, as presented in chapter 4 of [MTRV], and in Section 7.7 above. The conditions (or division system axioms) of chapter 4 are easily verified for this rectangular Cartesian product domain Ω and $\mathcal{I}(\Omega)$, even though the factor domain structure is a bit more complex than usual. (See Example 55 below. Also, the definition of the gauge γ can be modified or adapted as described in Section 7.4.)

The definition of the integral of h in chapter 4 of [MTRV] follows the usual pattern for -complete (or Henstock) integration. A function h is integrable on Ω, with integral $\int_\Omega h = \beta$, if, given $\varepsilon > 0$, there exists a gauge γ such that, for every γ-fine division $\mathcal{D}$ of Ω, the corresponding Riemann sum satisfies

$$\left| (\mathcal{D}) \sum h(\chi, \mathbf{I}[\mathbf{N}], \mathbf{N}) - \beta \right| < \varepsilon. \qquad (9.33)$$

The range of integration theory in chapter 4 of [MTRV] is then available; such as a theory of variation or outer measure, Fubini's theorem, and limit theorems (for "taking limits under the integral sign").

The objective of Part II of this book is to give a precise mathematical meaning to Feynman's integral over histories (9.14) in his theory of quantum electrodynamics:

$$\int_\Omega \omega(\chi)\, d\chi \; = \; \int_\Omega e^{\frac{i}{\hbar}\left(\int_T \left(\int_{\mathbf{R}_{\mathcal{F}}} \lambda(x(s)),\, A(y,s),\, \phi(y,s))\, dy \right) ds \right)} d\chi,$$

where λ is a lagrangian function for the system whose sampling format is given by (9.25). (Any wave amplitude or normalization factor $\eta(\cdot)$ is omitted by Feynman from this integrand; though, at least at the level of intuition, the form of such a factor, and the need for it, have to be borne in mind in developing the theory.)

The Riemann sum inequality (9.33) provides the required mathematical meaning in terms of the integration theory of [MTRV], summarized in Chapter 7 above.

In [MTRV], and in this book, integrands h are generally of the sampling kind. This means that the form of $h(\chi, \mathbf{I}[\mathbf{N}], \mathbf{N})$ in (9.33) is $h(\chi(\mathbf{N}), \mathbf{I}(\mathbf{N}), \mathbf{N})$. In particular, for the scenarios of interest here, sampling function integrand h has the more familiar form of a product of point function f multiplied by a volume function $|\mathbf{I}(\mathbf{N})|$ (or by some other additive volume function $k(\mathbf{I}(\mathbf{N}))$),

$$h(\chi(\mathbf{N}), \mathbf{I}(\mathbf{N}), \mathbf{N}) = f(\chi(\mathbf{N}), \mathbf{N}) |\mathbf{I}(\mathbf{N})|. \tag{9.34}$$

Thus, from (9.30), a sampling integrand depends on a finite number of real values, namely,

$$(A_{yt}, \phi_{yt}) \quad \text{and} \quad x_t$$

for $y \in N_{\mathcal{F},t}$ and $t \in N_T$; along with the corresponding boundary points of associated cells $I_{A_{yt}}$, $I_{\phi_{yt}}$, and I_{x_t}; and also including the (finite number of) elements of $N_{\mathcal{F},t}$ (for $t \in N_T$) and of N_T. Despite the complexity of the integration structures, "finiteness" is inbuilt and can be deployed to advantage in carrying out calculations or estimates of integral values, as in Example 48.

Sampling reduces the complexity of (8.31). If, in addition, the sampling values $\mathbf{N}$ are held constant (so each term of the Riemann sum in (9.33) has the same $\mathbf{N}$), then the integrand is *cylindrical*. Furthermore, if the integrand h has constant value in each cell $\mathbf{I}[\mathbf{N}]$ of some fixed partition of Ω, then h is not just cylindrical, it is a *step function*; and the Riemann sum is easier to calculate.

Cylindrical functions are used in [F1] and [FH]. These functions are useful in reducing infinite-dimensional domains to finite-dimensional ones. Integration theory (including this book and [MTRV]) also uses step functions. See, for instance, the numerical calculation of $\int_\Omega \cdots d\chi$ in Section 9.7.

Example 55 *The Henstock integral of [MTRV], chapter 4, is based on the "axioms" DS1, DS2, ..., DS8 of pages 111–113. In terms of an integral over histories, DS1 and DS2 require the points χ, cells $\mathbf{I}[\mathbf{N}]$, association $(\chi, \mathbf{I}[\mathbf{N}])$, and gauges γ which are defined above for the domain Ω:*

$$\Omega = \left(\mathbf{R} \times (\mathbf{R} \times \mathbf{R})^{\mathbf{R}}\right)^T = \left(\mathbf{R}_\mu \times (\mathbf{R}_A \times \mathbf{R}_\phi)^{\mathbf{R}_{\mathcal{F}}}\right)^{T'}.$$

For DS3 it must be established that, for any γ, there exists a γ-fine division of Ω, and a method for proving this has been demonstrated in Chapter 7. For DS4, given two gauges γ^p as defined by (9.29), $p = 1, 2$, there exists a gauge $\gamma < \gamma^p$ for $p = 1, 2$. In notation of [MTRV] this can be written $\gamma = \gamma^1 \wedge \gamma^2 < \gamma^p$, $p = 1, 2$. For $\gamma^p = (\mathbf{L}^p, L^p, \delta^p)$, let $\mathbf{L} = \mathbf{L}^1 \cap \mathbf{L}^2$, and, for each $t \in \mathbf{L}$ let $L = L^1 \cap L^2$, and $\delta_t = \min\{\delta_t^1, \delta_t^2\}$ (in summary form, with obvious abbreviations). Axiom DS5 follows, essentially, from DS3. Similarly for DS6. DS7 is called the diagonalization axiom: if, for each $\chi \in \Omega$, there is a gauge $\gamma(\chi)$, then there is a gauge γ such that, if $(\chi, \mathbf{I}[\mathbf{N}])$ is γ-fine then $(\chi, \mathbf{I}[\mathbf{N}])$ is $\gamma(\chi)$-fine. This follows from the construction in (9.29). (As demonstrated by the proofs in sections 4.12 and 4.13 of [MTRV] (pages 165–178), diagonalization of gauges is how limit theorems such as the monotone convergence theorem are proved.) DS8 is the joint divisions axiom on which Fubini's theorem is based. This property follows directly from the geometry of the component or factor cells in (9.28), and the fact that any division of Ω contains only a finite number of terms.

9.4   Review of Point-Cell Structure

Section 9.3 provides the basic ingredients for a "-complete" interpretation of integral-over-histories in the interaction of charged particle with electromagnetic field. We now address Riemann sum calculations based on these ingredients.

In preparation, this section reviews the system of points, cells, and gauges presented in Section 9.3.

Variables t and y belong to sets T and $M_{\mathcal{F}}$, respectively; T being the interval of time during which the interaction takes place, and $M_{\mathcal{F}}$ being the region of space in which the electromagnetic field is located. The latter is essentially three-dimensional, but has a simplified presence above as a one-dimensional domain of space, with $M_{\mathcal{F}}$ taken to be the set of all real numbers $y \in M_{\mathcal{F}} = \mathbf{R}_{\mathcal{F}} = \mathbf{R}$.

The charged particle μ in motion occupies successive point-locations $x_t^{(\mu)}$, $= x_t$, at successive times $t \in T$. It is assumed to be interacting with the electromagnetic field $\mathcal{F}$, so $x_t = y \in M_\mu$, the latter being, physically, a proper subset of the region $M_{\mathcal{F}}$ of space in which the field $\mathcal{F}$ extends. But, again for simplicity, take $M_\mu = \mathbf{R}_\mu = \mathbf{R}$; even though this makes domains $\mathbf{R}_\mu$ and $\mathbf{R}_{\mathcal{F}}$ identical—an identification which is "non-physical".

The electromagnetic field extends through a region $M_{\mathcal{F}}$ of space. The electric current passes through that region, but does not "fill" the space occupied by the field. The particle (or electrical current) component S_μ is generated in M_μ, as is the current-field "excitation" component $S_{\mu\mathcal{F}}$; see (8.48).

In dealing with the particle component of the interactive system, the history of the particle μ requires point-cell pairs $\left(x^{(\mu)}, I^{(\mu)} \right)$ with $x^{(\mu)} = x_T = (x_t)_{t\in T}$, which is a succession of values y in $\mathbf{R}_\mu \cap \mathbf{R}_{\mathcal{F}}$; and

$$I_t^{(\mu)} \subset \mathbf{R}_\mu, \qquad I^{(\mu)}(N_T) \subset \mathbf{R}_\mu^{N_T}, \qquad I^{(\mu)} = I^{(\mu)}[N_T] \subset \mathbf{R}_\mu^T;$$

with gauge $\gamma^{(\mu)} = \left(L^{(\mu)}, \left(\delta_t^{(\mu)} \right)_{t\in N_T} \right)$, $N_T \supseteq L^{(\mu)} \left(x_T^{(\mu)} \right)$, and $I_t^{(\mu)}$ being $\delta_t^{(\mu)} (x_t)$-fine for $t \in N_T$. Gauge $\gamma^{(\mu)}$ is absorbed into the gauge $\gamma = (\mathbf{L}, L, \delta)$ defined by (9.29) above and is implicit in the latter.

In association with particle history $x_T^{(\mu)}$, there is electromagnetic field history $\varsigma, = \varsigma_{\mathcal{F},T}$, given by (9.9), (9.10), and (9.11), with

$$\varsigma = (\varsigma(t))_{t\in T} = \left((A_{yt}, \phi_{yt})_{y\in\mathbf{R}_{\mathcal{F}}} \right)_{t\in T} \in \left((M_A \times M_\phi)^{M_{\mathcal{F}}} \right)^T ; \quad \text{or simply}$$

$\varsigma \in \left((\mathbf{R}_A \times \mathbf{R}_\phi)^{\mathbf{R}_{\mathcal{F}}} \right)^T$. The history of the interacting charge-field system is $\chi =$

$$= (\chi(t))_{t\in T} = (x_t, \varsigma_t)_{t\in T} = \left(x_t, \left((A_{yt}, \phi_{yt})_{y\in\mathbf{R}_{\mathcal{F}}} \right) \right)_{t\in T}$$

$\in \left(M_\mu \times (M_A \times M_\phi)^{M_{\mathcal{F}}} \right)^T$; or simply $\chi \in \Omega = \left(\mathbf{R}_\mu \times (\mathbf{R}_A \times \mathbf{R}_\phi)^{\mathbf{R}_{\mathcal{F}}} \right)^T$ with the usual simplifications.

Each of the interacting system variables x_t, A_{yt}, and ϕ_{yt} is allowed to vary independently in the integral over histories. The latter two can be regarded as

$$\varsigma(y,t) \;=\; (A(y,t),\, \phi(y,t)) \;=\; (A_{yt},\, \phi_{yt}),$$

in specifying the electromagnetic field $\mathcal{F}$ throughout the region $M_{\mathcal{F}}$ (or $\mathbf{R}_{\mathcal{F}}$) for duration T of time ($y \in M_{\mathcal{F}}$, $t \in T$). Time t links all three variables. Location y and time t link the two variables in $\varsigma(y,t)$. Interaction variable x_t is itself a y-value, but—physically—having a much smaller range of values ($x_t \in M_\mu$, $M_\mu \subset M_{\mathcal{F}}$, $y \in M_{\mathcal{F}}$).

The variability of the numbers A_{yt} and ϕ_{yt} in the integral over histories $\int_\Omega \cdots$ means that, for each given $t \in T$, and for all $y \in M_{\mathcal{F}}$ (or $\mathbf{R}_{\mathcal{F}}$), their domains M_A (or $\mathbf{R}_A$) and M_ϕ (or $\mathbf{R}_\phi$) must be partitioned by intervals $I_{A_{yt}}$ and $I_{\phi_{yt}}$, respectively.

Sets $M_{\mathcal{F}}$, M_A, and M_ϕ relate to the variable values y, A, and ϕ, respectively; and, in principle, are three distinct sets of real numbers. Each of the numbers y, A, and ϕ is regarded above as belonging to a continuum, so $M_{\mathcal{F}}$, M_A, and M_ϕ can be thought of as real intervals—or simply as $\mathbf{R}$, in which case the three variable domains are treated here as coinciding.

In finance, time ticks and basis points are discrete. If variable values in quantum mechanics and field theory are discrete, then this too has implications for integrals over histories. (In principle they should be easier to analyse. But this is a question of physics, not mathematics.)

The oscillation function $w(\chi) = \exp\left(\frac{\iota}{\hbar}S\right)$ depends on the action functional $S(\chi)$; which, in turn, involves integrals $\int_T \cdots dt$ and $\int_{\mathbf{R}_{\mathcal{F}}} \cdots dy$. (These integrals are replaced by sampling sums (9.25) in the -complete theory.)

The elements t and y are "labels" for dimensions (of Cartesian product domain Ω) in $\int_\Omega w(\chi)\,d\chi$. But in the construction of the integrand $w(\chi)$ (for any given χ), the variables t and y are integration/Riemann sum variables for the functional $S(\chi)$. Thus the numbers y and t have several different roles in the integral over histories.

9.5   Calculating Integral Over Histories

Section 9.3 provides the technical basis for integrals-over-histories on domain Ω using the -complete system—the theory of integration set out in chapter 4 of [MTRV]. Here are some questions:

- Are there particular functions or integrands to which the theory of Section 9.3 is applicable?

- If there are such, how in practice can such integrals be used?

- While Section 9.3 provides a definition, can such an integral be evaluated?

- As a well defined mathematical entity (regardless of whether it can be evaluated as a real or complex number), can such an integral be used as an element of some larger physical scenario or theory?

As discussed in Section 9.1, the Feynman theory of the quantum phenomena in the interaction of electric current with an electromagnetic field involves the integral (9.2) or (9.14), $\int_\Omega e^{\frac{\iota}{\hbar} S(\chi)} d\chi$, =

$$= \int_\Omega \exp\left(\frac{\iota}{\hbar}\left(S_\mu(x) + S_{\mu\mathcal{F}}(x,(A,\phi)) + S_\mathcal{F}(A,\phi)\right)\right) \mathcal{D}x\,\mathcal{D}A\,\mathcal{D}\phi, \qquad (9.35)$$

where $\Omega = \left(\mathbf{R}_\mu \times (\mathbf{R}_A \times \mathbf{R}_\phi)^{\mathbf{R}_\mathcal{F}}\right)^T$. In the -complete system of integration, that construction involves Riemann sums[3]

$$(\mathcal{D}) \sum e^{\frac{\iota}{\hbar} S(\chi)} |\mathbf{I}| = (\mathcal{D}) \sum \exp\left(\frac{\iota}{\hbar}\left(S_\mu + S_{\mu\mathcal{F}} + S_\mathcal{F}\right)\right) |\mathbf{I}[\mathbf{N}]| \qquad (9.36)$$

where $\mathcal{D}$ is a γ-fine division of Ω; and where history action wave amplitudes $\eta(\mathbf{N})$ have been omitted for the moment.

The existence of such a γ-fine division $\mathcal{D}$ for any gauge γ is the foundation on which this approach to integration works. The definition of gauge and γ-fineness in (9.29), (9.31), and (9.32) differs from that in theorem 4 of [MTRV], and from the versions in Section 7.7 above. But the differences in the latter case are superficial, and the proof of existence of γ-divisions differs only superficially in this case, and can profitably be undertaken by the reader.

The definition of the integral on Ω is based on the general definition of the Henstock integral given in chapter 4 of [MTRV]. Suppose h is a function of associated point-cell pairs in Ω, $h = h(\chi, \mathbf{I})$. Inherent in the meaning of

$$\mathbf{I} = \mathbf{I}[\mathbf{N}] \subset \Omega = \left(\mathbf{R}_\mu \times (\mathbf{R}_A \times \mathbf{R}_\phi)^{\mathbf{R}_\mathcal{F}}\right)^T$$

is the role of sampling dimension sets $N_T \in \mathcal{N}(T)$ and $N_\mathcal{F} \in \mathcal{N}(\mathbf{R}_\mathcal{F})$. A relationship between these sets is indicated by notation

$$\mathbf{I} = \mathbf{I}[\mathbf{N}] = \mathbf{I}[N_T : N_\mathcal{F}],$$

where $\mathbf{N} = N_T : N_\mathcal{F}$ is a composition of sampling points as expressed in (9.28). An integrand h will generally contain explicitly elements $t \in T$ and $y \in \mathbf{R}_\mathcal{F}$, so integrand h in Ω can be written

$$h = h(\chi, \mathbf{I}) = h(\chi, N_T, N_\mathcal{F}, \mathbf{I}[N_T : N_\mathcal{F}]) = h(\chi, \mathbf{N}, \mathbf{I}[\mathbf{N}]).$$

Example 56 *The -complete structure for integral over histories is rather complicated. There are various ways in which it can be approached. As an assessment of the concepts, consider a point-sampling interpretation of the construction of Riemann sums for -complete integration—an interpretation which makes no explicit reference to intervals or cells.*

Suppose the domain of integration is a real interval $[a, b]$, with associated point-cell pairs (x, I) in which x is a vertex of I; and suppose $h(x, I)$ is an

---

[3] As already pointed out, analogy with the single-particle system of Section 8.2 prompts the conjecture that the integrand should also include a factor $\eta(\mathbf{N})$ corresponding to $\alpha(N)$ in (8.17), and to Feynman's "normalizing factor" A in (8.10).

integrand. Suppose $\delta(x) > 0$ is a gauge defined for $a \leq x \leq b$. If a finite set $\mathcal{P}$ of numbers x is chosen, then $\mathcal{P}$ can be regarded as a sample of points of the domain $[a, b]$,

$$\mathcal{P} = \{x_1, x_2, \ldots, x_n\}, \qquad a \leq x_1 < x_2 < \cdots < x_{n-1} < x_n \leq b,$$

giving a partition of $[a, b]$ whose elements are vertices of cells I in $[a, b]$:

$$[a, x_1], \quad]x_1, x_2], \quad]x_2, x_3], \ldots,]x_{n-2}, x_{n-1}], \quad]x_{n-1}, x_n], \quad]x_n, b].$$

(If $a = x_1$ and/or $b = x_n$, then remove the empty cell(s).) In -complete integration each cell $]x_{j-1}, x_j]$ requires further choice of tag point (or associated point) $\bar{x}_j$; involving selection of another finite sample of points $\{\bar{x}_j : j = 1, \ldots, n\}$, with $x_{j-1} \leq \bar{x}_j \leq x_j$ (—in this book the choice is limited to $\bar{x}_j = x_{j-1}$ or $\bar{x}_j = x_j$) resulting in a linked pair of finite samples

$$\mathcal{P} = \{x_1, \ldots, x_n\}, \qquad \bar{\mathcal{P}} = \{\bar{x}_1, \ldots, \bar{x}_n\},$$

called a division of the domain $[a, b]$. Given a gauge $\delta(x) > 0$ $(a \leq x \leq b)$, a division $\mathcal{D} := (\mathcal{P}, \bar{\mathcal{P}})$ is δ-fine if

$$x_j - x_{j-1} < \delta(\bar{x}_j), \qquad j = 1, 2, \ldots, n;$$

Existence of such δ-fine divisions $\mathcal{D}$, $= (\mathcal{P}, \bar{\mathcal{P}})$, is the basis of -complete integration; and the integration can be expressed in terms of finite samples of points without explicit reference to cells or intervals.

This provides an alternative way of setting up -complete integrals, referring only to finite samples of points in the domain. The division system axioms DS1 to DS8 in chapter 4 of [MTRV] are expressed in terms of cells or intervals. But if the domain of integration Ω consists of (products of) $\mathbf{R}$, or of real interval subsets of $\mathbf{R}$, then cells (finite- or infinite-dimensional), can be replaced by points, specified as (finite or infinite 'tuples of) real numbers. In other words -complete integration can be done without mentioning cells, only gauges and finite samplings of points.

In the case of $\Omega = \mathbf{R}^T$ with T infinite (an interval of time, say), then the definition of the -complete integral requires further sampling of times $N \subset T$. That is to say, a Riemann sum for the -complete integral of a function h in $\mathbf{R}^T$ involves a finite-sample-triple

$$\mathcal{D} := \left(\mathcal{P}, \bar{\mathcal{P}}, \mathcal{P}_N \right)$$

from $\Omega \times \Omega \times T$; where $\mathcal{P}_N$ is a finite collection of finite samples N. If integrand h depends on cells I, these need not be explicitly mentioned since they have alternative specification in terms of finite samples of points $x \in \mathcal{P}$ and $t \in N$; so a Riemann sum of integrand h can be expressed in terms of $\mathcal{D} = \left(\mathcal{P}, \bar{\mathcal{P}}, \mathcal{P}_N \right)$,

$$(\mathcal{D})h(\bar{x}, N, I[N]) = (\mathcal{P}, \bar{\mathcal{P}}, \mathcal{P}_N) \sum h(\bar{x}, N, (x)),$$

with each cell I delimited by points (x).

Example 7 in Section 3.1 describes an integrand which is independent of cells and consists only of points.[4] Section 9.7 below uses finite samples of points to

---

[4]This rather strange integrand is re-visited in Example 58.

evaluate an integral of the kind that arises in chapter 9 of [FH]:

$$\int_\Omega \exp\left(\frac{\iota}{\hbar}\left(S_\mu(x) + S_{\mu\mathcal{F}}(x,(A,\phi)) + S_{\mathcal{F}}(A,\phi)\right)\right) \mathcal{D}x\,\mathcal{D}A\,\mathcal{D}\phi, \quad \text{or} \quad \int_\Omega f(\chi)|\mathbf{I}|.$$

The Riemann sum inequality (9.33) gives meaning or definition to the integral over histories $\int_\Omega f(\chi)|\mathbf{I}|$, (from which a conjectured integrand factor $\eta(\mathbf{N})$ is omitted). But definition (9.33) does not establish existence of the integral, nor, subject to integral evaluation, its value as a known complex number β. Given initial and final conditions or states $\chi_{\tau'}, = \mathbf{x'}$, and $\chi_\tau, = \mathbf{x}$, as in (9.16), does $\int_\Omega f(\chi)|\mathbf{I}|$ actually exist? And if it does exist, what is its value? Are there any practical methods for evaluating integrals of this kind? Apart from that, are there any other useful consequences of definition (9.33)?

Like any other integral, $\int_\Omega f(\chi)|\mathbf{I}|$ probably does not generally exist unless some restrictions, conditions, or hypotheses are imposed on integrand f. In the case of a single particle subject to conservative mechanical force, the path integral is

$$\psi(\xi,\tau) \;=\; \int_{\mathbf{R}^T} \exp\left(\frac{\iota}{\hbar}S_\mu(x(N))\right)|I[N]|$$

of Section 8.2 above. As in (8.10), it was shown in [MTRV] that this integral requires an additional factor $\alpha(N)$, which Feynman called the *normalization factor* in [F1] and in [FH]; and which, in Section 8.2 above, is called (hypothetically) the *amplitude* for the action wave of a path.

Whether or not the latter conjecture is physically realistic, the comparison with the path integral suggests that some factor $\eta(\mathbf{N})$, $= \eta(N_T : N_{\mathcal{F},T})$ (corresponding to $\alpha(N)$), may be required for the integral over histories to exist as wave function

$$\psi(\mathbf{x},\tau) = \int_\Omega \eta(\mathbf{N})\exp\left(\frac{\iota}{\hbar}S(\chi(\mathbf{N}))\right)|\mathbf{I}|, \;=\; \int_\Omega \eta(\mathbf{N})\,\omega(\chi(\mathbf{N}))\,|\mathbf{I}[\mathbf{N}]|. \quad (9.37)$$

The analogy from Section 8.2 then suggests that any such "normalization factor" $\eta(\mathbf{N})$ should be called (conjecturally) an *amplitude for a history action wave*, where the *oscillation* $\omega(\chi)$ (or $\omega(\chi(\mathbf{N}))$) for the wave is $\exp\left(\frac{\iota}{\hbar}S(\chi(\mathbf{N}))\right)$.

Just like more familiar integrals, a step function approximation to (9.37) can be undertaken, in which the integrand $f(\chi(\mathbf{N}),\mathbf{N})$, $= \eta(\mathbf{N})\,\omega(\chi(\mathbf{N}))$, is assigned a constant value in cells $\mathbf{I}$ of a fixed partition $\mathcal{P}$ of Ω. This is done in Section 9.7 below.

Constant functions are even easier to integrate than step functions. For instance, if $f(\chi) = 0$ for all $\chi \in \Omega$, then, for every division $\mathcal{D}$ of Ω, $(\mathcal{D})\sum f(\chi)|\mathbf{I}| = 0$, and $\int_\Omega f(\chi)|\mathbf{I}| = 0$. If $f(\chi)$ equals a non-zero constant β for all $\chi \in \Omega$, then, for every division $\mathcal{D}$ of Ω,

$$(\mathcal{D})\sum f(\chi)|\mathbf{I}| \;=\; \beta \sum |\mathbf{I}|.$$

If each of the component domains $M_{\mathcal{F}}$, M_μ, M_A, and M_ϕ is the unit interval $]0,1[$, it is fairly easy to show that $\int_\Omega f(\chi)\,|\mathbf{I}|$ exists and equals β. But if any of M_μ, M_A, or M_ϕ is unbounded, then the Riemann sums diverge.

On the other hand, in Section 8.2 above, instead of volume function $|I[N]|$, integrator function $G(I[N])$ takes finite values in the unbounded domain $\mathbf{R} \times \mathbf{R} \times \cdots$. (See Example 11 above, and also chapter 7 of [MTRV]. The exponential $e^{\iota y^2}$ in G and g ensures vanishingly small contributions for large y, and this is reinforced when y^2 is divided by the very small quantity $\hbar$.)

Even though $\mathbf{R} = {]-\infty, \infty[}$ has been used for convenience, the variability of x, A, ϕ is small.

9.6 Integration of a Step Function

A definition of Feynman's integral over histories (9.35) is provided in (9.33). The latter uses Riemann sums over divisions $\mathcal{D} = \{(\chi, \mathbf{I})\}$ of the domain Ω of system histories χ.

The good thing about a Riemann sum is that it amounts to just adding up a finite number of terms. The bad thing in this case is that the domain Ω is a structured product of product domains—a somewhat complicated structure as it happens.

In any case it is desirable at this stage to actually do a Riemann sum calculation for Feynman's integral-over-histories. While the complicated parts of (9.33) are needed for -complete integration theory, the method itself can be demonstrated without too much difficulty when certain simplifications are applied (in addition to the simplifications already made earlier).

As suggested in Section 9.5, and in Examples 44 and 47, regular partitions are easier. Also, gauge restrictions can be omitted,[5] and simpler versions of the component or factor domains can be used. Take each of

$$T, \quad \mathbf{R}_\mu, \quad \mathbf{R}_{\mathcal{F}}, \quad \mathbf{R}_A, \quad \mathbf{R}_\phi$$

to be the unit interval $]0, 1[$, denoted by R. Thus initial and final times τ', τ are 0 and 1, respectively. Denote the resulting space by

$$\Upsilon = \left(R \times (R \times R)^R\right)^R, \text{ to avoid confusion with } \Omega = \left(\mathbf{R}_\mu \times (\mathbf{R}_A \times \mathbf{R}_\phi)^{\mathbf{R}_{\mathcal{F}}}\right)^T.$$

Whether $\mathbf{I}$ is a cell of Υ or of Ω, we continue to use notation $I_t^{(\mu)}$ or I_{x_t}, $I_{yt}^{(A)}$ or $I_{A_{yt}}$, and $I_{yt}^{(\phi)}$ or $I_{\phi_{yt}}$, to denote the factor cells of $\mathbf{I}$.

To illustrate Riemann sum construction, each of the factor domains R of Υ could be assigned two partition points, $\frac{1}{3}$ and $\frac{2}{3}$, giving the same partition for each domain R:

$$]0, 1[= \left]0, \frac{1}{3}\right] \cup \left]\frac{1}{3}, \frac{2}{3}\right] \cup \left]\frac{2}{3}, 1\right[. \tag{9.38}$$

---

[5]When integrating step functions, gauges may be needed to ensure conformance; see Example 45.

So, using the notation of (9.30), a cell in R corresponding to I_{x_t}, $I_{A_{yt}}$, or $I_{\phi_{yt}}$ of domains $\mathbf{R}_\mu$, $\mathbf{R}_A$, or $\mathbf{R}_\phi$, respectively, consists of one of

$$\left]0, \frac{1}{3}\right] \quad \text{or} \quad \left]\frac{1}{3}, \frac{2}{3}\right] \quad \text{or} \quad \left]\frac{2}{3}, 1\right[.$$

In the case of T ($= R$) and $\mathbf{R}_{\mathcal{F}}$ ($= R$), the partition points are the sampling points

$$N_T = \left\{\frac{1}{3}, \frac{2}{3}\right\}, \qquad N_{\mathcal{F},\frac{1}{3}} = \left\{\frac{1}{3}, \frac{2}{3}\right\}, \qquad N_{\mathcal{F},\frac{2}{3}} = \left\{\frac{1}{3}, \frac{2}{3}\right\}.$$

With this kind of partitioning, (9.38) gives a regular partition $\mathcal{P}$ of Υ.

For any given history $\chi \in \Omega$ (or Υ), the sampling points

$$\mathbf{N} = N_T : N_{\mathcal{F},T} = \left(N_T, (N_{\mathcal{F},t})_{t \in N_T}\right) \tag{9.39}$$

are used as in (9.21), (9.23), and (9.24) to calculate a sampling version of the action S of the interaction of charged particle and field; with $|\imath|$, $|\jmath|$ replaced by expressions $t_j - t_{j-1}$ ($t_j \in N_T$) and $y_p - y_{p-1}$ for $y_p \in N_{\mathcal{F},t}$ and $t \in N_T$.

The details of the calculation of $S(\chi(\mathbf{N}))$ in $\omega(\chi(\mathbf{N}))$ are omitted here, since the exact details of the action functional S have not been specified in Section 9.2 above, which provides only an outline or sketch of some aspects of the physics theory in [FH].

Nevertheless in Υ, or in Ω, the calculation of the sampling version $S(\chi(\mathbf{N}))$ involves just a finite number of arithmetical steps as indicated in (9.23), and (9.24). So for elements t and y of (9.39), the amount of arithmetic in the calculation of $\omega(\chi(\mathbf{N}))$ is small, especially for regular partitions $\mathcal{P}$.

In the domain Υ replace the integrand h by $\omega(\chi(\mathbf{N}))\,|\mathbf{I}|$ in definition (9.33) of the -complete integral, where $\mathbf{I}(\mathbf{N})$ is

$$\prod_{t \in N_T}\left(I_{x_t} \times \left(\prod_{y \in N_{\mathcal{F},t}}\left(I_{A_{yt}} \times I_{\phi_{yt}}\right)\right)\right) \tag{9.40}$$

—as in (9.30). Cells $\mathbf{I}[\mathbf{N}]$ have unrestricted components of the form $R = {]}0,1{[}$ corresponding to $t \in {]}0,1{[}\backslash N_T$ and to $y \in {]}0,1{[}\backslash N_{\mathcal{F},t}$. These are omitted since, for divisions $\mathcal{D}$ of Υ, components $R = {]}0,1{[}$ do not affect the result of the multiplication of factors

$$|\mathbf{I}| = |\mathbf{I}[\mathbf{N}]| = |\mathbf{I}(\mathbf{N})| = \prod_{t \in N_T}\left(|I_{x_t}|\left(\prod_{y \in N_{\mathcal{F},t}}\left(|I_{A_{yt}}|\,|I_{\phi_{yt}}|\right)\right)\right) \tag{9.41}$$

in the calculation of the Riemann sum $(\mathcal{D})\sum\omega(\chi(\mathbf{N})|\mathbf{I}(\mathbf{N})|$. In order to get a more explicit display of the "fractions" in (9.41), here is a less concise representation of (9.40). First, write N_T and $N_{\mathcal{F},t}$ in expanded form:

$$
\begin{aligned}
N_T &= \{t_1, \ldots, t_{n-1}\}, \\
N_{\mathcal{F},t_1} &= \{y_{1,t_1}, \ldots, y_{m_1,t_1}\}, \\
N_{\mathcal{F},t_2} &= \{y_{1,t_2}, \ldots, y_{m_2,t_2}\}, \\
&\;\;\vdots \qquad \vdots \\
N_{\mathcal{F},t_{n-1}} &= \{y_{1,t_{n-1}}, \ldots, y_{m_{n-1},t_{n-1}}\}.
\end{aligned}
$$

The corresponding cells are

$$
\begin{aligned}
\mathbf{I}(\mathbf{N}) \;=\; &\left(I_{t_1}^{(\mu)} \times \cdots \times I_{t_{n-1}}^{(\mu)}\right) \times \\
&\times \left(\left(I_{1,t_1}^{(A)} \times I_{1,t_1}^{(\phi)}\right) \times \cdots \times \left(I_{m_1,t_1}^{(A)} \times I_{m_1,t_1}^{(\phi)}\right)\right. \times \\
&\times \left(I_{1,t_2}^{(A)} \times I_{1,t_2}^{(\phi)}\right) \times \cdots \times \left(I_{m_2,t_2}^{(A)} \times I_{m_2,t_2}^{(\phi)}\right) \times \\
&\qquad\qquad \vdots \\
&\times \left. \left(I_{1,t_{n-1}}^{(A)} \times I_{1,t_{n-1}}^{(\phi)}\right) \times \cdots \times \left(I_{m_{n-1},t_{n-1}}^{(A)} \times I_{m_{n-1},t_{n-1}}^{(\phi)}\right)\right) \\
\;=\; &\prod_{j=1}^{n-1} \left(I_{t_j}^{(\mu)} \times \left(I_{1,t_j}^{(A)} \times I_{1,t_j}^{(\phi)}\right) \times \cdots \times \left(I_{m_j,t_j}^{(A)} \times I_{m_j,t_j}^{(\phi)}\right)\right). \qquad (9.42)
\end{aligned}
$$

Then $\bigcup\{\mathbf{I}[\mathbf{N}] : \mathbf{I}[\mathbf{N}] \in \mathcal{P}\} = \Upsilon$ when $\mathcal{P} = \{\mathbf{I}[\mathbf{N}]\}$ is a partition (regular or otherwise) of Υ; and, by combining the fractions in (9.41),

$$
(\mathcal{P}) \sum |\mathbf{I}[\mathbf{N}]| \;=\; 1.
$$

For the simplified integral-over-histories, the oscillation factor of the integrand in sampling form is $\omega(\chi(\mathbf{N})) \;=\; \omega(\chi(N_T : N_{\mathcal{F},T})) \;=\;$

$$
\;=\; \omega(\chi(N_T, N_{\mathcal{F},t_1}, \ldots, N_{\mathcal{F},t_{n-1}}; \mathbf{x}', \mathbf{x}; \tau', \tau)).
$$

with $t_0 = \tau'$ and $t_n = \tau$. This includes dependence on parameters $\mathbf{x}'$ and $\mathbf{x}$, which are the initial and final states of the system, specified by values for

$$
x(\tau'), \;\; x(\tau), \;\; (A(y,\tau'), \;\; \phi(y,\tau'))_{y \in R_{\mathcal{F}}}, \;\; (A(y,\tau), \;\; \phi(y,\tau))_{y \in R_{\mathcal{F}}},
$$

as described in (9.16).

In sampling form, $\omega(\chi(\mathbf{N}))$ is obtained from the action function $S(\chi)$ for system histories χ by the method indicated in (9.23) and (9.24). As explained in chapter 9 of [FH], the details of the functional S are determined by the physics of the interacting system.

This establishes the cells $\mathbf{I}$ of integrator function $|\mathbf{I}|$, along with the parameters and calculations needed to determine the required values of point-function integrand $\omega(\chi(\mathbf{N}))$. Then, for any division $\mathcal{D} = \{(\chi, \mathbf{I})\}$ of Υ, the Riemann sum estimate of the integral over histories is $(\mathcal{D}) \sum \omega(\chi(\mathbf{N}))|\mathbf{I}[\mathbf{N}]|$.

9.7   Regular Partition Calculation

A relatively easy way to display $(\mathcal{D}) \sum \omega(\chi(\mathbf{N}))|\mathbf{I}[\mathbf{N}]|$ explicitly is to use a regular partition $\mathcal{P}_\Upsilon$ of the simplified domain Υ consisting of factor domains

$$]0, 1[\;=\; T \;=\; R_{\mathcal{F}} \;=\; R_\mu \;=\; R_A \;=\; R_\phi;$$
$$\Upsilon \;=\; \left( ]0, 1[\times (]0, 1[\times]0, 1[)^{]0, 1[} \right)^{]0, 1[}.$$

with initial and final times $\tau' = 0$ and $\tau = 1$; and with sets of real numbers specifying initial and final system states $\boldsymbol{\chi}'$ and $\boldsymbol{\chi}$. Using domains $]0, 1[$ instead of $]-\infty, \infty[$ is not as extreme an over-simplification as it might seem. Intuitively, the variability in field parameters and particle positions amounts to the most minute of tremors. Also, the Fresnel factors $e^{\frac{i}{\hbar} y^2}$ in the integrand ensures negligible contributions outside a small range of values.[6]

For simplicity use only a single, fixed sample point $\frac{1}{2}$ in each of five factor domains, so (9.42) reduces to

$$N_T \;=\; \left\{ \frac{1}{2} \right\}, \qquad N_{\mathcal{F}, \frac{1}{2}} \;=\; \left\{ \frac{1}{2} \right\}, \qquad \mathbf{I}(\mathbf{N}) \;=\; I_{\frac{1}{2}}^{(\mu)} \times \left(I_{A_{\frac{1}{2}, \frac{1}{2}}} \times I_{\phi_{\frac{1}{2}, \frac{1}{2}}} \right),$$

where each of $I_{\frac{1}{2}}^{(\mu)}$, $I_{A_{\frac{1}{2}, \frac{1}{2}}}$, $I_{\phi_{\frac{1}{2}, \frac{1}{2}}}$ is one of

$$\left] 0, \frac{1}{2} \right], \qquad \left] \frac{1}{2}, 1 \right[;$$

giving $2^3 = 8$ different cells $\mathbf{I}[\mathbf{N}]$ composing a regular partition $\mathcal{P}_\Upsilon$ of Υ, with each cell having volume

$$|\mathbf{I}[\mathbf{N}]| \;=\; |\mathbf{I}(\mathbf{N})| \;=\; \left| I_{\frac{1}{2}}^{(\mu)} \right| \left| I_{A_{\frac{1}{2}, \frac{1}{2}}} \right| \left| I_{\phi_{\frac{1}{2}, \frac{1}{2}}} \right| \;=\; \frac{1}{8}.$$

Accordingly, for the Riemann sum $(\mathcal{P}_\Upsilon) \sum \omega(\chi(\mathbf{N}))|\mathbf{I}[\mathbf{N}]|$ over the regular partition $\mathcal{P}_\Upsilon$, there are 8 sample histories $\chi(\mathbf{N})$ to be chosen; one for each cell $|\mathbf{I}[\mathbf{N}]|$; with corresponding values $\omega(\chi(\mathbf{N}))$ to be calculated in order to evaluate the Riemann sum.

Note that $\mathcal{P}_\Upsilon$ is not required to be γ-fine at this stage; so it is sufficient, for instance, that each sample history $\chi(\mathbf{N})$ be some easily determined internal point of $\mathbf{I}(\mathbf{N})$. In this case the only "random" or variable co-ordinate of $\chi(\mathbf{N})$ is the one for which $t = \frac{1}{2} \in N_T$ and $y = \frac{1}{2} \in N_{\mathcal{F}, \frac{1}{2}}$. For simplicity let that random co-ordinate be the left hand boundary point of each of the two choices of $I_{\frac{1}{2}}^{(\mu)}$, $I_{A_{\frac{1}{2}, \frac{1}{2}}}$, $I_{\phi_{\frac{1}{2}, \frac{1}{2}}}$, respectively, for each $\mathbf{I}[\mathbf{N}] \in \mathcal{P}_\Upsilon$. Thus

$$x\left(\frac{1}{2} \right) = 0 \text{ or } \frac{1}{2}, \qquad A\left(\frac{1}{2}, \frac{1}{2} \right) = 0 \text{ or } \frac{1}{2}, \qquad \phi\left(\frac{1}{2}, \frac{1}{2} \right) = 0 \text{ or } \frac{1}{2}.$$

---

[6]See figures 6.1–6.4, pages 269 and 302 of [MTRV].

$I_{x_{t_i}}$	$I_{A_{ji}}$	$I_{\phi_{ji}}$	$\|\mathbf{I}\|$	$\chi_{xA\phi}$	$\omega^{(xA\phi)}$
$]0,\tfrac{1}{2}]$	$]0,\tfrac{1}{2}]$	$]0,\tfrac{1}{2}]$	$\tfrac{1}{8}$	$(0,0,0)$	ω^{000}
$]0,\tfrac{1}{2}]$	$]0,\tfrac{1}{2}]$	$]\tfrac{1}{2},1[$	$\tfrac{1}{8}$	$(0,0,\tfrac{1}{2})$	ω^{001}
$]0,\tfrac{1}{2}]$	$]\tfrac{1}{2},1[$	$]0,\tfrac{1}{2}]$	$\tfrac{1}{8}$	$(0,\tfrac{1}{2},0)$	ω^{010}
$]0,\tfrac{1}{2}]$	$]\tfrac{1}{2},1[$	$]\tfrac{1}{2},1[$	$\tfrac{1}{8}$	$(0,\tfrac{1}{2},\tfrac{1}{2})$	ω^{011}
$]\tfrac{1}{2},1[$	$]0,\tfrac{1}{2}]$	$]0,\tfrac{1}{2}]$	$\tfrac{1}{8}$	$(\tfrac{1}{2},0,0)$	ω^{100}
$]\tfrac{1}{2},1[$	$]0,\tfrac{1}{2}]$	$]\tfrac{1}{2},1[$	$\tfrac{1}{8}$	$(\tfrac{1}{2},0,\tfrac{1}{2})$	ω^{101}
$]\tfrac{1}{2},1[$	$]\tfrac{1}{2},1[$	$]0,\tfrac{1}{2}]$	$\tfrac{1}{8}$	$(\tfrac{1}{2},\tfrac{1}{2},0)$	ω^{110}
$]\tfrac{1}{2},1[$	$]\tfrac{1}{2},1[$	$]\tfrac{1}{2},1[$	$\tfrac{1}{8}$	$(\tfrac{1}{2},\tfrac{1}{2},\tfrac{1}{2})$	ω^{111}

Table 9.1: Elements of calculation of Riemann sum over a regular partition.

With this selection, each of the 8 sample histories $\chi(\mathbf{N})$ in the Riemann sum $(\mathcal{P})\sum\omega(\chi(\mathbf{N}))|\mathbf{I}[\mathbf{N}]|$ is a vertex of the corresponding cell $\mathbf{I}[\mathbf{N}]$, on the boundary of $\mathbf{I}[\mathbf{N}]$ but not contained in it. The eight sample values for $\omega(\chi(\mathbf{N}))$ in the Riemann sum are complex numbers which can be denoted ω^{pqr} for $p,q,r = 0,1$.

In Example 48 for comparison, the terms of a Riemann sum for the path integral of a step function are listed in (8.1) and (8.2). We wish to provide a similar listing of Riemann sum values for an integral over histories.

Table 9.1 provides this list of terms of a Riemann sum over the regular partition $\mathcal{P}_\Upsilon$. Note that each χ listed has initial and final components $\boldsymbol{\chi}'$ and $\boldsymbol{\chi}$, respectively; but only the variable components of χ are shown in Table 9.1.

In Example 48 of Section 8.7, Table 8.2 corresponds to Table 9.1; and (8.36) provides a list of numerical values of ground state action S_0 for a simplified single particle system. (If values for potential energy of the mechanical force are provided, then the interaction can also be included.) From these numerical values, the oscillation terms can be evaluated for Riemann sum calculation of a path integral.

The corresponding numerical evaluations of action $S = S_\mu + S_{\mu\mathcal{F}} + S_\mathcal{F}$ can be done for the sampling data in Table 9.1 using the calculations indicated by (9.24). If only ground state oscillation is required, then the numerical calculations can be done for $S_0 = S_\mu + S_\mathcal{F}$; so

$$\omega^{(xA\phi)} = \exp\left(\frac{\iota}{\hbar}S^{(xA\phi)}\right) \quad \text{or} \quad \exp\left(\frac{\iota}{\hbar}S_0^{(xA\phi)}\right)$$

for "excited" or ground states, respectively. Either way, with amplitudes $\eta(\mathbf{N})$ and $\bar{\eta}$ omitted, the Riemann sum estimate for the wave function $\psi(\chi,\tau) = \int_\Omega \omega(\chi)|\mathbf{I}|$ is

$$\sum_x\left(\sum_A\left(\sum_\phi \omega^{(xA\phi)}\left|I_{\phi_{ji}}\right|\right)\left|I_{A_{ji}}\right|\right)\left|I_{x_{t_i}}\right|, \quad = \frac{1}{8}\sum_{p=0}^{1}\left(\sum_{q=0}^{1}\left(\sum_{r=0}^{1}\omega^{pqr}\right)\right). \qquad (9.43)$$

This relatively straightforward calculation is expedited by the regularity of partition $\mathcal{P}_T$, which places component cell factors in parallel arrangement, so terms
can be added in convenient order. The iterated sum (9.43) corresponds to the
iterated integral

$$\int_{\mathbf{R}_\mu} \left(\int_{\mathbf{R}_A} \left(\int_{\mathbf{R}_\phi} e^{\frac{t}{\hbar}\left(\Sigma_{\mathcal{P}_T}\left(\Sigma_{\mathcal{P}_{\mathcal{F}}}\lambda(x(t),\,(A(y,t),\,\phi(y,t)))|j|\right)|\imath|\right)}\mathcal{D}\phi \right)\mathcal{D}A \right)\mathcal{D}x$$

of (9.25) above. Also, if history action wave amplitudes $\eta(\mathbf{N}) = \eta^{pqr}$ can be
found, then the oscillation terms ω^{pqr} can be replaced by amplitude-times-
oscillation products $\eta^{pqr}\omega^{pqr}$ in Table 9.1.

9.8   Integrand for Integral over Histories

Part II of this book set itself the objective of providing a "-complete" structure for the integration aspect of Feynman's integral-over-histories in quantum
electrodynamics. It sought to define the $\int_\Omega$ part of $\int_\Omega \exp\left(\frac{t}{\hbar}S(\chi)\right)d\chi$, with
amplitude/normalization factors $\eta(\cdot)$ omitted.[7] With some simplifications, the
Riemann sum inequality in (9.33) provides a version of $\int_\Omega$ which accords with
the physical and mathematical reasoning in [FH]. There is a basic numerical
outline in Sections 9.6 and 9.7 above.[8]

The issue of normalization factors A^{-1}, corresponding to history action wave
amplitude η, is left unresolved by Feynman in [FH]. Despite this apparent gap
in the theory of integration over histories, [FH] is able to deduce an extensive theory of physical phenomena from the analysis of action wave oscillation
$\exp\left(\frac{t}{\hbar}S\right)$, with wave amplitude omitted.

On the other hand, in his PhD thesis [10], and in [F1] and [FH], Feynman
includes normalization factor A^{-1} in order to be able to deduce Schrödinger's
equation for his integral over particle paths. So, in this case at least, the normalization factor/action wave amplitude is required in order to make the mathematical theory work in physics.

To sum up, a theory of $\int_\Omega$ has been provided without specifying the integrand
in detail. This section offers some conjectures, on the margins of physics, and
based on mathematical analogy, as to how an appropriate integrand might be
found for the integral over histories. Since all that is offered here is supposition
and conjecture, the stages and components of the analogy are merely outlined.

The analogy is a comparison of the path integral theory of [MTRV] with the
integral-over-histories of this book. These are based on action waves

$$\alpha(N_T)\omega(S^{(\mu)}(x_T(N_T))), \qquad \eta(\mathbf{N})\omega(S^{(\mu,\mathcal{F},\mu^{\mathcal{F}})}(\chi(\mathbf{N}))),$$

respectively. To track the comparison or analogy, here is a summary review of
notation and symbols for the two systems. With $T =]\tau',\,\tau[$ a single particle

---

[7]They are omitted because they are not known.

[8]Gauges γ do not appear in the numerical work of Section 9.7. For discussion of this, see
comments appended to Example 48.

system (μ) has paths $x_T \in \mathbf{R}^T$ with initial and final states

$$\xi' = x(\tau') = x_{\tau'}, \qquad \xi = x(\tau) = x_\tau,$$

respectively. A charged particle-field system $(\mu, \mathcal{F}, \mu\mathcal{F})$ has histories

$$\chi_T = (\chi(s))_{s \in T} = (x_s, \varsigma(s))_{s \in T} = \left(x_s, (A_{ys}, \phi_{ys})_{y \in M_{\mathcal{F}}}\right)_{s \in T},$$

with initial and final states

$$\boldsymbol{\chi}' = \chi_{\tau'} = (\xi', s') = (\xi', (A', \varphi')) = \left(x_{\tau'}, (A_{y\tau'}, \phi_{y\tau'})_{y \in M_{\mathcal{F}}}\right),$$

$$\boldsymbol{\chi} = \chi_\tau = (\xi, s) = (\xi, (A, \varphi)) = \left(x_\tau, (A_{y\tau}, \phi_{y\tau})_{y \in M_{\mathcal{F}}}\right),$$

respectively. At time τ the symbol $\mathcal{A}$ represents a set of numbers $A_{y\tau} \in M_A$ for $y \in M_{\mathcal{F}}$. Similarly for $\mathcal{A}'$, φ, and φ'. As before, we can take $T = \,]0, t[$, so $\tau' = 0$ and $\tau = t$. The wave functions (or aggregate action waves) for the two systems in their ground states are, respectively,

$$\psi_0(\xi, t) = \int_{\mathbf{R}^T} \alpha(N_T) \omega\left(S_0^{(\mu)}(x_T(N_T))\right) dx_T,$$

$$\psi_0(\boldsymbol{\chi}, t) = \int_\Omega \eta(\mathbf{N}) \omega\left(S_0^{(\mu, \mathcal{F})}(\chi(\mathbf{N}))\right) d\chi. \qquad (9.44)$$

Intuitively, it would seem that $\int_\Omega \cdots d\chi$ should include integration of the variables $A_{y\tau'}$, $A_{y\tau}$, $\phi_{y\tau'}$, $\phi_{y\tau}$; so $d\chi$ should include factors $dA_{y\tau'} \, dA_{y\tau} \, d\phi_{y\tau'} \, d\phi_{y\tau}$. If that is the case, Table 9.1 needs additional columns, which could be profitably supplied by the reader. Also, the only fixed terminal value in the particle-field interaction would then be the particle location $x(\tau) = \xi$. In that case $\psi_0(\boldsymbol{\chi}, t)$ should be interpreted as $\psi_0(\xi, t)$ in (9.44), and likewise for $\psi(\boldsymbol{\chi}, t)$ below (so ξ replaces $\boldsymbol{\chi}$. It is not immediately clear from [FH] which approach is intended.[9]

Returning to (9.44), both of the systems are investigated for interaction effects—due to a conservative force with potential energy $V(x_s, s)$ in the first case; and due to the interaction of electric current and electromagnetic field in the second case. When the interaction components of action are included, the wave functions become, respectively,

$$\psi(\xi, t) = \int_{\mathbf{R}^T} \alpha(N_T) \omega\left(S^{(\mu)}(x_T(N_T))\right) dx_T,$$

$$\psi(\boldsymbol{\chi}, t) = \int_\Omega \eta(\mathbf{N}) \omega\left(S^{(\mu, \mathcal{F}, \mu\mathcal{F})}(\chi(\mathbf{N}))\right) d\chi.$$

For the single-particle system (μ), the action for the excited state is $S = S_0 + S'$ in (8.19) and (8.21) of Section 8.2, as in chapter 7 of [MTRV]. So

$$S = S_0(x(N_T)) + S'(x(N_T)) = \frac{\mu}{2} \sum_{j=1}^n \frac{(x_j - x_{j-1})^2}{s_j - s_{j-1}} - \sum_{j=1}^n V(x_{j-1})(s_j - s_{j-1})$$

---

[9]See, for instance, (9.1) in chapter 9 of [FH].

and $\psi(\xi, t)$ becomes

$$\int_{\mathbf{R}^T} \left(e^{\frac{i}{\hbar} S'} \right) \left(\alpha(N_T) \omega \left(S^{(\mu)}(x_T(N_T)) \right) \right) dx_T, \; = \; \int_{\mathbf{R}^T} e^{\frac{i}{\hbar} S'(x(N))} G(I[N]),$$

using (8.22) and (8.23), in which $\alpha(N_T)$ is a "natural" component part of Fresnel integrator G. The latter representation is possible because of the Fresnel form of the ground state oscillation $\omega_0 = \omega(S_0(x(N_T)))$ and the resulting "additivity" expressed in lemmas 12 and 13 of [MTRV], page 262.

This in turn leads to a probabilistic interpretation of the wave function $\psi(\xi, t)$ as "marginal density of expectation" ([MTRV] page 345); since, in augmented form, the integrator G is a (complex-valued) probability distribution function in the augmented domain $\mathbf{R^T}, = \mathbf{R}^{]0,t]}$ (or $\mathbf{R}^{]\tau',\tau]}$).

The proposed analogy or conjecture for the integral over histories follows on from the preceding interpretation of the integral over paths.

According to chapter 9 of [FH], the action functional S for the interaction of charged particle (μ) with electromagnetic field $(\mathcal{F})$ is the combination of ground state action from (μ) and $(\mathcal{F})$ with a further interaction contribution $S' = S_{\mu\mathcal{F}}$, giving $S^{(\mu, \mathcal{F})} + S_{\mu\mathcal{F}}$; or, in accordance with (9.1) above,

$$S \; = \; (S_\mu + S_\mathcal{F}) + S_{\mu\mathcal{F}} \; = \; (S_\mu(\mathbf{x}) + S_\mathcal{F}(\mathbf{A}, \phi)) + S_{\mu\mathcal{F}}(\mathbf{x}, \mathbf{A}, \phi) \; = \; S_0 + S',$$

the sum of ground state particle-field action $S_0(\chi(\mathbf{N}))$ and "excitation" action $S'(\chi(\mathbf{N}))$. To proceed in analogy with the path integral theory of [MTRV] (as summarized above), a construction of the following kind can be attempted. For $T =]0, t[$ write wave function

$$\psi(\mathbf{x}, t) \; = \; \int_\Omega e^{\frac{i}{\hbar} S'(\chi(\mathbf{N}))} \left(\eta(\mathbf{N}) e^{\frac{i}{\hbar} S_0(\chi(\mathbf{N}))} \right) d\chi.$$

With augmented domain $\mathbf{T} =]0, t]$ and Ω^+ correspondingly augmented, find a distribution function $\mathbf{G}(\mathbf{I}[\mathbf{N}])$ on Ω^+; so that

$$\int_{\Omega^+} \eta(\mathbf{N}) e^{\frac{i}{\hbar} S_0(\chi(\mathbf{N}))} d\chi \; = \; \int_{\Omega^+} \mathbf{G}(\mathbf{I}[\mathbf{N}]) \; = \; 1;$$

and, in the unaugmented domain of histories Ω (with diminished version of $\mathbf{G}$),

$$\psi(\mathbf{x}, t) \; = \; \int_\Omega e^{\frac{i}{\hbar} S'(\chi(\mathbf{N}))} \mathbf{G}(\mathbf{I}[\mathbf{N}]).$$

In path integral theory, the integrator function $G(I[N])$ for $\mathbf{R}^T$ is obtained from the ground state action functional for mechanical particle motion as summarized in Sections 8.2 and 8.3. If the comparison with integration over histories is valid, then a similar approach might be taken for the construction of an analogous cell function $\mathbf{G}(\mathbf{I}[\mathbf{N}]$ in the domain of histories Ω.

That brings in the corresponding lagrangian and action functionals of Section 9.2, such as (9.6) and (9.13); including their sampling versions as indicated in (9.24). Fresnel integral lemmas 12 and 13 of [MTRV] (page 262) deliver the rather convenient solution $G(I[N])$ in the path integral case. It is not clear that there is such a convenient analytic device available to deal with the ground state lagrangian of the particle-field system in Ω.

To proceed, the precise form of these functionals has to be worked out. According to [FH] there are subtle and difficult questions of physics involved; so the issue has been bypassed in this book.

Fresnel integration resolves the convergence or existence issue in the path integral theory of [MTRV]. In [FH] and [F1] this aspect of the subject is given the designation Gaussian integral, and is repeatedly referred to in both sources as the key to convergence (or existence) of both path integrals and integrals over histories.

There is a difference between Gaussian and Fresnel integrals—real- and complex-valued, respectively. In graphical terms, the difference is expressed in figures 6.2 and 6.3 (page 302 of [MTRV]), with figure 6.4 providing a "bridge" between the two. The properties of Gaussian integrals are easier to establish, Fresnel integrals being somewhat more subtle (see Section 11.1 below). But the claims in [FH1] and [FH] are verifiable, at least in regard to path integrals.

In regard to integrals over histories, the concluding section 9-8 of [FH] (dealing with the particle-field interaction contribution $S_{\mu\mathcal{F}}$ to the action functional S) says:

> "The integral over $\mathbf{A}$ and ϕ is then easily performed because it is a gaussian integral."

Though this assertion is grounds for optimism,[10] a means of verifying it directly is not immediately evident in [FH]. So the proposed integrator $\mathbf{G}(\mathbf{I}[\mathbf{N}])$ above is speculative.

9.9 Action Wave Amplitudes

Section 9.8 above noted the emphasis that Feynman placed on the Gaussian (or Fresnel) factors in the action functionals on which path integrals and integrals over histories are based. These factors are particularly helpful to the cylindrical path integrands on which Feynman based his theory, involving iterations of successive one-dimensional integrals $\int_{-\infty}^{\infty} \cdots$. Such integrals converge (or exist) for Gaussian integrands.

But if these iterated integrals are not placed in the context of a multidimensional domain such as $\mathbf{R}^T$, it is difficult to establish the normalization factors A^{-1} (designated action wave amplitudes in this book) which make the theory physically meaningful.

At an even more basic level the Feynman theory originates in the Principle of Least Action, in which a path $\bar{x}_T$ selects an oscillation $\omega(\bar{x}) = \exp\left(\frac{i}{\hbar} S(\bar{x})\right)$ for which $S(\bar{x})$ is stationary or minimal, and which contains information about the physical system as a whole. The calculus of variations version of $\bar{x}_T$ does not

---

[10]A similar quote, from section 3-5 of [FH], is given above, page 246.

seem to deliver any direct approach to the factors A^{-1}. The path integral theory (unlike the calculus of variations) includes information about paths "adjacent to" $\bar{x}_T$ which turn out to be physically important.

But it is not necessary to restrict the theory to cylindrical integrands. In particular, the Riemann sum approach presented in this book points towards step function estimates which are easier to understand and to calculate, and which are not restricted to Gaussian/Fresnel integrands.[11]

In this book a mathematical theory of integration over histories $\int_\Omega$ has been provided without specifying integrands in detail. This section offers some further conjectures, based on mathematical analogy, as to how action wave amplitudes might be found, or at least estimated.

In [MTRV] the action wave amplitude/normalization factor $\alpha(N)$ is arrived at by a route different from Feynman's. The analogy or similarity between Feynman's path integral theory on the one hand, and the older mathematical theory of Brownian motion on the other hand, was noticed fairly early in the chronology of the subject. This similarity is rooted in the more basic relationship between the integral of Fresnel and the Gaussian integral, described in detail in chapter 6 of [MTRV]; leading on to a single, unified mathematical theory of Brownian motion and path integration, which is described as *c-Brownian motion* in chapter 7 of [MTRV].

The "natural" emergence of $\alpha(N)$ and $\bar\alpha$ is rehearsed in Example 49 above, in which the sampling form $S_0(x(N))$ of the ground state action for the single particle mechanical system leads to a functional $g(x(N))$ and distribution function $G(I[N])$ on which the c-Brownian theory is based, including the amplitude expressions $\alpha(N)$ and $\bar\alpha$.

Taking this as its starting point, [MTRV] develops a theory of path integration whose validity in physics is demonstrated by a derivation of Schrödinger's equation (section 7.16, pages 345–353), and Feynman's perturbation diagrams (pages 366–381). For the latter, with action S resolved into ground state component $S_0(x(N))$ and excitation component $S'(x(N))$, a "random variable" $\omega'(X(N))$ emerges from

$$\omega'(x(N)) \;=\; \exp\left(-\frac{\iota}{\hbar}\sum_{i=1}^{n} V(x(s_{j-1}))(s_j - s_{j-1})\right);$$

and, with distribution function $G(I[N^+])$ derived from the ground state action $S_0(x(N))$, the resulting wave function $\psi(x(\tau),\tau)$ is described in page 372 of [MTRV] as a *marginal density of expectation* because of the relation

$$\int_{\mathbf{R}} \psi(\xi,\tau)\,d\xi \;=\; \int_{\mathbf{R}^{]\tau',\tau]}} \omega'(x(N))G(I[N^+]) \;=\; \mathrm{E}\left[\omega'(X(N))\right]$$

---

[11]The fact that step functions are piece-wise constant is not inconsistent with quantum mechanics: "*These difficulties* [of classical physics] *have necessitated a modification of some of the most fundamental laws of nature and have led to a new system of mechanics, called quantum mechanics, since its most striking ... differences from the old mechanics apparently show a discontinuity in certain physical processes and a discreteness in certain dynamical variables.*" P.A.M. Dirac [25] page 1.

(with $\xi = x(s_n) = x(\tau)$).

This expresses the theory in probability terms (just like its Brownian motion analogue); so other probabilistic ideas and methods may also be relevant.

For instance, "marginal" calculations are calculations on a single variable of functions of many (including infinitely many) variables. Their role in estimating path amplitudes $\alpha(N)$ and $\bar{\alpha}$ are illustrated in Examples 49 and 50 above.

This role is analogous to the use of frequencies and relative frequencies in calculation of expected value, as in Table 1.4 on page 27 of [MTRV]. Just as "relative amplitude" $\frac{\alpha(N)}{\bar{\alpha}}$ appears in Examples 49 and 50, probabilities appear in Table 1.4 as

$$\frac{\text{frequency}}{\text{total frequency}}, \qquad \frac{\text{marginal frequency}}{\text{total frequency}}, \qquad \frac{\text{total marginal frequency}}{\text{total frequency}}.$$

By analogy, it is conceivable that methods like those of Examples 49 and 50 could be extended to the theory of -complete integration over histories, including amplitude factors $\eta(\mathbf{N})$ and $\bar{\eta}$.

For the path integral theory, with mechanical action $S = S_0 + S'$, the sampling version of $\exp\left(\frac{\iota}{\hbar} S_0\right)$ includes factors

$$\cdots \exp\left(\frac{\iota}{\hbar} \frac{m}{2} \frac{(x_i - x_{i-1})^2}{s_i - s_{i-1}}\right) \cdots$$

which are Fresnel integrals (similar to Gaussian), and which, in successive integrations on variables x_i, give rise to factors

$$\cdots \sqrt{2\pi \iota \hbar m^{-1}(s_i - s_{i-1})} \cdots$$

from which path amplitudes $\alpha(N)$ are obtained. For the interaction of charged mass and electromagnetic field, action is $S = S_\mu + S_{\mathcal{F}} + S_{\mu\mathcal{F}}$ where S_μ and $S_{\mathcal{F}}$ are ground state action for particle and field, respectively. Assuming that ground state action for the charged particle-field system is the sum of the separate ground state actions, then, for the charged particle-field system,

$$S_0 = S_\mu + S_{\mathcal{F}}, \qquad\qquad S' = S_{\mu\mathcal{F}}.$$

If the analogy with the single particle system holds, the ground state oscillation

$$\omega_0(\chi(\mathbf{N})) = \exp\left(\frac{\iota}{\hbar}\left(S_\mu(N_T) + S_{\mathcal{F}}(\mathbf{N})\right)\right) = \exp\left(\frac{\iota}{\hbar} S_\mu(N_T)\right) \exp\left(\frac{\iota}{\hbar} S_{\mathcal{F}}(\mathbf{N})\right)$$

will, when integrated, involve amplitudes $\eta(\mathbf{N})$; and since the factors $\exp\left(\frac{\iota}{\hbar} S_\mu\right)$ and $\exp\left(\frac{\iota}{\hbar} S_{\mathcal{F}}\right)$ are independent of each other, the first factor may deliver, in some form, a recognizable contribution $\alpha(N_T)$ to the amplitude $\eta(\mathbf{N})$.

With details of $S_{\mathcal{F}}$ left unspecified, the other components of $\eta(\mathbf{N})$ and $\bar{\eta}$ remain to be worked out. If the mathematical analogy between path integrals and integrals over histories is valid, values for $\eta(\mathbf{N})$ and $\bar{\eta}$ can be sought as follows.

The path integral theory of [MTRV] uses ground state action S_0, equal to the contribution S_μ in the charge-field system of quantum electrodynamics. In sampling format, the oscillation factor $\omega\left(S_0(x(N_T))\right)$ for the particle path integral theory has factors

$$\exp\left(\frac{\iota\mu}{2\hbar}\left(\frac{(x(s'')-x(s'))^2}{s''-s'}\right)\right)$$

for sample times $s'' = t_i$, $s' = t_{i-1}$ in $N_t = \{t_1,\ldots,t_{n-1}\}$ with $\tau' = t_0$ and $\tau = t_n$. The terms ι, μ, $\hbar$, and time increments $s'' - s'$ form the components of $\bar{\alpha}$ (with $s'' = \tau$, $s' = \tau'$), and of $\alpha(N_T)$ (with $s'' = t_j$, $s' = t_{j-1}$), so amplitudes $\bar{\alpha}$ and $\alpha(N_T)$ can, in knowledge of the path integral theory of [MTRV], be assembled directly from the components of the ground state action expression S_0.

But what if S_0 is known, but the path integral theory of [MTRV] is unknown to us? Is there any operation or calculation by which the amplitude factors α can be deduced from the oscillation $\omega_0 = \omega(S_0)$? This is the situation we face regarding integration over histories for charge-field interaction—ground-state action $S_0 = S_\mu + S_\mathcal{F}$ is obtained from knowledge of physical properties, but "corrections"/normalization/amplitudes $\eta(\mathbf{N})$ and $\bar{\eta}$ are not obvious, and the issue is not addressed in [FH].

Evaluations of incremental Fresnel integrals in chapter 6 of [MTRV] give

$$\int_{\mathbf{R}]s',s''[}\omega\left(S_0(x(N))\right)|I[N]| = \sqrt{2\pi\iota\hbar\mu^{-1}(s''-s')}\exp\left(\frac{\iota\mu}{2\hbar}\frac{(x(s'')-x(s'))^2}{s''-s'}\right),$$

so for $s'' = \tau$ and $s' = \tau'$, with fixed $x(s'') = x(s') = 0$, the integral on the left works out at $\sqrt{2\pi\iota\hbar\mu^{-1}(s''-s')}$. which is the correct value of $(\bar{\alpha})^{-1}$; while $s'' = t_j$ and $s' = t_{j-1}$ give the component factors $\sqrt{2\pi\iota\hbar\mu^{-1}(t_j-t_{j-1})}$ of $(\alpha(N_T))^{-1}$. Thus $\int \omega_0$ is a way of calculating

$$\bar{\alpha} = \left(2\pi\iota\hbar\mu^{-1}(\tau-\tau')\right)^{-\frac{1}{2}}, \qquad \alpha(N_T) = \prod_{j=1}^{n}\left(2\pi\iota\hbar\mu^{-1}(t_j-t_{j-1})\right)^{-\frac{1}{2}}.$$

In fact it is not necessary to resort to infinite-dimensional integration $\int_{\mathbf{R}]s',s''[}$. The following is sufficient. With $x(s') = x(s'') = 0$, choose s with $s' < s < s''$, so $N = \{s\}$ and

$$S_0(x(N)) = \exp\left(\frac{\iota\mu}{2\hbar}\left(\frac{(0-x(s))^2}{s''-s}+\frac{(x(s)-0)^2}{s-s'}\right)\right).$$

Then, with $\mathbf{R}^{\{s\}} = \mathbf{R}$ and $y = x(s)$, the following one-dimensional Fresnel integral gives the same result:

$$\int_{\mathbf{R}}\omega\left(S_0(x(N))\right)|I(N)| = \int_{\mathbf{R}}\omega\left(S_0(y)\right)dy$$

$$= \sqrt{2\pi\iota\hbar\mu^{-1}(s''-s')}\exp\left(\frac{\iota\mu(0-0)^2}{2\hbar(s''-s')}\right)$$

$$= \sqrt{2\pi\iota\hbar\mu^{-1}(s''-s')}.$$

Thus the action wave amplitudes for a single particle system can be determined by a single Fresnel integration.

Provided the analogy between path integration and integration over histories is valid, it may be that amplitudes $\bar{\eta}$ and $\eta(\mathbf{N})$ can be determined similarly. The construction in Section 9.6 shows how the factor spaces in Ω can be arranged for this purpose.

And if the assertion in [FH]—that the ground state action $S_{\mathcal{F}}$ of the electromagnetic field has Gaussian (or Fresnel) format—is valid, then, like $\bar{\alpha}$ and $\alpha(N_T)$, a closed analytic formulation of $\bar{\eta}$ and $\eta(\mathbf{N})$ may be obtainable. Otherwise, numerical estimates can be produced by the step function approximation method of Sections 9.6 and 9.7.

9.10   Probability and Wave Functions

Here are some extracts from a discussion [167] by physicists which includes probabilistic interpretations of the wave function of quantum mechanics. According to S. Weinberg,

> *In quantum mechanics the state of a system is not described by giving the position and velocity of every particle and the values and rates of change of various fields, as in classical physics. Instead, the state of any system at any moment is described by a wave function, essentially a list of numbers, one number for every possible configuration of the system. If the system is a single particle, then there is a number for every possible position in space that the particle may occupy. This is something like the description of a sound wave in classical physics, except that for a sound wave a number for each position in space gives the pressure of the air at that point, while for a particle in quantum mechanics the wave function's number for a given position reflects the probability that the particle is at that position. ...*

> *There is a rule of quantum mechanics, known as the Born rule, that tells us how to use the wave function to calculate the probabilities of getting various possible results in experiments. For example, the Born rule tells us that the probabilities of finding either a positive or a negative result when the spin in some chosen direction is measured are proportional to the squares of the numbers in the wave function for those two states of the spin.*

> *The introduction of probability into the principles of physics was disturbing to past physicists, but the trouble with quantum mechanics is not that it involves probabilities. We can live with that. The trouble is that in quantum mechanics the way that wave functions change with time is governed by an equation, the Schrödinger equation, that does not involve probabilities. It is just as deterministic as Newton's*

equations of motion and gravitation. That is, given the wave func-
tion at any moment, the Schrödinger equation will tell you precisely
what the wave function will be at any future time. There is not even
the possibility of chaos, the extreme sensitivity to initial conditions
that is possible in Newtonian mechanics. So if we regard the whole
process of measurement as being governed by the equations of quan-
tum mechanics, and these equations are perfectly deterministic, how
do probabilities get into quantum mechanics?

J. Bernstein commented:

The probabilistic interpretation of the Schrödinger wave function was
introduced by Max Born in a very brief note in 1926. He considered
the collision of an electron with a target and studied the wave func-
tion that represented the electron after the collision. It is a function
of position and he said that the wave function determined the probab-
ility that the electron would occupy that position. Later he modified
it to say that it was the square of the wave function and still later
he said the absolute value determined the probability.

It must be acknowledged that debate of this nature, between physicists, is about
physics. But from a mathematical perspective the following questions arise.

- Is the wave function ψ a probability?

- Is the square ψ^2 of the wave function a probability?

- Is the absolute value $|\psi|$ of the wave function a probability?

From the mathematical perspective of [MTRV] the wave function of quantum
mechanics is none of these things. It is proposed instead that it is a **density
function.**[12]

Consider the wave function $\psi(\xi, \tau)$ of a particle subject to a conservative
force with potential energy $V(y)$ at location y; with initial time $\tau' = 0$ and
initial position $\xi' = 0$. For simplicity assume particle mass $m = 1$. In the ground
state, with force zero and $V = 0$, the wave function $\psi = \psi_0$ is given by the path
integral

$$\psi_0(\xi, \tau) = \int_{\mathbf{R}^T} \exp\left(\frac{\iota}{\hbar} S_0(x(N))\right) dx(N)$$

$$= \int_{\mathbf{R}^T} G(I[N]) = \frac{\exp\left(\frac{\iota}{\hbar}\frac{\xi^2}{2\tau}\right)}{\sqrt{2\pi\iota\tau}}, \qquad (9.45)$$

by theorem 202 of [MTRV], page 324. (This value of the wave function is confirmed as satisfying the appropriate version of Schrödinger's equation—theorem

---

[12]Section 7.2 of [MTRV] (pages 322–325) uses the designation *marginal density of expectation.*

221, page 351 of [MTRV]. See also Section 8.2 above.) Then the ground state wave function ψ_0, its square ψ_0^2, and its absolute value $|\psi_0|$ are, respectively,

$$\frac{\exp\left(\frac{\iota}{\hbar}\frac{\xi^2}{2\tau}\right)}{\sqrt{2\pi\iota\tau}}, \qquad \frac{\exp\left(\frac{\iota}{\hbar}\frac{\xi^2}{\tau}\right)}{2\pi\iota\tau}, \qquad \frac{1}{\sqrt{2\pi\tau}}.$$

None of these quantities is a probability in the usual sense of the word; the first two are complex numbers, while the third one—though a non-negative real number—is greater than 1 if τ is greater than $(2\pi)^{-1}$.

In contrast, [MTRV] says that the wave function is a density. What does that mean? The following example provides illustrations of various kinds of density functions, including **probability density** and **density of expectation**.

Example 57 *The function $\phi(y) = \frac{1}{\sqrt{2\pi}}e^{-\frac{1}{2}y^2}$ is a density function for the standard normal distribution function Φ, and*

$$\Phi(J) = \int_J \phi(y)\,dy = \int_J \frac{1}{\sqrt{2\pi}}e^{-\frac{1}{2}y^2}\,dy, = \int_J \frac{1}{\sqrt{2\pi}}e^{-\frac{1}{2}y^2}|I|,$$

(the latter being in -complete notation). Using the theory of -complete integration, $\Phi(I)$ is the indefinite integral of $\phi(y)|I|, = \frac{1}{\sqrt{2\pi}}e^{-\frac{1}{2}y^2}|I|$, with

$$1 = \int_{\mathbf{R}} \Phi(I) = \int_{\mathbf{R}} \frac{1}{\sqrt{2\pi}}e^{-\frac{1}{2}y^2}|I|.$$

Also they are variationally equivalent so, writing $h(I) = \Phi(I) - \phi(y)|I|$,

$$\int_{\mathbf{R}} h(I) = \int_{\mathbf{R}} |h(I)| = 0.$$

The function Φ is additive on disjoint cells I, with $(\mathcal{P})\sum \Phi(I) = 1$ for every[13] partition $\mathcal{P}$ of $\mathbf{R}$; so, in -complete terms, Φ is a probability function.[14] In contrast, the function $\frac{1}{\sqrt{2\pi}}e^{-\frac{1}{2}y^2}|I|$, though integrable, is not additive.

More generally, if $H(I)$ is the indefinite integral of $f(y)|I|$, we can say that $f(y)$ is a density function for H.

Suppose $f(Y)$ is a contingent observable, $f(Y) \simeq f(y)[\mathbf{R}, \Phi]$. Then $f(Y)$ is a random variable (or absolute random variable) if, respectively,

$$E[f(Y)] = \int_{\mathbf{R}} f(y)\Phi(I) = \int_{\mathbf{R}} f(y)\left(\frac{1}{\sqrt{2\pi}}e^{-\frac{1}{2}y^2}\right)|I|,$$

$$E[|f(Y)|] = \int_{\mathbf{R}} |f(y)|\Phi(I) = \int_{\mathbf{R}} |f(y)|\left(\frac{1}{\sqrt{2\pi}}e^{-\frac{1}{2}y^2}\right)|I|.$$

We can then say that $f(y)\phi(y)$ and $|f(y)|\phi(y)$, or

$$f(y)\left(\frac{1}{\sqrt{2\pi}}e^{-\frac{1}{2}y^2}\right), \qquad |f(y)|\left(\frac{1}{\sqrt{2\pi}}e^{-\frac{1}{2}y^2}\right),$$

---

[13]Note that $\Phi(I)$ is well defined even if I is unbounded.

[14]See [MTRV]. In classical probability theory, the interval function $\Phi(I)$ generates a probability measure on the sigma-algebra of Borel subsets of $\mathbf{R}$. The -complete version is simpler.

are **density functions for the expectations** E[f(Y)] *and* E[|f(Y)|], *respectively.*

Now suppose $y = (y_1, \ldots, y_m)$ and $I = I_1 \times \cdots \times I_m$ are points and cells of $\mathbf{R}^m$, with

$$\phi(y) = \prod_{j=1}^{m}\left(\frac{1}{\sqrt{2\pi}}e^{-\frac{1}{2}y_j^2}\right), \qquad \Phi(J) = \int_J \phi(y)|I| = \int_J \phi(y)\,dy$$

where $J \in \mathcal{I}(\mathbf{R}^m)$ is a cell or interval of $\mathbf{R}^m$ and interval volume $|I| = \prod_{j=1}^m |I_j|$, the product of one-dimensional interval lengths. (If I has a component of infinite length, then, in the -complete theory, we can take $|I| = 0$.) As in the one-dimensional case, $\Phi(I)$ is a (m-dimensional, mutually independent, joint normal) probability distribution function on domain $\mathbf{R}^m$, and $\phi(y)$ is a probability density function for Φ. A random variable $f(Y)$ in $\mathbf{R}^m$ has

$$\mathrm{E}[f(Y)] = \int_{\mathbf{R}^m} f(y)\Phi(I) = \int_{\mathbf{R}^m} f(y)\phi(y)\,dy,$$

and $f(y)\phi(y)$ is a **density function of the expectation.** Now consider points $y' = (y_1, \ldots, y_{m-1})$ and cells $I' = I_1 \times \cdots \times I_{m-1}$ of $\mathbf{R}^{m-1}$, so

$$y = (y', y_m) \in \mathbf{R}^{m-1} \times \mathbf{R} = \mathbf{R}^m,$$
$$I = I' \times I_m \subset \mathbf{R}^{m-1} \times \mathbf{R} = \mathbf{R}^m;$$

with probability distribution function and probability density function

$$\Phi(I) = \Phi(I')\Phi(I_m), \qquad \phi(y) = \phi(y')\phi(y_m),$$

respectively. Suppose $f(Y) = f(Y', Y_m)$ is a random variable in $\mathbf{R}^m$. Then

$$\mathrm{E}[f(Y)] = \int_{\mathbf{R}_m} f(y)\Phi(I)$$
$$= \int_{\mathbf{R}^{m-1}\times\mathbf{R}} f(y', y_m)\Phi(I')\Phi(I_m)$$
$$= \int_{\mathbf{R}}\left(\int_{\mathbf{R}^{m-1}} f(y', y_m)\Phi(I')\right)\Phi(I_m) \qquad (9.46)$$
$$= \int_{\mathbf{R}}\left(\int_{\mathbf{R}^{m-1}} f(y', y_m)\Phi(I')\phi(y_m)\right)dy_m, \qquad (9.47)$$

where (9.46) uses Fubini's theorem ([MTRV], page 160), and (9.47) uses variational equivalence of $\Phi(I_m)$ and $\phi(y_m)|I_m|$ (theorem 44, [MTRV] page 148). Write

$$\varphi(y_m) = \left(\int_{\mathbf{R}^{m-1}} f(y', y_m)\Phi(I')\right)\phi(y_m).$$

Then, by (9.47),

$$\mathrm{E}[f(Y)] = \int_{\mathbf{R}} \varphi(y_m)\,dy_m, \qquad (9.48)$$

so $\varphi(y_m)$ is a **marginal density function of the expectation** E[f(Y)]. The term "marginal" is customary in statistics and probability, in situations such as this where y_m is "on the margin" of $(y_1, \ldots, y_{m-1}, y_m)$.

The "marginal density" interpretation can be applied to wave functions, with $\mathbf{R^T} = \mathbf{R}^T \times \mathbf{R}$ (instead of $\mathbf{R}^m = \mathbf{R}^{m-1} \times \mathbf{R}$), where $\mathbf{T} =]0, \tau]$, $T =]0, \tau[$. But first consider some relevant aspects of distribution functions $G(I[N])$ in $\mathbf{R^T}$.

The Brownian cell function $G(I[N])$ is real-valued in (5.8) above (just as the cell function Φ is); whereas, in its quantum mechanical form (8.23), it is complex-valued.[15] But, in either case, G is additive on disjoint cells $I[N] \in \mathcal{I}(\mathbf{R^T})$, with $\sum G(I[N]) = 1$ over partitions of $\mathbf{R^T}$. Therefore, in the meaning and terminology of [MTRV], **real- or complex-valued $G(I[N])$ is a distribution function in domain $\mathbf{R^T}$**. Also, in either case, the function (8.22),

$$g(y(N)) \;=\; \prod_{j=1}^{n} \left(\frac{\exp\left(\frac{c(y_j - y_{j-1})^2}{s_j - s_{j-1}}\right)}{\sqrt{\frac{\pi}{-c}(s_j - s_{j-1})}} \right) \quad \text{(with } c = -\frac{1}{2} \text{ or } \frac{\imath m}{2\hbar}, \text{ respectively)},$$

is a density function for the distribution function G, with $g(x(N))|I[N]|$ variationally equivalent to $G(I[N])$;

$$h(x, N, I[N]) := g(x(N))|I[N]| - G(I[N]) \implies \int_{\mathbf{R^T}} |h(x, N, I[N])| = 0$$

in both cases. In [MTRV], and in this book, the sample times $N \in \mathcal{N}(\mathbf{T})$ should always include $s_n = \tau$.[16] A wave function $\psi(\xi, \tau)$ of quantum mechanics (and a particle/probability density function $p(\xi, \tau)$ of Brownian motion) is obtained by integration on the "diminished" domain $\mathbf{R}^T$, $= \mathbf{R}^{\mathbf{T} \setminus \{\tau\}}$ (corresponding to $\mathbf{R}^{m-1}$ in Example 57 above), omitting a factor $|I_\tau|$ in each of the integrands $g(x(N))|I[N]|)$ and $G(I[N])$.

This corresponds to omission of $\int_{\mathbf{R}} \cdots dy_m$ in Example 57; and the resulting "diminished" integrator $G(I[N^-])$ is a density function[17] for the distribution function G in $\mathbf{R^T}$, in the sense that, for any cell $J[N]$, $G(J[N]) =$

$$= \int_{J[N^-] \times J_\tau} G(I[N^-]) dx_\tau \;=\; \int_{J_\tau} \left(\int_{J[N^-]} G(I[N^-]) \right) dx_\tau \;=\; \int_{J_\tau} G(J[N^-]) dx_\tau.$$

Now suppose $f(X(N))$ is a G-random variable in domain $\Omega = \mathbf{R^T}$. Then

$$\mathrm{E}[f(X(N))] \;=\; \int_{\mathbf{R^T}} f(x(N)) G(I[N])$$
$$=\; \int_{\mathbf{R}} \left(\int_{\mathbf{R^T}} f(x(N)) G(I[N^-]) \right) dx_\tau,$$

and, as in Example 57, $f(x(N))G(I[N^-])$ is a **marginal density function for the expectation**. Let $V(y, s)$ be the potential energy for a conservative force acting on the system, and take

$$f(x(N)) \;=\; \exp\left(c \sum_{j=1}^{n} V(x_{j-1}, s_{j-1})(s_j - s_{j-1}) \right).$$

---

[15]In [MTRV] notation $G_c(I[N])$ is used, with real or complex value of c selecting Brownian motion or quantum mechanics.
[16]Even if τ is not included in N, in gauges $\gamma = (L, \delta_\mathcal{N})$ can be required to have $\tau \in L(x_\mathbf{T})$ for all $x_\mathbf{T} \in \mathbf{R^T}$.
[17]Equation (9.45) has N instead of N^- because the latter is implied by the context.

Then, writing $s_n = \tau$, $x_n = x(s_n) = \xi$, $E[f(X(N))] =$, respectively,

$$= \int_{\mathbf{R}} p(\xi, \tau) d\xi \quad \left(\text{for } c = -\frac{1}{2}\right), \qquad = \int_{\mathbf{R}} \psi(\xi, \tau) d\xi \quad \left(\text{for } c = -\frac{\iota m}{2\hbar}\right).$$

Since the latter integrations $d\xi$ take place on only a single variable x_τ of $x(N)$, the wave function ψ is a **marginal density function,** as is the Brownian particle density $p(\xi, \tau)$. Details of proof are in section 7.8 of [MTRV]; and theorem 220 (page 350) shows that $p(\xi, \tau)$ satisfies a Fokker-Planck/Smoluchowski equation, while $\psi(\xi, \tau)$ satisfies Schrödinger's equation. (A single generalized diffusion equation represents all of them.)

The term "expectation" is valid for the case $c = -\frac{1}{2}$, since, in that case $G(I[N])$ is non-negative, and is a probability distribution function. But for imaginary number c, $G(I[N])$ is complex-valued. (Traditionally, only non-negative, additive functions with sum 1 can be considered to be probabilities yielding expected values.)

In [MTRV] the term "potentiality" was used to describe probability-like expressions which do not meet all the requirements of classical or standard probability theory, and in order to avoid confusion.

But as the discussion in [167] shows, the idea of probability has, for better or worse, inserted itself into quantum mechanics even when its exact meaning in that subject is disputed by physicists. Rather than introducing a new term (such as "potentiality" instead of "probability"), another standard way of dealing with such issues is to extend or broaden the meaning of the original term, even if this means dropping some supposedly defining characteristics.

For instance, [167] declares that *"probabilities must all be positive numbers, and add up to 100 percent"*, which is a mathematical statement: if probability is thought of in terms of relative frequency, an event either occurs or does not occur. There is no such thing as "negative occurrence". In other words, occurrence contributes positively to relative frequency (probability), while non-occurrence contributes zero.

In those terms there can be no such thing as negative forms of relative frequency or probability. On the other hand:

> *The particles of atomic physics exhibit properties of* interference *or cancellation, just as waves do. If a beam of particles is directed toward some measuring device, then the number of particles arriving at the device can be counted. Extra particles can make a positive contribution, producing a more intense beam. But the particles may also manifest wave properties, and, depending on their phase, the interaction of waves can produce modulation or cancellation rather than intensification. In other words extra particles can reduce rather than increase the intensity of the beam. Thus a "probability calculus" of such particles may require distribution functions which take negative as well as positive values.*
> (From section 2.16 (*"Negative Probability"*) of [MTRV].)

To sum up, chapter 7 of [MTRV] uses a general version of the function $G(I[N])$ which includes as special cases the real-valued $G(I[N])$ of Brownian motion (5.10), and the complex-valued $G(I[N])$ of quantum mechanics (8.23). (A wave can be described by real-valued expressions such as $\alpha \cos \varpi t$; or, for ease of calculation, by the real part of $\alpha e^{\iota \varpi t}$. The latter modification means that the corresponding "probability" distribution function G has complex values.)

In terms of mathematics (as distinct from physics), [MTRV] effected a union between the classical mechanics of Brownian motion and the quantum mechanics of Feynman's path integral theory. A single "ground state probability" measure $G(I)$ is defined (see (5.8) and (8.23)) on cells I composed of potential events in either system, classical or quantum. The cell function G is, respectively, non-negative or complex-valued. In either case G is finitely additive on disjoint cells, with $\int_{\mathbf{R}^T} G(I[N]) = 1$. Therefore, for mathematical reasons, G can be designated as a probability; and the corresponding wave function of quantum mechanics is a probability density function.

Part III

Appendices

Chapter 10

Appendix 1: Integration

This book is based on the -complete system of integration, originated by R. Henstock and J. Kurzweil. Why not use the better-known Lebesgue integral?[1]

While Lebesgue's dominated convergence theorem is a crucial pillar of modern analysis, there are certain areas of the subject where this theorem is deficient. Deeper criteria for convergence of integrals are described in this Chapter.

We learn calculus in secondary school: first, differentiation of functions, and later integration as the inverse or opposite of differentiation: the integral is the anti-derivative or primitive function, from which *definite integrals* can be easily deduced.

In more advanced mathematics courses we learn Riemann's definition of definite integrals which enables us to integrate more functions. The Riemann definition does not make use of differentiation. Instead it is similar to the ancient method of finding areas and volumes "by exhaustion"—that is, the *quadrature* method of estimating an area or volume by dividing it up into pieces which are more easily estimated, and then taking the aggregate of the pieces.

Specialists in mathematical analysis go on from this to study the Lebesgue method of integration. Why? Again, one of the stock answers to this question is that the Lebesgue method enables us to integrate functions which cannot be integrated by more familiar methods such as the calculus integral and the Riemann integral.

The Dirichlet function is sometimes mentioned. In the unit interval $[0,1]$ this function takes value one at the rational points, and zero at the other points. The Dirichlet function is not the derivative of some other function, so it cannot be integrated by the method we learn at school in calculus lessons. Also, it cannot be integrated by Riemann's method. But the Lebesgue integral of the Dirichlet function exists: the definite integral of the Dirichlet function on the unit interval has value 0.

---

[1] E.J. McShane [111] devised a gauge version of the Lebesgue integral. (This is not the Lebesgue-complete integral of [123], page 34.) In [111, 113] McShane gave an account of stochastic integrals in Riemann terms which differs from the approach of [MTRV] and this book.

Does this matter? Apart from some specialists and experts, does anybody have any real use for the Dirichlet function, and does anybody really care whether or not it is amenable to calculus? It cannot be pictured as lines in a graph, nor does it have a straightforward expression as polynomial, trigonometric or exponential formula. Unlike the area and volume calculations of antiquity, and unlike the calculus of Newton and Leibniz which explained the world in mechanical terms, what difference does the Dirichlet function make to anyone outside the narrow and rarified world of a tiny number of people in pure mathematics?

Likewise many of the other arcane and exotic functions, such as the Devil's Staircase, invented during the long nineteenth century gestation of Lebesgue integration, measure theory and set theory. These are functions whose counterintuitive and challenging qualities we can admire and wonder at, but which were described by Hermite and Poincaré as unwelcome monsters causing mayhem in the rich and fertile garden of mathematical analysis.

"Does anyone believe that the difference between the Lebesgue and the Riemann integrals can have physical significance, and that whether, say, an airplane would or would not fly could depend on this difference? If such were claimed, I should not care to fly in that plane." (Richard W. Hamming [66]).

This critique is understandable, but unhistorical. By the early nineteenth century, the rich and fertile garden was on the verge of becoming a barren and dangerous wilderness—and not because of trespassing monster-functions.

10.1   Monstrous Functions

In the tradition of Newton and Leibniz, Fourier series representation of functions opened up the analysis of wave motion, crucial to an understanding of sound, light, electricity and other physical phenomena. But strange and paradoxical things could happen when the integral of a function was obtained by integrating its Fourier series term by term. Certain questions could no longer be avoided. To what extent, and under what conditions, is a function identical to its corresponding Fourier series representation? When is the integral of a function equal to the series obtained by integrating the terms of the corresponding Fourier series?

This boils down to whether a convergent series of integrable functions has integrable limit, and whether the integral of the limit is equal to the limit of the integrals whenever the latter limit exists.

Such issues motivated decades of investigation of the notion of integration, until a satisfactory resolution was found in the convergence theorems—uniform, monotone, and dominated—of Lesbesgue integration theory. In particular the dominated convergence theorem tells us that if a sequence of integrable functions f_j converges to f, and if the sequence satisfies $|f_j| < g$ for all j, where g is integrable, then f is integrable and $\int f_j$ converges to $\int f$ as $j \to \infty$.

The integral here is the definite Lebesgue integral on some domain. But the theorem holds for functions which are integrable in the older and more familiar senses of Riemann, Cauchy, and Newton/Leibniz, since, broadly speaking, functions which are integrable in the latter senses are, *a fortiori*, integrable in the Lebesgue sense.

From this point onwards somebody—not necessarily expert in Lebesgue's integration—who is contentedly doing some familiar integral operations, and who encounters some issue of convergence such as term-wise integration of a Fourier series, can proceed in safety if a dominant integrable function g can be found for the convergence.

This is the practical significance of Lebesgue's theory. It is a reason why "it is safe to fly in airplanes", so to speak. It is why the fertile garden did not turn into a barren wilderness. And the "monster-functions" were in reality guard dogs that played their part in protecting the garden.

10.2 A Non-monstrous Function

But is this the end of the story? Did Lebesgue's 1901 and 1902 papers [100, 101] give the last word on the subject?

Here is a sequence of non-monstrous functions formed by combining some familiar polynomial and trigonometric expressions. For $j = 2, 3, 4, \ldots$, let

$$f_j(x) = 2x \sin \frac{1}{x^2} - \frac{2}{x} \cos \frac{1}{x^2} z \text{ if } \frac{1}{j} \le x \le 1, \text{ and } = 0 \text{ if } 0 \le x < \frac{1}{j}.$$

An impression of function f_j can be obtained from Figure 10.3 below, which contains the graph of $f(x) = 2x \sin \frac{1}{x^2} - \frac{2}{x} \cos \frac{1}{x^2}$ for $0 < x \le 1$. Figure 10.2 resembles Figure 10.3 in the neighbourhood of $x = 0$. The difference between the two is $2x \sin x^{-2}$, whose graph is in Figure 10.1.

But the values of the latter function are very small in the neighbourhood of $x = 0$, and that is why its graph is included here. Note that the vertical scale of Figure 10.1 is much more magnified than the vertical scales of Figures 10.2 and 10.3.

Figure 10.4 is the graph of the primitive function or anti-derivative of f, which will play a big part in the following discussion.

Each f_j has a single discontinuity (at $x = \frac{1}{j}$), but is differentiable at every other point. Each f_j is integrable (in the sense of Riemann and Lebesgue), and the sequence f_j is convergent at each x to $f(x) = 2x \sin \frac{1}{x^2} - \frac{2}{x} \cos \frac{1}{x^2}$, $f(0) = 0$, whose graph is Figure 10.3.

The limit function f has a single discontinuity (from the right) at $x = 0$; and it has a primitive function $F(x) = x^2 \sin x^{-2}$ $(x > 0)$, $F(0) = 0$ (Figure 10.4); so in fact f has a definite integral $F(1) - F(0) = \sin 1$ on the domain $[0, 1]$ **provided** the Newton-Leibniz definition of the integral is used.

But f is unbounded on $[0, 1]$ and therefore is not Riemann integrable on $[0, 1]$. And, though clearly non-monstrous, the function f is **not** Lebesgue integrable. See below for discussion of this point.

On the face of it, this example indicates a step backwards, so to speak, where the old school method of Newton/Leibniz is actually more effective than more modern methods. Lebesgue's theory of the integral threw up anomalies of this kind, and accordingly investigation of the theory continued through the twentieth century.

To recapitulate, Lebesgue's dominated convergence theorem can be said to be the cutting edge of modern integration theory. But it fails to encompass the convergence of sequences such as f_j and $\int_0^1 f_j(x)\,dx$. The graph in Figure 10.3 suggests $2x^{-2}$ as a conceivable candidate for dominating function g for the terms $|f_j|$, $j = 1, 2, 3, \ldots$, at least in a neighbourhood of the critical point $x = 0$. But $2x^{-2}$ is not integrable in a neighbourhood of 0, and it seems that the dominated convergence theorem is not applicable here.

This failure must sound some alarm bells, because while many working mathematicians can get by without the Lebesgue integral, we cannot really do without convergence theorems which allow us, for instance, to perform routine operations on Fourier series; or, more generally, to safely find the integral of the limit of a sequence of integrable functions by taking the limit of the corresponding sequence of integrals.

The purpose of this chapter is to dip into some aspects of modern integration theory in order to introduce Theorems 9, 10, and 11 below, which are delicate enough to deal with, for instance, the convergence of the functions f_j above and their integrals; while also covering the ground already covered by the convergence theorems of Lebesgue's theory.

For ease of reference, here again are the sequence f_j and related functions, $j = 1, 2, 3, \ldots$:

$$
f(x) \;=\; \begin{cases} 2x \sin \frac{1}{x^2} - \frac{2}{x} \cos \frac{1}{x^2} & \text{if } 0 < x \le 1, \\ 0 & \text{if } x = 0, \end{cases} \tag{10.1}
$$

$$
f_j(x) \;=\; \begin{cases} 2x \sin \frac{1}{x^2} - \frac{2}{x} \cos \frac{1}{x^2} & \text{if } \frac{1}{j} \le x \le 1, \\ 0 & \text{if } 0 \le x < \frac{1}{j}, \end{cases} \tag{10.2}
$$

$$
F(x) \;=\; \begin{cases} x^2 \sin \frac{1}{x^2} & \text{if } 0 < x \le 1, \\ 0 & \text{if } x = 0, \end{cases} \tag{10.3}
$$

$$
F_j(x) \;=\; \begin{cases} x^2 \sin \frac{1}{x^2} & \text{if } \frac{1}{j} \le x \le 1, \\ 0 & \text{if } 0 \le x < \frac{1}{j}. \end{cases} \tag{10.4}
$$

Figures 10.3 and 10.4 are, respectively, the graphs of the functions f and F. The graphs of f_j and F_j are easily substituted—just insert a horizontal line segment at height 0 from $x = 0$ to $x = \frac{1}{j}$.

The function f has a single discontinuity at $x = 0$, while F is continuous. The reader should verify that F is differentiable, including at the point $x = 0$ (from the right), and that $F'(0) = 0, = f(0)$. The fact that $F'(x) = f(x)$ for $x > 0$ is easily verified.

This establishes that f, $= \lim f_j$, has a primitive or anti-derivative, and F is the calculus- or Newton-indefinite integral of f. Also the definite integral on

$[0,1]$, in the calculus or Newton-Leibniz sense, is

$$\int_0^1 f(x)\,dx \;=\; F(1) - F(0) \;=\; \sin 1 - 0 \;=\; \sin 1.$$

For each j both f_j and F_j have a discontinuity at $x = 1/j$, but provided $x \neq 1/j$, we have $F'(x) = f_j(x)$. Thus, for each j, f_j is Riemann and Lebesgue integrable on $[0,1]$, but not calculus- or Newton-integrable on $[0,1]$. However f_j **is** calculus-integrable on $[j^{-1}, 1]$ for each j, and

$$\int_0^1 f_j(x)\,dx \;=\; \int_{\frac{1}{j}}^1 f_j(x)\,dx \;=\; F(1) - F\!\left(\frac{1}{j}\right) \;=\; \sin 1 - \frac{1}{j^2}\sin j^2,$$

provided we interpret $\int_0^1$ as a Riemann (or Lebesgue) integral. Thus, as $j \to \infty$,

$$\int_0^1 f_j(x)\,dx \;\to\; \int_0^1 f(x)\,dx \tag{10.5}$$

provide we interpret the left hand integrals in the sense of Riemann or Lebesgue, and the right hand one as a calculus or Newton/Leibniz integral.

Unless we can interpret it in some other way, (10.5) is false as it stands, since we cannot ascribe the same meaning to the symbol $\int_0^1$ in the left- and right-hand terms. In fact we will establish later that (10.5) is valid for an adapted[2] version of the Riemann integral; and that the convergence—including integrability of the limit function f—though unrelated to any kind of dominated convergence, satisfies a new kind of Riemann sum convergence criterion which goes beyond the Lebesgue dominated convergence theorem.

To recapitulate, for $0 \leq x \leq 1$ the function f_j is bounded and continuous—except for a discontinuity at $x = j^{-1}$. Also it is differentiable except at $x = j^{-1}$. It has anti-derivative $F_j(x)$—except at $x = j^{-1}$.

By familiar standards, f_j is Riemann integrable and Lebesgue integrable on domain $]0,1]$. But, as discussed above, its limit function f is **not** Lebesgue (or Riemann) integrable.

There are theorems which tell us when to expect Lebesgue integrability of the limit function of Lebesgue integrable functions. If the convergence of the functions f_j to the function f is uniform, monotone, or dominated by a Lebesgue integrable function, then Lebesgue integrability of the functions f_j implies Lebesgue integrability of their limit function f, with

$$\lim_{j\to\infty}\left(\int_0^1 f_j(x)\,dx\right) \;=\; \int_0^1 \left(\lim_{j\to\infty} f_j(x)\right)dx \;=\; \int_0^1 F(x)\,dx.$$

Inspection of the graphs indicates that convergence of functions f_j is not uniform, monotone or dominated. And, even though each f_j is Lebesgue integrable, the limit function f is not Lebesgue integrable—as demonstrated below.

---

[2]That is, the Riemann-complete integral, also know as the generalized Riemann or Henstock-Kurzweil integral.

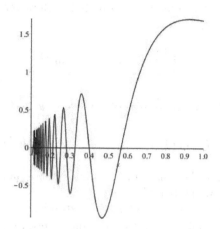

Figure 10.1: $2x \sin x^{-2}$

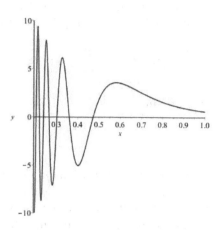

Figure 10.3: $2x \sin x^{-2} - 2x^{-1} \cos x^{-2}$

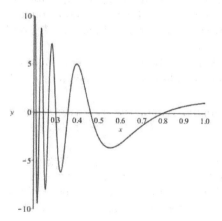

Figure 10.2: $2x^{-1} \cos x^{-2}$

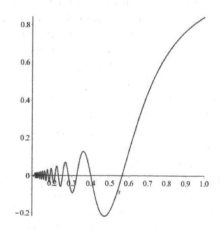

Figure 10.4: $x^2 \sin x^{-2}$

Is this a big problem for the garden of mathematics, or is it just a minor incursion, not by a monster, but by an atypical creature which is easily contained? This chapter seeks to provide some perspective.

The names of Denjoy, Perron, Kolmogorov and others are associated with twentieth century efforts [61] to pursue the implications of problems such as the convergence of functions f_j and their integrals. This chapter deals with the Riemann sum approach of R. Henstock and J. Kurzweil.

10.3 Riemann-complete Integration

Kurzweil came to this subject through his investigations of differential equations. Henstock was interested in convergence issues in integration. Independently, each of them focussed on careful construction of Riemann sums.

A partition $\mathcal{P}$ of a domain such as $[0,1]$ is a set of points $u_0 < u_1 < u_2 < \cdots < u_n = 1$. Now identify $\mathcal{P}$ with the corresponding set of disjoint intervals I:

$$\mathcal{P} = \{[u_0, u_1], \,]u_1, u_2], \,]u_2, u_3], \ldots, \,]u_{n-1}, u_n]\} = \{I\}.$$

For each $I \in \mathcal{P}$ let $|I|$ denote the length $u_i - u_{i-1}$ of I. Given a function f defined on $[0,1]$, and given a partition $\mathcal{P}$, a Riemann sum for f is

$$\sum \{f(x)|I| : I \in \mathcal{P}\}, \quad \text{or} \quad (\mathcal{P}) \sum f(x)|I|,$$

where, for each I, the point x of the term $f(x)|I|$ satisfies $u_{i-1} \le x < u_i$. Call the point x of $f(x)|I|$ the *evaluation point*.

The integral $\int_0^1 f(x)\,dx$ of f on the domain $[0,1]$, denoted by α, exists if α satisfies a condition of the following form.

Given $\varepsilon > 0$, partitions $\mathcal{P}$ can be chosen such that

$$\left| \alpha - (\mathcal{P}) \sum f(x)|I| \right| < \varepsilon.$$

This inequality is reminiscent of the Riemann integral of f, but it is not the full definition. More is required here. There must be some rule for selecting the partitions $\mathcal{P}$ that can be admitted in the inequality.

For Riemann integration, the rule is that, given $\varepsilon > 0$ there exists a constant $\delta > 0$ such that, for every partition $\mathcal{P}$ for which $|I| < \delta$ for each $I \in \mathcal{P}$, the above inequality[3] holds. Denote such a rule by γ, and denote a partition $\mathcal{P}$ which satisfies an appropriate instance of the rule γ by $\mathcal{P}_\gamma$.

An integral constructed from such a rule can be identified by notation $^\gamma\!\int$. Then the definition of the integral $\alpha = \,^\gamma\!\int_0^1 f(x)\,dx$ is as follows. There is a number α for which, given any $\varepsilon > 0$, there exists a corresponding[4] instance $\gamma(\varepsilon)$ of γ such that every partition $\mathcal{P}_{\gamma(\varepsilon)}$ satisfies

$$\left| \alpha - (\mathcal{P}_{\gamma(\varepsilon)}) \sum f(x)|I| \right| < \varepsilon. \tag{10.6}$$

---

[3]If f is continuous on $[0,1]$ then its Riemann integral exists there.

[4]For the ordinary Riemann integral, read "there exists a corresponding number $\delta > 0$".

We will omit the $^\gamma$ in $^\gamma\!f$, and allow the context to demonstrate which version of the integral is being discussed.

While the Riemann sum rule for ordinary Riemann integration is "$|I| < \delta$", the innovation of Kurzweil and Henstock was to replace the constant δ by variable $\delta(x) > 0$, where x is the *evaluation point* in the term $f(x)|I|$ of the Riemann sum in (10.6). To distinguish this from the Riemann integral, call it the *Riemann-complete* integral. (It is also called the Henstock-Kurzweil integral, generalized Riemann integral, and gauge integral. The term *gauge* is applied to the function $\delta(x)$.) Clearly, every Riemann integrable function is also integrable in the Riemann-complete sense.

A Stieltjes-type definition of the integral of a function f can be expressed as follows. Suppose $f(x)$ and $g(x)$ are point functions defined on the domain $[0, 1]$. The Riemann-Stieltjes integral of f with respect to g is got by replacing the length function $|I|$ by the increment function $g(I) = g(u_i) - g(u_{i-1})$ in the above definitions. A standard result is that if f is continuous and g is monotone (or has bounded variation), then the Riemann-Stieltjes integral $\int_0^1 f(x)\,dg$ exists. If the constant $\delta > 0$ in the definition is replaced by the function $\delta(x) > 0$ of the Riemann-complete construction, call the resulting integral the *Stieltjes-complete* integral of f with respect to g.

The fundamental result on which -complete integration is based is *Cousin's lemma*. This states that, if a gauge $\delta(z) > 0$ is given for a domain such as $[0, 1]$, then there exist a positive integer n, points $z_1, \ldots, z_n$, and disjoint cells (intervals) $[0, u_1] = I_1$, $]u_1, u_2] = I_2, \ldots,]u_{n-1}, 1] = I_n$ partitioning the domain $[0, 1]$, such that

$$u_{j-1} \leq z_j \leq u_j, \qquad u_j - u_{j-1} < \delta(z_j), \qquad (u_0 = 0,\ u_n = 1,\ j = 0, 1, \ldots, n).$$

In this book z_j is restricted to u_{j-1} or u_j, not[5] $u_{j-1} \leq z_j \leq u_j$. The point z_j is the associated point (or tag point) of I_j. The collection $\{(z_j, I_j)\}$ of associated point-cell pairs is a division of the domain, and is δ-fine if each (z_j, I_j) is δ-fine.

The Riemann sum consists of terms $f(x_j)(u_j - u_{j-1})$. If the evaluation point x_j of the integrand f is the tag point z_j of the associated cell I_j, the resulting integral is *Riemann-complete*. If $x_j \neq z_j$, then the resulting integral is *Burkill-complete*.

To see how the Riemann-complete integral matches the calculus or Newton/Leibniz integral, suppose a point function $f(x)$ has an anti-derivative or primitive function $F(x)$ or $0 \leq x \leq 1$, so its definite integral, in the Newton/Leibniz sense is $F(1) - F(0)$. Proceeding as follows, it is easy to deduce that f is Riemann-complete integrable.

If $\mathcal{P}$ is a partition of $[0, 1]$ with partition points u_i, $0 = u_0 < u_1 < \cdots < u_n = 1$,

---

[5] $z_j = u_{j-1}$ or u_j implies $u_{j-1} \leq z_j \leq u_j$. Riemann sums satisfying the first condition also satisfy the second. So if f is integrable under the first arrangement, it is also integrable under the second, but not conversely. Nevertheless, the first condition on evaluation points in Riemann sums is sufficient for the purposes of this book.

and if $u_{i-1} \le x_i \le u_i$, then

$$
\begin{aligned}
(\mathcal{P})\sum f(x)|I| &= \sum_{i=1}^{n} f(x_i)(u_i - u_{i-1}) \\
&= \sum_{i=1}^{n} \left(f(x_i)(x_i - u_{i-1}) + f(x_i)(u_i - x_i) \right), \\
F(1) - F(0) &= \sum_{i=1}^{n} F(u_i) - F(u_{i-1}) \\
&= \sum_{i=1}^{n} \left((F(x_i) - F(u_{i-1}) + (F(u_i) - F(x_i)) \right)
\end{aligned}
$$

Let $\varepsilon > 0$ be given. Then, for each x, $0 < x < 1$, there exists a number $\delta(x) > 0$ such that, for $|x - a| < \delta(x)$,

$$
\left| \frac{F(x) - F(a)}{x - a} - f(x) \right| < \varepsilon. \tag{10.7}
$$

Now choose a partition $\mathcal{P}$ so that each term $f(x_i)(u_i - u_{i-1})$ satisfies

$$
x_i - u_{i-1} < \delta(x_i), \qquad u_i - x_i < \delta(x_i).
$$

The existence of such partitions is a consequence of the Heine-Borel theorem. For such a partition (10.7) implies

$$
\begin{aligned}
|(F(x_i) - F(u_{i-1})) - f(x_i)(x - u_{i-1})| &< \varepsilon(x_i - u_{i-1}), \tag{10.8} \\
|(F(u_i) - F(x_i)) - f(x_i)(u_i - x_i)| &< \varepsilon(u_i - x_i); \tag{10.9}
\end{aligned}
$$

with corresponding inequalities for $x = 0$ and $x = 1$. Writing $\alpha = F(1) - F(0)$ and $\beta = (\mathcal{P})\sum f(x)|I|$,

$$
\begin{aligned}
|\alpha - \beta| &= \left| \sum_{i=1}^{n} (F(u_i) - F(x_i)) + (F(x_i) - F(u_{i-1})) \ - \right. \\
&\qquad\qquad \left. - \sum_{i=1}^{n} (f(x_i)(x_i - u_{i-1}) + f(x_i)(u_i - x_i)) \right| \\
&\le \sum_{i=1}^{n} \{|(F(x_i) - F(u_{i-1}) - f(x_i)(x_i - u_{i-1})| \ + \\
&\qquad\qquad + \ |(F(u_i) - F(x_i)) - f(x_i)(u_i - x_i)|\} \\
&< \sum_{i=1}^{n} \{\varepsilon(x_i - u_{i-1}) + \varepsilon(u_i - x_i)\} \\
&= \varepsilon \sum_{i=1}^{n} (x_i - u_{i-1} + u_i - x_i) \ = \ \varepsilon,
\end{aligned}
$$

so $\beta = \alpha$. Thus the Riemann-complete integral of f exists and equals the Newton/Leibniz definite integral $F(1) - F(0)$.

In the case that f is given by (10.1), the Newton/Leibniz and Riemann-complete integrals exist. But it has been asserted above that the Lesbesgue integral does not exist.[6] This assertion is verified in 10.11 below.

---

[6] In that case the Riemann integral of f does not exist either.

The definition of the Lebesgue integral of a function can be provided in various equivalent ways. (See [138] or [61], for example.) Given a real-valued, measurable function f defined on an arbitrary measurable space S, with measure μ defined on a family of measurable subsets of S, [123] shows how to define the Lebesgue integral of f on S as a Riemann-Stieltjes integral in the set of real numbers $\mathbf{R}$. In fact, writing

$$g(x) \;=\; \mu\big(f^{-1}(]-\infty,x])\big), \qquad\qquad (10.10)$$

the Lebesgue integral $\int_S f\,d\mu$ is the Riemann-Stieltjes integral $\int_{-\infty}^{\infty} f\,dg$; the latter being also Stieltjes-complete integrable. If $S \subseteq \mathbf{R}$ and μ is Lebesgue measure in $\mathbf{R}$, and if the Lebesgue integral of f exists, then the Riemann-complete integral of f exists and the two integrals are equal. Every Lebesgue integrable function is integrable in the Riemann-complete sense (see theorem 120, [MTRV] page 246).[7]

A key point is that the Lebesgue integral is an absolute integral, while the Riemann-complete integral is non-absolute. Writing $f_+(x) = f(x)$ if $f(x) \ge 0$, with $f_+(x) = 0$ otherwise, and $f_-(x) = f(x) - f_+(x)$, absolute integrability implies that f is Lebesgue integrable if and only if both f_+ and f_- are Lebesgue integrable.[8] We use this point to demonstrate that the function f defined by (10.1) is not Lebesgue integrable.

With that in mind, Figure 10.3 provides an indication of how the function f defined by (10.1)

- fails to be Lebesgue integrable, while

- its Riemann-complete integral exists.

In fact Figure 10.3 shows that, in neighbourhoods of $x = 0$, the graph of f oscillates increasingly rapidly, in positive loops (f_+ above the x-axis) and negative loops (f_- below the x-axis) whose amplitude (or height/depth) increases without limit as $x \to 0$. This creates the intuition, or expectation, that the sum of areas of the positive loops diverges to $+\infty$, while the sum of areas of the negative loops diverges to $-\infty$.

But if, instead of treating positive and negative loops separately, we add up their areas in their natural sequence, then positive and negative areas will tend to cancel each other out, and the resulting sequence of net values may converge.[9] The latter is what happens in the Riemann sum construction of the Riemann-complete integral of f.

The following discussion seeks to confirm these points. In any interval, not including zero, but with small values of x, Figure 10.1 shows that the contribution from the term $2x \sin x^{-2}$ to the area under the graph of f is vanishingly small

---

[7]A "Lebesgue-complete" integral is described in [123], definition 8.2, page 34.

[8]This restriction does not apply to the Riemann-complete integral of f, which does **not** require the Riemann-complete integrability of f_+ and f_-.

[9]For example, with $b_i = (-1)^i i^{-1}$, the series $\sum_{i=1}^{\infty} b_i$ converges, but the series consisting of only the positive terms (or only negative terms) diverges; so $\sum_{i=1}^{\infty} |b_i|$ diverges.

in neighbourhoods of $x = 0$, while the corresponding contribution from the term $2x^{-1}\cos x^{-2}$ in f is relatively large. Therefore, disregarding the term $2x\sin x^{-2}$, the zeros of (10.1) can, for present purposes, be estimated approximately as

$$x = \sqrt{\frac{2}{(2n+1)\pi}} \quad \text{as } x \to 0 \text{ (or integer } n \to \infty).$$

Accordingly we may estimate that, for large, even values of n,

$$\int_{\sqrt{\frac{2}{(2n+3)\pi}}}^{\sqrt{\frac{2}{(2n+1)\pi}}} f_+(x)\,dx \quad \text{is approximately} \quad \frac{2}{\pi}\left(\frac{1}{2n+1} + \frac{1}{2n+3}\right)$$

while, for large and odd values of n,

$$\int_{\sqrt{\frac{2}{(2n+3)\pi}}}^{\sqrt{\frac{2}{(2n+1)\pi}}} f_-(x)\,dx \quad \text{is approximately} \quad \frac{2}{\pi}\left(\frac{1}{2n+1} + \frac{1}{2n+3}\right).$$

Writing

$$a_n = \frac{2}{\pi}\left(\frac{1}{2n+1} + \frac{1}{2n+3}\right),$$

each of the two series

$$a_2 + a_4 + a_6 + \cdots, \qquad a_1 + a_3 + a_5 + \cdots$$

diverges, so it is clear that f is not Lebesgue integrable in $[0,1]$. But it is easy to see that the series

$$-a_1 + a_2 - a_3 + a_4 - \cdots$$

is non-absolutely convergent, even if we did not already know, from existence of the primitive function $F(x)$ for $0 \le x \le 1$, that f is Riemann complete integrable.

This is because the Riemann-complete convergence is obtained from the cancellation effects produced by successively summing the positive and negative parts in their natural sequence.

Convergence of the Riemann sums can be ensured by choosing $\delta(x)$ as follows. When x lies between adjacent roots $\sqrt{\frac{2}{(2n+1)\pi}}$ and $\sqrt{\frac{2}{(2n+3)\pi}}$ let

$$\delta(x) < \min\left\{x - \sqrt{\frac{2}{(2n+3)\pi}}, \sqrt{\frac{2}{(2n+1)\pi}} - x\right\}.$$

If x is one of the roots $\sqrt{\frac{2}{(2n+1)\pi}}$, take

$$\delta(x) < \min\left\{\sqrt{\frac{2}{(2n+1)\pi}}, \sqrt{\frac{2}{(2n+1)\pi}} - \sqrt{\frac{2}{(2n+3)\pi}}\right\}. \tag{10.11}$$

And when $x = 0$ let $\delta(0) > 0$ be arbitrary. Any partition corresponding to this definition of $\delta(x)$ ($0 \le x \le 1$) will contain a term with $f(0) = 0$, and the terms for non-zero x will each contain an arbitrarily close estimate of the area of the corresponding positive or negative loop in Figure 10.3. This provides the required cancellation and convergence of Riemann sums, since the alternating loops are monotone decreasing in area as x approaches 0.

In the case of the Lebesgue integral this cancellation effect is absent, and convergence fails.

This establishes that, just as there are Lebesgue integrable functions that are not Riemann integrable, there are Riemann-complete integrable functions that are not Lebesgue integrable.

10.4   Convergence Criteria

Anybody experienced in the theory of integration will be aware that most of the preceding discussion covers fairly well-worn ground which has already been worked through in many excellent publications, such as [6] and many others.

But at the outset of this chapter it was stated that, while Lebesgue's dominated convergence theorem is a crucial pillar of modern analysis, there are certain areas of the subject where this theorem is deficient. The sequence $\{f_j\}$ of (10.2) demonstrates that the dominated convergence theorem provides no illumination in this particular instance of convergent non-absolute integrals. This section addresses the deficit.

There are convergence conditions and criteria which encompass and surpass the dominated, monotone and uniform convergence theorems of standard integration theory. These are the convergence criteria of Theorems 9, 10, and 11 below. They are valid for Riemann-complete integrals (which include integrals of the Newton/Leibniz, Riemann, and Lebesgue kinds). Measurability of the integrand functions is not assumed.

Theorem 9 *Suppose f_j is integrable on $[0,1]$ and $f_j(x)$ converges to $f(x)$ for $x \in [0,1]$. Then g is integrable if and only there exist a ball B_1 of arbitrarily small radius and, for $x \in [01]$, a gauge $\delta(x)$ and integers $p = p(x)$ depending on B_1 so that*

$$(\mathcal{D}) \sum f_j(x)|I| \in B_1$$

for all choices of $j = j(x) > p(x)$ in the Riemann sum, and for all δ-fine $\mathcal{D}$.

Theorem 10 *Suppose f_j is integrable on $[0,1]$ and $f_j(x)$ converges to $f(x)$ for $x \in [0,1]$. Then $\int_0^1 f_j(x)dx$ converges as $j \to \infty$ if and only if there exist a ball B_2 of arbitrarily small radius, a corresponding integer $q = q(B_2)$, and a gauge δ_j depending on j so that*

$$(\mathcal{D}_j) \sum f_j(x)|I| \in B_2$$

for each $j > q$ and for each δ_j-fine division $\mathcal{D}_j$ of domain $[0,1]$.

Theorem 11 *Suppose f_j is integrable on $[0,1]$ and $f_j(x)$ converges to $f(x)$ for $x \in [0,1]$. Given the existence of $\int_0^1 f$ and $\lim_{j\to\infty}\int_0^1 f_j$, the two are equal if and only if*

$$B_1 \cap B_2 \neq \varnothing.$$

Proofs can be found in [MTRV], chapter 4, pages 174–178. (Note that theorem 63 on page 175 is false. Details of this are in Chapter 11 below.)

For anybody more familiar with the classical integration theorems on passage to a limit, these theorems or criteria may appear somewhat indigestible at first sight.[10]

Their starting point is that a convergent sequence of functions f_j is given. These functions are assumed to be Riemann-complete integrable, which is a weaker assumption than Lebesgue or any other kind of integrability. There is no assumption of properties like continuity or measurability.

To answer questions about convergence of the corresponding sequence of Riemann-complete integrals, and about the Riemann-complete integrability of the limit function, from previous experience of integration we might be led to expect some condition, not about Riemann sums, but only about the functions f_j—such as monotonicity, domination by a fixed integrable function g; or the like. But nothing like this appears in the above convergence criteria. Instead we have various statements about Riemann sums.

However, setting aside for a moment the conception of integral as primitive function, or anti-derivative, the original and more durable meaning of integral involves slicing up (partitioning), followed by summation, followed by taking a limit of the sums.

From this perspective, it may be less of a surprise that Riemann sums appear in the formulation of conditions for limits of integrals, since Riemann sums are central to the concept of integral.

Consider Theorem 9. Given integrability of the terms f_j in the sequence, this theorem addresses the integrability of the limit function f, which, essentially, involves the question of convergence of Riemann sums $\sum f(x)|I|$.

As initial exploration, consider sequences of Riemann sums $\sum f_j(x)|I|$, $j = 1, 2, 3, \ldots$. We know that, for each x, the sequence of values $f_j(x)$ converges to $f(x)$. We also know that, for each j, Riemann sums of the form $\sum f_j(x)|I|$ converge. Can we somehow put these two facts together to deduce, as an immediate consequence, convergence of Riemann sums $\sum f(x)|I|$?

The answer is no. But the answer is yes if the terms f_j satisfy some conditions such as $|f_j| < g$ where g is integrable. If we want a condition expressed in the form of a condition on Riemann sums, clearly something more delicate than convergence of Riemann sums $\sum f_j(x)|I|$ is required.

For instance, with $\varepsilon > 0$ given, the condition we need is **not** that, for all j greater than some $j_0 = j_0(\varepsilon)$, every Riemann sum $\sum f_j(x)|I|$ will be contained within some ball $B_1(\varepsilon)$ of the form $]\beta - \varepsilon,\ \beta + \varepsilon[$.

This is too crude for our purpose. The convergence of $f_j(x)$ to $f(x)$ may be very fast at some points x, and very slow at other points x. This behaviour is provided for in Theorem 9, by choosing, **not** $j_0 = j_0(\varepsilon)$, but $j_0 = j_0(\varepsilon, x)$ which is different for each x.

---

[10]The formulation here is taken from [76]. More loosely, the criteria state that is that there is a family of balls with a common centre so that, for any radius, there corresponds a gauge... etc. Expressed this way, Theorem 11 is trivial: the integral of f and the limit of the integrals are equal if the centres of the balls coincide. See also [78].

This formulation is sufficient, and necessary, for integrability of the limit function f. Once this point is established, the criteria of Theorems 10 and 11 are fairly obvious, and less subtle.

But are these conditions of Theorems 9, 10 and 11 "workable" in the way that the dominated convergence condition $|f_j| < g$ is?

After all, Riemann sums are fine for defining the meaning of the integral of a function. But when we actually want to find the value of an integral we do not typically work with Riemann sums. Instead we revert to the integral as primitive or anti-derivative, using the substitution method or integration by parts. Or we use some less direct method, such as solving a related differential equation. Or a myriad of other *ad hoc* methods.[11]

To respond to such questions, and to demonstrate that Riemann sums can actually be of use here, take, for example, the sequence f_j defined in (10.2):

$$ f_j(x) \;=\; \begin{cases} 2x \sin \frac{1}{x^2} - \frac{2}{x} \cos \frac{1}{x^2} & \text{if } \ \frac{1}{j} \le x \le 1, \\ 0 & \text{if } \ 0 \le x < \frac{1}{j}. \end{cases} $$

Remember, for each j the function f_j is Riemann integrable and Lebesgue integrable, but not Newton/Leibniz integrable, and their limit function f (from (10.1)) is Newton/Leibniz integrable but not Riemann or Lebesgue integrable:

$$ f(x) \;=\; \begin{cases} 2x \sin \frac{1}{x^2} - \frac{2}{x} \cos \frac{1}{x^2} & \text{if } \ 0 < x \le 1, \\ 0 & \text{if } \ x = 0. \end{cases} $$

For each j, f_j is Riemann-complete integrable.[12] This discussion of the convergence criteria of Theorems 9, 10 and 11 is set in the context of Riemann-complete integrability.

The subject of the first criterion is the (Riemann-complete) integrability of the limit function f, and it is established by examining Riemann sums of the form

$$ (\mathcal{D}) \sum f_{j(x)}(x)|I|, \quad \text{or} \quad (\mathcal{D}) \sum f_{j(x)}(x)(v-u) $$

where $I =]u, v]$ and $x = u$ or v. We already know, by various means, including a direct investigation of the Riemann sums $\sum f(x)|I|$, that $f(x)$ is (Riemann-complete) integrable on $[0, 1]$.

The function $f(x)$ is the limit of functions $f_j(x)$. Is it possible to further confirm the integrability of f by direct examination, not just of $\sum f(x)|I|$, but of Riemann sums $\sum f_{j(x)}(x)|I|$ involving functions f_j instead of f, where the factor f_j in the sum has variable index $j = j(x)$, depending on the element x of the division $\mathcal{D} = \{(x, I)\}$ used to construct the Riemann sum?

This is the essence of Theorem 9. And according to Theorem 9 the answer to this question should be yes. Given $\varepsilon > 0$, and with a suitable gauge $\delta(x)$,

---

[11]Which is **not** to say that Riemann sums are "merely" a device of fundamental theory, and nothing else. Versions of them have other uses. Such as the ancient techniques of quadrature; or computer programs for estimating numerical values of an integral. Simpson's rule is another example.

[12]As is f, from earlier discussion. But for present purposes we wish to **deduce** Riemann-complete integrability of f from Theorem 9.

provided factors $f_{j(x)}(x)$ are chosen appropriately, we should be able to demonstrate that the value of each Riemann sum $(\mathcal{D}_\delta) \sum f_{j(x)}(x)|I|$ will lie within some ball B_1 of arbitrarily small radius ε.

Since we are already convinced of the integrability of f in this case, what we are really trying to do here is to get a sense of the behaviour of sums $\sum f_{j(x)}(x)|I|$. So, given the integrability of f, write $\alpha = \int_{]0,1]} f(x)dx$ and choose a gauge δ so that, for every δ-fine division $\mathcal{D}$ of $]0,1]$,

$$|(\mathcal{D}) \sum f(x)|I| - \alpha| \; < \; \varepsilon, \quad \text{or} \quad (\mathcal{D}) \sum f(x)|I| \; \in \; B(\alpha, \varepsilon),$$

the ball with centre α and radius ε.

Consider any one of these Riemann sums $(\mathcal{D}) \sum f(x)|I|$, corresponding to a particular δ-fine division $\mathcal{D} = \{(x, I)\}$. For each x with $(x, I) \in \mathcal{D}$, choose

$$r(x) \; \geq \; \frac{1}{x}.$$

Then, by definition of f_j, if $j = j(x) \geq r(x)$,

$$f_j(x) \; = \; f_{j(x)}(x) = f(x),$$

so, for all choices $j(x) \geq r(x) = r(x, \varepsilon)$,

$$(\mathcal{D}) \sum f_{j(x)}(x)|I| \; = \; (\mathcal{D}) \sum f(x)|I| \; \in \; B(\alpha, \varepsilon),$$

as required by Theorem 9.

Generally speaking the convenient equation $f_j(x) = f_{j(x)}(x) = f(x)$ will not be available. But if, with suitable choices of $j = j(x)$, the differences

$$f_j(x) - f(x), \; = \; f_{j(x)}(x) - f(x),$$

can make sufficiently small contributions to the Riemann sum, then it may be plausible that

$$(\mathcal{D}) \sum f_{j(x)}(x)|I| \; \in \; B'(\varepsilon) \quad \text{implies} \quad (\mathcal{D}) \sum f(x)|I| \; \in \; B''(\varepsilon),$$

so f is integrable. This is the intuitive content of Theorem 9. (A proof can be found in [MTRV], page 175.)

Next consider Theorem 10. The preceding remarks are concerned with the integrability of $\lim_{j \to \infty} f_j$. The fundamental assumption is that each function f_j in the sequence $\{f_j\}$ is integrable. In this case with $f_j(x)$ given by (10.2) the anti-derivatives (10.4) are the sequence $\{F_j\}$, giving a sequence of integrals

$$\int_{]0,1]} f_j(x)dx \; = \; F_j(1) - F_j(j^{-1}) \; = \; \sin 1 - F_j\left(j^{-1}\right),$$

which can be denoted by α_j. Note that continuity of F implies $F_j\left(j^{-1}\right) \to 0$ as $j \to \infty$. In this case it is already clear that the sequence of integrals $\int_{]0,1]} f_j(x)dx$ converges as $j \to \infty$, the limit being $\sin 1$. What we want from this example is confirmation that Theorem 10 makes sense intuitively.

The convergence of a sequence of integrals is the subject of Theorem 10, and it is again expressed in terms of Riemann sums. The criterion states that there is a ball B_2 of arbitrarily small radius and a corresponding integer q, so that for each $j \geq q$, a gauge $\delta_j(x)$ can be found such that every δ_j-fine division $\mathcal{D}_j$ the resulting Riemann sum $(\mathcal{D}_j) \sum f_j(x)|I|$ is contained in B_2.

This will be satisfied if, for instance, we can show that, with $\varepsilon > 0$ given, there is a number β so that, for every δ_j-fine division $\mathcal{D}_j$,

$$\left| (\mathcal{D}_j) \sum f_j(x)|I| - \beta \right| < \varepsilon$$

whenever $j \geq q$. Unlike Theorem 9, j is the same for each term of any particular Riemann sum in this case.

Again, this is easy to demonstrate since we already know in this case that the integrals $\int_{]0,1]} f_j(x)dx$ converge to the value $\sin 1$ as $j \to \infty$. Just take

$$\beta = \sin 1 = \lim_{j \to \infty} g(1) - g(j^{-1}) = \int_{]0,1]} f_j(x)dx,$$

and choose q so that $j \geq q$ implies

$$\left| \beta - \int_{]0,1]} f_j(x)dx \right| < \varepsilon.$$

For each $j \geq q$ choose a gauge $\delta_j(x)$ $(0 < x \leq 1)$ so that, for any δ_j-fine division $\mathcal{D}_j$ of $]0,1]$,

$$\left| \int_{]0,1]} f_j(x)dx - (\mathcal{D}_j) \sum f_j(x)|I| \right| < \varepsilon.$$

Then, by the triangle inequality,

$$\left| \beta - (\mathcal{D}_j) \sum f_j(x)|I| \right| < 2\varepsilon, \quad \text{or} \quad (\mathcal{D}_j) \sum f_j(x)|I| \in B_2 = B(\beta, 2\varepsilon)$$

for all $j \geq q = q(\varepsilon)$ and all δ_j-fine divisions $\mathcal{D}_j$ of $]0,1]$. In other words, the criterion of Theorem 10 confirms the convergence of the sequence of integrals

$$\left\{ \int_{]0,1]} f_j(x)dx \right\};$$

and this illustrates the intuitive content of Theorem 10.

Finally, the question arises whether the integral of the limit

$$\int_{]0,1]} \lim_{j \to \infty} f_j(x)dx$$

equals the limit of the integrals

$$\lim_{j \to \infty} \int_{]0,1]} f_j(x)dx.$$

For the sequence f_j of (10.2), we already know by direct evaluation that these two quantities have the same value, namely $\sin 1$. But how does this stand with the criterion of Theorem 11, which requires that B_1 and B_2 have a non-empty intersection? Since

$$B_1 \;=\; B(\sin 1, \varepsilon) \quad \text{and} \quad B_2 \;=\; B(\sin 1, 2\varepsilon),$$

this criterion is obviously satisfied, with $B_1 \subset B_2$. Again, the intuitive content of Theorem 11 is clear in the context of this example.

Example 7 uses "slightly monstrous" Burkill integrands $h_n(s, I)$ and their limit $h(s, I)$ (based on the Dirichlet function) which, for $I = \,]s', s'']$, depend on the right hand boundary point s'' of I, but are independent of $|I| = s'' - s'$. The monotone convergence theorem was used to deduce integrability of the limit function h from the integrability of h_n, $n = 1, 2, 3, \dots$. Unlike the sequence of trigonometric functions f_j above, the convergence of h_n to h is absolute. Nonetheless, Theorem 9 can be applied (as alternative to the monotone convergence theorem) to deduce the integrability of the limit function h.

Example 58 *For ease of reference, here are the integrands h_n and h. Let Q denote the rational numbers in the domain $]0, 1[$, enumerated as*

$$Q \;=\; \{r_1, r_2, r_3, \dots\} \quad \text{and write} \quad Q_n \;=\; \{r_1, r_2, \dots, r_{n-1}\}, \quad n = 1, 2, 3, \dots.$$

For $I = \,]s', s''] \subseteq \,]0, 1]$, and associated pairs (x, I) with $x = s'$ or $x = s''$, define

$$h_n(x, I) \;:=\; \begin{cases} \frac{1}{2^j} & \text{if } s'' = r_j, \ 1 \le j \le n, \\ 0 & \text{otherwise;} \end{cases}$$

$$h(x, I) \;:=\; \begin{cases} \frac{1}{2^j} & \text{if } s'' = r_j, \ j = 1, 2, 3, \dots, \\ 0 & \text{otherwise.} \end{cases}$$

Theorems 9, 10, and 11 above refer to Riemann-complete integration of functions $f_j(x)|I|$, $f(x)|I|$; while h_n and h are Burkill-complete integrands. Theorems 62, 64, and 65 in [MTRV] deal with convergence of Burkill-type integrands. Theorem 62 can be applied to h_n, h. Here is a suitable statement of this theorem: Suppose, for $n = 1, 2, 3, \dots$, the integrable functions $h_n(x, I)$ converge to $h(x, I)$. Then h is integrable if and only if there exist a ball B of arbitrarily small radius, and, correspondingly, a gauge $\delta(x) > 0$ and integers $m = m(x, I)$ depending on B so that

$$(\mathcal{D}) \sum h_{n(x,I)}(x, I) \;\in\; B$$

for all choices $n = n(x, I) \ge m(x, I)$ in the Riemann sum, and for all δ-fine divisions $\mathcal{D}$.

To apply this, let $\varepsilon > 0$ be given, and choose $m = m_\varepsilon$ so that

$$1 - \sum_{j=1}^{n} \frac{1}{2^j} \;<\; \varepsilon \quad \text{for all} \ n \ge m_\varepsilon.$$

Referring back to the construction in Example 7, choose a gauge $\delta(x) > 0$ which conforms to $P_{m,} = P_{m_\varepsilon}$. This ensures that, for all δ-fine divisions $\mathcal{D} = \{(x, I)\}$

of $]0,1]$, $\mathcal{D}$ contains (r_j, I) for $j = 1, 2, \ldots, m_\varepsilon$. Therefore, for all $m = m(x, I) \geq m_\varepsilon$, and for all δ-fine divisions $\mathcal{D}$ of $]0,1]$,

$$1 - (\mathcal{D})\sum h_{m(x,I)}(x, I) \;\leq\; 1 - \sum_{j=1}^{m_\varepsilon} \frac{1}{2^j} \;<\; \varepsilon;$$

so $(\mathcal{D})\sum h_{m(x,I)}(x, I)$ is contained in a ball of centre 1 and radius ε. Integrability of the limit function h then follows from theorem 62 of [MTRV]. Theorems 64 and 65 of [MTRV] can be used to confirm that $\int_0^1 h(x, I) = 1$.

10.5   "I would not care to fly in that plane"

Does it really matter whether aviation designers work out their aerodynamic equations using old-fashioned Riemann integrals or the latest sophisticated Lebesgue integrals?

Probably not much. But it matters a lot if the value $22/7$ for π were hard-wired into every computer in the world. Or if the wrong value for elasticity of O-rings at freezing temperature was used in space shuttle design. And it certainly matters whether our aviation designers make tricky, unjustifiable calculations involving, for instance, term-by-term integration of Fourier series.

So, thanks to the intellectual diligence of the nineteenth century, not to mention its monster-functions, we have the dominated convergence theorem to keep the garden of mathematics safe and fertile—and, indeed, to keep airplanes flying safely.

But do we really need anything more than the dominated convergence theorem for absolutely convergent integrals? Why bring up the convergence criteria of Theorems 9, 10, and 11? Is the sequence $\{f_j\}$ of (10.2) just an exceptional one-off, or is it representative of something more significant? If the latter, where are all these non-absolute integrals?

In fact they are very widespread. Modern stochastic calculus [127, 137] is based on integrals for which absolute convergence fails, but which may converge weakly or, in some cases, non-absolutely. These are described in Part I of this book, and in [MTRV].

Quantum mechanics can be formulated in terms of path integrals which also fail to converge absolutely. Famously, the dominated convergence theorem does not work for these integrals, and, as described in [MTRV], the non-absolute convergence criteria must be invoked.

"*Does anyone believe ... I would not care to fly in that plane.*" A healthy scepticism must be welcomed. But what is certain is that, while integration is central to mathematical analysis, there are no certain and definite ways of tackling any problem of integration, and even a beginning student has to exercise imagination and ingenuity. From the ancient methods of quadrature, to the methods of Newton/Leibniz, Cauchy, Riemann, Lebesgue, Denjoy, Perron, Kolmogorov, Kurzweil, or Henstock, it is a mistake to disregard any resource or insight that can be called upon.

Chapter 11

Appendix 2: Theorem 63

Theorem 63 of [MTRV] (page 175) states the following: *In domain $\mathbf{R}^T$ suppose*

- $f_j(x, N) \to f(x, N)$ *as $j \to \infty$,*

- $h(x, N, I[N])$ *is VBG**,*

- *and suppose $f_j(x, N)h(x, N, I[N])$ is integrable on $\mathbf{R}^T$ for each j.*

Then $f(x, N)h(x, N, I[N])$ is integrable on $\mathbf{R}^T$.

This is false, as Example 59 below (a counter-example) shows. If the functions f_j and f of Example 59 are inserted into the argument of pages 175–176, it breaks down.

Unlike other errors and typos[1] in [MTRV], there is no correction or replacement for theorem 63. And since this book is a supplement or continuation of [MTRV], it is appropriate and instructive to revisit it here.

11.1 Fresnel's Integral

There are additional reasons for paying some attention to this point. Saks' *Theory of the Integral* [141] was a basic text for the first half of the twentieth century, and is still a fundamental resource. From beginnings in the 1950s, J. Kurzweil and R. Henstock independently moved the subject forward. But while the description "gauge integral" is applied indiscriminately to the work of both, they had different orientations and priorities.

Henstock's major works ([70, 75, 76]) can be viewed as a contribution to the Saks' agenda. Just as Lebesgue's work on convergence of integrals, such as the dominated convergence theorem, established a new age of integration theory, Henstock's work in the subject is exemplified in his radically new convergence theorems which appeared in fully developed form in [76], in 1991.

---

[1]These are detailed in [**website**], along with discussion of theorem 63.

In [MTRV] these convergence theorems[2] are expressed in a form suitable for Feynman integrals, as theorems 62, 64, and 65 (pages 174–178). Since they are central to Feynman's non-absolute integration, and since theorem 63 sits amongst them, it is best to give the latter some attention here. (The fallacious theorem 63 is due to the present author, not Henstock.) Another reason for revisiting it is closer scrutiny of the Fresnel and Gaussian integrals.

Example 59 *In domain* **R** *suppose* $f_j(x) = 1$ *for* $0 \le x \le j$ *and* $= 0$ *otherwise. Suppose* $h(x, I) = |I|$, *so h is VBG* * in* **R**. *Then, for each* x, $f_j(x) \to f(x)$ *as* $j \to \infty$, *where* $f(x) = 0$ *for* $x < 0$ *and* $f(x) = 1$ *for* $x \ge 0$. *Also, for each* j, $f_j(x)|I|$ *is integrable on* **R**, *with integral value* j, *but* $f(x)|I|$ *is not integrable on* **R**, *in contradiction to theorem 63.*

The "proof" of theorem 63 (page 175) is invalid. All it does is to "prove" that the identically zero function f is integrable with respect to the integrator function h. But the zero function is integrable with respect to any integrator h, regardless of whether or not h is VBG*. (In Example 59, the domain **R** can be resolved into a countable union of subsets X_j, $= [j, j+1[$, in each of which the function $h(x, I)$, $= |I|$, is bounded with variation 1; and is therefore VBG*.)

Theorem 63 is subsequently used to "prove" theorem 142 (page 269), lemma 28 (page 299), theorem 211 (page 332), and theorem 217 (page 338). Accordingly, these results have to be re-examined for valid proofs.

The presentation in this book is mostly discursive. But at this point, in order to deal with theorem 63, a more technical approach is adopted using the language, notation, methods, and "mentality" of [MTRV].

Since theorem 63 is false, those results which refer to it in [MTRV] need to be reconsidered. Regarding theorem 142, if $J \in \mathcal{I}(\mathbf{R}^T)$ has no unbounded components then both $\varphi_{nc}(x)|I|$ and $\varphi_{nc}(I)$ are integrable on J, with the same integral value, and the two functions are variationally equivalent in the domain J. The revised theorem 142 below implies that these statements are true when J is replaced by $\mathbf{R}^n$.

The condition "$a \le 0$, $b \ge 0$, $c = a + \iota b \ne 0$" is used in chapter 6 of [MTRV] for generalized Gaussian-Fresnel integrals. But when $a = 0$ the condition $b > 0$ can usually be replaced by $b \ne 0$, as $b > 0$ has no greater efficacy than $b < 0$. Therefore, replace "$a \le 0$, $b \ge 0$, $c = a + \iota b \ne 0$" by "$a \le 0$, $c = a + \iota b \ne 0$".

Theorem 142 *For any positive integer* n, *with* $a \le 0$, $c = a + \iota b \ne 0$, *the function* $\varphi_{nc}(x)|I|$ *is integrable on* $\mathbf{R}^n$, *with*

$$\int_{\mathbf{R}^n} \varphi_{nc}(x)|I| \; = \; \left(\sqrt{\frac{\pi}{-c}} \right)^n . \tag{11.1}$$

The expression on the left, an integral on domain $\mathbf{R}^n$, is obtained as a limit of Riemann sums of combinations of various n-dimensional expressions. For

---

[2]See also Theorems 9, 10, and 11 above.

instance, the expression $|I|$ is a volume in $\mathbf{R}^n$ obtained by multiplying together the lengths $|I_j|$ of the one-dimensional components I_j of $I = I_1 \times \cdots \times I_n$.

The expression on the right is equal to $\int_{-\infty}^{\infty} \exp(cy^2)dy$ multiplied by itself n times, and can be thought of as a product of extended Riemann integrals.

Experience with integrals in bounded one-dimensional domains might suggest that some seemingly obvious manipulations of the -complete kind will give this result. But a counter-example[3] in [118] shows that this will not always work in unbounded domains $\mathbf{R}^n$ with $n > 1$, so a proof on the following lines is required. The notation and references are those of [MTRV].

Proof of theorem 142. It is slightly easier to give this proof for $\varphi_{nc}(I)$ instead of $\varphi_{nc}(x)|I|$. For $r = 1, 2, 3, \ldots$, let

$$J^r = \ ]-r, r]^n = \ ]-r, r] \times \cdots \times \ ]-r, r].$$

By Fubini's theorem, and by repeated applications of theorem 133,

$$\int_{J^r} \varphi_{nc}(I) = \left(\int_{-r}^{r} e^{cy^2} dy \right)^n \rightarrow \left(\sqrt{\frac{\pi}{-c}} \right)^n$$

as $r \to \infty$. Let $\varepsilon > 0$ be given. Choose r_ε so that $r \geq r_\varepsilon$ implies

$$\left| \int_{J^r} \varphi_{nc}(I) - \left(\sqrt{\frac{\pi}{-c}} \right)^n \right| < \varepsilon.$$

Choose a gauge $\delta(x)$ in $\mathbf{R}^n$ as follows. For $x \in \bar{J}^r$ let δ conform to J^1 and to $J^r \setminus J^{r-1}$ for each $r > 1$. If x has an infinite component $x_j = +\infty$ let $\delta(x) = (r_\varepsilon)^{-1}$; and if x has an infinite component $x_j = -\infty$ let $\delta(x) = -(r_\varepsilon)^{-1}$. (We need not worry about cases where one component $x_j = \infty$ and a different component $x_{j'} = -\infty$.) For associated pairs (x, I) where x has an infinite component, use the convention that $\varphi_{nc}(I) = 0$. (This convention is convenient, but not essential, for the following argument.) Let $\mathcal{D}$ be a δ-fine division of $\mathbf{R}^n$. The non-zero terms of the Riemann sum $(\mathcal{D}) \sum \varphi_{nc}(I)$ can be separated into two groups $\mathcal{D} = \mathcal{D}' \cup \mathcal{E}''$ as follows. There is $r' \geq r_\varepsilon$ such that r' is the maximum value of r for which $\mathcal{D}'$ is a full δ-fine division of $J^{r'}$; and there is $r'' \geq r' \geq r_\varepsilon$ for which $\mathcal{E}'' = \mathcal{D} \setminus \mathcal{D}'$ is a partial division of $J^{r''} \setminus J^{r'}$. If $r'' = r'$ then $\mathcal{E}''$ is empty and

$$(\mathcal{D}) \sum \varphi_{nc}(I) = (\mathcal{D}') \sum \varphi_{nc}(I) = \int_{J^{r'}} \varphi_{nc}(I) = \left(\int_{-r'}^{r'} e^{cy^2} dy \right)^n$$

and the proof is complete. Therefore suppose $r'' > r'$. Let $\mathcal{D}$ be a δ-fine division of $\mathbf{R}^n$. Choose $\mathcal{E}^*$ so that $\mathcal{D}'' = \mathcal{E}'' + \mathcal{E}^*$ is a δ-fine division of $J^{r''}$, so

$$(\mathcal{D}'') \sum \varphi_{nc}(I) = \int_{J^{r''}} \varphi_{nc}(I) = \left(\int_{-r''}^{r''} e^{cy^2} dy \right)^n$$

---

[3] When $n > 1$ a function which is integrable on $\mathbf{R}^n$ in an "extended Riemann" sense is not necessarily integrable in the -complete sense.

and

$$\left| \int_{Jr'} \varphi_{nc}(I) - \left(\sqrt{\frac{\pi}{-c}}\right)^n \right| < \varepsilon, \quad \left| \int_{Jr''} \varphi_{nc}(I) - \left(\sqrt{\frac{\pi}{-c}}\right)^n \right| < \varepsilon.$$

This implies

$$\left|(\mathcal{E}'') \sum \varphi_{nc}(I)\right| < 2\varepsilon, \quad \text{and} \quad \left|(\mathcal{D}) \sum \varphi_{nc}(I) - \left(\sqrt{\frac{\pi}{-c}}\right)^n \right| < 3\varepsilon,$$

which gives the required result. ○

Next, consider lemma 28, page 299 of [MTRV], which purports to establish that, if Q is a countable dense set of real numbers (for instance, if Q consists of the rational numbers $\{s_1, s_2, s_3, \ldots\}$ in the interval $]0, t]$), then the set D of those $x \in \mathbf{R}^Q$, which fail to be continuous everywhere relative to the elements s of Q, form a $G_{\iota b}$-null set in $\mathbf{R}^Q$:

$$\int_{\mathbf{R}^Q} 1_D(x) G_{\iota b}(I[N]) = \int_{\mathbf{R}^Q} 1_D(x) g_{\iota b}(x(N))|I[N]| = 0$$

whenever $b \neq 0$; or, with $C = \mathbf{R}^Q \setminus D$ (the set of $x \in \mathbf{R}^Q$ which are continuous relative to Q),

$$\int_{\mathbf{R}^Q} 1_C(x) G_{\iota b}(I[N]) = \int_{\mathbf{R}^Q} 1_C(x) g_{\iota b}(x(N))|I[N]| = 1.$$

Lemma 28 states that the integral exists, and, in justification, refers to theorem 63, which is false. Therefore it is necessary to provide some proof.

For insight into this, consider Fresnel's integral ([MTRV], section 6.2, pages 261–265), with $\Re$ and $\Im$ denoting real and imaginary parts, respectively,

$$\frac{2}{\sqrt{\iota\pi}} \int_0^\infty e^{\iota y^2} dy = 1, \quad \Re\left(\frac{2}{\sqrt{\iota\pi}} \int_0^\infty e^{\iota y^2} dy\right) = 1, \quad \Im\left(\frac{2}{\sqrt{\iota\pi}} \int_0^\infty e^{\iota y^2} dy\right) = 0.$$
(11.2)

Therefore, using $\frac{1}{\sqrt{\iota}} = -\frac{1}{\sqrt{2}} - \iota\frac{1}{\sqrt{2}}$,

$$\frac{2}{\sqrt{\pi}}\left(-\frac{1}{\sqrt{2}} - \iota\frac{1}{\sqrt{2}}\right) \int_0^\infty \left(\cos y^2 + \iota \sin y^2\right) dy = 1, \quad \text{giving}$$

$$\sqrt{\frac{2}{\pi}} \int_0^\infty \left(\sin y^2 - \cos y^2\right) dy = 1, \quad \sqrt{\frac{2}{\pi}} \int_0^\infty \left(\sin y^2 + \cos y^2\right) dy = 0$$

$$\int_0^\infty \sin y^2 = \frac{\sqrt{2\pi}}{4}, \quad \int_0^\infty \cos y^2 = -\frac{\sqrt{2\pi}}{4}, \quad (11.3)$$

when real and imaginary parts of Fresnel's integral are separated.

The convergence behaviour of $\int_0^\infty \cos y^2 dy$ (or $\int_0^\infty \sin y^2$) is indicated in figure 6.3 ([MTRV] page 302). Broadly speaking, the graph of $\sin y^2$ (and of $\cos y^2$)

consists of cycles, each of which consists of a pair of half-cycles which, respectively, bound (with the horizontal axis) alternately positive and negative areas. The absolute values of these areas are strictly decreasing as y increases, so the succession of areas constitutes a non-absolutely convergent alternating series.

Thus the improper Riemann integral $\int_0^\infty \sin y^2$ (and $\int_0^\infty \cos y^2$) converges; and, with $\varepsilon > 0$ given, there exists $\beta_\varepsilon > 0$ for which $\beta > \beta_\varepsilon$ implies

$$\left| \int_\beta^\infty \sin y^2 dy \right| < \varepsilon.$$

It is useful to try to establish some relationship between ε and β_ε. A change of variable makes it easier to establish upper and lower bounds for the "tail" $\int_\beta^\infty \sin y^2 dy$. To examine the convergence in greater detail, let $u = y^2$, so

$$\int_\beta^\infty \sin y^2 dy = \frac{1}{2} \int_{\beta^2}^\infty \frac{\sin u}{\sqrt{u}} du.$$

The graph of $\sin u$ consists of paired half-cycles on values $(2n+1)\pi \le u \le (2n+2)\pi$ in which $\sin u$ is non-positive, and values $(2n+2)\pi \le u \le (2n+3)\pi$ in which $\sin u$ is non-negative. An overestimate of the area of one of the full cycles

$$\int_{(2n+1)\pi}^{(2n+3)\pi} \frac{\sin u}{\sqrt{u}} du$$

is given by

$$\int_{(2n+1)\pi}^{(2n+2)\pi} \frac{\sin u}{\sqrt{(2n+2)\pi}} du + \int_{(2n+2)\pi}^{(2n+3)\pi} \frac{\sin u}{\sqrt{(2n+2)\pi}} du, \quad = 0, \quad (11.4)$$

where the first integral (negative half-cycle) replaces $u^{-\frac{1}{2}}$ by the smallest value taken by $u^{-\frac{1}{2}}$ in the domain $[(2n+1)\pi, (2n+2)\pi]$, and the second integral (positive half-cycle) replaces $u^{-\frac{1}{2}}$ by the largest value taken by $u^{-\frac{1}{2}}$ in the domain $[(2n+2)\pi, (2n+3)\pi]$.

The results in (11.2) are based on Fresnel's integral (theorem 133 ([MTRV] page 261),

$$\int_{-\infty}^\infty e^{\iota b y^2} dy = \sqrt{\frac{\pi}{-\iota b}} = \sqrt{\iota} \sqrt{\frac{\pi}{b}} = -\sqrt{\frac{\pi}{2b}} - \iota \sqrt{\frac{\pi}{2b}}$$

provided $b > 0$, giving

$$\int_{-\infty}^\infty \cos by^2 dy = -\sqrt{\frac{\pi}{2b}}, \qquad \int_{-\infty}^\infty \sin by^2 dy = -\sqrt{\frac{\pi}{2b}}.$$

for $b > 0$. Inspection of figure 6.3 ([MTRV] page 302) suggests that these non-absolute integrals should converge also for $b < 0$. In that case,

$$\cos by^2 = \cos |b| y^2, \qquad \int_{-\infty}^\infty \cos by^2 dy = -\sqrt{\frac{\pi}{2|b|}};$$

so the latter is valid for $b \neq 0$ (that is, for b positive or negative). Similarly, $\sin by^2 = -\sin|b|y^2$, giving

$$\int_{-\infty}^{\infty} \sin by^2\, dy = -\int_{-\infty}^{\infty} \sin|b|y^2\, dy = -\left(-\sqrt{\frac{\pi}{2|b|}}\right) = \sqrt{\frac{\pi}{2|b|}};$$

so, for $b < 0$,

$$\int_{-\infty}^{\infty} e^{\iota by^2}\, dy = \sqrt{\frac{\pi}{2|b|}}(1+\iota) = \sqrt{\frac{\pi}{|b|}}e^{\iota\frac{3\pi}{4}}.$$

11.2   Theorem 188 of [MTRV]

Example 59 above shows that theorem 63 ([MTRV] page 175) is false. Therefore valid replacement proofs must be provided for those results in [MTRV] which cite theorem 63. Replacement results are now given for lemma 28 (page 299 of [MTRV]) and theorem 188 of (page 300 of [MTRV]). The replacements that follow are designated as Theorem 188A, Theorem 188B, Theorem 188C, and Theorem 188D.

Theorem 188A *For sufficiently large $\beta > 0$,*

$$\left|\frac{1}{\sqrt{\iota\pi}}\int_{|y|>\beta} e^{\iota y^2}\, dy\right| < \frac{8\sqrt{2}}{\sqrt{\pi\beta}}.$$

Proof. Suppose n in (11.4) is such that β^2 satisfies

$$(2n-1)\pi \leq \beta^2 < (2n+1)\pi.$$

An overestimate of the real number $\int_{\beta^2}^{(2n+1)\pi} u^{-\frac{1}{2}}\sin u\, du$ is then given by

$$\int_{2n\pi}^{(2n+1)\pi} \frac{\sin u}{\sqrt{2n\pi}}\, du, = \frac{2}{\sqrt{2n\pi}},$$

in which any negative contribution $\int_{(2n-1)\pi}^{\beta^2}$ (if there is one) is omitted, and in which $u^{-\frac{1}{2}}$ is replaced by the largest value of $u^{-\frac{1}{2}}$ in the domain of the integral. With n chosen by $(2n-1)\pi \leq \beta^2 < (2n+1)\pi$,

$$\int_{\beta}^{\infty} \sin y^2\, dy = \int_{\beta^2}^{\infty} \frac{\sin u}{\sqrt{u}}\, du < \frac{2}{\sqrt{2n\pi}},$$

since each of the full cycles in $u > (2n+1)\pi$ has upper bound 0. A similar calculation provides a lower bound:

$$\frac{-2}{\sqrt{2n\pi}} < \int_{\beta^2}^{\infty} \frac{\sin u}{\sqrt{u}}\, du = \int_{\beta}^{\infty} \sin y^2\, dy.$$

Thus, for $(2n-1)\pi \le \beta^2 < (2n+1)\pi$,

$$\frac{-2}{\sqrt{2n\pi}} < \int_{\beta^2}^{\infty} \frac{\sin u}{\sqrt{u}} du = \int_{\beta}^{\infty} \sin y^2 dy < \frac{2}{\sqrt{2n\pi}}, \qquad (11.5)$$

with similar bounds for $\int_{\beta}^{\infty} \cos y^2 dy$. Since $\sqrt{(2n+1)\pi} - \sqrt{(2n-1)\pi} \to 0$ as $n \to 0$, continuity implies that, for β sufficiently large,

$$\frac{-2}{\sqrt{\beta}} \le \int_{\beta}^{\infty} \sin y^2 dy < \frac{2}{\sqrt{\beta}}, \qquad \frac{-2}{\sqrt{\beta}} \le \int_{\beta}^{\infty} \cos y^2 dy < \frac{2}{\sqrt{\beta}}. \quad (11.6)$$

With $\int_{|y|>\beta} = \int_{-\infty}^{-\beta} + \int_{\beta}^{\infty}$,

$$\left| \int_{|y|>\beta} \sin y^2 dy \right| < \frac{1}{\sqrt{\beta}}, \qquad \left| \int_{|y|>\beta} \cos y^2 dy \right| < \frac{4}{\sqrt{\beta}}, \qquad (11.7)$$

giving

$$\left| \int_{|y|>\beta} (\sin y^2 - \cos y^2) dy \right| < \frac{8}{\sqrt{\beta}}, \quad \left| \int_{|y|>\beta} (\sin y^2 + \cos y^2) dy \right| < \frac{8}{\sqrt{\beta}}; \quad (11.8)$$

so, referring to (11.3),

$$\left(\Re \frac{1}{\sqrt{\iota\pi}} \int_{|y|>\beta} e^{\iota y^2} dy \right)^2 < \frac{64}{\pi\beta}, \quad \left(\Im \frac{1}{\sqrt{\iota\pi}} \int_{|y|>\beta} e^{\iota y^2} dy \right)^2 < \frac{64}{\pi\beta}. \quad (11.9)$$

Thus

$$\sqrt{\left(\Re \frac{1}{\sqrt{\iota\pi}} \int_{|y|>\beta} e^{\iota y^2} dy \right)^2 + \left(\Im \frac{1}{\sqrt{\iota\pi}} \int_{|y|>\beta} e^{\iota y^2} dy \right)^2} < \sqrt{\frac{64 \times 2}{\pi\beta}}. (11.10)$$

Other useful versions of this inequality: for $a > 0$ and for sufficiently large β,

$$\left| \sqrt{\frac{a}{\iota\pi}} \int_{|y|>\beta} e^{\iota a y^2} dy \right| < 8\sqrt{\frac{2}{\pi\beta}},$$

$$\left| \int_{|y|>\beta} e^{\iota a y^2} dy \right| < 8\sqrt{\frac{2}{a\beta}},$$

$$\left| \int_{|y| \ge \sqrt{a}\beta} e^{\iota(ay^2+by+c)} dy \right| \le 8\sqrt{\frac{2}{a\beta}} \qquad (11.11)$$

where b and c are real numbers. The result follows from (11.10). $\quad\bigcirc$

The main point to be taken from this is that, provided absolute value lines $|\cdots|$ are kept outside the integral sign, the function $e^{\iota y^2}$ behaves like a Gaussian function e^{-y^2}, in that total area in the tail of the function can be arbitrarily small, as demonstrated in figure 2.4 (page 74 of [MTRV]), figures 6.2 and 6.3 (page 302), and, for multi-dimensional versions, in figure 6.1 on page 269.

This is the non-absolute feature of Feynman integrals. For real-valued Gaussian-type functions, bounds on integral values are found by calculating Riemann sums of absolute values. But when $|\cdots|$ has to be kept outside the integral functional, or the Riemann sum estimate, this means that a different kind of calculus or analysis is required for these non-absolute integrals.

Given $\tau \in Q$, let $S = \{\tau_j\}$ be a sequence of elements of Q such that $\tau_j \to \tau$ as $j \to \infty$. For simplicity, let τ_j be monotone increasing from the left, so $\tau_j < \tau$, $j = 1, 2, 3, \ldots$. As in [121] section 6.9 ([MTRV] pages 288–303), let

$$A_r^{j\tau} = \left\{ x : x \in \mathbf{R}^Q, \ |x(\tau) - x(\tau_j)| > \frac{1}{r} \right\}.$$

If the (α, θ) condition is required (see [MTRV] page 290), then the format is

$$A_{\alpha\theta r}^{j\tau} = \left\{ x : x \in \mathbf{R}^Q, \ \frac{|x(\tau) - x(\tau_j)|}{\theta \, |\tau - \tau_j|^\alpha} > \frac{1}{r} \right\},$$

but, for simplicity, only sets $A_r^{j\tau}$ are considered here. Extension to the (α, θ) format is straightforward—see [121], [MTRV] page 290.

With $c = a + \iota b$ take $a = 0$, $b = 1$ for simplicity, so $c = \iota$ and the Gaussian integrand $G_c(I[N])$ is $G_\iota(I[N])$. ○

Theorem 188B *Writing* $\kappa = 8\sqrt{2}\,\pi^{-1}$,

$$\left| \int_{\mathbf{R}^Q} \mathbf{1}_{A_r^{j\tau}}(x) G_\iota(I[N]) \right| \ < \ \kappa\sqrt{r(\tau - \tau_j)}.$$

Proof. Using theorems 159 and 160 ([121], page 280 of [MTRV]; and also [website]), and writing $z = x(\tau) - x(\tau_j)$,

$$\int_{\mathbf{R}^Q} \mathbf{1}_{A_r^{j\tau}}(x) G_\iota(I[N]) \ = \ \frac{1}{\sqrt{\pi\iota(\tau - \tau_j)}} \int_{|z| > \frac{1}{r}} e^{\iota \frac{z^2}{\tau - \tau_j}} \, dz.$$

Write $y = \frac{z}{\sqrt{\tau - \tau_j}}$, so $dz = \sqrt{\tau - \tau_j}\, dy$. Then $|z| > \frac{1}{r}$ implies $|y| > \frac{1}{r\sqrt{\tau - \tau_j}}$, and

$$\left| \int_{\mathbf{R}^Q} \mathbf{1}_{A_r^{j\tau}}(x) G_\iota(I[N]) \right| \ = \ \left| \int_{|z| > \frac{1}{r}} e^{\iota \frac{z^2}{\tau - \tau_j}} \, dz \right| \frac{1}{\sqrt{\pi(\tau - \tau_j)}}$$

$$= \ \left| \int_{|y| > \frac{1}{r(\tau - \tau_j)}} e^{\iota y^2} \, dy \right| \sqrt{\tau - \tau_j}\, \frac{1}{\sqrt{\pi(\tau - \tau_j)}}$$

$$< \ \frac{1}{\sqrt{\pi}} \times 8 \sqrt{\frac{2}{\pi \frac{1}{r(\tau - \tau_j)}}} \ = \ \frac{8\sqrt{2}}{\pi} \sqrt{r(\tau - \tau_j)},$$

using Theorem 188A. ○

The (α, θ)-format can easily be substituted here, and the ensuing follow-up arguments will be valid provided $\alpha < \frac{1}{2}$ and $\theta > 0$. As in [121] ([MTRV] page

291), B^τ denotes the set of $x \in \mathbf{R}^Q$ which are discontinuous at τ. The auxiliary sets

$$B_{kr}^\tau = \bigcup_{j=k}^{\infty} A_r^{j\tau}, \quad B_r^\tau = \bigcap_{k=1}^{\infty} B_{kr}^\tau, \quad B^\tau = \bigcup_{r=1}^{\infty} B_r^\tau$$

are used. Proofs are similar to those in [MTRV], except that the latter use "absolute integral" principles which are not applicable to the non-absolute Fresnel-based integrand $G_\iota(I[N])$.

Let $\varepsilon > 0$ be given. For given r and for $j = 1, 2, 3, \ldots$ choose τ_j so that $\sqrt{r(\tau - \tau_j)} < \varepsilon 2^{-j}$. Then

$$\left| \int_{\mathbf{R}^Q} 1_{B_{kr}^\tau}(x) G_\iota(I[N]) \right| \leq \sum_{j=k}^{\infty} \left| \int_{\mathbf{R}^Q} 1_{A_r^{j\tau}}(x) G_\iota(I[N]) \right| < \sum_{j=k}^{\infty} \frac{\varepsilon}{2^j} = \frac{\varepsilon}{2^k}.$$

For each k, $B_r^\tau \subset B_{kr}^\tau$, so for each k and for arbitrary $\varepsilon > 0$,

$$\left| \int_{\mathbf{R}^Q} 1_{B_r^\tau}(x) G_\iota(I[N]) \right| < \frac{\varepsilon}{2^k},$$

giving $\int_{\mathbf{R}^Q} 1_{B_r^\tau}(x) G_\iota(I[N]) = 0$ for $B_1^\tau \subset B_2^\tau \subset B_3^\tau \subset \cdots$. $\quad\bigcirc$

Theorem 188C The function $1_{B^\tau}(x) G_\iota(I[N])$ is integrable and

$$\int_{\mathbf{R}^Q} 1_{B^\tau}(x) G_\iota(I[N]) = 0.$$

Proof. The functions $1_{B_r^\tau}(x) G_\iota(I[N]) \to 1_{B^\tau}(x) G_\iota(I[N])$ as $r \to \infty$. But because of the oscillation of $G_\iota(I[N])$ from positive to negative, the convergence is not monotone; nor is it dominated by any integrable function. Therefore some other method of establishing the integrability of the limit must be sought. Suppose $x \in B^\tau$ (so x is discontinuous at τ). Then there exists a subsequence $(j_{p(x)})$ of the sequence $j = 1, 2, 3, \ldots$, so $\left(\tau_{j_{p(x)}} \right)$ is a subsequence of (τ_j); and there exists $r(x)$; such that $j_{p(x)} \geq r(x)$ implies

$$1_{B_{j_{p(x)}}^\tau}(x) G_\iota(I[N]) = 1_{B^\tau}(x) G_\iota(I[N]) = G_\iota(I[N]).$$

It has been established that, for each r, $\int_{\mathbf{R}^Q} 1_{B_r^\tau}(x) G_\iota(I[N])$ exists and equals 0. Thus, for each r there exists a gauge $\gamma_r = (L_r, \delta_r)$ so that, for every γ_r-fine division $\mathcal{D}_r$, and every partial division $\mathcal{E}_r \subseteq \mathcal{D}_r$,

$$\left| (\mathcal{D}_r) \sum 1_{B_r^\tau}(x) G_\iota(I[N]) \right| < \frac{\varepsilon}{2^r}, \quad \left| (\mathcal{E}_r) \sum 1_{B_r^\tau}(x) G_\iota(I[N]) \right| < \frac{\varepsilon}{2^{r-1}}.$$

(The latter inequality follows from theorem 17, page 131 of [MTRV].) For each $x \in B^\tau$ there exists $j_p, = j_{p(x)}, \geq r(x)$, such that $x \in B_{j_p}^\tau = B_{j_{p(x)}}^\tau$. Define a gauge $\gamma = (L, \delta)$ by

$$L(x) = L_{j_{p(x)}}, \qquad \delta = \delta_{j_{p(x)}},$$

so, for this x and for any $N \in \mathcal{N}(Q)$, $\delta(x, N) = \delta_{j_{p(x)}}(x, N)$. For $x \notin B^\tau$ let γ be arbitrary.[4] Let $\mathcal{D}$ be any γ-fine division of $\mathbf{R}^Q$. Then

$$
\begin{aligned}
\left| (\mathcal{D}) \sum \mathbf{1}_{B_r^\tau}(x) G_\iota(I[N]) \right| &= \left| \sum_k \left\{ (\mathcal{E}_{j_{p(x)}}) \sum \mathbf{1}_{B_{j_{p(x)}}^\tau}(x) G_\iota(I[N]) : j_{p(x)} = k \right\} \right| \\
&\leq \sum_k \left\{ \left| (\mathcal{E}_{j_{p(x)}}) \sum \mathbf{1}_{B_{j_{p(x)}}^\tau}(x) G_\iota(I[N]) \right| : j_{p(x)} = k \right\} \\
&< \sum_k \frac{\varepsilon}{2^{k-1}} \ < \ 2\varepsilon,
\end{aligned}
$$

where $\sum_k$ is a finite sum which aggregates those terms $(x, I[N]) \in \mathcal{D}$ for which $j_{p(x)} = k$, $k = 1, 2, 3, \dots$. The inequality holds for all γ-fine $\mathcal{D}$, so theorem 62 ([MTRV], page 174) can be applied. ◯

A similar proof establishes that B, the set of x which are discontinuous relative to any countable, dense set Q, is G_ι-null in $\mathbf{R}^Q$. The proof goes as follows.

Theorem 188D The function $\mathbf{1}_{B^\tau}(x) G_\iota(I[N])$ is integrable and

$$
\int_{\mathbf{R}^Q} \mathbf{1}_{B^\tau}(x) G_\iota(I[N]) \ = \ 0.
$$

Proof. Let $\{t_1, t_2, t_3, \dots\}$ be an enumeration of Q. Let $\varepsilon > 0$ be given. If $x \in B$ then there exists $j, = j(x)$, such that $x \in B^{t_j}$; and by Theorem 188C there exists a gauge $\gamma_j, = \gamma_{j(x)}, = (L_j, \delta_j)$, such that, for every $\gamma_{j(x)}$-fine division $\mathcal{D}_j = \{(y, I[N])\}$ of $\mathbf{R}^Q$, and every partial division $\mathcal{E}_j \subset \mathcal{D}_j$,

$$
\left| (\mathcal{D}_j) \sum \mathbf{1}_{B^{t_j}}(y) G_\iota(I[N]) \right| < \frac{\varepsilon}{2^j}, \qquad \left| (\mathcal{E}_j) \sum \mathbf{1}_{B^{t_j}}(y) G_\iota(I[N]) \right| < \frac{\varepsilon}{2^{j-1}}.
$$

Define a gauge $\gamma = (L, \delta)$ as follows. For any given $x \in B$, let

$$
L(x) \ = \ L_{j(x)}, \qquad \delta(x, N) \ = \ \delta_{j(x)}(x, N) \quad \text{for each} \quad N \in \mathcal{N}(Q).
$$

For $x \notin B$ let $L(x)$ and $\delta(x, N) > 0$ be arbitrary. Suppose $\mathcal{D} = \{(x, I[N])\}$ is a γ-fine division of $\mathbf{R}^Q$. By construction, each $(x, I[N]) \in \mathcal{D}$ is $\gamma_{j(x)}$-fine, and, for any positive integer k the set of $(x, I[N])$ for which $j(x) = k$ is either an empty set, or is a member of a partial division $\mathcal{E}_{j(x)}$ of $\mathbf{R}^Q$. Therefore, using theorem 17 (page 131 of [MTRV]) again,

$$
\begin{aligned}
\left| (\mathcal{D}) \sum \mathbf{1}_B(x) G_\iota(I[N]) \right| &= \left| \sum_k \left\{ (\mathcal{E}_{j(x)}) \sum \mathbf{1}_{B^{t_{j(x)}}}(x) G_\iota(I[N]) : j(x) = k \right\} \right| \\
&\leq \sum_k \left\{ \left| (\mathcal{E}_{j(x)}) \sum \mathbf{1}_{B^{t_{j(x)}}}(x) G_\iota(I[N]) \right| : j(x) = k \right\} \\
&< \sum_k \frac{\varepsilon}{2^{k-1}} \ < \ 2\varepsilon,
\end{aligned}
$$

and theorem 62 ([MTRV] page 174) can again be applied. ◯

---

[4]This uses the definition of γ in chapter 4 of [MTRV]. The alternative definition of γ in Section 7.5 above can equally be used.

11.3   Some Consequences of Theorem 63 Fallacy

These consequences include the following.

- Theorem 211 (pages 331–332 of [MTRV] is false as it stands; see counter-example below. It may be valid under additional hypotheses, such as *BM-continuity* of the integrand $f(x_{M_r})$. BM-continuity is relatively "flat" or step-function-like behaviour—see pages 360–366 of [MTRV].

- Theorem 217 (page 338) is false as it stands. Continuity of the integrand f_U is not enough. For instance, a sample function of Brownian motion is typically continuous, but such a function may have oscillatory features which, as an integrand, can tend to interfere with the oscillatory cancellations of the integrator G_c. Again, some form of BM-continuity (pages 360–366) of the integrand f_U may ensure convergence of the integral with respect to G_c.

- Theorem 217 is itself cited on subsequent pages, as follows.

- The first paragraph of section 7.14 (pages 339–341) depends on theorem 217, and is therefore meaningless; though the rest of that section makes some useful points.

- Theorem 219 (page 345) is a consequence of theorem 217, and is false.

- Page 361: Line 1 (citing heorem 217) is false; likewise line 6.

- Page 377: Line −8 repeats the false assertion of theorem 219.

- Page 486: Line 23 mentions theorem 217, though the sentence is basically all right.

Example 60 *Here is a counter-example for theorem 217 (pages 338 of [MTRV]). Let $T =]0,1]$, with*

$$\mathbf{R}^T \;=\; \{x(t): x(0) = 0,\ x(t) = x_t \in \mathbf{R},\ 0 < t \le 1\}.$$

Let $r = 1$ and let $M_r = \{\tau\}$ where $0 < \tau < 1$. Let

$$f(x_{M_r}) \;=\; f(x_\tau) \;=\; \sqrt{\iota\pi\tau}\, e^{-\iota \frac{x_\tau^2}{\tau}}.$$

Then f satisfies the hypotheses of theorem 211; so if this theorem is valid the integral

$$\int_{\mathbf{R}^T} f(x_{M_r}) g_c(x(N))|I[N]|, \;=\; \int_{\mathbf{R}^T} \left(\sqrt{\iota\pi\tau}\, e^{-\iota \frac{x_\tau^2}{\tau}} \right) \prod_{j=1}^{n} \left(\frac{e^{\iota \frac{(x(t_j)-x(t_{j-1}))^2}{t_j-t_{j-1}}}}{\sqrt{\iota\pi(t_j - t_{j-1})}} |I_j| \right),$$

exists, and then theorem 160 (page 280) and theorem 172 (page 286) imply that

$$\int_{\mathbf{R}^T} f(x_{M_r}) g_c(x(N)) |I[N]| \;=\; \int_{\mathbf{R}} \left(\sqrt{\iota \pi \tau}\, e^{-\frac{\iota x_\tau^2}{\tau}} \right) \left(\frac{e^{\frac{\iota x_\tau^2}{\tau}}}{\sqrt{\iota \pi \tau}} \right) dx_\tau.$$

But the latter expression has the form of a one-dimensional integral,

$$\int_{\mathbf{R}} \left(\sqrt{\iota \pi \tau}\, e^{-\frac{\iota x_\tau^2}{\tau}} \right) \left(\frac{e^{\frac{\iota x_\tau^2}{\tau}}}{\sqrt{\iota \pi \tau}} \right) dx_\tau, \;=\; \int_{\mathbf{R}} e^{-\frac{\iota x_\tau^2}{\tau}} e^{\frac{\iota x_\tau^2}{\tau}} dx_\tau, \;=\; \int_{\mathbf{R}} 1\, dx_\tau,$$

which diverges. Therefore $f(x_{M_r})$ *is not* G_ι*-integrable, so theorem 217 must be false.*

Chapter 12

Appendix 3: Option Pricing

An analytic method for pricing American call options is provided, followed by an empirical method for pricing Asian call options; with the objective of further practical illustration of some of the ideas and methods of Part I of this book.

12.1 American Options

Suppose a share has market value $z(s)$ at time s, and suppose an option contract between two parties is entered into at time $s = 0$, so that one of the parties to the contract acquires the right, but not the obligation, to purchase the share, from the other party, for a fixed price κ at any time t, $0 < t \leq \tau$. The option expires at time τ, so there is no right to purchase the share at times $t > \tau$. These are the features of an American call option.

An American call option can be exercised at any time during the term or lifetime of the contract. This differs from a European call option which can only be exercised on the date of termination of the contract.

Here are some questions about American call options. What is the initial value (or purchase price) $w(0)$ of the option contract at time 0? At any time t, $0 < t \leq \tau$, what is the market value $w(t)$ of the (unexercised) call option?

If time $t > \tau$, then $w(t) = 0$ since the option can no longer be exercised. At time $t = \tau$, the American call (if not previously exercised) has value

$$w(\tau) = \max\{z(\tau) - \kappa, 0\},$$

the same as the corresponding European call option.

At time $t < \tau$, the American call option and the corresponding European call option have non-zero value[1] even if $z(t) < \kappa$. The difference between the two is that, for $t < \tau$, the European call cannot be exercised even if $z(t) > \kappa$.

If $z(0)$ is the observed value of the underlying share at the time of writing the contract, the purchase price $w(0)$ and the exercise price κ of this contract

---

[1] If κ is excessively large the option may be effectively worthless. See Figure 12.5 below.

Gauge Integral Structures for Stochastic Calculus and Quantum Electrodynamics, First Edition. Patrick Muldowney.
© 2021 John Wiley & Sons, Inc. Published 2021 by John Wiley & Sons, Inc.

must satisfy

$$w(0) + \kappa \ \geq \ z(0). \tag{12.1}$$

The reason for this is that, if $w(0) < z(0) - \kappa$, the option can be exercised immediately, the share can be purchased for the amount κ, and then re-sold for amount $z(0)$, realizing a riskless profit of $z(0) - \kappa - w(0)$.

For $t < \tau$ the American call option will not be exercised at time t if $z(t) < \kappa$. What if $z(t) > \kappa$? In that case, and disregarding the outlay required to purchase the option, the holder of the option will obtain immediate profit of amount $z(t) - \kappa$ by purchasing the share for the option exercise price κ and immediately re-selling the share for the market price $z(t)$. If $t = \tau$ that is what is expected to happen, since there are no further opportunities to purchase the share at option exercise price κ.

But if $t < \tau$, and even if $z(t) > \kappa$, then, depending on various factors, the holder of the American call option may decide not to exercise in case they expect the underlying share price to rise further in the time interval $]t, \tau]$. If that were to happen, then the profit (or *payoff*) $z(s) - \kappa$ to be obtained by exercising the option at the later date $s > t$ would be greater than what would be obtained by exercising at time t.

This introduces a new element of unpredictability. If time t is the "present", the share price $z(t)$ has a known or observed value. Call this value κ_t. Even if $\kappa_t > \kappa$, the option holder, in anticipation of further increase in the price of the underlying share, may decline to exercise the option at time t.

On the face of it, in advance of exercising, the optimal time t at which to consider exercising the American call is the time $\tilde{t}$ for which the share price $z(t)$ is maximal, $0 < t \leq \tau$; so $x(\tilde{t}) > x(s)$ for $s \neq \tilde{t}$.

However, the time value of money has to be factored into this argument. The time value of money is measured by the risk-free interest rate ρ. The following formulation takes account of time value of any potential option payoff.

If $\tau \geq \tilde{t} > s$ and $z(s) > z(\tilde{t}) > \kappa$, it is more advantageous to exercise the American call at the earlier time s if

$$(z(s) - \kappa) \, e^{\rho(\tilde{t}-s)} \ > \ z(\tilde{t}) - \kappa.$$

At time $s = 0$ the optimal[2] exercise time ς is unpredictable. It is possible that the holder of an American call will exercise the option in advance of the optimal time ς, or will do so after the optimal point has passed. Or will not exercise at all.

Assuming "equal likelihood", each of these possibilities is just as likely as the other. On this reasoning, the exercise time (or date) t of an American call option is a random variable, with expected value equal to ς.

To explore further, assume that $t < \tau$ is the "present" so the share price $z(t) = \kappa_t$ is known. Suppose that, even if $\kappa_t > \kappa$, the option holder, in anticipation of further increase in the price of the underlying share, declines to exercise the

---

[2] Likewise, at times $s > 0$ the optimal exercise time ς is unknown, $s \leq \varsigma \leq \tau$.

option at time t. Assume that the likelihood of the option holder making this decision is equal to the probability that $z(s) > \kappa_t$ for some $s \in]t, \tau]$. That is,

$$z(s) - \kappa \ > \ z(t) - \kappa \ = \ \kappa_t - \kappa.$$

But it is necessary to factor in the time value of money here also. The option holder will decline to exercise if they expect that the discounted value of the term on the left hand side of the following inequality exceeds the right hand side; so this condition becomes:

$$(z(s) - \kappa)\, e^{-\rho(t-s)} \ > \ z(t) - \kappa \ = \ \kappa_t - \kappa.$$

If t is given, and if s is the "present", with $s < t$, similar considerations apply to any decision to exercise the American call during the period $]s, t]$.

This argument implies certain assumptions about the motivation and behaviour of option holders/buyers as a body, in regard to future prices of an underlying share. But the future market value of the underlying share is said to be "correctly" established by similar motivation and behaviour of those buying and selling the share. So there seems to be some justification for such assumptions.

Take time 0 to be the "present". Suppose s is a given "future" time, $0 < s < \tau$; so it is not known whether $z(s)$ is going to be less than or greater than the exercise price κ. Consider an option exercise decision at the future time s.

For any exercise date s $(0 < s \leq \tau)$ denote the discounted difference function for the option by

$$p(s, z(s)) \ := \ (z(s) - \kappa)\, e^{-\rho s}. \tag{12.2}$$

If time $t > 0$ is the present and $0 < t < s \leq \tau$, then the relevant difference function is discounted from time s to time t:

$$p_t(s, z(s)) \ := \ (z(s) - \kappa)\, e^{-\rho(s-t)}. \tag{12.3}$$

Take time 0 to be the present, so option exercise decisions are based on (12.2). The option is exercisable (or "in the money") if $z(s) > \kappa$ for some $s \in]0, \tau]$; and it is optimally exercisable at time $s = s(z_T)$ if, for some other time s' before or after s, the payoff, discounted from time s' to time 0, is not greater than the payoff at time $s(z_T)$ discounted to time 0. The optimal exercise time $s(z_T)$, if exercise[3] is feasible, is variable for different histories z_T.

The graph or history z_T is a potential occurrence of a joint-basic observable Z_T if there is a distribution function F for which[4]

$$Z_T \ \simeq \ z_T \big[\mathbf{R}_+^T, F\big].$$

In that case, for $0 < s \leq \tau$ the discounted payoff function values $p(s, z_T)$ are potential occurrences of a contingent observable

$$p(s, Z_T) \ \simeq \ p(s, z_T)\big[\mathbf{R}_+^T, F\big],$$

---

[3]Exercise may never be feasible if the exercise price κ is set too high. See Figure 12.5 below.

[4]It is sometimes assumed that F is the geometric Brownian distribution function $\mathcal{G}$.

whose expected value is

$$q(s) := \mathrm{E}\left[p(s, Z_T)\right] = \int_{\mathbf{R}_+^T} p(s, z_T) F(I[N]). \tag{12.4}$$

If the exercise date s is selected, then s is fixed in this integration. Therefore the point-integrand p is a cylinder function, so the integral reduces to a one-dimensional integral[5] on $\mathbf{R}_+ = \mathbf{R}_+^{\{s\}}$.

Choose s so that $q(s)$ is maximized[6] and denote the chosen s by ς. Then $0 \le \varsigma \le \tau$, and, if time 0 is the present, then, at time 0, it is predicted that the American call option will be exercised at time ς.

The question also arises as to whether, if the "present" is some time t later than 0, the predicted exercise date ς_t will be different.

The objective here is to consider ways of finding a purchase price $w(t), = w_t$, for a call option at any time t, $0 \le t < \tau$; in particular the price $w(0), = w_0$, payable by the initial purchaser of the option. Consider the latter problem first, so $t = 0$ is the present.

For a European call option the exercise date is τ, and the initial value of the option is obtained by means of the risk-neutral pricing formula (8.46) of [MTRV], page 440:

$$e^{-\rho\tau} \int_{\mathbf{R}_+^T} f_\tau(z_T) \bar{F}(I[N]),$$

where $\bar{F}$ is the risk-neutral martingale distribution function discussed in sections 8.14 and 8.15, pages 436–440 of [MTRV], and $f_\tau(z_T) = \max\{z(\tau) - \kappa, 0\}$.

For the corresponding American call option, the predicted exercise date is ς, which may be less than τ. But apart from that, the pricing argument is the same and gives initial value of the American call as

$$e^{-\rho\varsigma} \int_{\mathbf{R}_+^T} f_\varsigma(z_T) \bar{F}(I[N]),$$

where $f_\varsigma(z_T) = \max\{z(\varsigma), 0\}$. The integral on $\mathbf{R}_+^T$ can be replaced by one on $\mathbf{R}_+^{]0,\varsigma]}$. Either way, since the integrand is cylindrical the integral reduces to a one-dimensional integral on $\mathbf{R}_+, = \mathbf{R}_+^{\{s\}}$.

When it is assumed that the underlying asset price process Z_T is geometric Brownian with $F = \mathcal{G}^{\mu\sigma}$, then section 8.16 ([MTRV] pages 440–444) shows that $\bar{F} = \mathcal{G}^{\rho\sigma}$, where σ is the volatility, μ is the growth rate of the process, and ρ is the risk-free interest rate. In that case the initial value of the American call is given by (8.53) of [MTRV] page 442, with τ replaced by ς.

How can the optimal exercise time ς be established? If the process distribution function F is an expression such as $\mathcal{G}^{\mu\sigma}$ then the expected discounted difference $q(s)$ of (12.4) has a particular form whose maximum value can be estimated; so the corresponding $s, = \varsigma$, can also be estimated.

---

[5]If F is the geometric Brownian distribution function $\mathcal{G}^{\mu\sigma}$ then the integral is the one given in line 5 of [MTRV] page 442, but with natural process growth rate μ in place of the risk-free growth rate ρ, and with $p(s, z_T)$ in place of $\max\{z_\tau - \kappa, 0\}$.

[6]Set aside, for the moment, the question of how to make this choice.

On the other hand, suppose F (or, at least, approximate values of F) is determined empirically, as indicated in [MTRV] section 9.11 page 488, and in Section 12.2 below. Then it may be possible to calculate empirically the expected discounted difference $q(s)$ of (12.4) using different values of s, and use the results to estimate ς, the exercise date which gives the largest expected discounted difference.

As an exercise in determining optimal exercise time ς, take $F = \mathcal{G}^{\mu\sigma}$, which is the traditional (but generally incorrect) assumption. Then, using lemma 33 (pages 309–310 of [MTRV]) with change of variable $z(s) = e^u$, (12.4) becomes

$$q(s) \;=\; \int_{\mathbf{R}_+^{]0,s]}} e^{-s}\,(z(s) - \kappa)\,\mathcal{G}^{\mu\sigma}(I[N]); \qquad (12.5)$$

so

$$q(s) \;=\; e^{-s}\left(\int_{-\infty}^{\infty} \frac{e^u}{\sigma\sqrt{2\pi s}} e^{-\frac{1}{2}\left(\frac{u-\mu}{\sigma\sqrt{s}}\right)^2} du - \kappa \int_{-\infty}^{\infty} \frac{1}{\sigma\sqrt{2\pi s}} e^{-\frac{1}{2}\left(\frac{u-\mu}{\sigma\sqrt{s}}\right)^2} du \right)$$

$$= \; e^{-s}\left(e^{\mu s + \frac{\sigma^2 s}{2}} - \kappa \right).$$

With $\mu = 0$, $\sigma = 0.5$, $\tau = 2$, and with exercise price $\kappa = 0.1$, 2, 1, and 5, Figures 12.1, 12.2, 12.4, 12.5, respectively illustrate the expected values $q(s)$ of the corresponding options if they are exercised at times s for $0 \le s \le 2$.

These graphs indicate that if exercise price κ is very low (Figure 12.1), the situation can be compared to that of (12.1), and it is best to exercise the option immediately. In other words, at time $t = 0$ essentially riskless profit is available "here and now"; so why wait around for potential increases in the underlying share price? After all, the latter could actually decrease.

If κ is a bit larger (Figure 12.2), at time $t = 0$ it is safest to plan on exercising the option at or near[7] the contract termination date $s = \tau$. Because, even if (as in the cases portrayed here) the growth rate μ of the underlying share price is zero, there is a "structural" mean growth rate of $\frac{1}{2}\sigma^2$ due to volatility. So the attitude of the investor in American call options at time $t = 0$ will lean towards late exercise.[8]

For intermediate values of κ (Figure 12.4) the optimal exercise time $\varsigma \approx 1$ is intermediate between 0 and 2. And for very large κ (Figure 12.5), the long position in the option contract appears unattractive—potential buyers may not be very interested in the long position in an option with such a high exercise price, and will not pay very much for it; so it can only be sold for some very small amount of money.

Figure 12.5 does not imply that an option with these parameters can never be in the money. The function $q(s)$ yields expected rather than actual values. But the diagram suggests that the option should be immediately exercised if

---

[7] Figure 12.3 gives a more detailed graph of $q(s)$ for this case. It shows that the maximal value of $q(s)$ occurs at about $s = 1.83$.

[8] Some commentary on American call options suggests that there is no merit at all in early exercise.

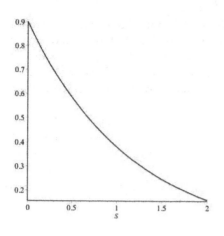

Figure 12.1: $\kappa = 0.1$

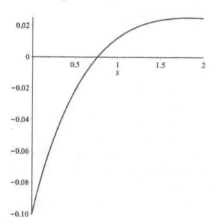

Figure 12.2: $\kappa = 1.1$

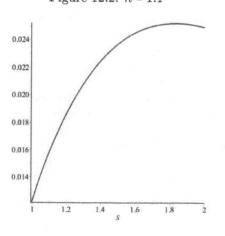

Figure 12.3: $\kappa = 1.1$, $t = 0$, $\tau = 2$

Figure 12.4: $\kappa = 1$

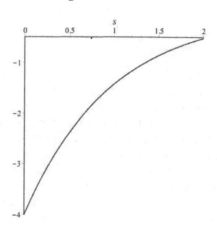

Figure 12.5: $\kappa = 5$

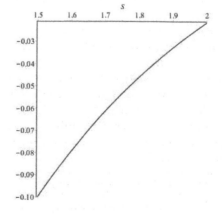

Figure 12.6: $\kappa = 1.1$, $t = 1.5$, $\tau = 2$

and when it achieves in-the-money status, without waiting around for further increases in underlying share price. Secondly, at time $t = 0$ the purchase price of such an option should be based on exercise at time $s = \tau$, not earlier, in order to allow for any possible growth in underlying share price.

The optimal exercise date ς can be estimated by methods such as the one demonstrated in Figures 12.1 to 12.5. Once ς is established, the Black-Scholes pricing formulae (8.53) of [MTRV] page 442 can be used, with ς replacing τ; giving the initial $(t = 0)$ value of the American call option as

$$w_0 = z_0 \Phi \left(\frac{\ln \frac{z_0}{\kappa} + \left(\rho + \frac{1}{2}\sigma^2 \right) \varsigma}{\sigma \sqrt{\varsigma}} \right) - \kappa e^{-\rho \varsigma} \Phi \left(\frac{\ln \frac{z_0}{\kappa} + \left(\rho - \frac{1}{2}\sigma^2 \right) \varsigma}{\sigma \sqrt{\varsigma}} \right) \qquad (12.6)$$

where Φ represents the standard normal distribution function. Now suppose the option, created at time $s = 0$, is unexercised at time t, $0 < t < \tau$. Suppose the holder of the option wishes to sell it at time t. What price should be charged for it? In other words, what is the value $w(t)$ (or w_t)?

Can it be argued that, at time t later than 0, the expected exercise date for the as yet unexercised option is ς, the exercise date obtained by maximising $q(s)$ in (12.2)? If that were the case, then formula (8.54) of [MTRV] page 442 could be used for w_t, with ς replacing τ; so w_t would be

$$z_t \Phi \left(\frac{\ln \frac{z_t}{\kappa} + \left(\rho + \frac{1}{2}\sigma^2 \right) (\varsigma - t)}{\sigma \sqrt{\varsigma - t}} \right) - \kappa e^{-\rho(\varsigma - t)} \Phi \left(\frac{\ln \frac{z_t}{\kappa} + \left(\rho - \frac{1}{2}\sigma^2 \right) (\varsigma - t)}{\sigma \sqrt{\varsigma - t}} \right) \qquad (12.7)$$

The trouble with this argument is that ς is not necessarily the actual exercise date. It is a date estimated at time $s = 0$ for the purpose of pricing the option at that time. The option holder may choose to exercise at some time s earlier or later than ς.

Indeed, if exercise is delayed, then t (which is the "present" in the new pricing scenario) may be greater than ς in (12.7). In that case (12.7) makes no sense.

However, with appropriate modification the pricing argument may be adapted for $t > 0$. At time $t \geq 0$ the estimated exercise date ς_t is the value of s $(t \leq s \leq \tau)$ which maximises

$$q_t(s) := \mathbb{E}\left[p_t \left(s, Z_T \right) \right] = \int_{\mathbf{R}_+^{]t,s]}} p_t(s, z_T) F(I[N]). \qquad (12.8)$$

If the distribution function F for the share price process Z is $\mathcal{G}^{\mu\sigma}$, (12.8) becomes

$$q_t(s) = \int_{\mathbf{R}_+^{]t,s]}} e^{-(s-t)} \left(z(s) - \kappa \right) \mathcal{G}^{\mu\sigma}(I[N]). \qquad (12.9)$$

As before, since the point-integrand p_t is cylindrical, this gives $q_t(s) =$

$$= e^{-(s-t)} \left(\int_{-\infty}^{\infty} \frac{e^u}{\sigma \sqrt{2\pi(s-t)}} e^{-\frac{1}{2}\left(\frac{u-\mu}{\sigma \sqrt{s-t}} \right)^2} du - \kappa \int_{-\infty}^{\infty} \frac{1}{\sigma \sqrt{2\pi(s-t)}} e^{-\frac{1}{2}\left(\frac{u-\mu}{\sigma \sqrt{s-t}} \right)^2} du \right)$$

$$\text{or} \qquad q_t(s) \;=\; e^{-(s-t)}\left(e^{\mu(s-t)+\frac{\sigma^2(s-t)}{2}} - \kappa\right).$$

With $\mu = 0$, $\sigma = 0.5$, $\kappa = 1.1$, $\tau = 2$, and $t = 1.5$, Figure 12.6 is the graph of $q_t(s)$ from $s = 1.5 = t$ to $s = 2 = \tau$. This option has the same parameters $(\mu, \sigma, \kappa, \tau)$ as the one described in Figures 12.2 and 12.3 above, except that it is entered into at time $t = 1.5$ instead of time 0. But whereas Figure 12.3 indicates an estimated exercise date $\varsigma \approx 1.83$, Figure 12.6 indicates an exercise date $\varsigma_{1.5} = 2$.

Thus the estimated value $w(t)$ (or w_t) of an American call option at time t $(0 \le t < \tau)$ is $w_t \;=\;$

$$= \; z_t \Phi\left(\frac{\ln\frac{z_t}{\kappa} + \left(\rho + \frac{1}{2}\sigma^2\right)(\varsigma_t - t)}{\sigma\sqrt{\varsigma_t - t}}\right) - \kappa e^{-\rho(\varsigma_t - t)} \Phi\left(\frac{\ln\frac{z_t}{\kappa} + \left(\rho - \frac{1}{2}\sigma^2\right)(\varsigma_t - t)}{\sigma\sqrt{\varsigma_t - t}}\right).$$

$$(12.10)$$

At time $t = \tau$ the value of an unexercised American call option is

$$w_\tau \;=\; \max\{z_\tau - \kappa, \, 0\}.$$

12.2 Asian Options

The preceding investigation uses distribution function $\mathcal{G}^{\mu\sigma}$. But the assumption that price processes Z follow a geometric Brownian (or lognormal) distribution is generally unreliable, as demonstrated in [MTRV] section 9.9, pages 479–485, and in the literature of this subject.

Also, both European and American options have exercise values which depend on the value taken by the underlying asset on the date of exercise of the option. In contrast, options described as Asian have value which depends, not just on the share value on the option exercise date, but on the average value of the share over the lifetime of the option:

$$w(\tau) \;=\; \max\{A(z_T) - \kappa, \, 0\},$$

where $A(z_T)$ is the average of the known or observed values taken by the share in advance of the exercise date τ, which, for this purpose, is the "present".

So if the lifetime of the option consists of one million "time ticks", by time τ the share value $z(s)$ will have been observed for each of one million values of $z(s)$, and, with $T = \,]0, \tau]$, $A(z_T)$ is the sum of a million values of $z(s)$ divided by a million.

However, at any time t earlier than τ, including time $t = 0$ when the option contract is created, the future values $z(s)$ $(0 < s \le \tau)$ are unknown and unpredictable. But if their joint distribution F is known, then $A(Z_T)$ is a joint-contingent observable, $A(Z_T) \simeq A(z_T)\big[\mathbf{R}_+^T, F\big]$, and likewise $f_\tau(Z_T) \simeq \max\{A(Z_T) - \kappa, \, 0\}\big[\mathbf{R}_+^T, F\big]$. The values w_t of the Asian option at times t, $0 \le t < \tau$ can then be deduced from (8.47) of [MTRV], page 440:

$$w_t \;=\; e^{-\rho(\tau - t)} \mathrm{E}^{\bar{F}}\big[f_\tau\big(\bar{Z}_T\big)\big] \;=\; e^{-\rho(\tau - t)} \int_{\mathbf{R}_+^{]t,\tau]}} f_\tau(z_T) \bar{F}(I[N]).$$

The defects of the hypotheses (assumption of geometric Brownian motion or lognormality of asset prices) underlying the preceding argument are outlined in section 9.9 (pages 479–485) of [MTRV]. Sections 9.10 and 9.11 of [MTRV] outline an empirical method for estimating the actual joint distribution functions for such processes.

This method is used below to estimate the value of an Asian call option on Glanbia shares, using the Maple program.

Both European and American options have exercise values which depend on the value taken by the underlying asset on the date of exercise of the option. In contrast, other kinds of option contract have value which depends, not just on the share value on the option exercise date, but on the values taken by the underlying asset at various times during the lifetime of the option contract.

For instance, an option of the Asian type depends on the average value of the share over the lifetime of the option:

$$w(\tau) = \max\{A(z_T) - \kappa, 0\},$$

where $A(z_T)$ is the average of the known or observed values taken by the share in advance of the exercise date τ.

At any time t earlier than τ, including time $t = 0$ when the option contract is created, the future values $z(s)$ $(0 < s \le \tau)$ are unknown and unpredictable. But if their joint distribution F is known, then $A(Z_T)$ is a joint-contingent observable,

$$A(Z_T) \simeq A(z_T)[\mathbf{R}_+^T, F],$$

and likewise $f_\tau(Z_T) \simeq \max\{A(Z_T) - \kappa, 0\}[\mathbf{R}_+^T, F]$. The values w_t of the Asian option at times t, $0 \le t < \tau$ can then be deduced from (8.47) of [MTRV], page 440:

$$w_t = e^{-\rho(\tau-t)}\mathbf{E}^F\left[f_\tau\left(\bar{Z}_T\right)\right] = e^{-\rho(\tau-t)}\int_{\mathbf{R}_+^{]t,\tau]}} f_\tau(z_T)\bar{F}(I[N]), \qquad (12.11)$$

in accordance with the risk-neutral martingale pricing arguments[9] of [MTRV] sections 8.13–8.19, pages 433–466, with

$$f_\tau(z_T) = \max\{A(z_T) - \kappa, 0\} \qquad (12.12)$$

for $z_T \in \mathbf{R}_+^T$. In the case of European and American options, f_τ depends on a single variable x_τ or x_ς with τ (or ς) fixed, so the latter integral reduces to an integral on $\mathbf{R}_+^T$ (or $\mathbf{R}_+^\varsigma$).

But the integral in (12.11) is infinite-dimensional. Note that if $t > 0$ is the "present", some values of $z_T, = \{z(s) : 0 < s \le \tau\}$, are known values, while the others are still in the future, and must be regarded as unpredictable observables. So

$$w_t = e^{-\rho(\tau-t)}\int_{\mathbf{R}_+^{]t,\tau]}} f_\tau\left((z_{T'}, z_{T''})\right)\bar{F}(I[N]),$$

---

[9]Unlike the assumption of lognormality of asset prices, these arguments seem to be fairly reliable.

Date	Price(£)	Date	Price(£)	···	Date	Price(£)
3/08/1991	0.800	3/15/1991	0.880	···	3/3/2011	3.751
3/11/1991	0.817	3/18/1991	0.878	···	3/4/2011	3.810
3/12/1991	0.836	3/19/1991	0.859	···	3/7/2011	3.782
3/13/1991	0.868	3/20/1991	0.859	···	3/8/2011	3.699
3/14/1991	0.885	3/21/1991	0.858	···	3/9/2011	3.692

Table 12.1: Glanbia share prices, 3 August 1991 to 3 September 2011.

where the values $z_{T'} \in \mathbf{R}_+^{]0,t]}$ are known and the values $z_{T''} \in \mathbf{R}_+^{]t,\tau]}$ (with $T'' =$ $]t,\tau]$) are potential occurrences of the joint-basic observable

$$Z_{T''} \simeq z_{T''}\left[\mathbf{R}_+^{]t,\tau]}, F\right].$$

To sum up,

$$w_0 = e^{-\rho(\tau)} \int_{\mathbf{R}_+^{]0,\tau]}} f_\tau(z_T)\bar{F}(I[N]) \quad \text{where} \quad f_\tau(z_T) = \max\{A(z_T) - \kappa\}.$$

$$(12.13)$$

Example 61 *Instead of assuming $F = \mathcal{G}^{\mu\sigma}$, pages 486–489 of [MTRV] suggest an empirical approach using the risk-neutral, no-arbitrage pricing argument of [MTRV] section 8.14, pages 436–438. The analysis can be applied to the Glanbia share price series [224] shown in Table 12.1. This table displays the first ten days and last five days of the Glanbia prices for the 5219 days of stock market listings from 3 August 1991 to 3 September 2011. The full listing is available as an Excel file at* Price Data *in* [website]. *Figure 9.16 on page 480 of [MTRV] has a graph of this share price series.*

Here is the proposed analysis of an Asian option on Glanbia stock:

- *Using the historic Glanbia data z_T, construct values for the empirical distribution $F(I[N])$ as in section 9.11 of [MTRV] pages 487–489.*

- *Estimate the growth rate μ of the price process. This can be done by, for example, calculating least squares regression for the Glanbia data z_T.*

- *Calculate a discounted version $z'(s) = e^{-\mu s}z(s)$ of the Glanbia price data.*

- *Using the growth rate estimate μ, adjust the distribution values F so that, with respect to amended values $\bar{F}$, the discounted prices $z'(s), w'(s)$ are occurrences of martingales $Z'_T \simeq z'_T[\mathbf{R}_+^T, \bar{F}]$, $W'_T \simeq w'_T[\mathbf{R}_+^T, \bar{F}]$:*

$$\mathrm{E}^{\bar{F}}[Z'_s] = z'(0), \qquad\qquad \mathrm{E}^{\bar{F}}[W'_s] = w'(0)$$

for all s, $0 < s \leq \tau$; with $z'(0) = z(0)$, $w'(0) = w(0)$.

- *Calculate an estimated value $w(0)$ by estimating the integral*

$$w(0) = w'(0) = \mathrm{E}^{\bar{F}}[W'_\tau] = \int_{\mathbf{R}_+^{]0,\tau]}} w'(\tau)\bar{F}(I[N]) = \int_{\mathbf{R}_+^{]0,\tau]}} f_\tau(z'_T)\bar{F}(I[N]).$$

For Asian options the latter integral depends on all occurrences $z(s)$, $0 < s \leq \tau$. So the integral is not cylindrical and, in order to estimate it, suitable regular partitions[10] of $\mathbf{R}_+^{]0,\tau]}$, $= \mathbf{R}_+^T$, can be used, along with step function estimates of the integrand.

Step function approximations are, in principle at least, easy enough to understand and to apply—even in the case of multi-dimensional integrals, when we do not have at our disposal the immense simplification that occurs when the integrand is a cylinder function whose integral happens to reduce to a one-dimensional integral, as in the cases of European and American options.

But, even when step function simplicity is present, it is still necessary to take into consideration the financial principles involved in risk-neutral valuation.

The main issue is converting the process distribution F to a martingale distribution $\bar{F}$. How can this be tackled in the case of the Glanbia price process, for instance?

Accepting that F cannot be taken to be the geometric Brownian function $\mathcal{G}$, [MTRV] sections 9.10 and 9.11 (pages 486–489) suggest using a counting or relative frequency method to estimate F for an observable $X_T \simeq x_T[\mathbf{R}^T, F]$.

This is problematic. Suppose an observable X has distribution function P on sample space $\mathbf{R}$, $X \simeq x[\mathbf{R}, P]$, so a P-measurable set of potential occurrences $A \subset \mathbf{R}$ has probability $P(A)$. The relative frequency idea implies that multiple occurrences of A, expressed as a fraction of total occurrences, provide an estimate of $P(A)$. Some justification of this estimate is provided by the Law of Large Numbers ([MTRV] section 5.9, pages 224–233).

A problem with this is that "multiple occurrences" of a random variable X (or of a process Z_T) are contrary to the usual mathematical understanding of a random variable, or of a process. A random variable/observable is generally understood as a mathematical description of a single future possible occurrence of something; that is, a single future observation or measurement which has multiple possible values, just one of which can actually occur.

So there are no multiple occurrences from which relative frequency can be calculated as estimate of probability.

On the other hand, multiple occurrences, and relative frequency, can be somehow extracted from the Law of Large Numbers by postulating, not a single random variable with multiple occurrences, but multiple random variables, each with a single occurrence.

In order to connect this line of reasoning to the "relative frequency approximates to probability" idea, this family of random variables must consist of statistically independent random variables.

Glanbia (or any other) historic share price data constitute a single occurrence z_T of a family of joint observables $\{Z_t : t \in T\}$, written as process $Z_T \simeq z_T[\mathbf{R}_+^T, F]$. The distribution function $F(I[N])$ for the process can be thought of as consisting of transition probabilities for joint occurrences

$$(z(t_j) \in I(t_j), \quad 1 \leq j \leq n), \qquad N = \{t_1, \ldots, t_n\}.$$

---

[10](r, q)-partitions are a bit cumbersome in practice, and other kinds of regular partitioning are used in the calculations below.

Can relative frequency be used to estimate $F(I[N])$? This question is addressed in [MTRV] pages 486–489. The idea is as follows.

The Glanbia share price series begins with the end-of-day price 0.8 on Friday 8 March 1991, and continues with the daily price for each day of 5-day weeks until Wednesday 9 March 2011 when the end-of-day price was 3.692. Thus the series contains over 5000 individual items.

The price transitions in $I[N]$ of (12.11) are possible transitions in the future. Assume that, at any stage in the "history" (past, present, or future) of the data, the likelihood of such a pattern of transitions is some more-or-less fixed amount for this particular series of share prices.

It is possible to count how many transitions in the historic or past data correspond to the set of possible future transitions represented by $I[N]$. To convert this to a relative frequency, divide this number by the total number of possible transitions corresponding to $I[N]$ in the past data. Provided the number of price transitions n in the event $I[N]$ is small compared to 5000, the preceding assumptions suggest[11] that this relative frequency could be used as an approximation to $F(I[N])$.

To perform risk-neutral valuation the (estimated) distribution function values $F(I[N])$ must be converted to (estimated) martingale distribution values $\bar{F}(I[N])$ for $I[N] \in \mathcal{I}(\mathbf{R}_+^T)$. How can this be done?

Consider a single share price z_t at some fixed time t, $0 < t \le \tau$. For cells $I_t =]u,v] \in \mathcal{I}(\mathbf{R}_+)$, methods such as those described above can be used to find an estimated distribution function $F_t(I_t)$ for the observable $Z_t \simeq z_t[\mathbf{R}_+, F_t]$,

$$P(z_t \in I_t) = F_t(I_t), \qquad \mu_t = \mathrm{E}[Z_t] = \int_{\mathbf{R}_+} z_t F_t(I_t) = \int_0^\infty z_t dP.$$

The basic observable Z_t can be regarded as a contingent observable in $\mathbf{R}_+^T$, with $f(Z_T) = Z_t$, $f(Z_T) \simeq f(z_T)[\mathbf{R}_+^T, F]$. Then f is a cylinder function, $f(z_T) = z_t$, t fixed; and

$$\mathrm{E}[Z_T] = \int_{\mathbf{R}_+^T} f(z_T) F(I[N]) = \int_{\mathbf{R}_+} z_t F_t(I_t) = \mu_t$$

since $F(I[N]) = F_t(I_t)$ for $N = \{t\}$, $I[N] = I_t$.

Given a price process $Z_T \simeq z_T[\mathbf{R}_+^T, F]$, risk-neutral pricing theory requires the construction of a process $\bar{Z}_T \simeq z_T[\mathbf{R}_+^T, \bar{F}]$ such that, under the distribution function $\bar{F}$, the discounted values $e^{-\rho s}z(s)$ are occurrences of a martingale, with

$$\mathrm{E}^{\bar{F}}\left[e^{-\rho s}\bar{Z}_s\right] = z(0), \quad \text{or} \quad \mathrm{E}^{\bar{F}}\left[\bar{Z}_s\right] = e^{\rho s}z(0)$$

for each $s \in T$, where $z(0)$ is the known initial price, and where, for present purposes, the risk-free interest rate ρ is taken to be constant for all s.

---

[11]This can be criticized from various angles. One benefit, however, is that there is no presumption of independence. Any conditioning or dependence is picked up by the counting procedure.

Example 62 *The construction of $\bar{F}_t$ can be illustrated as follows. Suppose there is an observable $X \simeq x[\mathbf{R}, H]$, with mean value $\mu = \mathrm{E}[X]$. Suppose we wish, by changing the values of H, to construct a different observable $\bar{X} \simeq \bar{x}[\mathbf{R}, \bar{H}]$ whose mean is some given number $\mathrm{E}^{\bar{H}}[\bar{X}] = \bar{\mu}, \neq \mu$. This can be done in two stages. First construct $X' \simeq x'[\mathbf{R}, H']$ with mean 0, then construct $\bar{X} = \bar{x}[\mathbf{R}, \bar{H}]$ with mean $\bar{\mu}$. (The elements $x, x', \bar{x}$ are not different numbers. Nor are they the same number. Each of them represents an arbitrary possible occurrence in $\mathbf{R}$.) For the first stage, suppose $I' =]u', v'] \in \mathbf{R}$. Write*

$$u = u' + \mu, \quad v = v' + \mu, \quad so \quad u' = u - \mu, \quad v' = v - \mu,$$

and, for $I_t =]u, v]$, define $H'(I') = H(I)$, so $\mathrm{E}^{H'}[X'] = 0$. The second stage is similar. For $\bar{I}_t =]\bar{u}, \bar{v}] \in \mathbf{R}$ write

$$u = \bar{u} + \mu - \bar{\mu}, \quad v = \bar{v} + \mu - \bar{\mu}, \quad so \quad \bar{u} = u - \mu + \bar{\mu}, \quad \bar{v} = v - \mu + \bar{\mu};$$

and, for $I_t =]u, v]$, define $\bar{H}(\bar{I}) = H(I)$, so $\mathrm{E}^{\bar{H}}[\bar{X}] = \bar{\mu}$, as required.

Returning to the construction of $\bar{Z}_T \simeq \bar{z}_T[\mathbf{R}_+^T, \bar{F}]$, Example 62 shows how to construct $\bar{F}_t$ for each $t \in T$, so, for $N = \{t\}$ the values $\bar{F}(I[N])$ can be found.

But if $\bar{Z}_T$ is to be defined as a *process* or *joint* observable, the *joint* distribution function $\bar{F}(I[N])$ must be defined, not just for singletons $N = \{t\}$, but for *all* $N \in \mathcal{N}(T)$ and all corresponding $I[N] \in \mathcal{I}(\mathbf{R}_+^T)$. The counting procedure addresses this by considering various patterns of transition.

The Glanbia daily price share data described in [MTRV] section 9.11 (pages 487–489) consists of 5225 daily end-of-day Glanbia share prices, beginning Friday 8 March 1991 and concluding Wednesday 9 March 2011.

Suppose the latter point in time is the present, and a 3 month call option contract on Glanbia shares is entered into at that moment. Since the share prices are reported daily on a 5-day week basis, take the 3-month term of the option to be 60 days, starting on morning of 10 March 2011 and expiring at end-of-day Wednesday 1 June 2011, a total of 5219 individual days/prices.

Suppose the option is the Asian type, with pay-off depending on the average share price over the term of the option.

To find an economic price for entering into this contract, it is useful to have estimates of the daily expected value of a share during the 60-day term of the option; likewise the daily standard deviation of the share price. Also, joint probability values $F(I[N])$ need to be estimated for the share price process during the term of the option.

Such estimates can be calculated by means of Maple computer software. Some code for this is given below, using historic Glanbia price data [57] from an Excel file in [website].

The first aim of the code is to produce estimates of expected share price value for each day of the term of the option—that is, the expected share price on 10 March 2011, on 11 March 2011, and so on, up to and including 1 June 2011; all of which are in the "future"—and the to estimate daily share price standard deviation for each day of the term of the option. (Line numbers 1., 2., 3., ... are not part of the code.)

Maple code for empirical distribution analysis:

```
 1. with(ExcelTools); with(Statistics);
```

2. $Q :=$ `Import`("glanbia.xls"):

3. $A1 :=$ `seq`$(Q[j][2], j = 2500..5010) : A2 := [A1] : A3 :=$`Vector`$(A2) :$

4. $J1 :=$`seq`$(j, j = 2500..5010); J2 := [J1]; J3 :=$`Vector`$(J2);$

5. $y :=$`ExponentialFit`$(J3, A3, t);$

6. `GlanbiaGraph` $:=$`[seq`$([t, Q[t + 7][2]], t = 2500..5000)];$

7. `plot`$($`[GlanbiaGraph`, $y], t = 2500..5000)$

8. $p := 7; q := p + 59;$

```
 9. for k from 1 to 5160 do
```

10. $A[k] :=$ `seq`$(Q[j][2], j = p..q);$

11. $B[k] := [A[k]]; p := p + 1; q := q + 1;$

12. $m[k] :=$ `Mean`$(B[k]); s[k] :=$ `StandardDeviation`$(B[k]):$

```
13. end do;
```

14. $m :=$ `seq`$(m[k], k = 1..5160); s :=$`seq`$(s[k], k = 1..5160);$

15. $S :=$ `seq`$(s[k], k = 5000..5160); T := [S]; s0 :=$ `Mean`(T)

16. $M[1] := 3.7*$`exp`$(0.006); M[2] := 3.7*$`exp`$(0.012); M[3] := 3.7*$`exp`$(0.018):$

```
17. for  k from 1 to 5219 do
```

18. $z[k] := Q[k + 6][2] :$

```
19. end do:
```

20. $N := 6; J :=$ `seq`$(j, j = -3..2);$ `seq`$(q, i = 1..3);$ `seq`$(b, i = 1..3);$

```
21. for i from 1 to 216 do
```

```
22. for j from 1 to 3 do
```

23. $q[j] :=$ `trunc`$((i - 1)/N^{(}j - 1)) + 1;$

24. $b[j] :=$ `mod`$(q[j] - 1, N) + 1;$

25. $p[i][j] := J[b[j]]$

```
26. end do
```

```
27.   end do
```

28. $cx := 3.5; w0 := 0;$

29. `Prob := 0;`

30. `for` i `from 1 to 216 do`

31. $f := 0;$

32. `for` k `from 4550 to 5000 do`

33. `if` $(z[k+20] >= m[k+20] + s[k+20] * p[i][1]$

34. `and` $z[k+20] < m[k+20] + (p[i][1]+1) * s[k+20])$

35. `and` $z[k+40] >= m[k+40] + s[k+40] * p[i][2])$

36. `and` $z[k+40] < m[k+40] + (p[i][2]+1) * s[k+40])$

37. `and` $z[k+60] >= m[k+60] + p[i][3] * s[l+60]$

38. `and` $z[l+60] < m[l+60] + (p[i][3]+1) * s[l+60])$

39. `then` $f := f + 1:$

40. `end if;`

41. `end do;`

42. $P := (1/50) * f;$

43. `Prob := Prob+`$P;$

44. $a1 := (M[1]+p[i][1] * s0 + M[2]+p[i][2] * s0 + M[3]+p[i][3] * s0)/3 - ex;$

45. `if` $a1 > 0$ `then` $a2 := a1;$ `end if;`

46. `if` $a1 <= 0$ `then` $a2 := 0;$ `end if;`

47. $a3 := a2$`*exp`$(-0.02);$

48. $w := a3 * P;$

49. $w0 := w0 + w:$

50. `end do;`

51. $w0;$

52. `evalf(Prob);`

The lines of Maple code above are numbered for the purpose of explanation. These numbers are not part of the program and should not be included in actual Maple code.

Lines 1–7 take the historic Glanbia data and import them into the program in order to perform Maple calculations on them. These data can be found in an Excel file in [**website**] and should be stored in the same folder in which the above Maple code is located, in order for this code to be able to access them. In Version 18 of Maple, the file with `Import` command must, on first use, be closed and then re-opened before `Import` will work successfully.

Lines 3 and 4 convert the share price data to a format suitable for line 5, using about half of the available prices—from day 2500 to day 5010. It is a matter of judgement which data to include.

Line 5 calculates a least squares regression for the price data, in the form $y = a\exp(\mu t)$. The idea here is that the prices follow some underling growth rate μ of a "proportional" character; with random variation of prices superimposed on this underlying trend. In other words, if the share price on day t is z_t, and disregarding the superimposed random variation of prices, then

$$\frac{z_{t+1}}{z_t} \text{ is constant,} \qquad \ln\left(\frac{z_{t+1}}{z_t}\right) = \mu.$$

In this case the exponential best fit calculated by Maple is approximately

$$y = 0.1e^{0.0007t},$$

so daily growth rate is approximately $\mu = 0.0007$. This translates into an annual growth rate of the share values amounting to about 20%.

Lines 7 and 8 of the Maple code produce a graph (y) of this underling growth, superimposed on a graph of the actual prices—see Figure 12.8.

For the purpose of pricing an Asian option entered into "today" (9 March 2011), when "today's" share price is 3.7, note that a risk-free daily growth rate of 0.0003. This is less than the underlying growth rate $\mu = 0.0007$. Thus the underlying trend[12] values of the share during the period of the Asian option (up to 1 June 2011, or 60 days in total) is taken to be

$$y = 0.1e^{0.0003t}.$$

Then, provided we can superimpose on these price trend values the appropriate amount of random variability, we can estimate the value of the Asian option by using (12.11).

Lines 8–13 are the first step in accomplishing this. This part of the program calculates successive 60-day average values of the historic share prices and also the corresponding standard deviations. The latter demonstrate the scale of the random variability or volatility in these share prices, and are a kind of indicator of the joint likelihood distribution $F(I[N])$ that "determines" the share price process $Z_T \simeq z_t[\mathbf{R}_+^T, F]$.

These "moving averages" and "moving standard deviations" are illustrated in Figures 12.7 and 12.9 at the end of this section.

The Maple program will also use these mean-and-standard-deviation data as "3σ" partition points in a regular partition of $\mathbf{R}_+^T$, T being the 60-day period or term of the Asian option, in order to estimate the option valuation (12.11).

---

[12]This is not the actual trend, just the hypothetical risk-free trend needed for martingale pricing of discounted values. Line 16 of the code shows these successive risk-free trend values at intervals of 20 days during the term of the option.

Line 15 takes an average $s0$ of the 60-day standard deviations. This is to be taken as the standard deviation of the daily share price for each day of the 60-day term of the Asian option—see line 44.

Line 16 computes three 20-day trend values $M[1], M[2], M[3]$ for the 60-day term of the option. But these are not "true" trend values; they are the hypothetical values showing a risk-free daily growth rate of 0.0003 needed to carry out the risk-neutral martingale valuation of (12.11).

Lines 17 to 19 ascribe the familiar notation $z[k]$ (or z_t) to the historic share price values. Lines 20–27 compute the permutations, with repetition, of six numbers $-3, -2, -1, 0, 1, 2$ taken three at a time. There are $6^3 = 216$ such permutations, which, for $i = 1$ to 216, are each given by

$$(p[i][1], \quad p[i][2], \quad p[i][3]).$$

One such permutation is $(1, -1, 2)$; that is

$$p[i][1] \;=\; 1, \quad p[i][2] \;=\; -1, \quad p[i][3] \;=\; 2.$$

With mean m and standard deviation s, these numbers are then used to construct 3σ partitioning intervals

$$[m - p[i][j] * s, \quad m - (p[i][j] + 1) * s[,$$

which, for $p[i][j] = -3, -2, \ldots, 2$, give six cells

$$[m + js, \; m + (j + 1)s[, \qquad j = -3, -2, -1, 0, 1, 2.$$

On any particular day the share price will usually be found to lie in one of these intervals. The probability of $|z_t - m| > 3s$ is small, and can be ignored in the option valuation calculation. Take the start of 9 March 2011 to be time $t = 0$, and 1 June 2011 to be $t = \tau$, and $T = \,]0, \tau]$, with $t = \tau_1 = 20$, $t = \tau_2 = 40$, and $t = \tau = 60$. For $j = 1, 2, 3$, write the interval I_j as

$$I_j \;=\; [M[j] + p[i][j] * s0, \; M[j] + (p[i][j] + 1) * s0[\tag{12.14}$$

where the values $s0$ and $M[j]$ are given by lines 15 and 16 of the Maple code.

Then, with $N = \{\tau_1, \tau_2, \tau\}$ the domain $\mathbf{R}_+^T$ of the option is partitioned by cylinders

$$I[N] \;=\; \prod_{j=1}^{3} I_j \times \mathbf{R}_+^{T \setminus M}. \tag{12.15}$$

There are 216 such cylindrical cells, corresponding to the 216 permutations

$$(p[i][1], \quad p[i][2], \quad p[i][3]).$$

This is a regular partition in $\mathbf{R}_+^T$. It does not fully exhaust the domain $\mathbf{R}_+^T$, but the probability that z_T is not in one of these cylindrical intervals is very small. Line 28 of the Maple program sets the exercise price of the option at $ex = 3.5$. The assignment $w0 := 0$ sets the initial value of the Riemann sum estimate of (12.11). The Riemann sum has 216 terms, one for each permutation $(p[i][j]; j = 1, 2, 3)$. Each term of the Riemann sum has value w (line 48), and the term values w are accumulated as $w0$ in the 216 iterations of lines 48 and 49.

The ultimate value returned by $w0$ is the value of the Riemann sum estimate of (12.11). For option exercise price $ex = 3.5$, the Maple program gives initial value $w0 = 1.31$. This is the w_0 or $w(0)$ of (12.11). The expression Prob of line 29 is used in the code to keep track of, or accumulate, the probabilities generated in the calculation. Total probability returned by line 43 is approximately 0.97, indicating that the 3σ regular partition of $\mathbf{R}_+^T$, though non-exhaustive, is almost full.

Using the historic Glanbia share price data, lines 30–43 of the program use a relative frequency-type argument to produce estimates of the joint transition probabilities $F(I[N]$ needed to calculate the Riemann sum estimates $w0$ of (12.11).

Line 32 shows that this program only uses 50 cycles or iterations of the 3×20-day transitions in the partitioning cells (12.15). For greater accuracy, it is easy to increase this sample size.

The relative frequency calculation for a single permutation of transitions $(p[i][j] : j = 1, 2, 3)$ is done in line 42. This value of P is the estimate of $F(I[N]$ for a single term of the Riemann sum for (12.11).

For $i = 1, \ldots, 216$, each permutation $(p[i][j] : j = 1, 2, 3)$ generates a single term of the Riemann sum for (12.11). The calculation of each such term is done in lines 44–48.

How are the values $F(I[N])$ converted to risk-neutral probabilities $\bar{F}(I[N])$? This is accomplished in line 44. The quantities $M[j] + p[i][j] * s0$ are the lower bounds of partitioning cells of the form (12.14) and (12.15), and these are taken as evaluation points for the function $A(z_T)$ in (12.11).

But these boundary points are unlike the corresponding boundaries in lines 33–38. The latter are growing at the daily trend growth rate $\mu = 0.0007$, whereas the bounds in line 44 are growing at the risk-free growth rate $\rho = 0.0003$.

So, in some sense, the probabilities $F(I[N])$ produced from lines 33-42 are, in line 44, applied to the "wrong" cells of $\mathbf{R}_+^T$. Except that this is the appropriate adjustment needed to produce the hypothetical $\bar{F}(I[N])$ in $\mathbf{R}_+^T$.

The point-integrand in (12.11) is the function (12.12), and this is calculated as $a2$ in lines 44–46. The discounted[13] value of (12.12) is calculated in line 47. Multiplying this by $P = \bar{F}(I[N])$ gives the value of a single term w of the Riemann sum. Summing over all the permutations, $i = 1, \ldots, 216$, gives the Riemann sum estimate of the option value at initial time $t = 0$: $w0 = w_0 = w(0) = 1.3$.

To test out the above Maple code, it is advisable to test its component parts individually in order to more easily detect and correct transcription and coding errors. Also, at the relevant points in the code, alternative parameters and

---

[13]For 60-day discounting, the daily rate 0.0003 becomes approximately $60 \times 0.0003 = 0.018$. Line 47 uses $e^{-0.02}$ instead of $e^{-0.018}$ for discounting at the risk-free rate.

estimation tactics can be experimented with.

In the above code the average $A(z_T)$ was estimated at only three 20-day intervals. For a 60-day option the Glanbia share price data in the Excel file permit up to 60 daily price values to be used to calculate $A(z_T)$. In practice, a reasonable balance must be struck between the demands of accuracy and the scale of computing power available. The latter quickly escalates when more accuracy is demanded.

[MTRV], page 480, has a twenty-year graph of the Glanbia share prices. Figure 12.8 is a graph of part of the data, with superimposed exponential regression graph $y = \exp(0.0007t)$.

The first average calculated by lines 8–13 of the Maple program is the mean of the prices for day 1 (8 March 1991) up to and including day 60. The second average is for day 2 (9 March 1991) up to and including day 61 of the data. The final average is the mean of the prices on day 5160 of the data up to and including day 5219 (9 March 2011).

This is a total of 5160 cycles, giving 5160 means and 5160 standard deviations. The latter are displayed in Figure 12.9.

Since the first moving average is the average for the first 60 days, it can be applied to day 30 or day 31, half-way through the first cycle. Likewise for each of the other moving averages; and also the standard deviations.

The graph of moving averages, Figure 12.7, is a "smoothed out" version of the graph (Figure 12.8) of the original Glanbia share price data. But it is not as smooth as the exponential regression graph for the data, in Figure 12.8.

Figure 12.10 shows how the initial value of the Asian option depends on the terminal exercise price. A low exercise price of 3.0 results in high option price of about 5.5. A high exercise price of 3.8 produces a low option price of about 0.02.

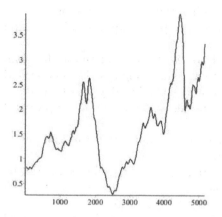

Figure 12.7: All moving averages

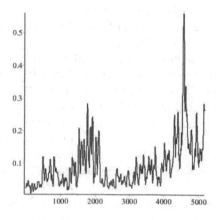

Figure 12.9: All moving SD's

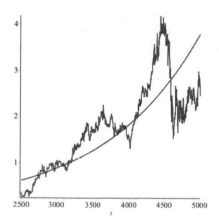

Figure 12.8: Exponential regression

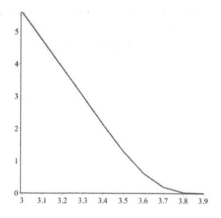

Figure 12.10: Option values

Chapter 13

Appendix 4: Listings

13.1   Theorems

Gauge Integral Structures for Stochastic Calculus and Quantum Electrodynamics, First Edition. Patrick Muldowney.
© 2021 John Wiley & Sons, Inc. Published 2021 by John Wiley & Sons, Inc.

13.2   Examples

Example number

13.3   Definitions

Definition

13.4   Symbols

Symbol

Bibliography

[1] Abbott, E.A., *Flatland: A Romance of Many Dimensions*, Seeley, London, 1884, `https://ned.ipac.caltech.edu/level5/Abbott/paper.pdf` (accessed 3 September 2019).

[2] Albeverio, S.A., Høegh-Krohn, R.J., and Mazzuchi, S., *Mathematical Theory of Feynman Path Integrals: An Introduction*, Springer-Verlag, Berlin Heidelberg, 2008.

[3] Alvarez, E.T.G., and Gaioli, F.H., *Feynman's proper time approach to QED* (2008), `https://arxiv.org/pdf/hep-th/9807132.pdf` (accessed 3 September 2019).

[4] Arfken, G., *Mathematical Methods for Physicists*, Academic Press, Orlando, 1985.

[5] Bartle, R.G., *Return to the Riemann integral*, American Mathematical Monthly 103(8) (1980), 625–632,
`http://classicalrealanalysis.info/documents/Bartle1996-2974874.pdf` (accessed 3 September 2019).

[6] Bartle, R.G., *A Modern Theory of Integration*, Wiley, Hoboken, 2001.

[7] Bernstein, J., A Palette of Particles, Harvard University Press, Massachusetts, 2013.

[8] Bonotto, E. de Mello, *A Equação de Black–Scholes com Ação Impulsiva*, PhD thesis, University of São Paolo, São Carlos, 2008.

[9] Bonotto, E. de Mello, Federson, M., and Muldowney, P., *A Feynman–Kac solution to a random impulsive equation of Schrödinger type*, Real Analysis Exchange 36(1) (2010-11), 107–148.

[10] Brown, L.M. (ed.), *Feynman's Thesis: A New Approach to Quantum Theory*, World Scientific, Singapore, 2005,
`https://www.academia.edu/6157188/`   (accessed 3 September 2019).

[11] Bullen, P.S., et al., *New Integrals*, Springer-Verlag, Berlin, 1980.

[12] Bullen, P.S., *Nonabsolute integration in the twentieth century*, American Mathematical Society Special Session on Nonabsolute Integration, Toronto, 23–24 September, 2000,
www.emis.de/proceedings/Toronto2000/papers/bullen.pdf
(accessed 3 September 2019).

[13] Burkill, J.C., *Functions of intervals*, Proceedings of the London Mathematical Society, 22(2) (1924), 275–310.

[14] Burkill, J.C., *The expression of area as an integral*, Proceedings of the London Mathematical Society, 22(2) (1924), 311–336.

[15] Chew, T.S., and Toh, T.L., *The Riemann approach to multiple Wiener integral*, Real Analysis Exchange 30(1) (2003–04), 275–289.

[16] Chung, K.L., and Williams, R.J., *Introduction to Stochastic Integration*, Birkhäuser, Boston, 1990.

[17] Cousin, P., *Sur les fonctions de n variables complexes*, Acta Mathematica 19 (1895), 1–62.

[18] Cyganowski, S., Kloeden, P., and Ombach, J., *From Elementary Probability to Stochastic Differential Equations with MAPLE*, Springer, Berlin, 2002.

[19] Daniell, P.J., *Integrals in an infinite number of dimensions*, Annals of Mathematics 20 (1919), 281–288.

[20] De Pauw, T., Pfeffer, W., *The divergence theorem for unbounded vector fields*, Transactions of the American Mathematical Society 359(12) (2007), 5915–5930.

[21] De Pauw, T., Pfeffer, W., *Distributions for which div v = F has a continuous solution*, Communications on Pure and Applied Mathematics 61(2) (2008), 230–260.

[22] Dias de Deus, J., Pimenta, M., Noronha, A., Peña, T., and Brogueira, P., *Introdução à Física*, McGraw-Hill, New York, 2000; *Introducción al la Física*, McGraw-Hill, New York, 2001.

[23] Dienes, P., *The Taylor Series*, Clarendon Press, Oxford, 1931.

[24] Di Piazza, L., and Maraffa, V., *The McShane, PU and Henstock integrals of Banach valued functions*, Czechoslovak Mathematical Journal 52 (2002), 609–633.

[25] Dirac, P.A.M., *The Principles of Quantum Mechanics*, Clarendon Press, Oxford, 1930, https://archive.org/details/in.ernet.dli.2015.177580/ (accessed 3 September 2019).

[26] Dirac, P.A.M., *The Lagrangian in quantum mechanics*, Physikalische Zeitschrift der Sowjetunion 3 (1933), 64–72.

[27] Dirac, P.A.M., *Quantum electrodynamics*, Scríbhinní Institiúid Árd-Léighinn Bhaile Átha Cliath A (1), 1943. (Communications of the Dublin Institute for Advanced Studies, Series A No. 1.)

[28] Dirac, P.A.M., *Lectures on Quantum Mechanics*, 1964, Dover edition, Dover, New York, 2001.

[29] F. Dyson, F.J., *The S-matrix in quantum electrodynamics*, Physical Review 75 (1949), 1736–1755.

[30] Dyson, F.J., *The Radiation Theories of Tomonaga, Schwinger, and Feynman*, Physical Review 75 (1949) 486–502.
http://web.ihep.su/dbserv/compas/src/dyson49b/eng.pdf
(accessed 3 September 2019).

[31] Emery, M., *Stochastic Calculus in Manifolds*, Springer-Verlag, Berlin, New York, 1989.

[32] Ehud, G., *Plenty of Room for Biology at the Bottom: An Introduction to Bionanotechnology*, Imperial College Press, London, 2007.

[33] Federson, F., *A equação de Schrödinger e a integral de Feynman*, São Carlos Institute of Physics, University of São Paulo, 2018.

[34] Federson, M., *A constructive integral equivalent to the integral of Kurzweil*, Czechoslovak Mathematical Journal 52(2) (2002), 365–367.

[35] Feynman, R.P., *The Principle of Least Action in Quantum Mechanics*, PhD thesis, Princeton University, 1942,
https://cds.cern.ch/record/101498/files/Thesis-1942-Feynman.pdf
(accessed 3 September 2019).

[36] Feynman, R.P., and Wheeler, J.A., *Interaction with the Absorber as the Mechanism of Radiation*, Reviews of Modern Physics 17 (1945), 157–181.

[37] Feynman, R.P., *A relativistic cutoff for quantum electrodynamics*, Physical Review 74 (1948), 939–946.

[38] Feynman, R.P., *Relativistic cutoff for quantum electrodynamics*, Physical Review 74 (1948), 1430–1438.

[39] Feynman, R.P., cited as [F1], *Space-time approach to non-relativistic quantum mechanics*, Reviews of Modern Physics 20 (1948), 367–387.

[40] Feynman, R.P., *The theory of positrons*, Physical Review 76 (1949), 749–759.

[41] Feynman, R.P., and Wheeler, J.A., *Interaction with the absorber as the mechanism of radiation*, Reviews of Modern Physics, 17 (1949), 157–181.

[42] Feynman, R.P., *Space-time approach to quantum electrodynamics*, Physical Review 76 (1949), 769–789.

[43] Feynman, R.P., *Mathematical formulation of the quantum theory of electromagnetic interaction*, Physical Review 80 (1950), 440–457.

[44] Feynman, R.P., *An operator calculus having applications in quantum electrodynamics*, Physical Review 84 (1951), 108–128.

[45] Feynman, R.P., *Quantum Electrodynamics*, Benjamin, New York, 1961.

[46] Feynman, R.P., and Hibbs, A.R., cited as [FH], *Quantum Mechanics and Path Integrals*, McGraw-Hill, New York, 1965;
emended edition: editor Styer, D.F., Dover, New York, 2010.

[47] Feynman, R.P., *The development of the space-time view of quantum electrodynamics*, Nobel Lecture, California Institute of Technology, 1965,
`https://www.nobelprize.org/prizes/physics/1965/feynman/lecture/`
(accessed 3 September 2019).

[48] Feynman, R.P., Leighton, R.B., Sands, M., *The Feynman Lectures on Physics*, Addison-Wesley, Boston, 1970,
`http://www.feynmanlectures.caltech.edu` (accessed 3 September 2019);
Lecture on the principle of least action:
`http://www.feynmanlectures.caltech.edu/II_19.html` (accessed 3 September 2019).

[49] Feynman, R.P., *QED: the Strange Theory of Light and Matter*, Princeton University Press, 1985;
based on Feynman's New Zealand Lectures on Quantum Electrodynamics:
`http://www.feynman.com/science/qed-lectures-in-new-zealand/`
(accessed 3 September 2019).

[50] Fokker, A.D., *Ein Invariante Variationssatz für die bewegung Mehrerer Elektrischer Massteilchen*, Zeitschrift für Physik 58 (1929), 386–393.

[51] Folland, G.B., *Quantum Field Theory: A Tourist Guide for Mathematicians*, American Mathematical Society, Providence, 2008.

[52] Forsyth, A. R., *Calculus of Variations*, Dover, New York, 1960.

[53] Fremlin, D.H., *The generalized McShane integral*, Illinois Journal of Mathematics 39(1) (1995), 39–67.

[54] Ganguly, D.K., Lee, P.Y., and Pal, S., *Henstock-Stieltjes integrals not induced by measure*, Real Analysis Exchange 26(2) (2000), 853–860.

[55] Gelfand, I.M., Yaglom, A.M., *Integration in function spaces*, Journal of Mathematical Physics 1 (1960), 48–69.

[56] Gill, T.L., and Zachary, W.W., *Functional Analysis and the Feynman Operator Calculus*, Springer, New York, 2016.

[57] [website] Glanbia, Smith-Glaxo-Kline, and Rio Tinto share price data, https://sites.google.com/site/stieltjescomplete/glanbia-smithglaxokline-and-rio-tinto-data (accessed 3 September 2019).

[58] Glimm, J., and Jaffe, A., *Quantum Physics: A Functional Integral Point of View*, Springer, New York, 1981.

[59] Goldstein, H., Poole, C., and Safko, J., *Classical Mechanics*, Addison-Wesley, Boston, 2002.

[60] Goldstine, H.H., *A History of the Calculus of Variations from the 17th through the 19th Century*, Springer-Verlag, New York, 1980.

[61] Gordon, R.A., *The Integrals of Lebesgue, Denjoy, Perron, and Henstock*, American Mathematical Society, Providence, 1994.

[62] Griffiths, D.J., Introduction to Quantum Mechanics, Pearson, Harlow, 2014.

[63] Griffiths, D.J., Introduction to Electrodynamics, Cambridge University Press, Cambridge, 2017.

[64] Grosche, C., and Steiner, F., *Handbook of Feynman Integrals*, Springer-Verlag, Berlin Heidelberg, 1998.

[65] Hamilton, W.R., *The Mathematical Papers of Sir William Rowan Hamilton (1805–1865), transcribed and edited by David R. Wilkins*, http://www.emis.de/classics/Hamilton/ (accessed 3 September 2019).

[66] Hamming, R., paraphrased in N. Rose's *Mathematical Maxims and Minims*, Raleigh NC, Rome Press Inc., 1988.

[67] Hardy, G.H., *Divergent Series* Clarendon Press, Oxford, 1949.

[68] Henstock, R., *Interval Functions and their Integrals*, PhD thesis, Birkbeck College, University of London, 1948, https://arxiv.org/abs/1702.08486 (accessed 3 September 2019).

[69] Henstock, R., *The efficiency of convergence factors for functions of a continuous real variable*, Journal of the London Mathematical Society 30 (1955), 273–286.

[70] Henstock, R., *Theory of Integration*, Butterworth, London, 1962, 1963.

[71] Henstock, R., *Linear Analysis*, Butterworth, London, 1968.

[72] Henstock, R., *Generalised integrals of vector-valued functions*, Proceedings of the London Mathematical Society 19 (1969), 509–536.

[73] Henstock, R., *Integration in product spaces, including Wiener and Feynman integration*, Proceedings of the London Mathematical Society 27 (1973), 317–344.

[74] Henstock, R., *Integration, variation and differentiation in division spaces*, Proceedings of the Royal Irish Academy 78A(10) (1978), 69–85.

[75] Henstock, R., *Lectures on the Theory of Integration*, World Scientific, Singapore, 1988.

[76] Henstock, R., *The General Theory of Integration*, Clarendon, Oxford, 1991.

[77] Henstock, R., editor P. Muldowney, *Lecture Notes of R. Henstock*, (1970), `https://arxiv.org/pdf/1602.02993.pdf` (accessed 3 September 2019).

[78] Henstock, R., *The Calculus and Gauge Integrals*, unfinished, c. 1992–3, Henstock Archive, University of Ulster, `https://arxiv.org/pdf/1608.02616.pdf`   (accessed 3 September 2019).

[79] Henstock, R., Muldowney, P., Skvortsov, V.A., *Partitioning infinite-dimensional spaces for generalized Riemann integration*, Bulletin of the London Mathematical Society, 38 (2006), 795–803.

[80] Itô, K., *Stochastic Integral*, Proceedings of Imperial Academy Tokyo 20 (1944), 519–524.

[81] Itô, K., and McKean, H.P., *Diffusion Processes and their Sample Paths*, Academic Press, New York, 1965.

[82] Jarrow, R., and P. Protter, *A Short History of Stochastic Integration and Mathematical Finance: The Early Years, 1880-1970*, Herman Rubin Festschrift, IMA Lecture Notes 45 (2004), 75–91.

[83] Jessen, B., *The theory of integration in a space of an infinite number of dimensions*, Acta Mathematica 63 (1934), 249–323, `https://projecteuclid.org/download/pdf_1/euclid.acta/1485888079` (accessed 3 September 2019).

[84] G. W. Johnson, G.W., and Lapidus, M.L., *The Feynman Integral and Feynman's Operational Calculus*, Oxford Science Publications, New York, 2000.

[85] Jorgensen, P.E.T., and Nathanson, E., *A global solution to the Schrödinger equation: From Henstock to Feynman*, Journal of Mathematical Physics 56 (9) (2015), 092102–092115, https://arxiv.org/pdf/1501.06226v1.pdf (accessed 3 September 2019).

[86] Karatzas, I., and Shreve, S.E., *Brownian Motion and Stochastic Calculus*, Springer-Verlag, New York, 1991.

[87] Karatzas, I., and Shreve, S.E., *Methods of Mathematical Finance*, Springer-Verlag, New York, 1998.

[88] Kanesawa, S., and Tomonaga, S., *On a Relativistically Invariant Formulation of the Quantum Theory of Wave Fields. IV.*, Progress of Theoretical Physics 3 (1948), 1–13.

[89] Kanesawa, S., and Tomonaga, S., *On a Relativistically Invariant Formulation of the Quantum Theory of Wave Fields. V.*, Progress of Theoretical Physics 3 (1948), 101–113.

[90] Koba, Z., Tati, T., and Tomonaga, S., *On a Relativistically Invariant Formulation of the Quantum Theory of Wave Fields. II.*, Progress of Theoretical Physics 2 (1947), 101–116.

[91] Koba, Z., Tati, T., and Tomonaga, S., *On a Relativistically Invariant Formulation of the Quantum Theory of Wave Fields. III.*, Progress of Theoretical Physics 2 (1947), 198–208.

[92] Koba, Z., and Tomonaga, S., *On Radiation Reactions in Collision Processes. I.*, Progress of Theoretical Physics 3 (1948), 290–303.

[93] Kolmogorov, A.N., *Grundbegriffe der Wahrscheinlichkeitreichnung*, Ergebnisse der Mathematik, Springer, Berlin, 1933 (*Foundations of the Theory of Probability*, Chelsea Publishing Company, New York, 1950).

[94] Kurzweil, J., *Generalized ordinary differential equations and continuous dependence on a parameter*, Czechoslovak Mathematical Journal (1957), 418–449.

[95] Kurzweil, J., *Nichtbsolut Konvergente Integrale*, Teubner-Texte zur Mathematik, Teubner, Leipzig, 1980.

[96] Kwok, Y.K., *Mathematical Models of Financial Derivatives*, Springer-Verlag, Singapore, 1998.

[97] Lagrange, J.L., *Analytical Mechanics*, (ed. Victor N. Vagliente, tr. Boissonnade, A.), Kluwer, Dordrecht, 1997.

[98] Lanczos, C., *The Variational Principles of Mechanics*, University of Toronto Press, 1949.

[99] Landau, L.D., and Lifshitz, E.M., Mechanics, *Course of Theoretical Physics*, Butterworth-Heinemann, London, 1976.

[100] Lebesgue, H., *Sur une généralisation de l'integrále définie*, Comptes Rendus de l'Academie des Sciences 132 (1901), 1025–1028.

[101] Lebesgue, H., *Integrále, longueur, aire*, Annali di Matematica Pura ed Applicata 7 (1902), 231–359.

[102] Lebesgue, H., *Leçons sur l'integration et la recherche des fonctions primitives*, Gauthier-Villars, Paris, 1904.

[103] Lee P.Y., and Vyborny, R., *The Integral: an Easy Approach after Kurzweil and Henstock*, Australian Mathematical Society Lecture Series 14, Cambridge University Press, Cambridge, 2000.

[104] Levy, C., *Measures from infinite-dimensional gauge integration*, pre-print, Institut de Mathématiques de Toulouse, Institut National Universitaire Champollion, 2019,   https://hal.archives-ouvertes.fr/hal-02308813 (accessed 9 October 2019).

[105] Loeb, P., and E. Talvila, *Lusin's theorem and Bochner integration*, Scientiae Mathematicae Japonicae Online, 10 (2004), 55–62.

[106] Lukashenko, T.P., Skvortsov, V.A., and Solodov A.P., *Generalized Integrals* (in Russian), URSS, Moscow, 2nd edition, 2011.

[107] Mandelstam, S., and Yourgrau, W., *Variational Principles in Dynamics and Quantum Theory*, Dover, New York, 1979.

[108] Mawhin, J., *Introduction á l'Analyse*, Cabay, Louvain-la-Neuve, 1983.

[109] Mazzuchi, S., *Mathematical Feynman Path Integrals and their Applications*, World Scientific, Singapore, 2009.

[110] McKean, H.P., *Stochastic Integrals*, Academic Press, New York, 1969.

[111] McShane, E.J., *A Riemann Type Integral that Includes Lebesgue–Stieltjes, Bochner and Stochastic Integrals*, Memoirs of the American Mathematical Society No. 88, Providence, 1969.

[112] McShane, E.J., *A unified theory of integration*, American Mathematical Monthly 80 (1973), 349–359.

[113] McShane, E.J., *Stochastic Calculus and Stochastic Models*, Academic Press, New York, 1974.

[114] Morse, P. M., and Feshbach, H., *Methods of Theoretical Physics Part I*, McGraw-Hill, New York, 1953.

[115] Muldowney, P., *A General Theory of Integration in Function Spaces*, Pitman Research Notes in Mathematics, Longman, Harlow, 1987.

[116] Muldowney, P., *Feynman's path integrals and Henstock's non-absolute integration*, Journal of Applied Analysis 6 (1) (2000), 1–24.

[117] Muldowney P., and Wojdowski, W., *Nonabsolute integration in Black–Scholes pricing*, Proceedings of AMS Special Session on Non-Absolute Integration, Toronto, 2000,
www.emis.de/proceedings/Toronto2000/papers/wojdowski.pdf
(accessed 3 September 2019).

[118] Muldowney, P., and Skvortsov, V.A., *Improper Riemann integral and Henstock integral in* $\mathbf{R}^n$, Mathematical Notes 78(2) (2005), 228–233; translated from *Nesobstvennyj integral Rimana i integral Henstoka v* $\mathbf{R}^n$, Matematicheskie Zametki 78(2) (2005), 251–258.

[119] Muldowney, P., *A Riemann approach to random variation*, Mathematica Bohemica 131(2) (2006), 167–188.

[120] Muldowney, P., *Henstock on random variation*, Scientiae Mathematicae Japonicae, No. 247 67(1) (2008), 51–69.

[121] Muldowney, P., cited as [MTRV], *A Modern Theory of Random Variation, with Applications in Stochastic Calculus, Financial Mathematics, and Feynman Integration*, Wiley, Hoboken, 2012.

[122] Muldowney, P., Supplement to [MTRV], 2019. Cited as [website],
https://sites.google.com/site/StieltjesComplete/Supplement
(accessed 3 September 2019).

[123] Muldowney, P., *Integration issues in probability*, Bulletin of the Irish Mathematical Society, 75 (2015), 21–44.
http://www.maths.tcd.ie/pub/ims/bull75/Muldowney.pdf
(accessed 3 September 2019).

[124] Muldowney, P., Ostaszewski, K., and Wojdowski, W., *The Darth Vader Rule*, Tatra Mountains Mathematical Publications 12 (2012), 53–63.

[125] von Neumann, J., *Mathematische Grundlagen der Quanten-mechanik*, Dover, New York, 1943.

[126] Norman, D., and Sanders, S., On the mathematical and foundational significance of the uncountable, Journal of Mathematical Logic 19(1) (2019), 1–40.

[127] Øksendal, B., *Stochastic Differential Equations*, Springer-Verlag, Berlin, 1985.

[128] Ostaszewski, K., *Henstock Integration in the Plane*, Memoirs of the American Mathematical Society, no. 353, Providence, Rhode Island, 1986.

[129] Ostaszewski, K., *The space of Henstock integrable functions of two variables*, International Journal of Mathematics and Mathematical Sciences 11 (1988), 15–22.

[130] Pedgaonkar, A., *Fundamental Theorem of Calculus for Henstock-Kurzweil Integral*, Bulletin of the Marathwada Mathematical Society 14(1) (2013), 71–80, https://sites.google.com/site/anilpedgaonkar/ (accessed 3 September 2019).

[131] Pedgaonkar, A., *Integrable Manifolds and Fundamental Theorem of Calculus*, Proceedings of National Conference on Recent Trends in Mathematics 1 (2016), 67–78, https://sites.google.com/site/anilpedgaonkar/ (accessed 3 September 2019).

[132] Pfeffer, W., *The Riemann Approach to Integration*, Cambridge University Press, London, 1993.

[133] Pfeffer, W., *The Lebesgue and Denjoy-Perron integrals from a descriptive point of view*, Ricerche di Matematica XLVIII (1999), 211–223.

[134] Pfeffer, W., *Derivation and Integration*, Cambridge University Press, Cambridge, 2001.

[135] Popov, V.N., *Functional Integrals in Quantum Field Theory and Statistical Physics*, Reidel, Dordrecht, 1983.

[136] Rao, M.M., *Measure Theory and Integration*, Marcel Dekker, New York, 2004.

[137] Ross, S., *An Introduction to Mathematical Finance*, Cambridge University Press, Cambridge, 1999.

[138] Royden, H.L., *Real Analysis*, Macmillan, New York, 1968.

[139] Rudin, W., *Real and Complex Analysis*, McGraw-Hill, New York, 1974.

[140] Sagan, H., *Introduction to the Calculus of Variations*, McGraw-Hill, New York, 1969.

[141] Saks, S., *Theory of the Integral*, Warszawa–Lwów; translated by L.C. Young, Stechert, New York, 1937, https://archive.org/details/theoryoftheinteg032192mbp/ (accessed 3 September 2019).

[142] Schiff, L.I., *Quantum Mechanics*, McGraw-Hill, New York, 1955.

[143] Schulman, L.S., *Techniques and Applications of Path Integration*, Wiley, New York, 1981.

[144] Schwabik, S., *Generalized Ordinary Differential Equations*, World Scientific, Singapore, 1992.

[145] Schwartz, L., *Review of [31]*, Bulletin of the American Mathematical Society 24(2) (1991), 451–466.

[146] Schweber, S.S., *QED and the men who made it: Dyson, Feynman, Schwinger, and Tomonaga*, Princeton University Press, Princeton, 1994.

[147] Schwinger, J.S., *On Quantum electrodynamics and the magnetic moment of the electron*, Physical Review 73 (1948), 416–417.

[148] Schwinger, J.S., *Quantum electrodynamics. I. A covariant formulation*, Physical Review 74 (1948), 1439–1461.

[149] Schwinger, J.S., *Quantum electrodynamics. II. Vacuum polarization and self-energy*, Physical Review 75 (1948), 651–679.

[150] Schwinger, J.S., *Quantum electrodynamics. III. The electromagnetic properties of the electron: Radiative corrections to scattering*, Physical Review 76 (1949), 790–817.

[151] Shreve, S.E., *Stochastic Calculus for Finance II*, Springer Finance Textbook Series, Springer, New York, 2004.

[152] Simon, B., *Functional Integration and Quantum Physics*, Academic Press, New York, 1981.

[153] Skvortsov, V.A., *Henstock integral in harmonic analysis*, Scientiae Mathematicae Japonicae, 67(1) (2008), 71–82.

[154] Skvortsov, V.A., and Solodov, A.P., *A variational integral for Banach-valued functions*, Real Analysis Exchange 24(2) (1998-99), 799–805.

[155] Skvortsov, V.A., and Solodov, A.P., *A descriptive characterization of the Denjoy–Bochner integral and its generalizations*, Moscow University Mathematics Bulletin 57 (2002), 36–39.

[156] Skworcow, W. (V.A. Skvortsov), and Sworowski, P., *Całki uogólnione*, Wydawnictwo Uniwersytetu Kazimierza Wielkiego, Bydgoszcz, 2010.

[157] Smirnov, V.A., *Evaluating Feynman Integrals*, Springer, Berlin, 2004.

[158] Solodov, A.P., *Riemann-type definition for the restricted Denjoy–Bochner integral*, Fundamentalnaya i Prikladnaya Matematika 7 (2001), 887–895.

[159] Sussman, G.J., and Wisdom, J., *Structure and Interpretation of Classical Mechanics*, MIT Press, Boston, 2001.

[160] Swartz, W., *Introduction to Gauge Integrals*, World Scientific, Singapore, 2001.

[161] Talvila, E., *Necessary and sufficient conditions for differentiating under the integral sign*, American Mathematical Monthly 108 (2001), 544–548.

[162] Tetrode, H., *Über den Wirkungszusammenhang der Welt: Eine Erweiterung der klassischen Dynamik*, Zeitschrift für Physik 10 (1922), 317–328.

[163] Thorber, N.S., and Taylor, E.F., *Propagator for the simple harmonic oscillator*, American Journal of Physics 66 (1998), 1022–1024.

[164] Tomonaga, S., *On a Relativistically Invariant Formulation of the Quantum Theory of Wave Fields*, Progress of Theoretical Physics 1 (1946), 27–42.

[165] Tomonaga, S., and Oppenheimer, J.R., *On Infinite Field Reactions in Quantum Field Theory.*, Physical Review 74 (1948), 224–225.

[166] Weinberg, S., *Lectures on Quantum Mechanics*, Cambridge University Press, Cambridge, 2013.

[167] Weinberg, S., *The trouble with quantum mechanics*, New York Review of Books, January 2017, responses April 2017,
`http://quantum.phys.unm.edu/466-17/QuantumMechanicsWeinberg.pdf`
(accessed 3 September 2019).

[168] Weinstock, R., *Calculus of Variations, with Applications to Physics and Engineering*, Dover, New York, 1974.

[169] Wiener, N., *Differential space*, Journal of Mathematics and Physics 2 (1923), 131–174.

[170] Wiener, N., *The average value of a functional*, Proceedings of the London Mathematical Society 23 (1924), 452–467.

[171] Wiener, N., *Generalised harmonic analysis*, Acta Mathematica 55 (1930), 117–258.

[172] Wiener, N., *Nonlinear problems in random theory*, Wiley, New York, 1958.

[173] Yeh, J., *Stochastic Processes and the Wiener Integral*, Dekker, New York, 1973.

[174] Yeh, J. *Real Analysis: Theory of Measure and Integration*, World Scientific, Singapore, 2006.

Index